Advances in
Carbohydrate Chemistry and Biochemistry

Volume 63

Advances in Carbohydrate Chemistry and Biochemistry

Editor
DEREK HORTON
The American University
Washington, DC

Volume 63

Amsterdam • Boston • Heidelberg • London • New York • Oxford
Paris • San Diego • San Francisco • Singapore • Sydney • Tokyo
Academic Press is an imprint of Elsevier

Academic Press is an imprint of Elsevier
Radarweg 29, PO BOX 211, 1000 AE Amsterdam, The Netherlands
Linacre House, Jordan Hill, Oxford OX2 8DP, UK
32 Jamestown Road, London NW1 7BY, UK
30 Corporate Drive, Suite 400, Burlington, MA 01803, USA
525 B Street, Suite 1900, San Diego, CA 92101-4495, USA

First edition 2010

ISBN: 978-0-12-380856-1
ISSN: 0065-2318

British Library Cataloguing in Publication Data
A catalogue record for this book is available from the British Library

Library of Congress Cataloging-in-Publication Data
A catalog record for this book is available from the Library of Congress

For information on all Academic Press publications
visit our website at books.elsevierdirect.com

Printed and bound in USA

10 11 12 13 10 9 8 7 6 5 4 3 2 1

CONTENTS

Multivalent Lectin—Carbohydrate Interactions: Energetics and Mechanisms of Binding

TARUN K. DAM AND C. FRED BREWER

Design and Creativity in Synthesis of Multivalent Neoglycoconjugates

YOANN M. CHABRE AND RENÉ ROY

LIST OF CONTRIBUTORS

C. Fred Brewer, Department of Molecular Pharmacology, Microbiology and Immunology, Albert Einstein College of Medicine, Bronx, New York, 10461, USA

Klaus Buchholz, Department of Technical Chemistry, Technical University, Braunschweig, D-38106, Germany

Yoann M. Chabre, Department of Chemistry, Université du Québec à Montréal, Montréal, Québec, H3C 3P8, Canada

Tarun K. Dam, Department of Molecular Pharmacology, Albert Einstein College of Medicine, Bronx, New York, 10461, USA

R. Colin Hughes, National Institute for Medical Research, Mill Hill, London, UK

Lennart Kenne, Department of Chemistry, Swedish University of Agricultural Sciences, Uppsala, Sweden

Olle Larm, ExThera AB, Karolinska Science Park, Stockholm, Sweden

Alf Lindberg, Department of Clinical Bacteriology, Karolinska Institutet, Huddinge University Hospital, Huddinge, Sweden

Susana D. Lucas, Centro de Química e Bioquímica/Departamento de Química e Bioquímica, Faculdade de Ciências da Universidade de Lisboa, Ed. C8, 5° Piso, Campo Grande, 1749-016, Lisboa, Portugal

Amélia P. Rauter, Centro de Química e Bioquímica/Departamento de Química e Bioquímica, Faculdade de Ciências da Universidade de Lisboa, Ed. C8, 5° Piso, Campo Grande, 1749-016, Lisboa, Portugal

René Roy, Department of Chemistry, Université du Québec à Montréal, Montréal, Québec, H3C 3P8, Canada

Miguel Santos, Centro de Química e Bioquímica/Departamento de Química e Bioquímica, Faculdade de Ciências da Universidade de Lisboa, Ed. C8, 5° Piso, Campo Grande, 1749-016, Lisboa, Portugal

Jürgen Seibel, Institute of Organic Chemistry, University of Würzburg, Am Hubland, D-97074, Würzburg, Germany

Nathan Sharon, Weizmann Institute of Science, Biological Chemistry Department, Rehovot, Israel

Nuno M. Xavier, Centro de Química e Bioquímica/Departamento de Química e Bioquímica, Faculdade de Ciências da Universidade de Lisboa, Ed. C8, 5° Piso, Campo Grande, 1749-016, Lisboa, Portugal

PREFACE

Synthesis has been a sustained area of major interest in carbohydrate science, and the preceding Volume 62 of this series featured three chapters focusing in detail on the construction of glycosidic linkages. Many established synthetic methods permit the elaboration of complex target molecules, but frequently involve the use of tedious protection–deprotection sequences and expensive or hazardous reagents. This current Volume 63 of *Advances* presents two articles that offer promise of useful methodologies for simplified procedures amenable to low-cost large-scale applications, using mild conditions and environmentally friendly materials.

Rauter and her coauthors Xavier, Lucas, and Santos (Lisbon) present here a detailed overview of the potential for heterogeneous catalysts in useful synthetic transformations of carbohydrates. Such silicon-based catalysts as zeolites are easy to handle and recover, are nontoxic, and can offer interesting possibilities for exercising stereo- and regio-control in many established carbohydrate transformations.

In the midst of wide-ranging research on the role of complex oligosaccharides in biological recognition processes, and the attendant focus on synthesis of such molecules, the important role of oligosaccharides in large-scale commercial applications is often overlooked. The contribution by Seibel and Buchholz (Wurzburg and Braunschweig) in this volume addresses in detail those tools of particular value for preparation of oligosaccharides that serve needs in the food, pharmaceutical, and cosmetic industries. Major emphasis is devoted to the use of readily available enzymes (glycosidases, glycosynthases, glucansucrases, fructansucrases) and abundant carbohydrate substrates, especially sucrose, and the applications of enzyme and substrate engineering. Particular attention is given to those large-scale applications of oligosaccharides that serve as sweeteners, as well as promising new medical applications.

Two complementary treatments deal with different aspects of the intense current interest in the biological recognition phenomena between carbohydrates and proteins. Dam and Brewer (New York) examine in detail the energetics and mechanisms of binding between lectins (carbohydrate-binding proteins) and the multivalent glycoprotein receptors on the surface of normal and transformed cells, as well as other types of carbohydrate receptors, including linear glycoproteins (mucins). The authors postulate a common ''bind-and-jump'' mechanism that involves enhanced entropic effects which facilitate binding and subsequent complex formation.

In an extensive and comprehensive survey, Chabre and Roy (Montreal) revisit the subject of neoglycoconjugates introduced three decades ago by Stowell and Lee in Volume 37 of this Series. It was then a nascent topic, and the Montreal authors now bring together in a single large article the vast new literature base that has subsequently evolved in the field of the ''glycoside cluster effect.'' The years have witnessed much creativity in the design and strategies of synthesis that have afforded a wide array of novel carbohydrate structures, and reflect the ongoing dynamic activity in this rapidly evolving area, even since the recent article by the Nicotra group in Volume 61 of this series. Elaborate nanostructures, now termed glycoclusters and glycodendrimers, feature arrays of carbohydrate epitopes joined via ligands onto a variety of scaffolds, including calixarenes, porphyrins, and such carbon nanostructures as fullerenes and nanotubes. Their synthesis and characterization is addressed in detail along with evaluation via such techniques as microarrays and other modern analytical techniques, for their potential in application to biological systems.

The contributions of two of the world's leading carbohydrate innovators are recognized in this issue. The work of Roger Jeanloz evolved from a classical background in synthesis and structure elucidation to wide applications in the biological area which have led him to be considered as the father of the subject now known as glycobiology. He contributed extensively to these *Advances*, with articles in Volumes 6, 11, 13, and 43 of the series, and his life and scientific work is the subject of the obituary article by Sharon (Israel) and Hughes (London). The article by Kenne, Larm, and Alf Lindberg (Stockholm) surveys the career of Bengt Lindberg, and especially notes Lindberg's development of the methodology for microscale analysis of carbohydrate structures that has permitted determination of the sequence structure of minute samples of oligo- and poly-saccharides from biological sources and has enabled the explosive growth of modern research in glycobiology. Lindberg was also a notable contributor to this series, with articles in Volumes 15, 31, 33, and 48 that document in Lindberg's classic terse style the evolution of structural methodology from early beginnings to sophisticated applications, in particular for bacterial polysaccharides.

The death on October 8, 2009 of Antonio Gómez Sánchez is noted with regret. He was one of the successors of the Seville carbohydrate school built up by Francisco García González, with whom he coauthored in Volume 20 a chapter on the reaction of amino sugars with 1,3-dicarbonyl compounds, a subject that was a major theme of Gómez Sánchez's research.

Sincere thanks are expressed to Professors Stephen Angyal and J. Grant Buchanan for their many years of advice and support as members of the Board of Advisors. Welcomed as a new member of the Board is Professor Todd Lowary.

Derek Horton

Washington, DC
November, 2009

ROGER W. JEANLOZ

1917–2007

Roger William Jeanloz, who passed away on September 12, 2007, 6 weeks short of his 90th birthday, was among the earliest pioneers in the field now known as Glycobiology. He made seminal contributions to the subject, and trained a number of leaders in the field. He was one of the prime organizers in the 1950s of the Glycosaminoglycan, Glycoprotein, and Glycolipid Group (known as the 4Gs), later named the Society of Complex Carbohydrates, and eventually the Society for Glycobiology, of which he served as President in 1974.

From Glycogen and Deoxy Sugars to Complex Carbohydrates

Jeanloz was born on November 3, 1917 in Berne, Switzerland, to a French mother and a Swiss–German father. He was brought up in French-speaking Geneva where he pursued classical studies emphasizing Greek and Latin at College Calvin. In 1936, he was awarded the B.S. degree from this College and in 1941 a Diploma in Chemical Engineering from the University of Geneva, where he studied organic chemistry and biochemistry. His keen interest in science and research began in 1941 when Kurt H. Meyer, well known for his pioneering studies on cellulose and starch, accepted him as a doctoral student. In Meyer's laboratory, Jeanloz investigated the enzymatic degradation of starch[1] and then the structure of muscle glycogen.[2] The latter studies were the subject of his D.Sc. thesis, completed in 1943. At the same time he served as head instructor for organic chemistry.

In 1943, after being awarded the D.Sc. degree, Jeanloz was appointed as Research Associate, first with Meyer and then with Tadeusz Reichstein, Nobel Laureate for his work on steroid hormones. With Reichstein he studied the chemistry of deoxy sugars, some of which are constituents of these hormones, and developed[3] a new method for the assay of these sugars. In 1946–1947 he spent 1 year in Canada as Research Associate at the University of Montreal, where he collaborated with D. A. Prins from the Research Division, Cleveland Clinic, Cleveland, OH, in the preparation of

ISBN: 978-0-12-380856-1
DOI: 10.1016/S0065-2318(10)63001-6.

a review of the chemistry of carbohydrates[4] for the prestigious *Annual Reviews of Biochemistry*, the first of many important ones he wrote during his scientific career. He then moved to the United States, where he worked for 1 year as a Senior Research Fellow in the National Institutes of Health laboratory of the noted carbohydrate chemist, Claude S. Hudson, where he became involved in the study of ribose and its derivatives[5–7], and he contributed with Hewitt Fletcher from the same laboratory a review[8] on the chemistry of ribose in the fledgling *Advances in Carbohydrate Chemistry*. The following 3 years were spent at the Worcester Foundation for Experimental Biology, then under the direction of the noted endocrinologist Gregory Pincus, working on steroid hormones, a subject on which he continued to collaborate with the Pincus group for several years thereafter.[9–14]

In 1951, Jeanloz was invited by Dr. Walter Bauer, Chief of the Medical Services and of the Arthritis Unit at the Massachusetts General Hospital, to become a member of the Robert W. Lovett Memorial Group for the Study of Crippling Diseases, and to organize a laboratory for the study of the chemical structure of the polysaccharides of connective tissue and of related biochemical problems. Ten years later, he was appointed as the Head of the newly formed Laboratory for Carbohydrate Research of the Lovett Group, and in 1969 as Professor of Biological Chemistry and Molecular Pharmacology at Harvard Medical School. He held these positions until his retirement, and continued to be active in research and teaching thereafter. It was in the Carbohydrate Research laboratory that Jeanloz, during half a century, made major contributions to our knowledge of the structure, biosynthesis, and function of complex carbohydrates, an area that now falls under the subject of glycobiology. Indeed, he may be considered as one of the early glycobiologists.

Most prominent among his contributions were the elucidation of glycosaminoglycan structures, chiefly by methylation analysis; the synthesis of rare amino sugars; establishment of the structure of the polysaccharide backbone of the bacterial cell-wall peptidoglycan; providing the foundation for structural studies of the carbohydrate moieties of N- and O-linked glycoproteins; synthesis of many of the glycopeptide constituents of glycoproteins; detailed structural analysis as well as synthesis of several of the dolichyl sugar phosphates involved in protein glycosylation; characterization of glycans accumulated in lysosomal storage diseases, and the action pattern of catabolic glycosidases. In addition, analysis of TA-3 glycoprotein initiated in his laboratory led to the most detailed investigation of any tumor-related glycoprotein examined at the time. His work has had implications in many areas of biology, among them biochemistry and bacteriology, immunology, and cancer research. A dominant feature of much of the this work was Jeanloz's emphasis on the importance of chemical synthesis of component parts of complex glycoconjugates in order to distinguish, for example, D or L enantiomers, ring structures and α or β anomers, in the overall structure of biologically active macromolecules.

Applications of Methylation Analysis

Before joining the Lovett Group, Jeanloz devoted considerable effort to understanding the periodate oxidation reaction, starting with an examination of its effects on glucosamine.[15] He and Enrico Forchielli then used it to examine the structure of hyaluronic acid[15,16] and chitin.[17] It rapidly became evident to him that this method, as was the case in the successful determination of the structure of simpler polysaccharides such as glycogen or starch, required supplementation with other ones, particularly the methylation technique. The latter technique had already been attempted in the field of glycosaminoglycans, but without positive results. The main difficulty resided in the fact that degradation of a methylated heteropolysaccharide, made up of repeating units of hexosamine and uronic acid with different linkages, could give rise to a large number of mono- di- and tri-*O*-methylated monosaccharides. It was therefore necessary to synthesize all of the reference substances, and to separate and identify them in artificial mixtures. Jeanloz was encouraged in this long and arduous task, which required gram quantities of starting materials, by the knowledge that if shown to be efficient, the method could be applied not only to the elucidation of the structure of the glycosaminoglycans but also to other classes of complex carbohydrates.

In the course of the 1950s Jeanloz synthesized 13 methylated derivatives of glucosamine and galactosamine (reviewed in Ref. 18). These included the 4-methyl[19] and 6-methyl[20] ethers of glucosamine hydrochloride and the 3,6- and 4,6-dimethyl ethers of this hexosamine.[21] Concurrently the 3-[22] 4-,[23] and 6-[24] monomethyl ethers of galactosamine hydrochloride, and its 3,6-[25] and 4,6-[26] dimethyl ethers were also prepared. Methylated derivatives of other monosaccharides were also synthesized subsequently, among them those of muramic acid, with Harold Flowers[27] and *N*-acetylmannosamine with Nasir-ud-Din.[28]

The methylated glucosamine and galactosamine derivatives in particular were in such demand by other laboratories as reference compounds that one research assistant worked full time solely to replenish supplies of these compounds. In this synthetic approach, which took 10 years to accomplish, Jeanloz was helped by several associates, including his wife Dorothy, and Pierre J. Stoffyn who joined his group in 1953 and stayed until 1961. One of the many results of their productive collaboration was the Stoffyn and Jeanloz method[29] for identification of hexosamines by ninhydrin degradation to the corresponding pentoses, which is still the simplest method to distinguish between glucosamine and galactosamine.

Connective Tissue Glycosaminoglycan Structure

The methylated derivatives served as the basis of elegant studies in which Jeanloz and his colleagues unequivocally established the structure of hyaluronan, dermatan sulfate, and the chondroitin sulfates. They also clarified many confusing issues, including the position of the sulfate groups in these polymers. In the course of this work, they proved that dermatan sulfate contained L-iduronic acid, a sugar not known before to occur in nature.

Application of the methylation technique to hyaluronan established the linkage at position 3 of the glucosamine☆ moiety and the linkage at position 4 of the glucuronic acid moiety.[30] Eventually, hyalobiouronic acid [β-GlcA-(1→3)-GlcNAc], the repeating unit of hyaluronan was synthesized; it was identical with one of the acid degradation products of the parent polysaccharide.[31,32] Jeanloz and his colleagues also isolated methyl derivatives of hyalobiouronic acid from the methylated polysaccharide.

Methylation analysis of chondroitin sulfate A showed that the sulfate is located at position 4 of the galactosamine moiety, that the *N*-acetyl-β-galactosamine residue is located at position 4 of the glucuronic acid moiety, and that galactosamine possesses a pyranose structure. This method gave also additional evidence for a β-(1→3) interglycosidic linkage between glucuronic acid and *N*-acetylgalactosamine residues, consistent with a repeating disaccharide unit of β-glucuronate-(1→3)-4-sulfo-*N*-acetyl-β-galactosamine-(1→4). Using the same technique, Jeanloz also investigated the structure of chondroitin sulfate C, and provided evidence that chondroitin sulfate C is an unbranched polysaccharide composed of alternate residues of glucuronic acid and *N*-acetylgalactosamine 6-sulfate residues β-glycosidically linked at positions (1→3) and (1→4), respectively (see Ref. 33 for a review).

Dermatan sulfate, also termed chondroitin sulfate B, a related glycosaminoglycan constituent of connective tissue, was known to be composed of galactosamine and a uronic acid, originally believed to be glucuronic acid but then claimed to be iduronic acid based largely on color reactions and paper chromatography. However, the D or L-enantiomer status of the latter monosaccharide was not clear. Jeanloz and Stoffyn unequivocally characterized the monosaccharide as L-iduronic acid by consecutive desulfation, reduction, and hydrolysis of the polysaccharide, followed by isolation of the crystalline 2,3,4-tri-*O*-acetyl-1,6-anhydro-β-L-idopyranose, which was shown to be identical to an authentic specimen synthesized from 1,2-*O*-isopropylidene-β-L-idofuranose.[34]

☆All sugars are of the D configuration, unless otherwise noted.

During this careful analytical work on chondroitin and dermatan sulfates, Jeanloz noted that the stoichiometry of the purified glycosaminoglycans indicated significant heterogeneity, in particular variable sulfation. It is now known that such variability in sulfation and uronic acid content significantly impacts on the ability of connective-tissue polysaccharides to bind a diverse range of biologically active molecules, including growth factors and protease inhibitors.

Mammalian Membrane Glycolipids

As early as 1955, Jeanloz felt that the methods devised for the study of connective-tissue glycosaminoglycans could be applied to other classes of glycoconjugates. Together with Sen-itiro Hakomori, who came to the Laboratory for Carbohydrate Research from the University of Sendai in Japan, he undertook the onerous task of purifying the substances conferring human erythrocyte blood-type specificity. Extraction of erythrocytes of human A and B blood type afforded small amounts of glycolipids ($\sim$1 mg/l of blood) possessing blood-group activity. Purification was obtained by adsorption on activated alumina and activated silica gel, followed by partition chromatography on cellulose. The active substances isolated from the two different blood types showed great similarities. Partial hydrolysis showed a resistant core composed of fatty acid, sphingosine, glucose, and galactose for the substances isolated from both blood types. In both substances, additional residues of either galactose or galactosamine constituents were located at, or near, the extremities of the carbohydrate chain, hinting at the possibility that these sugars were part of the blood type immunodeterminants.[35] Subsequently, they isolated from human cancerous tissue a glycolipid rich in fucose,[36] and together with Jerzy Koscielak and K. J. Bloch showed that this glycolipid cross-reacted immunologically with human blood-group substances.[37] In addition, the first definitive evidence of the immunogenicity of glycolipids was obtained by immunizing rabbits with a purified glycolipid (globoside) with the formation of a precipitating antigloboside antibody.[38] The antibody agglutinated and hemolyzed human red cells, provided that they were pretreated with trypsin.

Bacterial Cell-Wall Peptidoglycan

In the early 1960s Jeanloz became intrigued by peptidoglycan, the rigid constituent of bacterial cell walls that endows these microorganisms with their characteristic shape, whether round (cocci) or elongated (bacilli). At the time it was believed that the polysaccharide backbone of peptidoglycan consists of a repeating disaccharide in which *N*-acetylglucosamine is β-(1$\rightarrow$6) linked to muramic acid (3-*O*-D-lactyl-*N*-acetylglucosamine), and that linkage between the disaccharide units in the backbone is β-(1$\rightarrow$4). To establish unequivocally the structure of the disaccharide, the synthesis of muramic acid and its various derivatives, including the disaccharide β-GlcNAc-(1 → 6)-MurNAc was performed in the Jeanloz lab by Harold Flowers[39–41] and by Toshiaki Osawa.[42] These included various methyl ethers[43] and glycosides,[44] and acetyl derivatives[45] of muramic acid and its *galacto*-[46] and *manno*-[47] analogues, in the synthesis of several of which Pierre Sinaÿ was involved. Additionally, muramic acid 6-phosphate was synthesized[48] and the absolute configuration of the carboxy-ethyl (lactyl) side-chain of muramic acid was established.[49] In the course of these studies, the peptidoglycan polysaccharide backbone was shown to contain some disaccharide units having the *manno* (rather than *gluco*) analog of muramic acid.[50] Methods were developed quite early on for the large-scale preparation of bacterial cell walls from *Micrococcus lysodeikticus* (later renamed *M. luteus*)[51] and for the isolation of the natural disaccharide from lysozyme digests of the walls. Comparison of the synthetic disaccharide with the natural one revealed quite surprisingly that the two were not identical. The conclusion was that the natural disaccharide is β-(1 → 4)-linked, as is the polysaccharide backbone throughout.[52] This was confirmed by comparison of the natural disaccharide with synthetic β-GlcNAc-(1 → 4)-MurNAc. The bacterial polysaccharide is therefore chemically homologous to chitin.

One of the minor by products of the isolation procedure of the disaccharide from lysozyme digests of bacterial cell walls was the tetrasaccharide β-GlcNAc-(1→4)-β-MurNAc-(1→4)-β-GlcNAc-(1→4)-MurNAc. This structure is readily cleaved by lysozyme, and has proved to be extremely useful in other laboratories for the study of the mechanism of action of the enzyme.

Invertebrate Matrix Glycoconjugates

In the late 1960s, Jeanloz began an extended study of the extracellular matrix of invertebrates. Earlier, Jeanloz's colleague in the Lovett Unit, Jerry Gross, had pointed out that invertebrate matrices contained greater and more varied amounts of carbohydrate than vertebrate matrices. The Jeanloz laboratory, in particular Richard Katzman, together with Andy Kang of the Gross laboratory, set out to identify these poorly characterized glycoconjugates from three invertebrate phyla, the sea anemone *Metridium diocanthus*, the sea cucumber *Thyone briareus*, and a marine sponge *Hippospongia gossypina*.[53–55] The major oligosaccharide chain of the invertebrate collagens, present in higher amounts than in vertebrate collagens, was shown to be α-glucopyranosyl-(1 → 2)-β-galactopyranosyl-(1 → 5)-L-lysine, similar to the major glycan of vertebrate collagen. Asparagine-linked *N*-glycans containing glucosamine, fucose, and mannose were also present. Further studies showed the presence in sea cucumber matrix of two major acidic glycosaminoglycans, namely a chondroitin sulfate similar in structure to vertebrate chondroitins, and a novel polyfucose sulfate. This latter component was named thyonatan and characterized as an α-(1→2)-linked polyfucose-4-sulfate polysaccharide. The absence of an amino sugar distinguishes thyonatan from all known vertebrate connective-tissue glycoconjugates. Chondroitin and thyonatan were not found in sponges and sea anemone, leading to the speculation that acidic glycoconjugates play no crucial role in collagen fibril formation, contrasting with the prevailing idea at the time that regularly spaced acidic groups may act as a template upon which collagen fibrils are built up. Interestingly, arabinose was isolated from sponge connective tissue: its identity was confirmed by the preparation of a crystalline derivative of benzyl-1-phenylhydrazine.[56] Although arabinose is a common sugar in plants, it is very rare in animals. Katzman proposed that the arabinofuranosylnucleosides of uracil and thymidine, isolated in the 1950s from sponges, may serve in biosynthesis of arabinose-containing polysaccharides in these organisms. Interest in invertebrate matrix glycoconjugates has recently increased. Some of these matrices, for example the sea cucumber, have an ability to alter their stiffness in response to changes in pH or ionic concentrations, creating the possibility of deriving new dynamic neocomposites for biomedical applications.

Rare Amino Sugars

The synthetic studies of Jeanloz led to the solution of many other questions in carbohydrate chemistry, especially that of uncommon amino sugars. Resulting from these studies was the synthesis, mainly with Dorothy Jeanloz and with Sonia Tarasiejskaya of several amino sugars, some of which for instance, gulosamine,[57,58] had been shown to be constituents of aminoglycoside antibiotics, such as the streptothricins, and others that at the time were not known to occur in Nature. These included allosamine and talosamine, 3-amino-3-deoxyidose, and 3-amino-3-deoxygulose, as well as 2,4-diamino-2,4-dideoxyglucose.[59–61] Bacillosamine (2,4-diamino-2,4,6-trideoxyglucose), another rare amino sugar, was first isolated in the Jeanloz laboratory by Nathan Sharon from a bacterial polysaccharide that he prepared at the Weizman Institute, Rehovot.[62] This structure, determined later elsewhere, has recently been identified as a constituent of the carbohydrate–peptide linking group of glycoproteins of many eubacterial pathogens, and appears to be required for their virulence. It is therefore attracting considerable current interest.

Structural Analysis of Protein-Bound Oligosaccharides

A number of the central themes of current glycoconjugate research can be traced back to studies begun in the Jeanloz laboratory. One prominent example is the elucidation of the sequences of the carbohydrates linked *N*- or *O*-glycosylically to asparagine or serine/threonine residues respectively in glycoproteins. Little was known about these structures at the time: rapid methods have subsequently been developed, providing information that is crucial in glycomics, describing the variety of glycan structures in different organisms and tissues. Work was initiated in the late 1950s by the Jeanloz laboratory on two serum glycoproteins, alpha 1-acid glycoprotein and fetuin by Ed Eylar and Robert Spiro, respectively.[63–66] In this work, purified glycosidases, such as *Diplococcus pneumoniae* neuraminidase and β-*N*-acetylglucosaminidase, were used for the first time. Sequential removal of sugars from the nonreducing terminals of the glycans by the glycosidases led to rational, if partial, proposals for oligosaccharide sequence. Additional information was obtained by mild acid hydrolysis for fragment sequencing, methylation analysis, and sequential periodate oxidation (Smith degradation). For the first time, the common presence of *N*-acetyllactosamine at the termini of complex-type asparagine-linked glycans was demonstrated and shown unequivocally by analysis of the crystalline disaccharide. Jeanloz also suggested from studies with alpha 1-acid glycoprotein that the inner core, linked to asparagine directly, was composed of two *N*-acetylglucosamine and several mannose residues. As additional proof of the proposed structures Jeanloz and his colleagues, in particular Hari Garg, used chemical synthesis.[67] The carbodiimide coupling reagent was employed extensively for the synthesis of asparagine-linked compounds such as β-GlcNAc-N-Asn. Syntheses were also described of disaccharide fragments such as β-GlcNAc-(1$\rightarrow$4)-GlcNAc and trisaccharides such as α-Man-(1$\rightarrow$6)-β-GlcNAc-(1$\rightarrow$4)-GlcNAc. These were coupled to give the corresponding asparagine-linked derivatives. Subsequently, methods were developed for the synthesis of O-glycans, based on the core structure α-GalNAc1-Ser.[67]

Lipid Intermediates in Glycoprotein Biosynthesis

Long-chain isoprenoid alcohols, polyprenols and dolichols, were identified in the 1970s as intermediates in biosynthesis of bacterial polysaccharides as well as in the multistep process of protein glycosylation in plants and animals. The donor that initiates N-linked glycan synthesis is a $Glc_3Man_9GlcNAc_2$ structure attached to the lipid dolichol through a pyrophosphate linkage. The enzymes that build up a dolichyl pyrophosphate-intermediate containing two *N*-acetylglucosamine residues and the first five mannose residues utilize sugar nucleotides directly. Transfer of the last four mannose residues and the three glucose residues are mediated by dolichyl phosphate-linked sugars. Progress in this research was hampered by the low abundance of the phosphorylated dolichol intermediates in cell extracts. Jeanloz and his colleagues, in particular Chris Warren and Annette Hercovics and initially in a very productive collaboration with Jack Strominger's laboratory at Harvard, undertook the unequivocal chemical synthesis of these still poorly characterized molecules, and their utilization in biochemical studies for comparison with the natural compounds (see for example, Refs. 68–70). Although the work raised many difficult problems, a novel methodology was developed for the first facile synthesis of dolichol phosphate and pyrophosphate, and their sugar derivatives. Although it was generally assumed at the time that the lipid moiety in the lipid intermediates was dolichol, this had not been established as insufficient material was available for chemical characterization: similarly the anomeric identity of the sugar intermediates was not known. To address these points, the radioactively labeled lipid intermediates formed in tissues such as calf pancreas microsomes were characterized by comparison with the synthetic compounds. Using this approach, the formation in tissues was established of dolichyl-β-mannosyl phosphate, dolichyl-β-glucosyl phosphate and dolichyl-β-*N*-acetylglucosaminyl pyrophosphate. A lipid-linked di-*N*-acetylchitobiose structure formed in the microsomal fraction was shown unequivocally to be identical to synthetic di-*N*-acetylchitobiosyl-dolichyl pyrophosphate. In a technical *tour de force* the pyrophosphate-linked lipid intermediate containing a substantial part of the natural $Glc_3Man_9GlcNAc_2$ lipid intermediate was prepared.[71] In this work, the synthetic chemistry in the Jeanloz laboratory went hand in hand with the biochemistry, and made a major contribution to understanding of the mode of assembly of the asparagine-linked carbohydrates.

Lysosomal Glycosidase Deficiencies

In the late 1980s, Jeanloz and Chris Warren, in collaboration with Brian Winchester from the Institute of Child Health, University of London and Peter Daniel from the Kennedy Schriver of Mental Retardation, Boston, became interested in the lysosomal storage disease mannosidosis, as well as in the toxicosis induced in sheep by ingestion of foodstuffs containing the toxin swainsonine, a powerful mannosidase inhibitor.[72–75] In part this interest was prompted by the need for relatively large amounts of mannose-rich oligosaccharides for chemical synthesis of dolichol intermediates and glycopeptides. Urine from patients with no or low levels of lysosomal mannosidases is a convenient source of such oligosaccharides. The work unexpectedly revealed some interesting data on the patterns of glycan catabolism. The structures of oligosaccharides accumulated and excreted by mannosidosis patients, as well as from cats and cattle genetically affected by loss of lysosomal mannosidases, were characterized. In all cases, the structures were consistent with the incomplete breakdown of N-linked glycans or biosynthetic intermediates. There were clear interspecies differences, however. Compounds containing a core of di-*N*-acetylchitobiose, instead of a single *N*-acetylglucosamine residue were isolated from cattle and cats but were absent in human mannosidosis. Human tissues alone contain a lysosomal endo-β-*N*-acetylglucosaminidase, accounting for this difference. There were also major differences in the structures of the excreted oligomannose compounds, presumably reflecting differences in substrate specificity of lysosomal mannosidases in the three species. In human mannosidosis, where there is a complete deficiency of the major lysosomal mannosidase, the structure and levels of the stored oligosaccharides appeared to result from the action of a minor lysosomal mannosidase with specificity for an α-(1$\rightarrow$6) linkage. In feline mannosidosis, a severe deficiency of lysosomal mannosidases leads to storage of intact $Man_9GlcNAc_2$ oligosaccharides and the undegraded Man_3-$GlcNAc_2$ core of complex-type glycans. In further studies, the detection of abnormal catabolic products in placenta was found to offer a promising early diagnosis of lysosomal mannosidase deficiency.[76]

Tumor-Cell Glycoproteins

At the end of the 1960s, there was considerable interest in emerging evidence that altered cell surface glycosylation appeared to be common in malignant and premalignant epithelial and some nonepithelial cells. The presence at cell surfaces of large amounts of carbohydrate-containing macromolecules had previously been demonstrated by histochemical staining. However, the identity of these components and their relevance to carcinogenesis was poorly understood. At this time, the Jeanloz laboratory, in particular John Codington, began work on the glycoconjugates of tumor cells in collaboration with Barbara Sanford of the MGH Pathology Division.[77–80] Previous work had shown that some sub-lines of the TA3 tumor cell, derived from a spontaneous mammary adenocarcinoma of an A-strain mouse, have a surface membrane enriched in carbohydrate. It was also known that these sub-lines had lost strain specificity and were transplantable into strains of allogeneic mice, thus evading immunological rejection across major histocompatibility barriers. Sanford had shown further that successful transplantation of the TA3 sub-lines into allogeneic mice was markedly reduced by treatment of the cells with neuraminidase. The suggestion was that these cells escape the host defense system through cell surface antigen masking by highly sialylated cell-surface glycoconjugates. In 1972, the major sialylated glycoconjugate of TA3 cells was isolated in the Jeanloz laboratory and called epiglycanin. As shown a little later epiglycanin turned out to be a surface membrane-associated mucin containing a typical serine and threonine-rich tandem-repeat sequence, a transmembrane domain and a cytoplasmic domain. The extracellular domain was shown by Dirk van den Eijden and Codington to be heavily substituted with O-glycans, based on the core structure β-Gal-(1$\rightarrow$3)-GalNAc attached to serine or threonine residues, heavily substituted with sialyl residues. Immunoelectron microscopy showed that epiglycanin covers the TA3 cell surface with an extended filamentous coat. Crucially, this glycoprotein was found not to be present, or present in low amounts, in TA3 tumor sublines that cannot grow in allogeneic hosts, offering strong corroboration of the idea that cell surface epiglycanin sterically hinders cytotoxic T-cells from mounting an immune response against the tumor. Many types of cancer cells are now known to possess large concentrations of mucin-like substances at their cell surface. These, like epiglycanin, are characterized by the presence of multiple *O*-glycosylically-linked glycans and extended rod-like conformations. The evidence indicates that these structures, like epiglycanin, play roles in hindering immune surveillance mechanisms and increasing metastatic potential. Recently, the human orthologue of epiglycanin has been cloned and named Muc21. It is expressed in many malignant tissues, including lung adenocarcinomas and breast cancers, and is of much current interest.

Editor, Symposia Organizer, and Expert on Carbohydrate Nomenclature

The productivity of Jeanloz is exemplified by his bibliography of more than 400 publications. He authored many reviews, which are outstanding for their clarity, concise style, and interpretation. He also edited several books, the most notable of which is *The Amino Sugars*, a three-volume treatise coedited with Endre Balazs.[81] He was particularly concerned with the nomenclature of carbohydrates and was instrumental in resolving key issues and introducing many new and well-defined terms such as hyaluronan and glycosaminoglycan.[82,83] In his publications, he was always careful in using the full chemical nomenclature of carbohydrates. Thus, he favored the term 2-amino-2-deoxy-D-glucose for *N*-acetylglucosamine, and almost completely avoided the use of the abbreviation ''GlcN.'' In addition to his research and teaching activities at Harvard and elsewhere throughout the world, Jeanloz served on numerous committees and editorial boards; among others, he was one of the founders in the early 1960s of the journal *Carbohydrate Research*, the first of its kind devoted to the subject.

The community of carbohydrate scientists is especially indebted to him as a central figure in the setting up and organization of the International Symposia on Glycoconjugates. The first of these Symposia was convened in 1965 in Swampscott with Jeanloz as Co-Chairman. He was also the prime mover in encouraging his European colleagues to organize the second and third meetings in Lille in 1973 and in Brighton in 1975, respectively, and was active in the organization of the 4th Symposium on Glycoconjugates held at Wood's Hole. Together with John Gregory he edited the two-volume proceedings of this Symposium.[84] In 1974–1975, Jeanloz was president of the National Society of Complex Carbohydrates. He was also active for many years in the Division of Carbohydrate Chemistry of the American Chemical Society, of which he served as Chairman. His numerous achievements have brought him many national and international honors. These include the Medal of the Société de Chimie Biologique de France; Medal of the University de Liége; Prix Jaubert, University of Geneva; Claude S. Hudson Award, American Chemical Society; Docteur Honoris Causa, University of Paris; and recently the Alexander von Humboldt Award, Germany's most prestigious scientific award. Posthumously, the second edition of *Essentials in Glycobiology*, edited by Ajit Varki *et al.* and published in 2008, is dedicated to his memory and that of Rosalind Kornfeld ''pioneers in the elucidation of glycan structure and function.''

Dedicated Teacher, Avid Sportsman, and Inveterate Traveler

The scope of Jeanloz's activities, and the knowledge and enthusiasm that he imparted to his colleagues, all combined to inspire his many students. He was for many years a tutor with the Faculty of Arts and Sciences of Harvard University, retiring from this activity only a few months before his death. Because of his qualities, he was able to attract and in many cases keep able associates to his laboratory and frequently continued collaborations with them after they left. The excellent training they received from Jeanloz and the contacts and friendships made in his laboratory exercised decisive influence on their careers. His demonstration of the importance of applying the rigorous techniques of organic chemistry to the solution of biochemical problems was the most crucial point of this training.

The uniquely congenial atmosphere in the Jeanloz laboratory that is remembered with affection by those who worked with him was aided greatly by the presence of his charming wife, Dorothea, an organic chemist, who collaborated with him for many years. In spite of their busy schedules, Dorothea and Roger always took great interest in the well-being of their associates. Jeanloz's personal qualities and bon vivant character endeared him to all of his colleagues.

Jeanloz was an inveterate traveler, who taught students from about as many countries as he had visited, a total difficult to tally in any category. In his younger years he was a skilled alpinist, and made many of the classical climbs in the Swiss French Alps. He was also an expert skier, a sport he continued to enjoy into his more advanced years, often accompanied by his grandchildren. Jeanloz was a lover of classical music, and an enthusiastic gardener. His love for plants came early in his life when he brought bouquets of alpine flowers from mountain expeditions. His interest in sports can be traced to his childhood. At the age of 12, he started playing basketball, a sport that had been introduced into Switzerland by the American YMCA. He was a member of the Geneva team that was selected to represent Switzerland in international games. Later he became a keen tennis player, and was a familiar figure on the tennis courts of the MGH, and in the Longwood Cricket Club near the family home in

Newton, Massachusetts, where they raised their four children Danielle, Sylvie, Raymond, and Claude. For many of the later years Roger and Dorothy enjoyed time in their delightful second home in Tourettes-sur-Loup in the south of France, where they were often generous hosts to friends and colleagues.

In 1986, a Symposium was held to honor Roger Jeanloz as part of the 8th International Symposium on Glycoconjugates in Houston, Texas. For this occasion the eminent British scientist Albert Neuberger wrote: ''His (Jeanloz's) work is characterized by a mastery of all chemical and biological methods, by reliability of the results obtained, and by careful and balanced interpretation. There is hardly any topic in this wide field, which he has not touched, and which has not benefited from his research. Roger Jeanloz has made his laboratory one of the few outstanding centers in the field of complex carbohydrates. His combination of supreme chemical competence and biological understanding are the hallmark of his career.''

Roger Jeanloz will be remembered with great affection and utmost respect by his many friends, colleagues, and disciples wherever they are.

List of associates who worked with Jeanloz at the Laboratory of Carbohydrate Research:

E. Alpert, J. Alroy, C. Augi, L. S Azaroff, J. Badet, N. Baggett, E. A. Balasz, M. Beppu, A. K. Bhattacharaya, A. Bhattacharyya, V. P. Bhavanandan, P. Biely, K. C. Blieszner, G. Blix, D. J. Bloch, K. J. Bloch, K. J. Bloch Sr., C. H. Bolton, A. A. Bothner, M. C. Brown, B. Bugge, R. A. Byrn, G. F. Cahill, M. S. Choudhary, A.-M. Close, J. F. Codington, A. G. Cooper, T. Dalianis, P. F. Daniel, D. Daniels, D. M. Darby, H. R. Das, E. A. Davidson, P. Degand, M. R. Dick, F. DuBois, D. van den Ejnden, J. V. Ellard, N. A. Evans, E. H. Eylar, N. E. Fayaz-ud-Din, H. M. Flowers, R. Fricke, D. M. Frim, T. C. Fuller, C. Gansser, H. G. Garg, J. De Gasperi, M. C. Glick, D. Gminsky, A. M. Golovchenko, Y. Goussault, P. H. Gross, M. Gut, M. A. Gvalambor, P. Gyorgy, S. Hakomori, H. van Halbeeck, A. M. Halford, M. D. A. Halford, M. H. Halford, A. Hallen, V. B. Hatcher, F. Heatley, A. Herscovics, O. Hoshino, S. A. Houssain, R. C. Hughes, M. R. Jahnke, L. F. James, D. Jeanloz, A. Jimbo, M. Z. Jones, R. Kaifu, R. L. Katzman, R. D. Kilker, W. Klaffke, G. Klein, E. H. Kolodny, Y. Konami, J. Koscielak, J. J. Lamar, G. Lamblin, L. A. Lampert, D. W. Laske, R. D. Lasky, M. L. Laver, M. M. Lee, N. Lee, C. Levrat, M. Lhermitte, K. B. Linsley, E. Lisowska, C. M. Liu, I. Y. Liu, T. Matsumoto, M. D. Maxfield, J. W. McArthur, D. Medrek, C. Merser, M.-L. Milat, D. K. Miller, S. C. Miller, T. Mitvedt, K Miyai, Y. Mizuno, J. D. Moore, J. H. G. M. Mutsaers, S. Nakabayashi, T. E. Nash, E. Salomon, R. Naves, N. Nikrui, H. Von Nocolai, T. Osawa, P. D. Palmer, M. Parquet, P. Perchfelides, J. M. Petit, J. R. Poortmans, D. Power, E. S. Rachaman, S. S. Raghavan, A. M. C. Rapin, E. J. S Rathke, V. N. Reinhold, G. P. Roberts, A. A. Rossini, S. W.

Rostad, P. Roussel, S. Sadeh, M. M. El Sadek, B. H. Sanford, S. Santikarn, W. Sasak, D. M. Schmid, K. Schmid, A. S. Schmit, D. Schwarzenbach, G. O. H. Schwarzmann, J. F. Scott, M. Shaban, M. Shalev, N. Sharon, C. Silber, P. Sinaÿ, H. S. Slayte, M. Spinola, R. G. Spiro, G. F. Springer, D. K. Stearns, R. L. Stephens, P. F. Stoffyn, G. Strecker, S. Suzuki, Z. Tarasiejska-Glazer, P. Thoma, J. S. Tkacz, M. Tomoda, M. Trémèege, R. B. Trimble, B. Tuttle, Nasir-ud-Din, J. R. Vercellotti, A. Veyrières, J. F. G. Vliegenthart, R. Vrba, E. Walker, C. D. Warren, H. Wecyer, J. F. Wedgwood, N. R. Williams, B. Winchester, J. K. Wold, I. Yamashina, T. Yamazaki, J. Yoshikawa, N. Zamchek, U. Zehavi, F. Zilliken.

The work of only a few of the more than two hundred colleagues and collaborators of Jeanloz listed here could be detailed within the scope of this tribute.

NATHAN SHARON
R. COLIN HUGHES

REFERENCES

1. K.-H. Meyer, E. Preiswerk, and R. W. Jeanloz, *Helv. Chim. Acta*, 24 (1941) 1395–1409.
2. R. Jeanloz, Thèse de Docteur es Sciences Chimiques (1943) de l'Université de Genève.
3. R. Jeanloz, D. A. Prins, and T. Reichstein, *Experientia*, 1 (1945) 1–2.
4. D. A. Prins and R. W. Jeanloz, *Ann. Rev. Biochem.*, 17 (1948) 67–96.
5. R. Jeanloz, H. G. Fletcher, Jr., and C. S. Hudson, *J. Am. Chem. Soc.*, 70 (1948) 4052–4054.
6. R. Jeanloz, H. G. Fletcher, Jr., and C. S. Hudson, *J. Am. Chem. Soc.*, 70 (1948) 4055–4057.
7. R. W. Jeanloz, G. R. Barker, and M. V. Lock, *Nature*, 167 (1951) 42–43.
8. R. W. Jeanloz and H. G. Fletcher, Jr., *Adv. Carbohydr. Chem.*, 6 (1951) 135–174.
9. O. Hechter, R. P. Jacobsen, R. Jeanloz, H. Levy, C. W. Marshall, and G. Pincus, *Arch. Biochem.*, 25 (1950) 457–460.
10. O. Hechter, R. P. Jacobsen, V. Schenker, H. Levy, R. W. Jeanloz, W. Marshall, and G. Pincus, *Endocrinology*, 52 (1953) 679–691.
11. H. Levy, R. W. Jeanloz, C. W. Marshall, R. P. Jacobsen, O. Hechter, V. Schenker, and G. Pincus, *J. Biol. Chem.*, 203 (1953) 433–451.
12. R. W. Jeanloz, H. Levy, R. P. Jacobsen, O. Hechter, V. Schenker, and G. Pincus, *J. Biol. Chem.*, 203 (1953) 453–461.

13. A. S. Meyer, R. W. Jeanloz, and G. Pincus, *J. Biol. Chem.*, 203 (1953) 463–468.
14. H. Levy, R. W. Jeanloz, R. P. Jacobsen, O. Hechter, V. Schenker, and G. Pincus, *J. Biol. Chem.*, 211 (1954) 867–881.
15. R. W. Jeanloz and E. Forchielli, *J. Biol. Chem.*, 188 (1951) 361–369.
16. R. W. Jeanloz and E. Forchielli, *J. Biol. Chem.*, 190 (1951) 537–546.
17. R. Jeanloz and E. Forchielli, *Helv. Chim. Acta*, 33 (1950) 1690–1697.
18. R. W. Jeanloz, *Adv. Carbohydr. Chem.*, 13 (1958) 189–214.
19. R. W. Jeanloz and C. Gansser, *J. Am. Chem. Soc.*, 79 (1957) 2583–2585.
20. R. W. Jeanloz, *J. Am. Chem. Soc.*, 76 (1954) 558–562.
21. R. W. Jeanloz, *J. Org. Chem.*, 26 (1961) 905–908.
22. P. J. Stoffyn and R. W. Jeanloz, *J. Am. Chem. Soc.*, 76 (1954) 561–562.
23. R. W. Jeanloz and P. J. Stoffyn, *J. Am. Chem. Soc*, 76 (1954) 5682–5684.
24. P. J. Stoffyn and R. W. Jeanloz, *J. Am. Chem. Soc.*, 80 (1958) 5690–5692.
25. D. K. Stearns, R. G. Naves, and R. W. Jeanloz, *J. Org. Chem.*, 26 (1961) 901–905.
26. P. J. Stoffyn and R. W. Jeanloz, *J. Am. Chem. Soc.*, 76 (1954) 563–564.
27. R. W. Jeanloz and H. M. Flowers, *Carbohydr. Res.*, 2 (1966) 411–413.
28. Nasir-ud-Din, and R. W. Jeanloz, *Carbohydr. Res.*, 28 (1973) 243–251.
29. P. J. Stoffyn and R. W. Jeanloz, *Arch. Biochem.*, 52 (1954) 373–379.
30. R. W. Jeanloz, *Chimia*, 7 (1953) 292.
31. R. W. Jeanloz and H. M. Flowers, *J. Am. Chem. Soc.*, 84 (1962) 3030–3126.
32. H. M. Flowers and R. W. Jeanloz, *Biochemistry*, 3 (1964) 123–125.
33. R. W. Jeanloz, *Adv. Enzymol. Relat. Areas Mol. Biol.*, 25 (1963) 433–456.
34. P. J. Stoffyn and R. W. Jeanloz, *J. Biol. Chem.*, 235 (1960) 2507–2510.
35. S. I. Hakomori and R. W. Jeanloz, *J. Biol. Chem.*, 236 (1961) 2827–2834.
36. S. Hakomori and R. W. Jeanloz, *J. Biol. Chem.*, 239 (1964) PC3606–PC3607.
37. S. I. Hakomori, J. Koscielak, K. J. Bloch, and R. W. Jeanloz, *J. Immunol.*, 98 (1967) 31–38.
38. S. I. Hakomori, J. Koscielak, and R. W. Jeanloz, *Immunochemistry*, 5 (1968) 441–455.
39. H. M. Flowers and R. W. Jeanloz, *J. Org. Chem.*, 28 (1963) 1377–1379.
40. H. M. Flowers and R. W. Jeanloz, *J. Org. Chem.*, 28 (1963) 1564–1567.
41. H. M. Flowers and R. W. Jeanloz, *J. Org. Chem.*, 28 (1963) 2983–2986.
42. T. Osawa and R. W. Jeanloz, *Carbohydr. Res.* (1965) 181–186.
43. R. W. Jeanloz, D. M. Schmid, and P. J. Stoffyn, *J. Am. Chem. Soc.*, 79 (1957) 2586–2590.
44. R. W. Jeanloz, E. Walker, and P. Sinaÿ, *Carbohydr. Res.*, 6 (1968) 184–196.

45. T. Osawa, P. Sinaÿ, M. Halford, and R. W. Jeanloz, *Biochemistry*, 8 (1969) 3369–3375.
46. P. Sinaÿ and R. W. Jeanloz, *Carbohydr. Res.*, 10 (1969) 189–196.
47. P. Sinaÿ, M. D. Halford, M. S. Choudhary, P. H. Gross, and R. W. Jeanloz, *J. Biol. Chem.*, 247 (1972) 391–397.
48. Y. Konami, T. Osawa, and R. W. Jeanloz, *Biochemistry*, 10 (1971) 192–196.
49. A. Veyrières and R. W. Jeanloz, *Biochemistry*, 9 (1970) 4153–4159.
50. O. Hoshino, U. Zehavi, P. Sinaÿ, and R. W. Jeanloz, *J. Biol. Chem.*, 247 (1972) 381–390.
51. N. Sharon and R. W. Jeanloz, *Experientia*, 20 (1964) 253–254.
52. N. Sharon, T. Osawa, H. M. Flowers, and R. W. Jeanloz, *J. Biol. Chem.*, 241 (1966) 223–230.
53. R. L. Katzman and R. W. Jeanloz, *Science*, 166 (1969) 758–759.
54. R. L. Katzman and R. W. Jeanloz, *J. Biol. Chem.*, 248 (1973) 50–55.
55. R. L. Katzman and R. W. Jeanloz, in E. A. Balazs, (Ed.), *The Chemistry and Biochemistry of the Intracellular Matrix*, Vol. 1, pp. 149–159.
56. R. L. Katzman, E. Lisowska, and R. W. Jeanloz, *Biochem. J.*, 119 (1970) 17–19.
57. Z. Tarasiejska and R. W. Jeanloz, *J. Am. Chem. Soc.*, 79 (1957) 2660.
58. Z. Tarasiejska and R. W. Jeanloz, *J. Am. Chem. Soc.*, 79 (1957) 4215–4218.
59. R. W. Jeanloz, Z. Tarasiejska-Glazer, and D. A. Jeanloz, *J. Org. Chem.*, 26 (1961) 537–541.
60. R. W. Jeanloz and D. A. Jeanloz, *J. Org. Chem.*, 26 (1961) 537–546.
61. R. W. Jeanloz and A. M. C. Rapin, *J. Org. Chem.*, 28 (1963) 2978–2986.
62. N. Sharon and R. W. Jeanloz, *J. Biol. Chem.*, 235 (1960) 1–5.
63. E. H. Eylar and R. W. Jeanloz, *J. Biol. Chem.*, 237 (1962) 622–628.
64. E. H. Eylar and R. W. Jeanloz, *J. Biol. Chem.*, 237 (1962) 1021–1025.
65. R. G. Spiro, *J. Biol. Chem.*, 237 (1962) 646–652.
66. R. G. Spiro, *J. Biol. Chem.*, 239 (1964) 567–573.
67. H. G. Garg and R. W. Jeanloz, *Adv. Carbohydr. Chem. Biochem.*, 43 (1985) 135–201.
68. C. D. Warren and R. W. Jeanloz, *Methods Enzymol.*, 50 (1978) 122–137.
69. Y. Goussault, S. Nakabayashi, C. D. Warren, B. Bugge, and R. W. Jeanloz, *Carbohydr. Res.*, 179 (1988) 381–392.
70. J. F. Wedgewood, C. D. Warren, R. W. Jeanloz, and J. L. Strominger, *Proc. Natl. Acad. Sci. USA*, 71 (1974) 5022–5026.
71. C. D. Warren, I. Y. Liu, A. Herscovics, and R. W. Jeanloz, *J. Biol. Chem.*, 250 (1975) 8069–8078.

72. C. D. Warren, A. S. Schmit, and R. W. Jeanloz, *Carbohydr. Res.*, 116 (1983) 171–182.
73. S. Sadeh, C. D. Warren, P. F. Daniel, B. Bugge, L. F. James, and R. W. Jeanloz, *FEBS Lett.*, 163 (1983) 104–109.
74. C. D. Warren, P. F. Daniel, B. Bugge, J. E. Evans, L. F. James, and R. W. Jeanloz, *J. Biol. Chem.*, 263 (1988) 15041–15049.
75. R. DeGasperi, S. al Daher, P. F. Daniel, B. G. Winchester, R. W. Jeanloz, and C. D. Warren, *J. Biol. Chem.*, 266 (1991) 16556–16563.
76. C. D. Warren, J. Alroy, B. Bugge, P. F. Daniel, S. S. Raghavan, E. H. Kolodny, J. J. Lamar, and R. W. Jeanloz, *FEBS Lett.*, 195 (1986) 247–252.
77. J. F. Codington and R. W. Jeanloz, *Z. Klin. Chem. Klin. Biochem.*, 9 (1971) 61.
78. J. F. Codington, B. H. Sanford, and R. W. Jeanloz, *J. Natl. Cancer Inst.*, 51 (1973) 585–591.
79. J. F. Codington, K. B. Linsley, R. W. Jeanloz, T. Irimura, and T. Osawa, *Carbohydr. Res.*, 40 (1975) 171–182.
80. D. H. Van den Eijnden, N. A. Evans, J. F. Codington, V. Reinhold, C. Silber, and R. W. Jeanloz, *J. Biol. Chem.*, 254 (1979) 12153–12159.
81. R. W. Jeanloz and E. A. Balazs, *The Amino Sugars* (1966) Academic Press, New York, 1A Distribution and Biological Roles, 591 pp, Vol. 2A; Chemistry of the Amino Sugars, 827 pp, Vol. 2B; Metabolism and Interactions.
82. R. W. Jeanloz, *Arthritis Rheum.*, 3 (1960) 233–237.
83. E. A. Balazs, T. C. Laurent, and R. W. Jeanloz, *Biochem. J.*, 235 (1986) 903.
84. J. D. Gregory and R. W. Jeanloz, *Glyconjugate Research Proceeding of the Fourth International Symposium on Glycoconjugates* (1797) Academic Press, New York, Vol. 1, pp. 1–571; Vol. 2, pp. 578–1103.

Bengt Lindberg

BENGT LINDBERG

1919–2008

One of the most important carbohydrate chemists during the last century, Bengt Lindberg, passed away on March 20, 2008.

Bengt Lindberg was born on July 17, 1919, in Stockholm, the capital of Sweden, and lived in the southern part of this city throughout his life. There he had all his school education and graduated in 1937 with excellent grades from Södra Latin, an old, well-reputed high school close to his home. At school Lindberg was already interested in life science, especially mathematics, and he continued his education at Stockholms Högskola (nowadays Stockholm University), where he studied mathematics, chemistry, and physics. After he obtained his B.Sc. in 1942 and M.Sc. in 1944, he moved to the Royal Institute of Technology in Stockholm, where he specialized in organic chemistry. His supervisor was Professor Holger Erdtman, an internationally reputed scientist in natural product chemistry. Erdtman understood the potential of the young man and introduced Bengt to natural product chemistry, and became a mentor for his academic career, which continued as a life-long friendship. Bengt had high ambitions and he wanted to break new ground as an organic chemist. At the age of 24, on Erdtman's initiative, he started to work on carbohydrate chemistry. This was a virgin research field in Sweden in the 1940s and it was regarded as difficult and complicated. It was a challenge that suited the young and ambitious Bengt Lindberg. His first contribution within this field was a study on Zemplén's glycoside synthesis ''Untersuchungen über die Zemplénsche Glucosidsynthese'' published in 1944 (*Arkiv Kemi Mineral. Geol.*). Géza Zemplén was a former coworker of Emil Fischer and he discovered the usefulness of mercuric acetate in the synthesis of glycosides. This procedure was suitable for the preparation of alkyl glycosides and, with the application of proper amounts of alcohols, either α- or β-glycosides could be obtained.

The results from his first study on glycoside synthesis and from the studies on the transformation of ''acetochloro sugars'' by $TiCl_4$, described by Eugene Pacsu,

ISBN: 978-0-12-380856-1
DOI: 10.1016/S0065-2318(10)63002-8.

initiated another 12 publications dealing with the isomerization of acetylated glycosides by strong acids. In 1949, Bengt Lindberg published the first example of the synthesis of an α-D-glucopyranosyl-glycoside by converting β-gentiobiose octaacetate in chloroform, in the presence of $TiCl_4$, into β-isomaltose octaacetate. This first synthesis of an α-linked glucoside attracted much attention from the scientific community (*Nature*, 1949).

In 1950, Bengt Lindberg defended his thesis ''Studies on glycosides, especially the α/β transformation.'' At that time Bengt had more than 20 scientific publications and had already gained an international reputation as a carbohydrate chemist. He included only 11 publications in the thesis; all dealing with the transformation reaction. His formal thesis amounted to one page containing an abstract and a list of these 11 scientific publications. With his characteristic razor-sharp logic he informed the opponent and the dissertation committee that they will find backgrounds and discussions in the publications. ''I find it unnecessary to repeat myself by writing an extensive thesis.'' This demonstration annoyed the academic community, and Bengt Lindberg was not awarded the highest mark on his thesis. Later Bengt admitted that this behavior showed a lack of judgment from his side and he prohibited his own Ph.D. students to follow his example.

After his dissertation Bengt Lindberg was appointed associate professor at the Royal Institute of Technology in Stockholm, and in 1956 he became head of the wood chemistry department at the Svenska Träforskningsinstitutet (Swedish Forest Research Laboratory). In 1959, he was awarded the honorary title of Professor by the Swedish King.

During these years, he managed to recruit a large number of highly talented coworkers who later ended up in leading industrial positions and as professors in academic institutions. His group was most productive, and postdoctoral researchers from different countries joined the team. Structural studies on oligo- and polysaccharides from different natural sources were initiated. These studies resulted in a series of more than 30 papers dealing with, respectively, high and low molecular-mass carbohydrates from wood, algae, and lichen. To simplify the structural studies, different degradation, substitution, and chromatographic technologies were introduced. The analysis of different methylated sugars was improved when they could be separated on carbon columns (*Acta Chem. Scand.*, 1954) and a review ''Methods in Structural Polysaccharide Chemistry,'' which set the standard for structural studies on polysaccharides, was published in 1960 (*Advances in Carbohydrate Chemistry*). The methylation analysis was further improved, and the procedure was summarized in the article ''Hydrolysis of methylated polysaccharides'' (*Methods in Carbohydrate Chemistry*, 1965).

Even if the main focus on the research activities were directed towards structural studies on carbohydrates of natural origin, the synthesis of model substances, derivatization of oligo- and poly-saccharides, oxidation, and reduction of carbohydrates, and identification of the products all were performed during this time.

In 1965, Bengt Lindberg was invited to take the Chair in Organic Chemistry at Stockholm University, and there he started to create an organization working with carbohydrate chemistry that later was called the ''Lindberg group'' by the scientific community. The name was not his invention. The environment around this group was open-minded and creative, and it had a broad perspective on carbohydrate chemistry with emphasis on structural analysis and development of methods allowing faster analysis with smaller amounts of sample. When monosaccharide analysis by gas chromatography of alditol acetates became available, the group applied the method on partially methylated sugars (*Acta Chem. Scand.,* 1967). In combination with the Hakomori procedure (1964) for methylation of polysaccharides and the use of gas chromatography–mass spectrometry, a new concept for methylation analysis of carbohydrates was developed (*Angewandte Chemie*, 1970). The new method allowed identification of the monomers and their linkage positions in a polysaccharide within a few days and used only a few milligram of sample. Lindberg early understood the importance of the method, and looked for new challenges, finding these in the bacterial polysaccharides. These polysaccharides were available in only small quantities and they consisted of oligosaccharide repeating-units built up of different monosaccharides that defined the immunological properties of the bacteria. This was a difficult field, and it demanded the collaboration of synthetic and analytical chemists, as well as microbiologists and immunologists. This led to numerous projects with other scientists, primarily at the Karolinska Insitutet, but also on an international level. These projects unraveled the structures of complex bacterial polysaccharides acting as surface antigens in the capsule and lipopolysaccharide (O-antigens). The specificities of poly- and mono-clonal antibodies could be determined by using the synthetic di-, tri-, tetra-, and larger oligosaccharides. Coupling of the oligosaccharides to immunogenic carrier proteins made the elucidation and production of defined oligo- and poly-clonal antibody reagents possible. Elucidation of the structures of the capsular polysaccharides in *Haemophilus* type b and *Streptococcus pneumoniae* resulted in the development of glycoconjugate vaccines for prevention of bacterial meningitis, pneumonia, and septicemia. Thus such vaccines as Hib-titer, Act-Hib, Menactra, and Prevnar have reduced mortality in infants well above 95%.

To solve the more complicated structures, a continuing development of technologies as tools for the structure determination took place and several methods now commonly used were developed. Examples of these are the uronic acid degradation, the chromic acid oxidation, and the N-deacetylation–deamination degradation. Structural studies of the resulting fragments made it possible to determine anomeric configurations and sequences of individual monosaccharide units in the parent polysaccharides. Other methods were the use of chiral 2-octanol for determination of D and L configuration by gas chromatography of the acetylated 2-octyl glycosides. These technologies, which are still in use, were combined with such spectroscopic methods as NMR and MS. Bengt remained interested and also active in the carbohydrate field many years after retirement, and he continued to publish scientific papers until 2002.

In 1967, Bengt Lindberg was elected to the Royal Swedish Academy of Sciences and was a member of the Nobel committee for chemistry between 1974 and 1987. He presented the prize winners of 1979 (Herbert Brown and Georg Wittig) and the prize winner of 1984 (Bruce Merrifield). He received a number of Swedish awards for his scientific contributions, including the Celsius medal (1985). He was awarded the Haworth Memorial Medal from the Royal Society of Chemistry (1981) and was the first non-American to receive the Hudson Award from the American Chemical Society (1983).

He set high standards for his own and his collaborator's work, and emphasized the importance of innovation and work across scientific borders. Many scientists with different backgrounds visited and worked in his laboratory for various periods. They solved the structures of a range of complex carbohydrates. The ''Lindberg group'' had a unique atmosphere of creativity and hard, disciplined work. We look back on that time with gratitude and pleasure; as we feel that we were involved in something scientifically important.

The lifetime achievement of Bengt Lindberg has, among other things, resulted in ~400 scientific publications (some of these being among the most cited papers dealing with carbohydrates), a number of reviews, and more than 40 Ph.D.s providing input in the carbohydrate field. A number of his students have continued his scientific work as professors in the academic community, and many others have leading positions in governmental institutions and industry.

In his private life Bengt showed a keen interest in recreation in the wilderness, canoeing, hiking, cross-country skiing, and skating. Many hours were spent with coworkers at coffee breaks with discussions on equipment, hard weather conditions, and distances traversed, making the laboratory also a friendly and familiar environment. Bengt was a heavy smoker, but in spite of that he was physically fit and an excellent squash player. Bengt had also a tender heart under his tough surface (although he did not like to admit it), as well as a wry sense of humor. He was a family man and loved children, especially his granddaughters.

In summary, Bengt Lindberg was one of the most admired and cited chemists in Sweden during the twentieth century.

Bengt is survived by his wife Ethel, daughter Elsa, and his granddaughters Hanna and Emma.

LENNART KENNE
OLLE LARM
ALF LINDBERG

ZEOLITES AND OTHER SILICON-BASED PROMOTERS IN CARBOHYDRATE CHEMISTRY☆

BY AMÉLIA P. RAUTER, NUNO M. XAVIER, SUSANA D. LUCAS and MIGUEL SANTOS

Centro de Química e Bioquímica/Departamento de Química e Bioquímica, Faculdade de Ciências da Universidade de Lisboa, Ed. C8, 5° Piso, Campo Grande, 1749-016, Lisboa, Portugal

ABBREVIATIONS

AW, Acid-washed; Chol, Cholesterol; DMAP, 4-(Dimethylamino)pyridine; DMF, *N,N*-Dimethylformamide; DMTr, Di(*p*-methoxyphenyl)phenylmethyl; GalNAc, *N*-Acetylgalactosamine, 2-acetamido-2-deoxy-D-galactose; HMF, 5-Hydroxymethylfurfural, 5-(hydroxymethyl)-2-furaldehyde; INOC, Intramolecular nitrile oxide–alkene cycloaddition; Le^a, Lewis^a; Le^x, Lewis^x; MOM, Methoxymethyl; MP, *p*-Methoxyphenyl; MS, Molecular sieves; NIS, *N*-Iodosuccinimide; PCC, Pyridinium chlorochromate; PDC, Pyridinium dichromate; PMA, Phosphomolybdic acid; PMB, *p*-Methoxybenzyl;

☆This article is dedicated to Prof. Joachim Thiem on the occasion of his 68th birthday.

ISBN: 978-0-12-380856-1
DOI: 10.1016/S0065-2318(10)63003-X.

PPTS, Pyridinium *p*-toluenesulfonate; SE, 2-Trimethylsilylethyl; TBDMS, *tert*-Butyldimethylsilyl; TBDPS, *tert*-Butyldiphenylsilyl; TEMPO, 2,6,6-Tetramethyl-1-piperidinyloxy; Tf, Trifluoromethanesulfonyl; THF, Tetrahydrofuran; THP, Tetrahydropyran; TIPDS, 1,1,3,3-Tetraisopropyl-1,3-disiloxanyl; TMS, Trimethylsilyl.

I. Introduction

The use of aluminosilicate porous materials as heterogeneous catalysts in organic synthesis has attracted considerable attention since the introduction of synthetic zeolites by Union Carbide approximately 50 years ago.[1] Zeolites and related silicon-based materials have found a wide range of applications, allowing mild and convenient catalytic methodologies in organic transformations.[2,3] In terms of the so-called "green" chemistry, these materials offer the positive features of lack of toxicity, noncorrosiveness, easy recovery, and possibility of reuse. Moreover, they can provide stereo- and regio-control in chemical reactions as a result of their regular porous structure and the consequent shape-selective properties. This constitutes an important goal for organic synthesis, and for carbohydrate chemistry in particular, where there are important issues of reaction selectivity because of the multifunctionalized nature of the molecules involved.

Many essential transformations of carbohydrates regularly employ toxic and corrosive reagents, such as Lewis acids and strong mineral acids. Hence, the use of zeolites and related porous solids as catalysts in such reactions affords a practical and environmentally compatible alternative to the standard protocols.

In this chapter, we demonstrate the potential of such agents as catalysts/promoters in key steps for the derivatization of sugars. The most significant catalytic approaches in carbohydrate chemistry that use aluminosilicate porous materials, namely zeolites and montmorillonite clays, are reviewed and discussed. Silica gel is a porous solid silicate that has also been used for heterogeneous catalysis of organic reactions in general. We include here its usefulness as promoter and reagent support for the reactions under consideration.

II. Zeolites, Clays, and Silica Gel as Heterogeneous Catalysts and Their Use in Organic Synthesis

Natural zeolites have been known for over 200 years, since the discovery of the first zeolite mineral, stilbit, by Crønstedt.[4] However, the first synthetic zeolites were only developed in the 1960s, through research in the laboratories of the petroleum companies Union Carbide and Mobil Oil.[1,2a,h] Since then, remarkable progress in our knowledge of their structure and use in synthetic methodologies has been made, and a wide range of synthetic zeolites are known. These materials have became the most widely used catalysts in the petroleum refining industry and also perform an important and still increasing role in the synthesis of fine and specialty chemicals. There are several reviews and books concerning the structural aspects of zeolites, and their use in chemical processes, and particularly in organic synthesis.[2,3] A comprehensive survey of their structural characteristics, properties and applications, is therefore not our purpose. However, it is worth mentioning the main aspects that enable them to serve as catalysts of valuable and distinctive performance in organic transformations.

These microporous crystalline materials possess a framework consisting of AlO_4 and SiO_4 tetrahedra linked to each other by the oxygen atoms at the corner points of each tetrahedron. The tetrahedral connections lead to the formation of a three-dimensional structure having pores, channels, and cavities of uniform size and dimensions that are similar to those of small molecules. Depending on the arrangement of the tetrahedral connections, which is influenced by the method used for their preparation, several predictable structures may be obtained. The most commonly used zeolites for synthetic transformations include large-pore zeolites, such as zeolites X, Y, Beta, or mordenite, medium-pore zeolites, such as ZSM-5, and small-pore zeolites such as zeolite A (Table I). The latter, whose pore diameters are between 0.3

TABLE I
Dimensions of Zeolites Commonly Used in Catalysis

Zeolite (Structure Type)	Dimensions	Commercial Products
Small Pore Zeolites		
Zeolite A (LTA)	4.1 Å diameter pore, 11.4 Å diameter cavity	Linde types 4, 3, 5 Å
Medium Pore Zeolites		
ZSM-5 (MFI)	5.3×5.6 Å channel, 5.1×5.5 Å channel	CBV-3020, CBV 3024E, CBV-30014
Large Pore Zeolites		
Zeolite X/Y (FAU)	7.4 Å diameter pore, 11.8 Å diameter cavity	13X, CBV-100, CBV-400, CBV-500, CBV-720, HSZ-320NAA, HSZ-90HUA
Mordenite (MOR)	6.5×7 Å channel, 2.6×5.7 Å channel	HSZ-620HOA, CBV-10A, HSZ-690HOA
Zeolite beta (BEA)	7.6×6.4 Å channel, 5.5×5.5 Å channel	CP-811BL-25, CP-814E

and 0.45 nm, allow the absorption of only small molecules. It is typically used as drying agent, and includes the commercialized 3, 4, and 5 Å Linde-type molecular sieves (MS).

Zeolites possess high adsorption capacity, and their unique and regular pore and channel structure leads to shape selectivity, which can be of reactant, product, or transition-state type. Because of size restrictions inside the zeolite, some reactants or products are more able to enter or leave the zeolite cavities than others, and in a chemical reaction, the more appropriate transition states, also in terms of size or shape, will be preferentially formed. This feature can be used to control the direction and the stereo- or regio-chemical outcome of a given reaction, avoiding the formation of undesired isomers or side products. Another important feature, crucial for their catalytic activity, is their Brønsted and Lewis acidity. The Brønsted acidity is associated with bridging hydroxyl groups attached to a framework of oxygen atoms that link tetrahedral Si and Al atoms: Al(OH)Si. Lewis acidity is conferred by the metal ions that compensate for the anionic aluminum centers or by the aluminum centers deficient in oxygen.

Thus, zeolites may replace such environmentally unfriendly acid catalysts as $AlCl_3$ or H_2SO_4 in organic transformations, contributing to cleaner and safer methodologies. The strength and concentration of the acid sites can be modified by controlling the Si/Al ratio, and therefore the zeolite acidity can be adjusted for a particular application.

The chemical and thermal stability of these solid catalysts is also an important advantage for their use, making them resistant to higher reaction temperatures and to a variety of chemical attacks. Moreover, the ease of handling and recovering of the zeolite by a simple filtration, with the possibility of reusing it, are valuable features for a catalysis-based reaction.

Among the wide variety of organic reactions in which zeolites have been employed as catalysts, may be emphasized the transformations of aromatic hydrocarbons of importance in petrochemistry, and in the synthesis of intermediates for pharmaceutical or fragrance products.[5] In particular, Friedel–Crafts acylation and alkylation over zeolites have been widely used for the synthesis of fine chemicals.[6] Insights into the mechanism of aromatic acylation over zeolites have been disclosed.[7] The production of ethylbenzene from benzene and ethylene, catalyzed by HZSM-5 zeolite and developed by the Mobil–Badger Company, was the first commercialized industrial process for aromatic alkylation over zeolites.[8] Other typical examples of zeolite-mediated Friedel–Crafts reactions are the regioselective formation of *p*-xylene by alkylation of toluene with methanol over HZSM-5,[9] or the regioselective *p*-acylation of toluene with acetic anhydride over HBEA zeolites.[10] In both transformations, the *p*-isomers are obtained in nearly quantitative yield.

Zeolites have also been described as efficient catalysts for acylation,[11] for the preparation of acetals,[12] and proved to be useful for acetal hydrolysis[13] or intramolecular lactonization of hydroxyalkanoic acids,[14] to name a few examples of their application. A number of isomerizations and skeletal rearrangements promoted by these porous materials have also been reported. From these, we can underline two important industrial processes such as the isomerization of xylenes,[2] and the Beckmann rearrangement of cyclohexanone oxime to ϵ-caprolactam,[15] which is an intermediate for polyamide manufacture. Other applications include the conversion of *n*-butane to isobutane,[16] Fries rearrangement of phenyl esters,[17] or the rearrangement of epoxides to carbonyl compounds.[18]

Clay minerals are compositionally similar to zeolites.[19] They differ, however, in their crystalline structure, which, in the case of clays, consists of tetrahedral and octahedral layers. The arrangements of these layers give rise to four different main groups of clays, namely smectite, illite, vermiculite, and kaolinite.[19] Among these, montmorillonite clays, which belong to the smectite group, are the most useful clay catalysts in synthetic organic chemistry. Their structural unit consists of a sandwich of one octahedrally coordinated sheet of $[Al_2(OH)_6]$ groups between two sheets of tetrahedrally coordinated silicate $[SiO_4]^{4-}$ groups. A particular feature of montmorillonite relies on its propensity to exchange the central cation, either at the tetrahedral sites, that is substitution of Si^{4+} by Al^{3+}, or at the octahedral sites, in which substitution of Al^{3+} by Mg^{2+} may occur. The substitution of cations in the lattice by lower valent cations results in a residual negative charge in the lattice, which can be balanced by cations, presumably situated between the lattice layers. The interlayer cations contribute strongly to the Brønsted acid character of clays, mainly because of the dissociation of the intercalated water molecules coordinated to cations. The Lewis-acidity character results from the Al^{3+} ions at the crystal edges. The catalytic activity of montmorillonites thus relies on their high capacity for ion exchange, strong acidity and, like zeolites, on their high surface area and sorptive properties. These materials are also environmentally benign, and allow mild reaction conditions and simple procedures.

A diverse group of organic reactions catalyzed by montmorillonite has been described and some reviews on this subject have been published.[19] Examples of those transformations include addition reactions, such as Michael addition of thiols to α,β-unsaturated carbonyl compounds,[20] electrophilic aromatic substitutions,[19c] nucleophilic substitution of alcohols,[21] acetal synthesis[19b,22] and deprotection,[23] cyclizations,[19b,c] isomerizations, and rearrangements.[19b,24]

Another silicon-based porous solid which has been used in heterogeneous catalysis of organic reactions is silica gel. However, unlike zeolites and clays, silica gel is amorphous and is most frequently used as a relatively inert support material. Nevertheless, its preparation for catalytic processes has been patented.[25] The structure of this material is based on a rigid three-dimensional network of branched and cross-linked siloxane (Si–O–Si) chains, which result from polymerization of silicic acid.[26]

The silica gel network readily retains water molecules. Therefore, the presence of hydrates or silicic acid is ascribed to be the support reagents responsible for mediating a chemical transformation.[27]

Silica gel-based catalytic systems have been described as efficient promoters for a number of organic reactions.[28] Illustrative examples include the oxidative cleavage of double bonds catalyzed by silica-supported $KMnO_4$,[29] reaction of epoxides with lithium halides to give β-halohydrins performed on silica gel,[30] selective deprotection of *tert*-butyldimethylsilyl ethers catalyzed by silica gel-supported phosphomolybdic acid (PMA),[31] and synthesis of cyclic carbonates from epoxides and carbon dioxide over silica-supported quaternary ammonium salts.[32]

Taking into consideration the advantages of using the foregoing siliceous materials in organic transformations, it is not surprising that many synthetic procedures employing these catalysts have been developed in carbohydrate chemistry. The utility of these inorganic templates is demonstrated by their application in reactions that traditionally involve strong Brønsted or Lewis acids, such as glycosylation and acetonation. Several research groups have established new protocols for key transformations of carbohydrates, profiting from the active catalytic sites and the reaction selectivity associated with such catalysts. The most significant results in this field are presented and discussed in this survey.

III. Catalysis of Key Transformations in Carbohydrate Chemistry

1. Glycosylation

The term glycosylation stands for synthetic methodologies that permit linking the anomeric carbon of a sugar to other sugar units or other molecules. In 1893, Fischer reported his historical glycosylation procedure, consisting of the reaction of unprotected monosaccharides with simple alcohols in the presence of HCl.[33] Since then, numerous methodologies for glycoside synthesis have been developed.[34] Most of them require activation of the glycosyl donor, as by Lewis acids, heavy metal salts, or protic acids. As demonstrated in this chapter, zeolites, montmorillonite clays, and silica provide mild and efficient catalyst-based procedures for these types of reactions.

a. Zeolites.—*(i) Medium and Large-Pore Zeolites.* The first zeolite-catalyzed glycosylation procedure was reported by Corma *et al.*[35] and consisted of the Fischer glycosylation of D-glucose with butanol. Different acid zeolites were used for this purpose, including HY, H-beta, HZSM-5, H-mordenite, and mesoporous MCM-22 MSs. All of the catalysts proved to be efficient for the conversion of glucose into the corresponding butyl glucoside, giving first the glucofuranoside product **2**, which is then isomerized to the pyranoside **3** (Scheme 1). The zeolite efficiency was correlated to its acid strength and its channel system, showing that both aspects influenced the reaction. Thus, zeolite HY performed the best, attributable to its adequately large-pore system, whereas MCM-22, despite having higher acidity, gave a poorer outcome, which may be explained by its unidirectional system of channels, and the consequent restriction to the diffusion entry of the reactants. It was also shown that zeolites having higher Si/Al ratio are more efficient for the transformation. This parameter not only defines the number and strength of acid sites, but also determines the adsorption properties of the zeolite. The results obtained are therefore consistent with the fact that zeolites having a high density of acid sites are highly hydrophilic and apt to form and strongly adsorb the polar glucosides, thus decreasing the efficiency of the catalyst. Taking into account all of the studied factors, the best catalysts were considered to be those large-pore tridirectional zeolites having high Si/Al ratios, namely faujasite and H-beta. Moreover, the formation of oligosaccharides and alkyl oligosaccharide glycosides was not detected, because of constrains in the zeolites' framework, where bulkier transition states are limited by shape-selective effects. The roles of the crystallite size and the hydrophobic character of the H-beta zeolite were also studied for this transformation.[36] It was demonstrated that, when maximum reaction rate is desired, the influence of diffusion through the micropores has to be minimized. This can be achieved by increasing the pore diameter and decreasing the length of the micropores, that is, the crystallite size. Concerning the Si/Al ratio, an optimum must be found for which adsorption properties, and number and strength of acid-active sites, are best matched. Also it was concluded that a high hydrophobicity of the catalyst is fundamental for its activity.

In further work, the same research group showed that it was possible to effect transacetalation of the initial butyl glucosides **2, 3** with octanol and dodecanol over H-beta zeolites. Direct Fischer glucosylation, leading to the desired long-chain glucosides **4, 5**, was also possible (Scheme 1).[37]

The zeolite-catalyzed glucosylation of *n*-butanol was also investigated by Chapat *et al.*[38] using dealuminated HY faujasites. The thermodynamically more-stable butyl glucopyranosides were predictably the final reaction products, resulting from

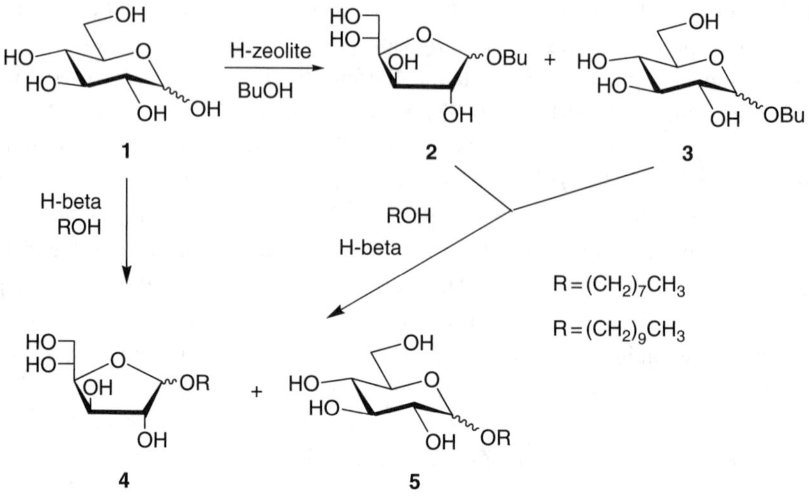

SCHEME 1. Acid-zeolite catalyzed formation of alkyl glucosides by Fischer glucosylation and by transacetalation of butyl glucosides.

isomerization of the intermediate furanosides. These results accord with those previously reported by Corma *et al.*[35,36] Zeolites having different Si/Al ratios were used to discover the optimal value, a balance between the number of acid sites and their strength. The performance of the HY zeolites was compared to that of HBEA zeolite in terms of selectivity. A higher β/α-anomer ratio (1.5) was obtained when HY zeolites were used, attributable to shape-selective properties of the catalysts for the transition states involved.

Although selectivity for the pyranosides has been found in the previously cited reports, our research group has demonstrated that selectivity for the kinetically controlled product, the furanoside derivative **7**, may be achieved by zeolite-mediated Fischer glycosylation of *N*-acetylgalactosamine (**6**) with methanol (Scheme 2) under reflux for 48 h, and the expected β-selectivity was obtained.[39] Large-pore zeolites (HY and HBEA) and the medium-pore HZSM-5 zeolite were tested in order to evaluate the influence of pore dimensions, the framework Si/Al ratio, and concentration of the acid sites, on the reaction efficiency and regioselectivity (Table II). For large-pore zeolites, the Brønsted acidity proved to be the determinant factor, and the HY zeolite (3.1) led to the highest conversion and best regioselectivity to the methyl furanoside, isolated as its acetylated (**8**) or isopropylidene (**9**) derivatives. Therefore, HY (13.2) and HBEA zeolites, which possess similar Brønsted acidity, led to identical conversion of starting material and furanoside/pyranoside ratio (2:1). When zeolite

i) ii) or i) iii)

6

7 $R^1=R^2=R^3=H$
8 $R^1=R^2=R^3=Ac$
9 $R^1=H$, R^2, $R^3=C(CH_3)_2$

10 $R^4=H$, $R^5=Me$ (only β)
11 $R^4=Ac$, $R^5=Me$ (only β)
12 $R^4=R^5=Ac$

i) CH_3OH, acid zeolite
ii) Ac_2O, Py
iii) $(CH_3)_2CO$, zeolite HY CBV 500

SCHEME 2. Fischer glycosylation of GalNAc catalyzed by acid zeolites.

TABLE II
Fischer Glycosylation of GalNAc with Methanol Followed by Acetylation (Compounds 8 and 11), as Mediated by Acid Zeolites

Zeolite (Si/Al Ratio)	External Surface Area ($m^2\ g^{-1}$)	Acid-Site Concentration ($\mu mol\ g^{-1}$)		Fischer Glycosylation Yield (%)		
		Brønsted	Lewis	8	11	12
HY (3.1)	37	670	247	67	31	2
HY (13.2)	64	327	74	32	14	52
HZSM-5 (13.3)	4	469	102	28	2	69
HBEA (12.5)	178	315	340	33	15	52

HZSM-5 was used, the starting material (**12**) was recovered in 69% yield, a result most probably attributable to its more confined-pore structure. However, the highest regioselectivity for the furanoside/pyranoside form (14:1) was obtained with this medium-pore zeolite. It could also be concluded that the external surface area of the zeolites and the Lewis acid sites (Table II), do not have any effect on yield/regioselectivity of this reaction.

Glycosylation using 1-*O*-acetyl sugars (glycosyl acetates) is a common and practical method for glycoside synthesis. These glycosyl donors are readily prepared and are generally activated by a Lewis acid. Helferich[40] introduced glycosyl acetates in 1933, and since then, several Lewis acids have been reported as promoters for glycosylation.[34a,41] H-beta zeolite was used by Aich and Loganathan[42] for the synthesis of aryl glycosides by reaction of a variety of phenols (**15a–e**) with β-D-glucose and β-D-galactose pentaacetates (**13**, **14**) under solvent-free conditions (Scheme 3). The corresponding β-glycopyranosides (**16a–e**, **17a–e**) were obtained in moderate

15a-e
H-beta zeolite
140 °C

13 $R^1 = H, R^2 = OAc$
14 $R^1 = OAc, R^2 = H$
R = H, CH_3, OCH_3, Cl, F
16a–e
17a–e
18a–e
19a–e

SCHEME 3. Phenols glycosylation promoted by H-beta zeolite.

yields (36–46%), along with the 2-deacetylated α-glycopyranosides (**18a–e**, **19a–e**), in 23–28% yield. Noteworthy is the fact that the latter by-products were obtained in a single-step reaction, in considerably higher yield than that previously described using multistep methods (< 12% overall yield[43]).

Subsequently, the same authors employed a similar methodology for the glycosylation of long-chain alcohols.[44] Different forms of beta zeolite (H-beta, Na-beta, Fe-beta, and Zn-beta) were used as catalysts for the reaction of β-D-galactose pentaacetate with cetyl alcohol, performed under conditions similar to those described in the previous work. The highest conversion was obtained over Fe-beta, which was therefore chosen for the subsequent glycosylation reactions with the other fatty alcohols. The major products were the alkyl glycopyranosides, whose yields ranged from 33% to 43%, and the C2-hydroxy α-glycopyranosides were obtained as secondary products in 18–19% yields.

1,2-Anhydro sugars are convenient glycosyl donors for the synthesis of 1,2-*trans*-glycosides, since the opening of the oxirane ring with nucleophiles proceeds with inversion of the configuration at the anomeric center.[45] One of the most significant methods using 1,2-anhydro sugars as glycosyl donors was reported by Danishefsky *et al.*[45a] who described the preparation of β-glycosides by reaction of alcohols with 1,2-α-anhydro sugars in the presence of $ZnCl_2$ at −78 °C.

Matsushita *et al.*[46] performed the glycosylation of cyclohexanol with 1,2-anhydro-3,4,6-tri-*O*-pivaloyl-α-D-glucopyranose (**20**) over silica gel and zeolites, including the mordenite-type zeolites HM, HY, and NaY. The HY zeolite presented the best catalytic performance, enabling the formation of the β-*O*-glycoside **21** in 79% yield, along with the diol **22** (Scheme 4). Stereoselectivity for the β-glycoside was obtained with most of the catalysts tested (68–72% yield). Zeolite NaY, having the weakest acidity among the solid acids, promoted the reaction only slightly.

Glycals have been widely used as glycosyl donors, especially for the synthesis of 2-deoxyglycosides.[45b,47] In addition, they can be readily converted into 2,3-unsaturated

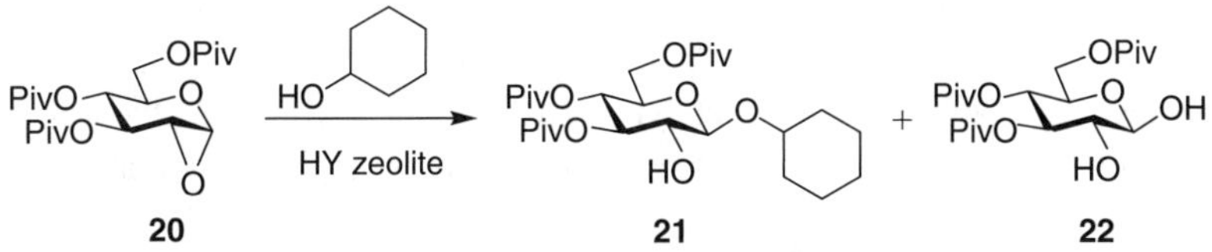

SCHEME 4. Glycosylation of cyclohexanol with 1,2-anhydro sugar **20** in the presence of HY zeolite.

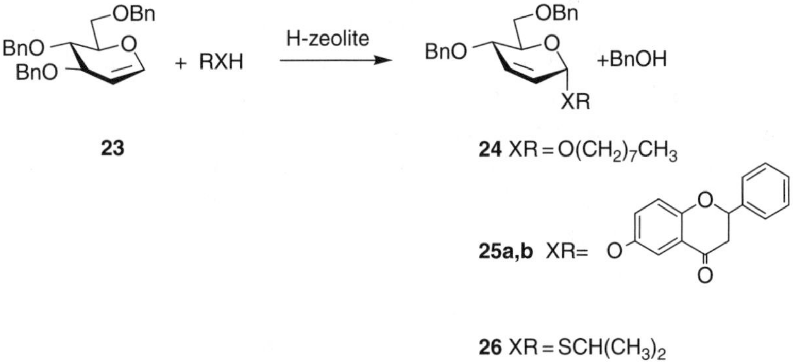

SCHEME 5. Ferrier rearrangement catalyzed by acid zeolites.

glycosyl derivatives by the Ferrier rearrangement.[48] This transformation involves migration of the glycal double bond with loss of an acyloxy group at the allylic position, under catalysis by Lewis acids. The intermediate delocalized carbenium ion then reacts with *O*-, *C*-, *N*-, or *S*-nucleophiles to give the corresponding 2,3-unsaturated glycosyl derivatives. Various catalysts have been used in this reaction, boron trifluoride etherate being the Lewis acid most frequently employed. Ferrier rearrangement of glycals possessing ether protection at the allylic position has only been reported for 3-*O*-methyl-D-glucal[49] and 3-*O*-benzyl-D-glucal,[50] under catalysis by boron trifluoride etherate. We have explored the potential of acid zeolites as promoters for the Ferrier rearrangement of 1,5-anhydro-2,5,6-tri-*O*-benzyl-2-deoxy-D-*arabino*-hex-1-enitol (nonpreferred trivial name: tri-*O*-benzyl-D-glucal, **23**).[39] Thus, reaction of **23** with octanol, 6-hydroxyflavanone, and propane-2-thiol, in the presence of HY (3.1), HY (13.2), HBEA, and HZSM-5 afforded only the corresponding 2,3-unsaturated α-glycosyl derivatives (**24–26**) (Scheme 5), in low to moderate yield. Yields increased with concentration of the Brønsted acid sites and with zeolite hydrophilicity. Hence, the best results were obtained when the reactions were performed over HY (3.1) zeolite, and the target molecules were obtained in moderate yield (**24** and **26** in 46% and 50% yields, respectively), while the glycosylflavonoid **25** was recovered in only 38% yield. The α-selectivity observed in these catalytic transformations arises from the well-known anomeric effect. This new methodology for synthesis of 2,3-unsaturated glycosyl derivatives involves zeolites-promoted debenzoxylation at position 3, thus expanding the synthetic use of the Ferrier rearrangement to benzyl-protected glycals.

(ii) Small-Pore Zeolites. Molecular sieves 3, 4, and 5 Å are small-pore zeolites (zeolites A) and have been employed, also in their acid-washed (AW) form, together with a variety of catalysts to ensure higher yields in glycosylation reactions. This cooperative effect has been observed with several catalysts, namely TMSOTf,[51–54] NIS/TfOH,[55–57] Tf_2O,[58] AgOTf,[55] $BF_3 \cdot Et_2O$,[55], or $H_3PW_{12}O_{40}$.[59] The role of the MS is mainly to stabilize the promoters, acting as a water and/or proton scavenger in the reaction mixture, and enhancing reaction rate, selectivity, and yield.

Posner and Bull[52] investigated the effect of the source and hydration level of the MS on the stereoselectivity of the dimerization of 1-hydroxy sugars into 1,1-linked disaccharides using TMSOTf as promoter. Molecular sieves 4 Å from Aldrich, Janssen, Lancaster, and Davison Chemical companies were tested and those from Davison gave the best results for the α,α-disaccharide both in yield and selectivity. The latter MS had significantly lower concentrations of calcium oxide and magnesium oxide and considerably higher concentrations of ferric oxide, sodium oxide, and sulfate ions, in comparison to the others, although it is not clear whether any of these variables are crucial for the observed α,α-selectivity. Moreover, the amount of adsorbed water in the latter MS did not seem to be an important factor, since the variation of its hydration level from 2.5% to16% gave no significantly different results, indicating that the role of the MS stays beyond that of water scavenger. Hence, the choice of an MS from a particular brand may have a major effect on the outcome of some chemical reactions.

The MS pore-size influences the efficiency of some reactions, as reported by Matsuo *et al.*[54] for glycosylation promoted by $TMSClO_4$, which had to be performed in the presence of 5 Å MS and did not proceed in the presence of 4 or 3 Å MS. Moreover, C-glycosylation of phenols, as promoted by several Lewis acids, (in particular by $Sc(OTf)_3$) was highly dependent on the MSs added. The highest yields and β-stereoselectivity were obtained in the presence of 5 Å MS, whereas 4 and 3 Å MS, among others, were much less effective.[60] Nagai *et al.*[59] used the heteropoly acid $H_3PW_{12}O_{40}$ as promoter for the glycosidation of sulfinyl glycosides with different alcohols and it was observed that, in the presence of 5 Å MS, the yields obtained increased from 68% to 84% and the α/β-selectivity also increased from 79/21 to 98/2, when compared to the glycosidation in the absence of MS.

The addition of AW 4 Å MS effected glycosylation of a trimethylsilylethyl glycoside with a peracetylated thioglycoside, with promotion by NIS/TfOH in almost quantitative yield (96–99%).[57] Triflic anhydride was also a suitable promoter in the presence of 4 Å MS for the glucosylation of glycosyl acceptors of medium or low reactivity, using 2,3,4,6-tetra-*O*-benzyl-*β*-D-glucopyranosyl fluoride, as reported by Wessel and Ruiz.[58] The ability of MSs to act as triflic acid scavengers also led to useful methodology for the formation of triflates under nonbasic conditions.[58]

Activation of 2,3,4-tri-*O*-acetyl-D-xylopyranosyl trichloroacetimidate with $BF_3 \cdot Et_2O$ in the presence of AW-300 MS, followed by reaction with azidosphingosine afforded the glycolipid *β*XylCer in 59% yield.[55] Nevertheless, $BF_3 \cdot Et_2O$-promoted glycosylation of phenolic hydroxy groups with 3,4,6-tri-*O*-acetyl-2-deoxy-2-*N*-phthalimido-*β*-D-glucopyranosyl trichloroacetimidate gave highly satisfactory results in the absence of MS, whereas in the presence of AW-300 MS the yields decreased surprisingly to less than 40%.[61] These contradictory results suggest that for the synthesis of glycosides by phenols glycosylation, the addition of AW MS may be counterproductive, probably because of their ability to promote C-glycosylation.

Disarmed glycosyl trichloro- and *N*-(phenyl)trifluoroacetimidates were efficiently activated by the system I_2/Et_3SiH for the rapid and high-yield glycosylation of primary and secondary saccharide acceptors, and it was observed that the presence of AW 4 Å MS was decisive for achieving high reaction yields.[62]

MSs are often described as acid or water scavengers, but Adinolfi *et al.*[63,64] later reported the use of 4 Å AW-300 MS to promote glycosidation of trifluoro- and trichloro-acetimidates in the absence of any other catalyst (Scheme 6). They found out that 4 Å AW-300 MS could lead to glycosidation, whereas regular 4 Å MS was totally ineffective. Nevertheless, to achieve good yields within a few hours, higher temperatures than those used with $Yb(OTf)_3$/AW-300 MS were required. Glycosidation

27 + 28 → (4 Å AW MS, Toluene, rt, 24 h) → 29, Yield > 90%; R = Ac, Bn

SCHEME 6. Glycosidation of a trifluoroacetimidate in the presence of AW MS.

proceeded in remarkably high yields with such nonpolar solvents as toluene or dichloroethane, while more-polar solvents gave lower yields. These results suggest that with the system $Yb(OTf)_3$/AW-300 MS, in ether or acetonitrile solvents, the activating power of AW-300 MS is negligible and the MS essentially acts as a drying agent.

This activation procedure is particularly attractive on account of its simplicity, to the avoidance of more-acidic promoters, and to the easy recovery and reuse of the MS. The mildness of this methodology allowed its successful application to the difficult regioselective carbinol glycosylation of 17β-estradiol.[64] In addition, the same group synthesized the nonreducing tetrasaccharide terminus of Globo H hexasaccharide, using 4 Å AW-300 MS for the stereocontrolled synthesis of 1,2-*trans*-glycosides bearing a participating group at position 2. Nevertheless, in the absence of such a group, the use of $Yb(OTf)_3$ with a suitable solvent is mandatory for either 1,2-*cis* or 1,2-*trans*-selectivity.[65]

Chandra and Brown[66] investigated the synthesis of α-ribonucleosides as mimetics of the lower axial ligand in coenzyme B_{12}. Reaction of unprotected indolines with 2,3-*O*-isopropylidene-5-*O*-trityl-D-ribofuranose in dry ethanol or dichloromethane was possible in the presence of 4 Å MS affording exclusively the desired indole α-ribonucleosides in 55–70% yield. This procedure is an effective alternative to the multistep route that requires protection of the indoline and use of expensive coupling agents as previously reported by the same group.[67]

b. Montmorillonite.—The application and utility of montmorillonite clays as catalysts in glycosylation reactions has been illustrated in the preparation of *O*- and *C*-glycosyl derivatives, and for promoting the allylic rearrangement of glycols. The first report of montmorillonite as a catalyst for glycosylation was published approximately 20 years ago by Florent and Monneret,[68] who used it for the glycosidation of 1-*O*-acetyl-2,3,6-trideoxyhexoses with methanol and benzyl alcohol. The corresponding β-glycosides were obtained stereoselectively in moderate yields (44–48%).

With respect to unprotected glycosyl donors, Fischer glucosylation of butanol and dodecanol in the presence of montmorillonites has been studied by Brochette *et al.*[69] The glucosylation of butanol was performed over K-10, KSF, and KSF/O montmorillonites, showing that the rate and the efficiency of the transformation depended on the relative acidity of the catalysts. Hence, the highest glucose conversion (88%) and the best product yield (78%) were achieved for the catalyst possessing the strongest acid character. The catalytic activity was ascribed to the transfer of acidity from the solid to the liquid phase and was detected by measuring the pH of the solvent after removal of the clay suspension by filtration. For the KSF/O-mediated O-glucosylation of butanol it was shown that the reaction rate is slightly accelerated by ultrasonic irradiation. In the case of dodecanol, the use of KSF/O led to substantial formation of oligomers, and the desired glycosides were obtained in only low yields (16%). Moreover, oligomerization was increased when ultrasound was employed.

Olivoses (2,6-dideoxy-D- and -L-*arabino*-hexoses) are components often present in bioactive glycosylated natural products, particularly in antitumor antibiotics of the aureolic acid family, some of which are used clinically. It is known that these sugar moieties are important for the bioactivity of these antibiotics and participate in the interaction with the cell target.[70] Jyojima *et al.*[71] employed montmorillonite K-10 as catalyst for the glycosidation of 3,4-di-*O*-protected olivoses (**30–34**) with alcohols (**35–40**) (Scheme 7). The reaction of 3,4-di-*O*-*tert*-butyldimethylsilyl (TBDMS)-olivose (**30**) and cyclohexylmethanol gave the corresponding olivosides in 94% yield and with α-selectivity (α/β ratio, 5.3/1). The nature of the protecting groups in the olivose was recognized as being essential for the stereochemical outcome of the glycosidation. Thus, montmorillonite K-10-catalyzed glycosidation of 3-*O*-acetyl-4-*O*-*tert*-butyldimethylsilyl-olivose (**33**) with cyclohexylmethanol proceeded with α-selectivity, although the β-olivoside was predominately obtained when the protecting groups at O-3 and O-4 of olivose were interchanged. This result may arise from the participating effect of the acetate at C-4. Glycosylation of the alcohols **35–40** further supported these findings, and occurred with good to excellent yield and stereoselectivity.

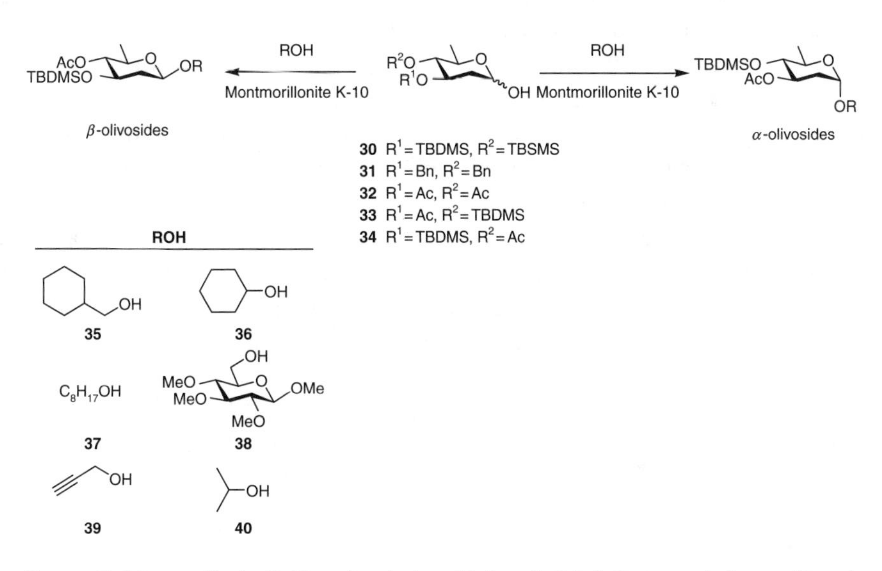

SCHEME 7. Montmorillonite K-10 catalyzed glycosidation of a 3,4-di-*O*-protected olivose with various alcohols.

Stereoselective synthesis of 2-deoxy-2-iodo-β-olivosides (**42**) was also accomplished by montmorillonite K-10-promoted glycosylation of various alcohols with a 2-deoxy-2-iodo-olivosyl fluoride **41** (Scheme 8).[72] The reaction conditions for the glycosylation of cyclohexylmethanol were investigated; use of dichloromethane as solvent in the presence of 5 Å MS (to avoid hydrolysis) provided better yields of the target glycoside. Under these conditions, the alcohols were efficiently glycosylated, affording the corresponding 2-deoxy-2-iodo-olivosides in yields ranging from 80% to 98%, with high β-stereoselectivity. Reduction of this intermediate iodide is feasible, leading to a useful synthetic route for the stereoselective synthesis of β-olivosides.

Conversion of glycals into 2,3-unsaturated glycosyl derivatives in the presence of K-10 montmorillonite has also been reported in the literature. Toshima *et al.*[73] described the use of this material as activator for O-glycosidation of glycals with several alcohols under mild conditions. For example, glycosidation of 3,4-di-*O*-acetyl-L-rhamnal (**43**) with alcohols **35–40** in dichloromethane, using 30% (w/w) of catalyst gave products of allylic rearrangement in high yield, with moderate to good α-stereoselectivity (Schemes 7 and 9).

Balasubramanian *et al.*[74,75] have shown that montmorillonite K-10 also induces the Ferrier rearrangement of glycals when conducted with microwave irradiation. Microwave heating enhanced the efficiency of the reaction by considerably decreasing the reaction time while increasing the yield of the product. By this methodology, glycosidation of 3,4,6-tri-*O*-acetyl-D-galactal and -D-glucal with alcohols and phenols led exclusively to the corresponding alkyl and aryl 2,3-unsaturated glycosyl derivatives, with major formation of the α anomer, attributable to the anomeric effect. Recently, montmorillonite K-10-supported bismuth(III) triflate was used as a catalytic system for the glycosylation of alcohols and thiols with 4,5-oxazoline derivatives of sialic acid.[76] The reactions also proceeded via Ferrier rearrangement to afford 3,4-unsaturated sialic acid derivatives in moderate yield.

Development of C-glycosylation methods has attracted great attention on account of the natural occurrence, biological profile, and usefulness of *C*-glycosyl derivatives as versatile chiral building blocks for the synthesis of optically active natural products.[77] Glycals are useful glycosyl donors for C-glycosylation, due to their ability to

AcO, AcO, O, F, I **41** —ROH / Montmorillonite K-10 / CH_2Cl_2, MS 5A→ AcO, AcO, O, OR, I **42**

SCHEME 8. Stereoselective β-glycosylation of various alcohols with a 2-deoxy-2-iodo-olivosyl fluoride catalyzed by Montmorillonite K-10.

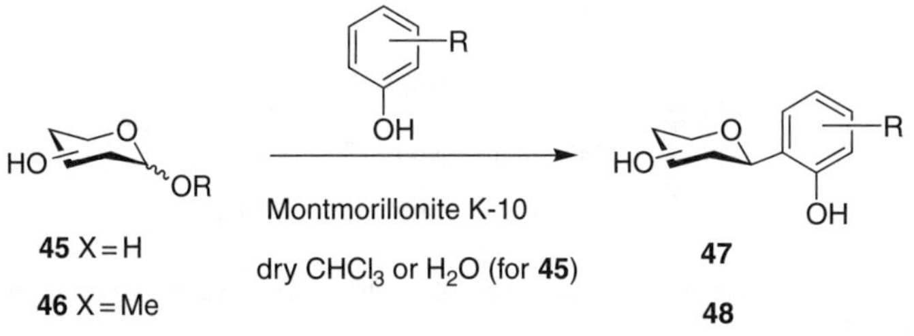

HO OR

45 X = H

46 X = Me

OH R

Montmorillonite K-10

dry $CHCl_3$ or H_2O (for **45**)

HO O R OH

47

48

SCHEME 9. Ferrier rearrangement by glycosidation of 3,4-di-*O*-acetyl-L-rhamnal with several alcohols catalyzed by Montmorillonite K-10.

react with *C*-nucleophiles to give 2,3-unsaturated *C*-glycosyl derivatives. This reaction normally takes place in the presence of such Lewis acids as boron trifluoride etherate or tin(IV) chloride, and occurs with high stereoselectivity. The most commonly used nucleophiles for C-glycosylation with glycals are organosilane reagents, especially allylsilane. Toshima *et al.*[78] reported the C-glycosylation of allyltrimethylsilane, vinyloxytrimethylsilane, or isopropenyl acetate with glycal acetates catalyzed by montmorillonite K-10 to give 2,3-unsaturated *C*-glycosyl derivatives in moderate to high yield and α-stereoselectivity.

C-Glycosyl aromatic compounds are commonly found as plant constituents and are present in a variety of biologically important natural products, including such antibiotics as angucyclin, pluramycin, and related families.[77,79] These possess C-glycosylically linked D-olivose moieties in their structure. Toshima *et al.*[80] also investigated the C-glycosylation of several phenol and naphthol derivatives with unprotected methyl olivosides (**46**) and the parent reducing sugars (**45**) in the presence of montmorillonite K-10 (Scheme 10). The unprotected sugars were efficiently linked to the glycosyl acceptors, affording stereoselectively the corresponding *β-C*-glycosyl derivatives (**47–48**) in yields ranging from 65% to 98%. In addition, montmorillonite K-10-catalyzed C-glycosylation reactions using anomerically unprotected sugars were carried out in water, giving yields and stereoselectivity similar to those obtained when dry chloroform was used. These unexpected results reinforce the usefulness of this environmentally friendly methodology.

c. Silica Gel.—The very first report on the use of silica gel as a promoter in glycosylation reactions was made by Matsushita *et al.*[46] 1,2-Anhydro-3,4,6-tri-*O*-pivaloyl-α-D-glucopyranose (**49a**) was allowed to react with cyclohexanol in the presence of silica gel, leading to the stereoselective formation of the *β* glucoside **50a**. However, epoxide hydrolysis also occurred and the reducing sugar was transformed into the corresponding methyl glucoside by treatment with an excess of methanol (Scheme 11).

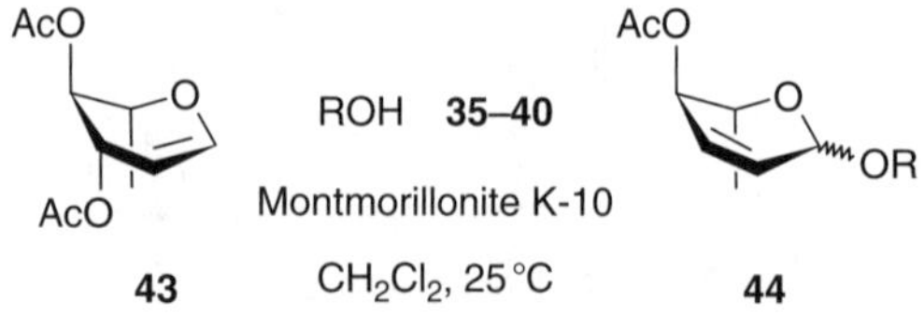

SCHEME 10. C-Glycosylation of aromatic alcohols with olivoses and olivosides in the presence of Montmorillonite K-10.

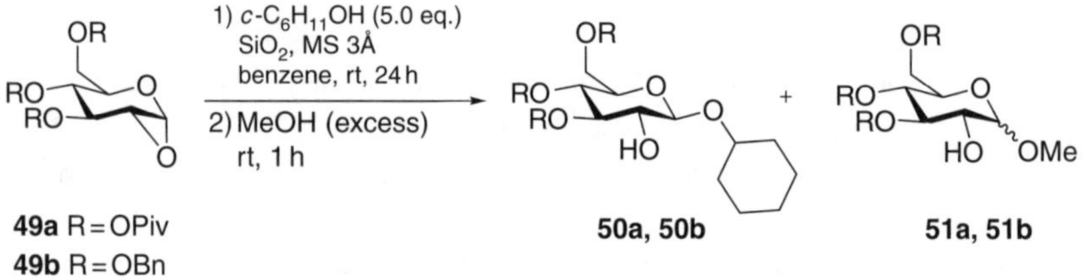

SCHEME 11. Glycosylation of a 1,2-anhydro sugar with cyclohexanol in the presence of silica gel.

Several sets of reaction conditions were compared and the best results were obtained when 3.0 wt eq. of SiO_2, and up to 9.0 wt eq. of 3 Å MS were used to transform both the pivaloyl- and the benzyl-protected glucosyl donors, to give **50a** and **50b** in 72% and 71% yields, respectively. The 3 Å MS were crucial when employed alongside silica, as they acted as water scavengers, minimizing the extent of hydrolysis and increasing the yield of the desired product. In addition, when **49a** reacted with cyclohexanol only, in the presence of 3 Å MS, no traces of the cyclohexyl β glucoside were obtained, confirming that 3 Å MS do not promote this reaction. Other catalysts were investigated for this reaction also, TiO_2–SiO_2 and the zeolites HM, HY, and NaY. Zeolite HM and TiO_2–SiO_2 gave good reaction yields but with a loss in stereoselectivity. The formation of **50a** in a stereocontrolled manner was achieved by using zeolites HY and NaY, although the yield was only 19% when the latter promoter was used. Thus, SiO_2 and HY zeolite were demonstrated to be the best catalysts, the advantage of the former being avoidance of possible side reactions attributable to its lower acidity.

Silica gel successfully catalyzed the stereoselective synthesis of several glucoside terpenoids. Treatment of **49a** with propan-2-ol, geraniol, the tetrahydropyranyl (THP) ether of coniferyl alcohol, and (−)-perillyl alcohol gave glucosides **52a–d** in good yields (Scheme 12). The acid-labile THP group was retained under these reaction

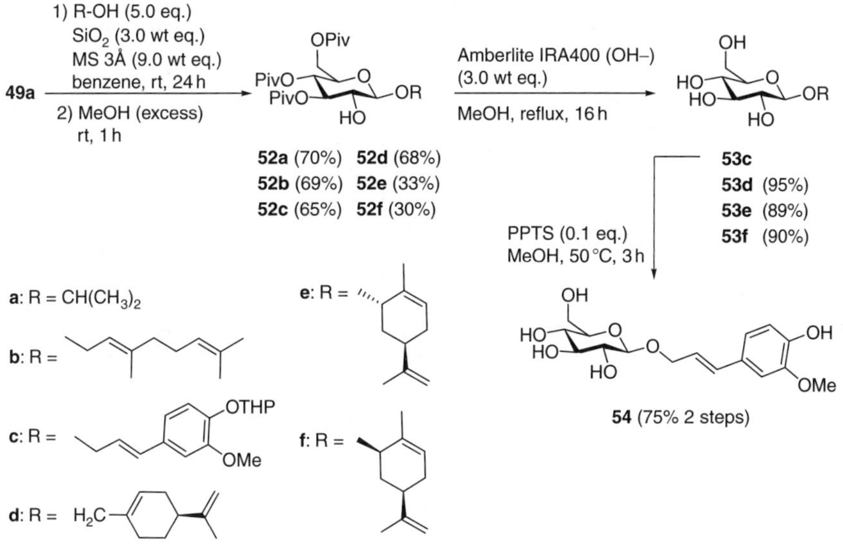

SCHEME 12. Synthesis of terpenoid glucosides via a stereoselective glucosidation of a 1,2-anhydro sugar catalyzed by silica gel.

conditions and, unlike other acid catalysts, silica gel led to good regioselectivity. However, glucosylation of the sterically hindered *trans*- and *cis*-carveol gave low yields of **52e** and **52f**, respectively. Further deprotection afforded the terpenoid glucosides **53d**, **53f,** and **54**.

Silica-supported reagents have been exploited as nontoxic, inexpensive, reusable, and environmentally acceptable catalysts for developing stoichiometric reaction methods in organic chemistry and specifically in carbohydrate chemistry. Apart from being easy to handle and to store, these systems allow facile workup, the catalyst being removed by simple filtration, and the reaction products isolated by chromatographic purification, if necessary.

Previously applied for the conversion of alcohols into the corresponding acetates,[81] perchloric acid immobilized on silica gel ($HClO_4$–SiO_2 200–400 mesh) was first reported in a glycosylation context by Agarwal *et al.*[82] to promote the Ferrier rearrangement. It proved to be a very efficient catalyst in the reaction of 3,4,6-tri-*O*-acetyl-D-glucal (**55**) with several primary, secondary, and allylic alcohols, as well as phenols and thiophenol, yielding the corresponding 2,3-unsaturated *O*/*S*-glycosides in good to excellent yields, with short reaction times and high α-selectivity (Table III). Moreover, the authors reported the formation of a 1-*C*-glucosyl derivative in the reaction with

TABLE III
Ferrier Reaction of Tri-*O*-acetyl-D-glucal with Different Nucleophiles in the Presence of $HClO_4$–SiO_2 Reagent[82]

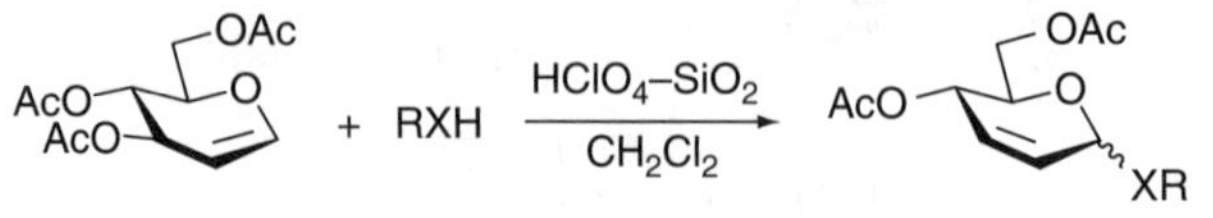

55 + RXH → ($HClO_4$–SiO_2, CH_2Cl_2) 56

XR	Time (min)	Yield (%)	α:β	XR	Time (min)	Yield (%)	α:β
OMe	180	82	6:1	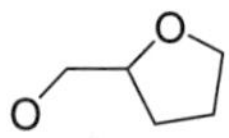 O–CH₂-tetrahydrofuranyl	45	84	7:2
OEt	180	81	15:1	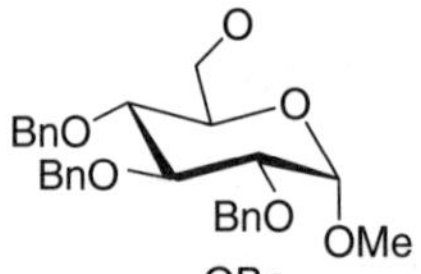 6-O-(methyl 2,3,4-tri-O-benzyl-glucopyranoside)	5	86	7:2
OBn	20	82	6:1				
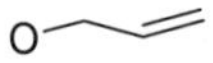 O-allyl	30	81	8:1	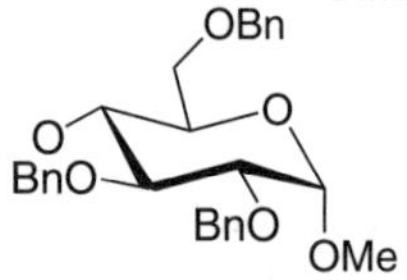 4-O-(methyl 2,3,6-tri-O-benzyl-glucopyranoside)	10	65	1:0
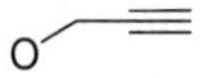 O-propargyl	20	92	5:1				
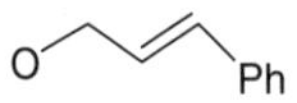 O-cinnamyl (Ph)	30	82	8:1	OPh	5	54	1:0
 O-CH₂CH=CHCH₂-OBn	20	95	6:1	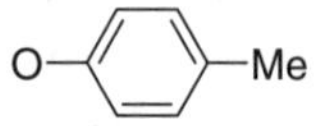 O-C₆H₄-Me	10	70	11:1
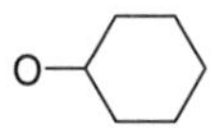 O-cyclohexyl	45	92	24:1	SPh	20	60	20:1
OChol	150	86	23:1	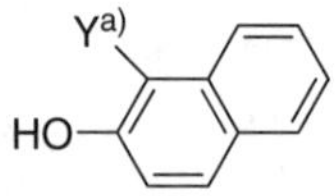 2-hydroxy-1-naphthyl (Y[a)], HO)	20	73	5:1

For RXH: Y=H and for the target molecule: Y=*C*-glycosyl.

β-naphthol in 73% yield with an *α*/*β* ratio of 5:1, probably resulting from rearrangement of the *O*-glycoside. Most reactions were complete within 1 h, except when MeOH, EtOH, and cholesterol were used as aglycons, and the amount of $HClO_4$–SiO_2 varied from 5 to 300/100 mg of starting sugar, according to the reactivity of the acceptor.

The results obtained by using this reagent system are comparable to those achieved using such alternative catalysts as $Sc(OTf)_3$, $Yb(OTf)_3$, CAN, and $BF_3 \cdot Et_2O$, and even better in some cases, in terms of yield, reaction time, and anomeric selectivity.

Misra *et al.*[83] also reported the Ferrier rearrangement of 3,4,6-tri-*O*-acetyl-D-glucal and -galactal with several alcohols, including monosaccharides, and thiols, bearing allyl, propargyl, isopropylidene, benzyl, and isopropyl groups. The corresponding 2,3-unsaturated *O*/*S*-glycosides were obtained in good yield after a few minutes as anomeric mixtures, with α/β ratios between 1:1 and 1:0. The $HClO_4$–SiO_2-promoted Ferrier rearrangement of glycals was also employed for the synthesis of 2,3-unsaturated *C*-glycopyranosyl compounds by reaction of acylated and alkylated glycals with trimethyl- and triethyl-silylated *C*-nucleophiles and with 1,3-diones and β-ketoesters, in acetonitrile at room temperature.[84] The reaction proceeded within 20–60 min, affording excellent yields of 2,3-unsaturated *C*-glycosyl derivatives, which were isolated as anomeric mixtures with α/β-selectivities between 20:0 and 1.5:1. The cleaner reaction profiles and simplicity in operation, with no need for special precautions in either handling the catalyst or in excluding moisture from the reaction medium, are other appealing aspects of this methodology for C-glycosylation reactions.

Synthesis of glycoconjugates promoted by $HClO_4$–SiO_2 was first described by Mukhopadhyay *et al.*[85] as a practical alternative to triflic acid for NIS-promoted glycosylation reactions. This fast and straightforward methodology allowed transformation of "disarmed" thioglycosides in the corresponding 1,2-*trans*-glycosides, with yields in the range of 74–87%. Moreover, the synthesis of a precursor to the N-linked glycan trimannoside core [3,6-di-*O*-(α-D-mannopyranosyl)-α-D-mannopyranoside], a natural ligand for the lectin concanavalin A, was accomplished by one-pot double glycosylation in 72% yield (Scheme 13). The same methodology was applied to glycosyl trichloroacetimidates[86] to afford several disaccharides, with yields in the range of 55–94%, and again the synthesis of the trimannoside **59** was successfully accomplished in 76% yield.

These results encouraged the authors to attempt the synthesis of more-complex oligosaccharides, such as the Le^x and Le^a derivatives, **65** and **66**, respectively. After formation of the $(1 \rightarrow 3)$-fucosyl linkage, as promoted by $HClO_4$–SiO_2, in a stereo- and regio-selective manner, the peracetylated galactosyl trichloroacetimidate was added to afford the desired Le^x derivative **65** in 62% yield. The same method was applied to obtain the Le^a derivative **66** in 59% (Scheme 14).[86]

The "on-column" synthesis approach was also explored, followed by *in situ* purification of the products for the glycosylation reactions using trichloroacetimidates. The authors primed the top of a standard silica chromatography column with perchloric acid immobilized on silica. After charging both reactants, dissolved in dry CH_2Cl_2, onto the

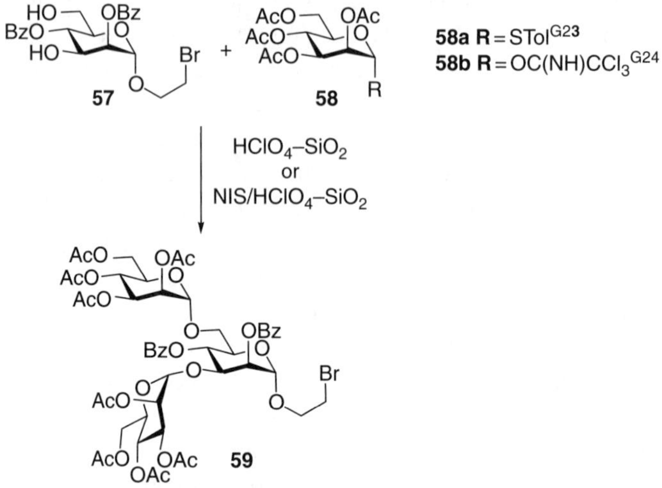

SCHEME 13. Synthesis of the trimanoside **59** by glycosidation of a thioglycoside or a trichloroacetimidate with a diol acceptor in the presence of $HClO_4$–SiO_2.

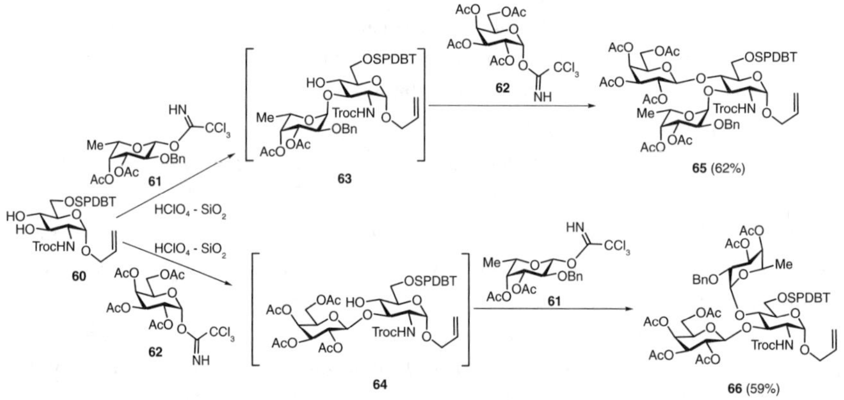

SCHEME 14. Synthesis of Le^x and Le^a derivatives by glycosylation in the presence of $HClO_4$–SiO_2.

column, they were allowed to react for 30 min and subsequently separated, affording the corresponding glycosides in comparable yields those obtained by solution-phase procedures. The same approach was applied in the preparation of glycosyl amino acids, affording the 1,2-*trans*-linked products in good yield, although the corresponding reducing sugar and unreacted amino acid were isolated as by-products.[86]

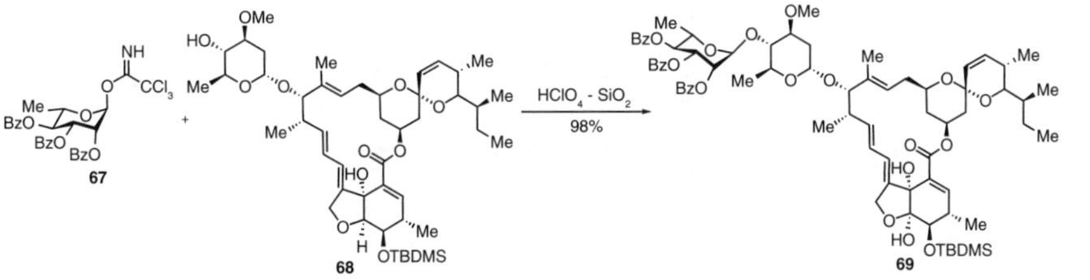

SCHEME 15. Synthesis of an Avermectin B1 analogue by glycosidation of a trichloroacetimidate donor catalyzed by $HClO_4$-SiO_2.

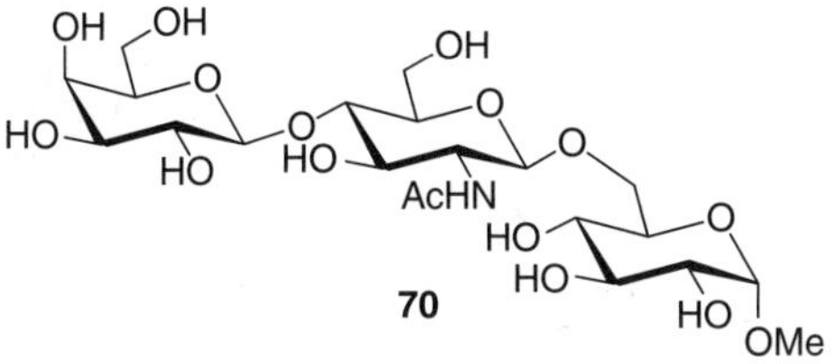

SCHEME 16. A methyl glycoside analogue of a trisaccharide fragment found in *E. coli*.

Glycosylation promoted by $HClO_4$–SiO_2 using trichloroacetimidates was also investigated by Du *et al.*[87] Using a 3–6% molar ratio of the promoter, the corresponding glycosides were obtained in 78–92% yield with excellent stereoselectivity. A variety of hydroxyl protecting groups such as isopropylidene, benzylidene, TBS, Bz, All, Tr, Bn, Ac, and other functional groups such as lactone, aldehyde, azide, and thioglycoside were found compatible under the glycosylation conditions studied and no side reactions resulting from migration or degradation were recorded, unlike the situation when other catalysts are used. In addition, this procedure allowed to the synthesis of the avermectin B1a analogue **69** (Scheme 15) in almost quantitative yield.

Synthesis of a disaccharide fragment related to the repeating unit of *Escherichia coli* O_{83}:K_{24}:H_3 O-antigen by glycosylation of methyl 2,3,4-tri-*O*-acetyl-*β*-D-galactopyranoside with methyl (ethyl 2,3,4-tri-*O*-acetyl-1-thio-*β*-D-glucopyranosid)uronate failed with NIS/$HClO_4$–SiO_2, probably because of the low reactivity of the donor. However, the methyl glycoside analogue **70** of a trisaccharide fragment, also found in *E. coli* O83:K24:H3, was synthesized in a 13-step procedure where the system $HClO_4$–silica was used in five steps, namely in detritylation, introduction of a benzylidene group and its subsequent removal, as well as in both NIS-promoted glycosylation reactions, all proceeding in good yields (Scheme 16).[88]

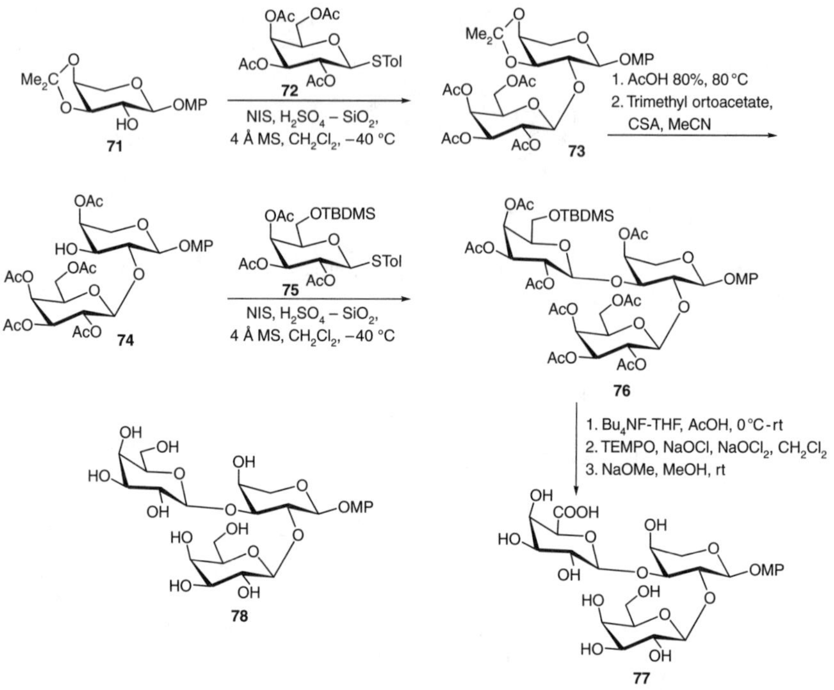

SCHEME 17. Synthesis of the glycon of Sokodoside B, involving H_2SO_4-SiO_2-promoted glycosylation steps.

The use of H_2SO_4 supported on silica gel was investigated as catalyst in Fischer-type glycosylations.[89] Reaction of D-glucose, D-galactose, D-mannose, *N*-acetyl-D-glucosamine, L-rhamnose, L-fucose, and maltose with propargyl alcohol was reported using the alcohol (5 eq.) in solvent-free conditions at 65 °C to furnish the desired glycosides in 69–83% yield with α,β-selectivity in the range of 10:1–1:0. This procedure was applied to the preparation of a variety of D-glucosides by reaction of D-glucose with allyl, aryl (benzyl, *p*-methoxybenzyl) and alkyl (octyl, dodecyl, 2-bromoethyl) alcohols, affording the desired glycosides in good yield (75–82%) and stereoselectively ($10:1<\alpha/\beta<13:1$).

The successful synthesis of trisaccharide **77** (Scheme 17), which is the glycon of the steroid glycoside Sokodoside B, isolated from *Erylus placenta*, was accomplished by Dasgupta *et al.*[90] Coupling of *p*-tolyl 2,3,4-tetra-*O*-acetyl-1-thio-β-D-galactoside (**72**) with α-L-arabinopyranoside (**71**), using NIS in the presence of H_2SO_4–SiO_2 afforded the corresponding disaccharide in 89% yield. After protective-group modification, reaction with the donor **75** in the presence of the same promoter gave the protected trisaccharide **76** in 83% yield. Selective deprotection of the primary alcohol, oxidation, and acetyl cleavage afforded the desired free trisaccharide **77**. Hence, the use of H_2SO_4–SiO_2 as catalyst for the NIS-promoted thioglycoside activation, instead of the corrosive TfOH or TMSOTf, proved to be a better choice.

Clean glycosylation required temperatures in the range of −30 to −40 °C, since partial loss of the TBDMS group occurred at temperatures higher than −10 °C. The authors took advantage of this observation and, anticipating the difference in reactivity among the 2- and 3- positions of the arabinose moiety, they attempted the one-pot sequential glycosylation. Thus, *p*-methoxyphenyl 4-*O*-acetyl-α-L-arabinopyranoside was first coupled with donor **75**, and then the second donor **72** was added. The sequential glycosylations were performed at −40 °C using NIS and H_2SO_4–SiO_2, and the acetylated trisaccharide, lacking the TBDMS group, was obtained after raising the temperature to ambient. The analogue of the trisaccharide of Sokodoside B (**78**) (Scheme 16) was also synthesized by a one-pot reaction of *p*-tolyl 2,3,4,6-tetra-*O*-acetyl-1-thio-*β*-D-glucopyranoside (2.5 eq.) with *o*-methylphenyl 4-*O*-acetyl-α-L-arabinopyranoside (1 eq.) at 10 °C using the same catalytic system, followed by complete deacetylation with NaOMe/MeOH, in 64% overall yield.[90]

Other glycoconjugates were synthesized by H_2SO_4–SiO_2-promoted glycosylation methodologies. The trisaccharides **79** and **80** (Scheme 18), which are related to the saponin isolated from *Centratherum anthelminticum*, were obtained in very good yield and with high stereoselectivity.[91] Glycosylation of a rhamnosyl acceptor at C-3 with the known *p*-tolyl 2,3,4,6-tetra-*O*-acetyl-1-thio-*β*-D-glucopyranoside, activated by NIS in the presence of H_2SO_4–SiO_2, afforded the disaccharide precursor of **79** in 91% yield. Subsequent attachment of the arabinosyl moiety was performed by activation of the corresponding disaccharide trichloroacetimidate at −40 °C, using H_2SO_4–SiO_2, to give the trisaccharide **79** in 89% yield.

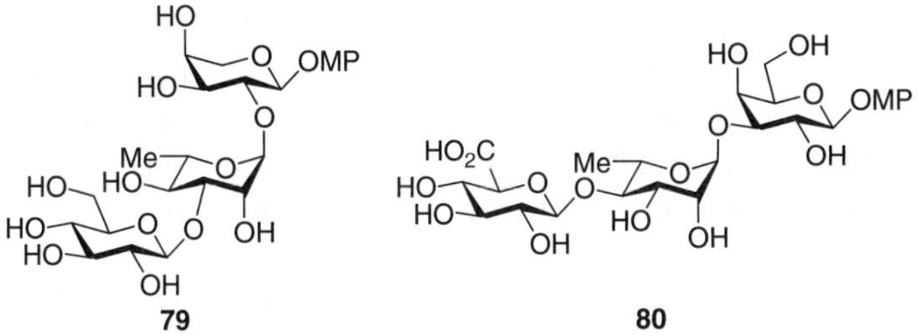

79 **80**

SCHEME 18. Trisaccharides related to the saponin isolated from *C. anthelminticum*.

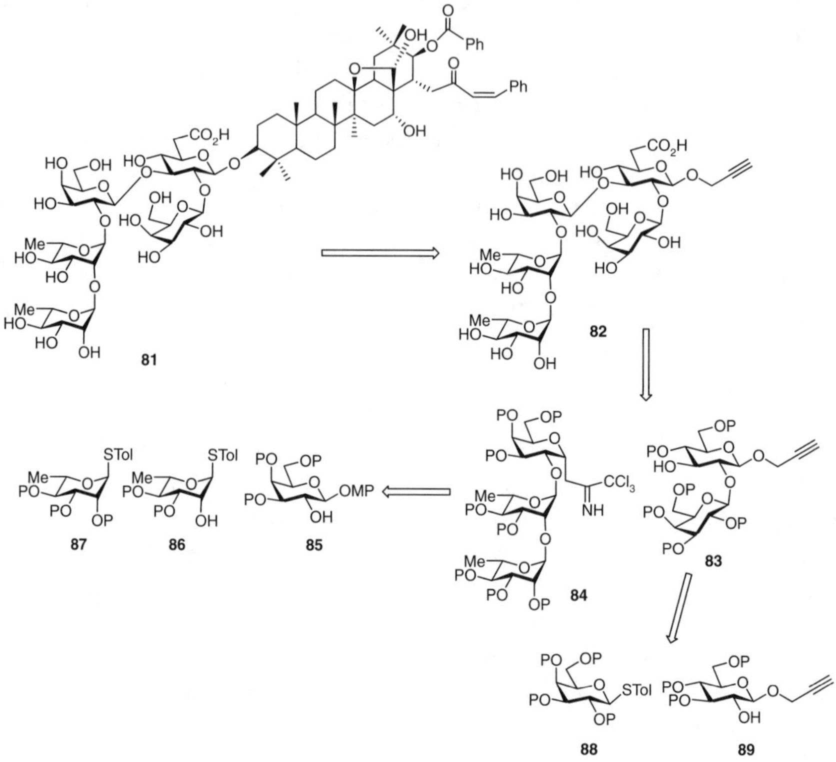

SCHEME 19. Retrosynthetic pathway for the glycon of the anti-Leishmania triterpenoid saponin **81**.

Trisaccharide **80** was synthesized by a similar approach, in which a rhamnoside was formed in 86% yield, transformed into its trichloroacetimidate, submitted to the second glycosylation step promoted by H_2SO_4–silica (84% yield), and subsequently deprotected to **80**.[91]

Subsequently, Rajput and Mukhopadhyay[92] successfully synthesized the glycone of the anti-leishmanial triterpenoid saponin **81**, isolated from *Maesa balansae* (Scheme 19). Glycosylations were formed by thioglycoside or glycosyl trichloroacetimidate activation using H_2SO_4–SiO_2 in conjugation with either NIS or without. For the synthesis of the target pentasaccharide, a 3 + 2 disconnection was planned, where the protected trisaccharide **84** was coupled to the disaccharide propargyl acceptor **83** under NIS/H_2SO_4–SiO_2 activation. The tri- and disaccharides **84** and **83** were synthesized from commercially available monosaccharides and the yields of the NIS-H_2SO_4–SiO_2 or H_2SO_4–SiO_2-assisted glycosylations were in the range of 82–87%.

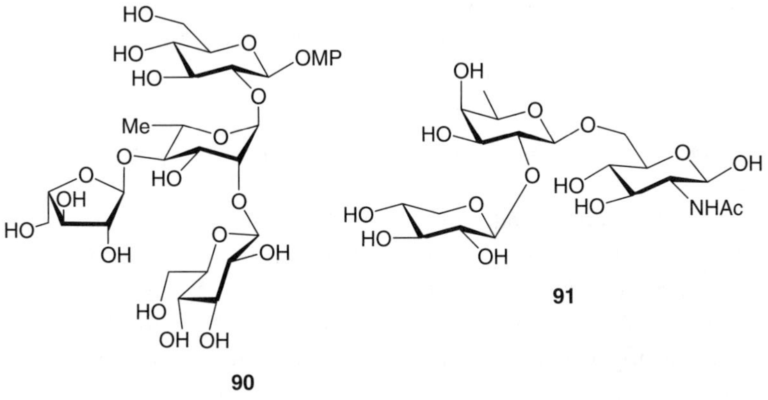

SCHEME 20. Tetra- and tri-saccharides **90** and **91**, contained in the antitumor agent Julibroside J_{28}.

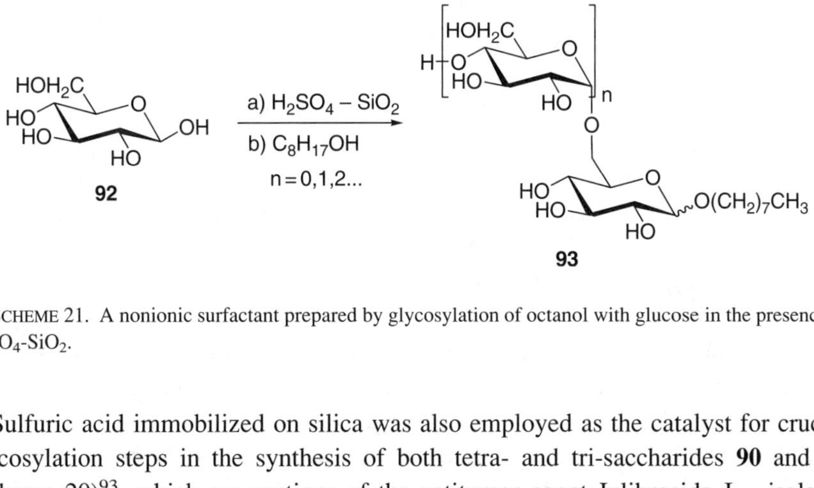

SCHEME 21. A nonionic surfactant prepared by glycosylation of octanol with glucose in the presence of H_2SO_4-SiO_2.

Sulfuric acid immobilized on silica was also employed as the catalyst for crucial glycosylation steps in the synthesis of both tetra- and tri-saccharides **90** and **91** (Scheme 20)[93], which are portions of the antitumor agent Julibroside J_{28}, isolated from *Albizia julibrissin*. Both compounds showed significant *in vitro* antitumor activity against HeLa, Bel-7402, and PC-3M-1E8 cancer cell-lines. The readily available D-glucose, L-rhamnose, and L-arabinose, or *N*-acetyl-D-glucosamine, D-fucose, and D-xylose were used as starting materials for the synthesis of **90** and **91**, respectively, using simple protecting-group manipulations. Glycosylations were promoted by NIS/H_2SO_4–SiO_2 or H_2SO_4–SiO_2 in very good yields (82–87%).

The octyl polyglucoside **93** (Scheme 21), a nonionic surfactant, having a degree of polymerization of 1.37, was also prepared in nearly quantitative yield by reaction of glucose with octanol in the presence of H_2SO_4–SiO_2.[94] The reaction was performed out at 110 °C and 20 mbar to remove the water formed and avoid glycoside hydrolysis.

2. Sugar Protection and Deprotection

The presence in carbohydrates of multiple hydroxyl groups of similar reactivity makes the chemo- and regio-selective manipulation frequently required quite difficult. For this reason, multistep protection–deprotection approaches are regularly employed in carbohydrate chemistry, and versatile techniques for these transformations are particularly helpful. The following section addresses this aspect, concentrating on the catalytic procedures that have been developed employing zeolites and related siliceous materials.

a. Zeolites.—*(i) Medium and Large-Pore Zeolites.* Acetylation is a common method for protecting hydroxyl groups of carbohydrates, in particularl to simplify the structural elucidation of many natural products containing saccharide units. Peracetylated sugars are also useful intermediates for the synthesis of glycosides, oligosaccharides, and other glycoconjugates. A standard acetylation procedure uses an excess of acetic anhydride in pyridine (or triethylamine), which serves as solvent and as a base. The inclusion of a cocatalyst, such as 4-(dimethylamino)pyridine (DMAP), is sometimes needed to increase the reaction rate and facilitate acetylation of more-hindered hydroxyl groups. Many other reagent systems have been introduced for this transformation; these include Lewis acids, Brønsted, acids and heterogeneous catalysts, in combination with acetic anhydride.[95]

Bhaskar and Loganathan[96] described O-peracetylation of monosaccharides, disaccharides, and methyl glycosides (**94**) with acetic anhydride under catalysis by acid zeolites. From the panel of zeolites tested (HY, HEMT, HZSM-5, HZSM-12, HZSM-22, and H-beta), the large-pore zeolite H-beta provided the best yields of the fully acetylated sugars, most of them being over 85% and up to 99%, with the pyranose forms **95** accounting for 66–100% of the reaction products (Scheme 22).

Subsequently, Gonçalves *et al.*[97] reported the acetylation of glycerol with acetic acid performed over different solid acids, including montmorillonite K-10 and such acid zeolites as HZSM-5 and HUSY. Among the siliceous porous materials examined, montmorillonite K-10 gave the best performance, with 96% conversion into the mono-, di- and tri-acetylated derivatives. When zeolites were used, the conversion was lower than with the other catalysts, giving a 30% conversion for HZSM-5 and only 14% for HUSY. However, selectivity for the primary monoacetylated product,

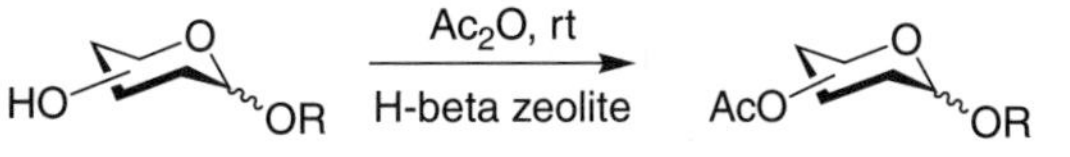

SCHEME 22. Peracetylation of monosaccharides, disaccharides and methyl glycosides in the presence of H-beta zeolite.

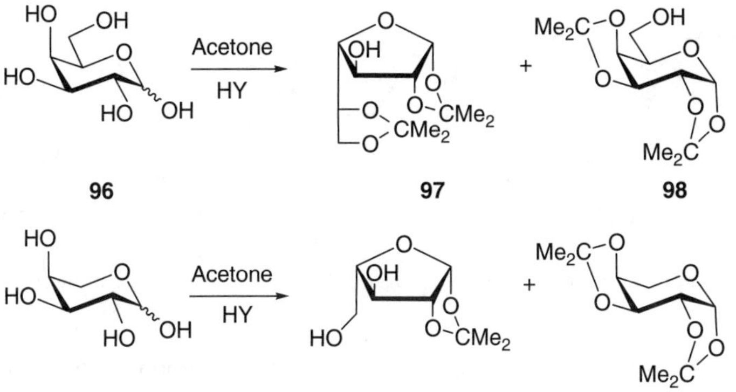

SCHEME 23. Acetonation of sugars catalyzed by the HY zeolite.

isolated in 79–83% yield, was achieved. The low efficacy observed for zeolites may be explained by space constrains and diffusion problems of the products within the pore system, favoring the formation of the monoacetylated derivative.

Isopropylidene acetals are convenient protecting groups in carbohydrate chemistry, particularly for the protection of 1,2- and 1,3-diols, and are readily formed by reaction of the diol with acetone or 2,2-dimethoxypropane under acidic conditions. Several protic and Lewis acids have been reported as catalysts for this purpose.[98]

Our group has demonstrated that zeolites are efficient catalysts for sugar isopropylidenation, offering the added value of directly promoting the major formation of the thermodynamically less stable furanose derivatives.[99] In the presence of HY zeolite, acetonation of D-galactose (**96**) and L-arabinose (**99**) furnished the corresponding furanose-derived diacetals **97** and **100**, in 40% and 37% yields, respectively, and the pyranoid diacetals **98** and **101** were obtained in lower yields, of 20% and 24%, respectively (Scheme 23). It is noteworthy that this methodology provided the first

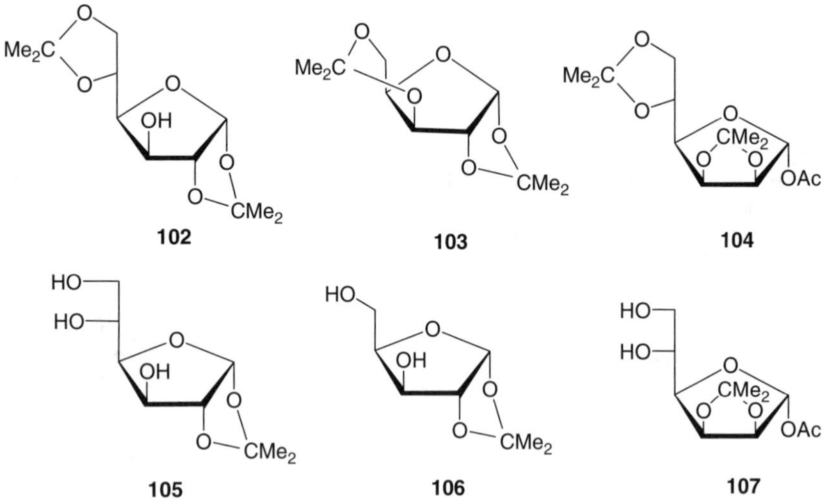

SCHEME 24. Selective deprotection of sugar di-*O*-isopropylidene acetals using acid zeolites.

straightforward synthesis of 1,2-*O*-isopropylidene-α-L-sorbopyranose, obtained in a moderate yield from L-sorbose.

Selective removal of one isopropylidene group from a diacetal may be achieved by a variety of procedures, most of them involving protic or Lewis acids.[100] Particularly common is the hydrolysis of the acetal engaging of the primary position of di-*O*-isopropylidene derivatives. Bhaskar *et al.*[101] studied the selective deprotection of di-*O*-isopropylidene acetals derived from D-glucose, D-xylose, and D-mannose, using acid zeolites and montmorillonite K-10. When **102** was submitted to acid hydrolysis in aqueous methanol, the best yields (85–96%) for the monoacetal **105** were obtained when H-beta and HZSM-5 zeolites were employed as catalysts (Scheme 24, Table IV). HY zeolite proved to be ineffective, whereas the yield obtained for the montmorillonite K-10-catalyzed reaction was low (22%). The zeolites found most effective were then used for the hydrolysis of the diacetal **103** and **104**, providing excellent yields for the desired corresponding monoacetals **106** and **107**.

(ii) Small-Pore Zeolites. While the ability to act as water scavenger is the most well-known property of MSs, it was found that they act as promoters for several protection and deprotection reactions of carbohydrates, including regioselective acetylation, diphenylmethylation, tritylation, and deacetylation, allowing mild transformations that ensure the stability of several functionalities present in the starting materials.

TABLE IV
Solid Acid-Catalyzed Selective Deprotection of 1,2:5,6-Di-*O*-isopropylidene-α-D-glucofuranose (102)

Catalyst	Solvent	Time (h)	Yield (%) of 105
HZSM-5	MeOH–H_2O (1:1)	16	92
	MeOH–H_2O (4:1)	48	85
	MeOH–H_2O (1:1)	2	96
H-beta	MeOH–H_2O (9:1)	2	94
	MeOH	4	88
Montmorillonite K-10	MeOH–H_2O (1:1)	48	22
HY	MeOH–H_2O (1:1)	48	–

Acetylation of carbohydrate hydroxyl groups with acetic anhydride using 4 Å MS as promoters was reported by Adinolfi *et al.*[102] Several saccharide polyols containing acid-labile functionalities were submitted to these reaction conditions, affording the corresponding acetates in good to excellent yield. The differential reactivity of primary and secondary alcohols allowed the regioselective protection of polyols by simply changing the reaction temperature.

In a later report,[103] the nucleosides inosine **108** and guanosine **109** (Table V) were peracetylated with acetic anhydride in the presence of a variety of silicon-based promoters, such as zeolite beta, NaY, ZSM-5, ZSM5-H, montmorillonite K-10 and KSF, 3, 4, 5 Å MSs and 13X (which is a large-pore MS), and $NaHSO_4.SiO_2$, under solvent-free conditions. Montmorillonites gave yields lower than 5%, zeolites, 3 and 5 Å MS promoted the reaction with moderate yields (45–52% for zeolites, and 50–55% for MSs). The more-basic 4 Å MS and 13X MS led to the highest yields (71–84%). Hence, enhancement of the basic properties of both 4 Å MS, 13X MS and 5Å MS was performed by ion exchange with potassium chloride, and the acetylation of several purine and pyrimidine nucleosides was investigated. The results demonstrated that potassium-exchanged MS is more effective than the corresponding commercial MS, leading to higher yields and shorter reaction times; the large-pore 13X/KCl MS being the most efficient catalyst. Acetylation of other nucleosides bearing such reactive functional groups as bromo, thiol, and acetal, was also successfully achieved. Recycled catalyst was used without any pretreatment and its catalytic activity was completely maintained. Under microwave irradiation the desired peracetylated products were also obtained.

Mild benzhydrylation (diphenylmethylation) of primary and secondary saccharide alcohols, and tritylation of the primary alcohols, were promoted by 4Å AW-300 MS at room temperature through a dehydration mechanism in the absence of any strong protic or Lewis acid. While protection of a single primary or secondary hydroxyl

TABLE V
Peracetylation of the Nucleosides Inosine 108 and Guanosine 109 with Acetic Anhydride in the Presence of Solid Catalysts

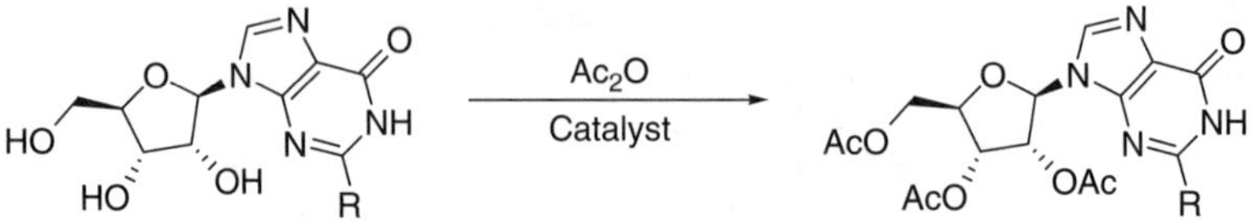

Ac_2O, Catalyst

108 R = H
109 R = NH_2

110 R = H
111 R = NH_2

Catalyst	Yield of 110 (%)	Yield of 111 (%)
None	40	42
Amberlyst-15		<5
I_2		39
$NaHSO_2$–SiO_2	36	35
Montmorillonite K-10	<5	
Montmorillonite K-10	<5	
Zeolite beta	45	45
Zeolite NaY	48	47
Zeolite ZSM-5	52	46
Zeolite ZSM-5-H	50	48
3 Å MS	52	55
5 Å MS	53	50
4 Å MS	71	74
13X MS	84	78

group was effected with good yields, regioselective benzhydrylation of the primary alcohol group of methyl 2,3-*O*-acetyl-α-D-glucopyranoside was less satisfactory, leading to a mixture of the mono- and diprotected saccharides in the ratio of ~3.5:1. It is noteworthy that tritylation was accomplished for the first time using the inexpensive triphenylmethanol, rather than the usual chlorotriphenylmethane, with good regioselectivity for the primary position. This methodology also allows the installation of ether protecting groups in the presence of such base-labile groups as esters.[104]

With respect to application of MS to deprotection reactions, Mizuno *et al.*[105] observed an unexpected *O*-acetyl cleavage in a building block of an N-glycopeptide by simple treatment with absolute methanol that had been stored over 3 Å MS. When acetates from other sugars were submitted to the same conditions, similar results were obtained, whereas no *O*-acetyl cleavage was observed when using absolute methanol that had not been dried over 3 Å MS. This deacetylation was evidently promoted by a methoxy active species generated by the 3 Å MS. Deacetylation and debenzoylation in several carbohydrates also took place upon treatment with methanol stored over 4 Å MS for 8 weeks at room temperature,[106] allowing selective deacylation in the presence of a benzyl ether function. To eliminate the waiting period necessary to form the alkali species in methanol, experiments using HPLC-grade methanol previously refluxed over powdered 4 Å MSs for 10 h gave the fully deacetylated products in excellent yields. In the case of O-peracetylated *N*-acetylglucosamine, the corresponding hemiacetal was detected in 90% yield after 2 h. However, a longer reaction time (48 h) was required to achieve complete O-deacetylation.

b. Montmorillonite.—The work reported on montmorillonite K-10-catalyzed sugar protection and deprotection reactions has shown that this porous solid can also provide mild and practical procedures, for acetylation, acetal protection, and deprotection, as well as for removal of other protecting groups frequently used in carbohydrate chemistry.

Acetylation of various mono- and disaccharides with acetic anhydride in the presence of montmorillonite K-10 was reported by Bhaskar and Loganathan.[107] Hexoses and pentoses underwent O-peracetylation in nearly quantitative yield, with major formation of the pyranose forms (75–100%) and α/β ratios ranging from 2.8/1 to 1.3/1. D-Galactose, D-xylose, L-arabinose, and L-fucose were the sugars that gave the highest yield of O-peracetylated furanose forms (25–14%), which were obtained in α/β ratio ranging from 2/1 to 1/2, depending on the starting material O-Peracetylation of the disaccharides maltose and cellobiose led mainly to the β anomer, whereas lactose and melobiose led mainly to the α anomer, the latter giving also furanose forms with α/β ratio (3:4).

The synthesis of isopropylidene acetal derivatives of monosaccharides and ribonucleosides has also been accomplished over montmorillonite K-10, in good yields, as reported by Asakura *et al.*[108] In further work, the same clay was employed to mediate removal of isopropylidene, silyl, and 4,4′-dimethoxytrityl (DMTr) groups from nucleoside and monosaccharide derivatives (Scheme 25).[109] Treatment of 2′,3′-*O*-isopropylideneuridine (**112**) with montmorillonite K-10 in methanol or aqueous methanol at different temperatures, revealed that 1:1 methanol–water used at 75 °C gave the highest yield of the parent uridine. These optimized reaction conditions were then applied to the nucleoside derivatives **113–119**, containing *tert*-butyldimethylsilyl (TBDMS) or *tert*-butyldiphenylsilyl (TBDPS) ether groups and 1,1,3,3-tetraisopropyl-1,3-disiloxanyl (TIPDS) groups. Deprotection of TBDMS- and TIPDS-containing nucleosides **114–118** was achieved in high yield (90–97%), although ether cleavage of 5′-*O*-TBDPS-2′-deoxyuridine (**119**) gave only traces of 2′-deoxyuridine, which is

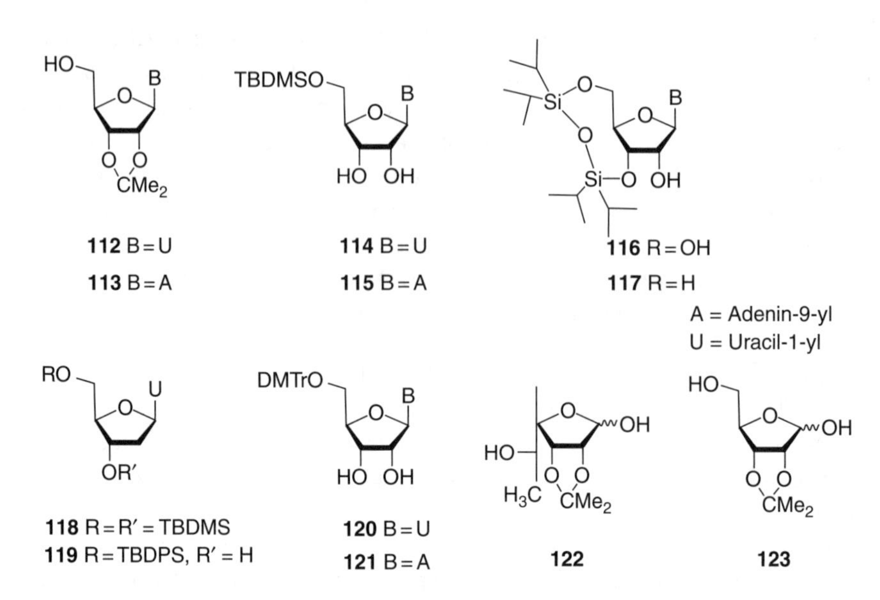

SCHEME 25. Monosaccharide derivatives containing protecting groups whose cleavage can be achieved by Montmorillonite K-10 in aqueous methanol.

in agreement with the requirement that stronger acidic conditions are needed for hydrolysis of these ethers. Removal of DMTr ether groups from nucleoside derivatives **120** and **121** was also successfully accomplished by this methodology. With respect to isopropylidene-protected monosaccharides, namely 1,2:5,6-di-*O*-isopropylidene-α-D-glucofuranose (**102**), the monoprotected 2,3-*O*-isopropylidene-α-L-rhamnofuranose (**122**), and 2,3-*O*-isopropylidene-D-ribofuranose (**123**), analogous treatment with montmorillonite K-10 gave the deprotected sugars in good to excellent yield.

c. Silica Gel.—Silica-supported reagents have gained much attention for protection and deprotection of carbohydrate functionalities, displaying significant advantages over other catalysts.

Acetylation of sugars was accomplished with $HClO_4$–SiO_2 using a stoichiometric quantity of protecting reagent and catalytic amounts of promoter under solvent-free conditions. O-Peracetylation of mono-, di- and tri-saccharides with acetic anhydride was successfully achieved within a few minutes in the presence of this catalyst. Most acetal functionalities remained unaffected under these reaction conditions, including glycosides, 1-thioglycosides, and sugar dithioacetals. Impregnation of $HClO_4$ in silica gel not only increased the effective surface area of the catalyst, but also excluded water from the reaction medium. However, use of hydrated sugars as starting materials led to a slight decrease in the yield, possibly because of partial consumption of the acetic anhydride by water.[110]

Chlorides of the rare earth metals erbium, lanthanum, europium, or neodymium, adsorbed on silica gel promoted regioselective acetylation of the primary hydroxyl groups of sugars using methyl orthoacetate as the acetylating agent. This methodology allowed synthesis of the primary monoacetylated derivatives in yields above 90% within 2–4 h. In addition, it was observed that those lanthanides having higher atomic numbers displayed greater catalytic effect.[111]

Mukhopadhyay *et al.*[112] developed a one-pot acetalation/ketalation–acetylation of sugars by simple addition of $HClO_4$–SiO_2 to a mixture of the starting material and benzaldehyde dimethyl acetal or 2,2-dimethoxypropane in acetonitrile, followed by the addition of acetic anhydride immediately after complete conversion of the starting sugar into the corresponding acetal/ketal. These reactions proceeded in 30–90 min, furnishing the peracetylated benzylidene or isopropylidene acetals in high yield. This procedure was extended to the synthesis of several S- and O-glycosides, such as the per-*O*-acetylated *O*-isopropylidene derivatives of L-rhamno-, L-arabino-, and D-galactopyranosides, and the per-*O*-acetylated *O*-benzylidene derivatives of D-gluco-, D-galacto-, and D-mannopyranosides. However, when applied to starting materials of the D-*manno* configuration, the solvent was replaced by DMF in order to avoid formation of the bis(benzylidene) product. Nevertheless, safety concerns limit the use of perchloric acid and H_2SO_4–SiO_2 was investigated as an alternative promoter for this one-pot reaction. The method was also successfully applied to the synthesis of per-*O*-acetylated *O*-benzylidene derivatives of various *S*- and *O*-glycosides having D-*gluco* and D-*galacto* configuration, and of per-*O*-acetylated *O*-isopropylidene derivatives of D-galacto-, D-manno-, L-rhamno-, and L-fucopyranosides, also retaining *S*- and *O*-acetal functionalities (Table VI). Furthermore, the yields obtained were reproducible when the catalyst was reused up to five times.[113]

Rajput and Mukhopadhyay[114] also reported the use of H_2SO_4–SiO_2 as promoter for the acetonation of reducing sugars (Table VII). While at room temperature no reaction of D-glucose with acetone occurred, after 3 h at 57 °C, the thermodynamically favored 1,2:5,6-di-*O*-isopropylidene-α-D-glucofuranose (**102**) was formed in 89% yield. When other hexopyranoses, namely D-galactose and D-mannose, were the starting materials used a slow increase of temperature led to the mixture of isopropylidene ketals in the pyranose and furanose forms, but the former could be obtained selectively by performing the reaction in a preheated oil bath. With the exception of L-arabinose, the aldopentoses submitted to the same reaction conditions, gave the corresponding isopropylidene acetals in the furanose form. Large-scale

TABLE VI

H_2SO_4–SiO_2-Promoted One-Pot Benzylidenation/Isopropylidenation-Acetylation, in Acetonitrile at Room Temperature[113]

Starting Material	Product	Time[a] (min)	Yield (%)
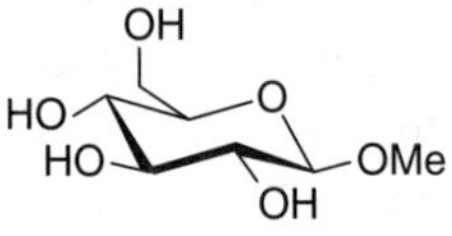 **124**	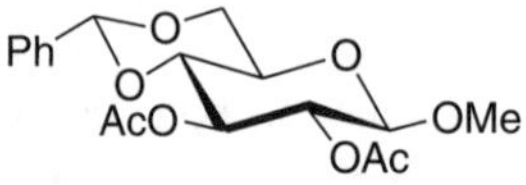 **125**	60	93
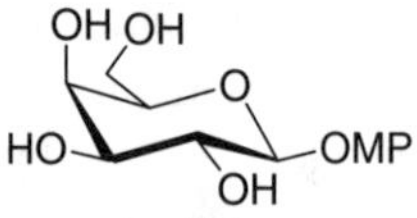 **126**	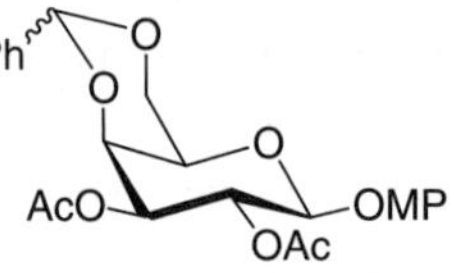 **127**	60	89
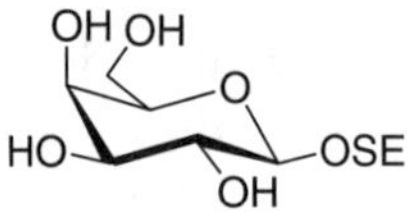 **128**	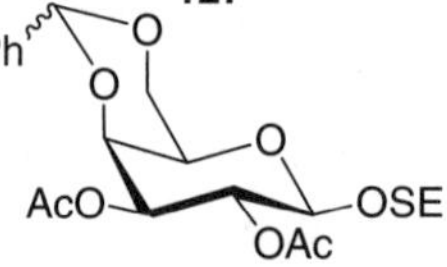 **129**	60	90
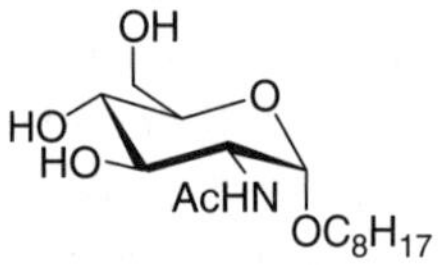 **130**	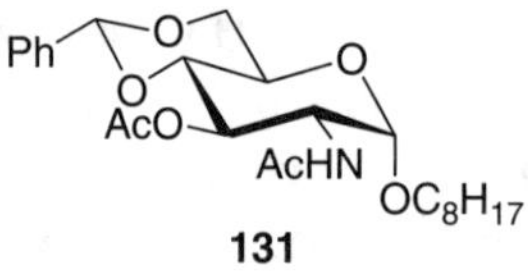 **131**	60	82
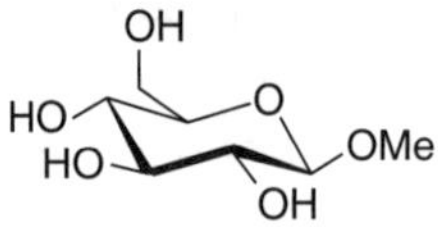 **132**	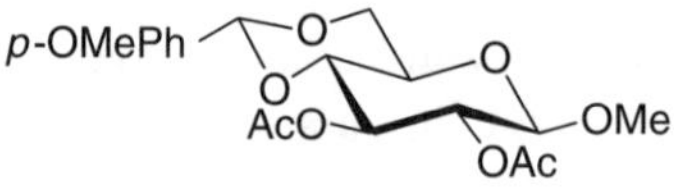 **133**	60	87
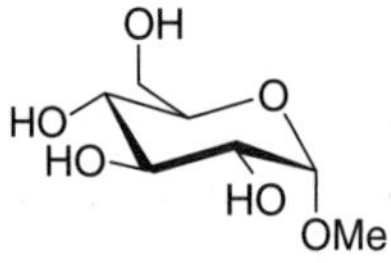 **134**	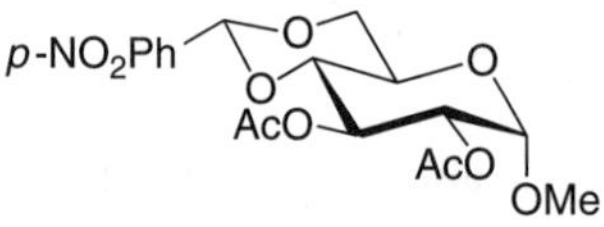 **135**	90	75
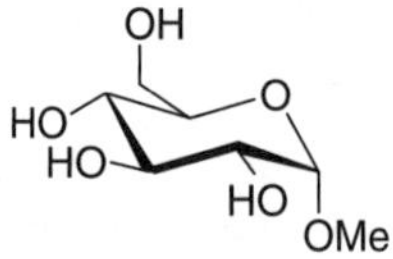 **134**	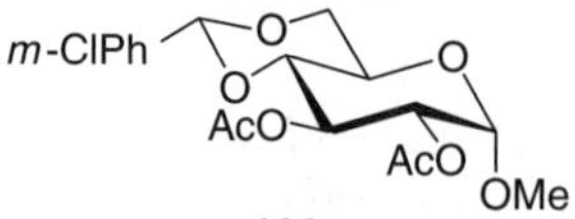 **136**	90	78

TABLE VI (*continued*)

Starting Material	Product	Time[a] (min)	Yield (%)
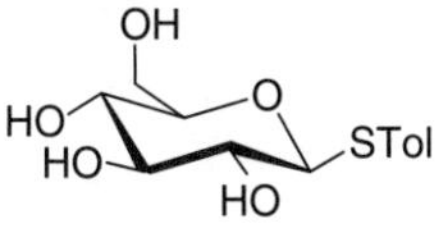 **137**	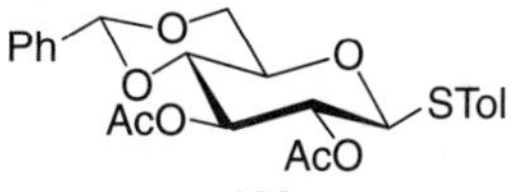 **138**	60	88
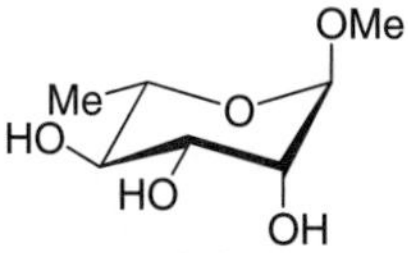 **139**	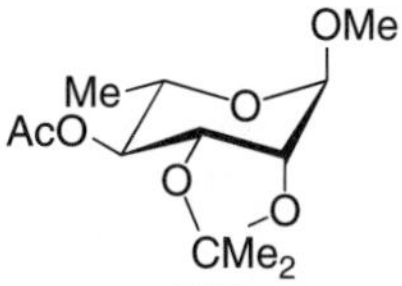 **140**	45	95
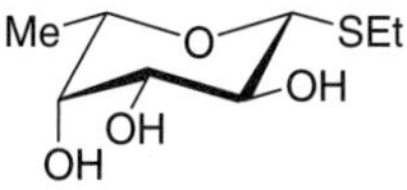 **141**	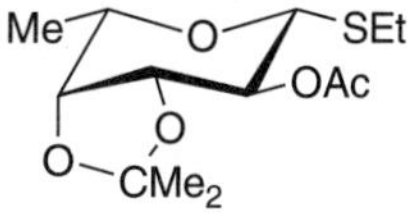 **142**	45	93
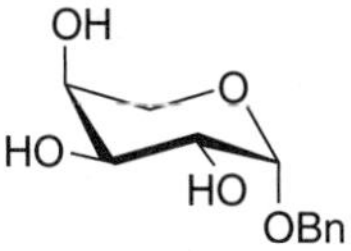 **143**	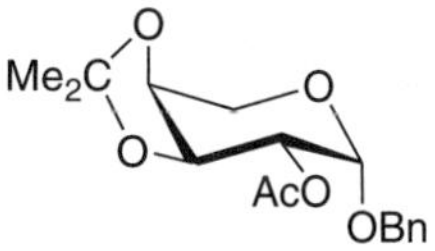 **144**	30	91
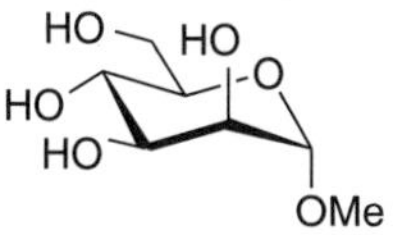 **145**	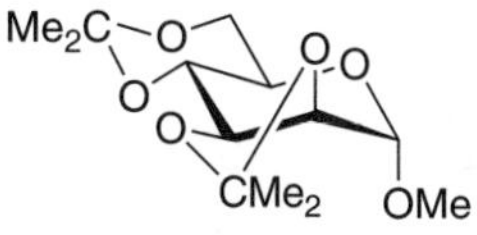 **146**	45	92

MP, *p*-methoxyphenyl; SE, 2-trimethylsilylethyl.
[a] Total time required for acetalation/ketalation and acetylation.

preparations are also feasible using this methodology, and the promoter can also be reused up to five times without loss in activity.

Sodium hydrogensulfate absorbed on silica ($NaHSO_4$–SiO_2) is also an acidic promoter for the introduction of benzylidene acetal groups (Scheme 26). Reaction of sugars of the D-*gluco*, D-*galacto*, or D-*manno* configurations bearing various functional and protecting groups (including azides, oxazolidinones, and *O*-/*S*-anomeric acetals), in acetonitrile with benzaldehyde dimethyl acetal in the presence of this promoter led successfully to the corresponding 4,6-*O*-benzylidene products in a very clean and

TABLE VII

Acetonation of Free Sugars Using H_2SO_4–SiO_2[114]

Starting Material	Product	Time (h)[a]	Yield (%)
D-Glucose	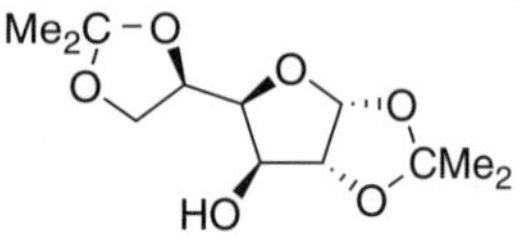 **102**	3	89
D-Galactose	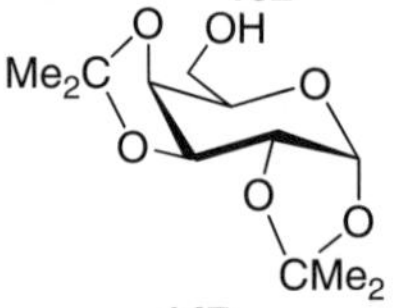 **147**	3	78
D-Mannose	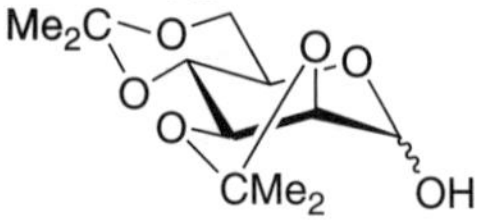 **148**	3	82
L-Rhamnose	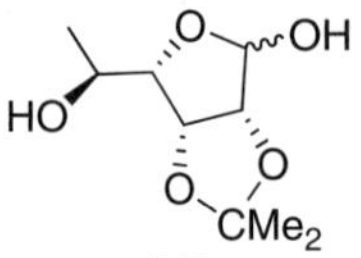 **149**	2	87
L-Arabinose	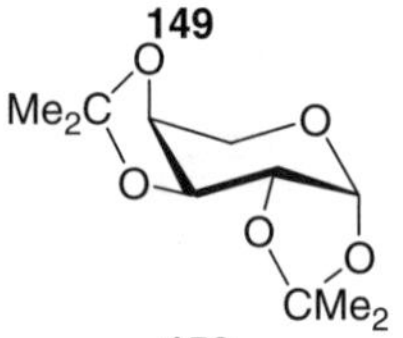 **150**	2	91
D-Xylose	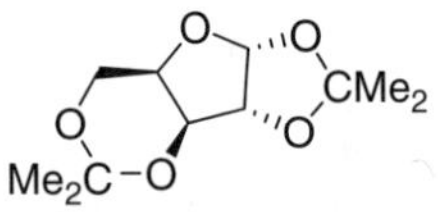 **103**	1.5	83
L-Sorbose	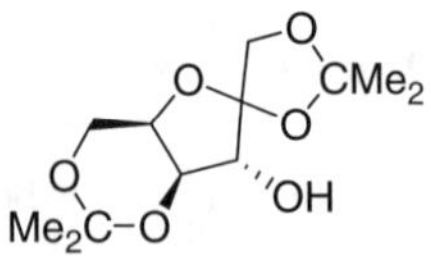 **151**	2	86

[a] All the reactions were started in a preheated oil bath so that the reflux started immediately after the addition of the catalyst.

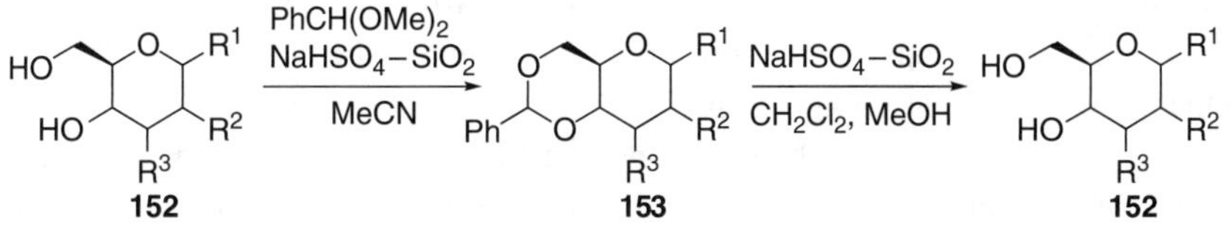

SCHEME 26. $NaHSO_4$–SiO_2 as an efficient promoter for benzylidene protection and deprotection of monosaccharides.

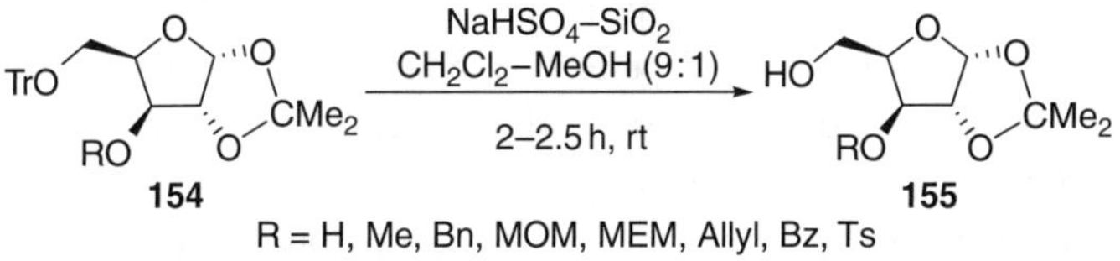

SCHEME 27. Removal of trityl group from 1,2-*O*-isopropyliden protected furanoses by $NAHSO_4$-SiO_2.

efficient manner. Although some of the groups are acid-labile (such as *p*-methoxybenzyl and epoxide), they remained unaffected. When phenyl 1-thio-α-D-mannoside was used as substrate, both mono- and diprotected benzylidene acetals were obtained, in a ratio between 3:1 and 4:1. The same system was also investigated for the benzylidene deprotection, and this transformation was successfully accomplished in excellent yield when benzylidene derivatives were treated with $NaHSO_4$–SiO_2 at room temperature, using a 1:4 methanol–dichloromethane as solvent.[115]

Clean removal of trityl ether groups from O-isopropylidenated furanoses by using $NaHSO_4$–SiO_2 was previously described by Das *et al.*[116] and afforded the corresponding alcohols at room temperature. The chemoselective deprotection was accomplished in yields above 91% within 2–2.5 h, leaving other protective-groups intact (Scheme 27).

4,6-Benzylidene acetal protecting groups were removed from sugar derivatives when absorbed in silica gel together with a 1:1 mixture of H_2O and AcOH and submitted to microwave irradiation for 7 min. The reaction occurred in an open vessel using a domestic microwave oven, and most of the deacetalated products were obtained in high yield, leaving other groups, namely the tosyl group and epoxides, intact.[117]

$HClO_4$–SiO_2 selectively cleaves terminal isopropylidene groups of the diisopropylidene derivatives of D-glucose, D-allose, and methyl D-gluconate in high yield.

Furthermore, several functionalities remained unaffected, namely the acid-labile TBDMS or PMB groups.[118] Deprotection yields were in the range of 85–95% when methanol was used at room temperature as the solvent, whereas acetonitrile or dichloromethane led to very sluggish or nonexistent reactions, respectively. Cleavage of primary trityl ethers was also accomplished using the same conditions in a very rapid and effective fashion. The trityl pyranosides and furanosides assayed were selectively deprotected in 2–3 h and yields higher than 85% were achieved. This reaction was also more efficient when conducted in methanol, which acts as a nucleophile to trap the generated trityl cation.

Removal of terminal *O*-isopropylidene groups using H_2SO_4–SiO_2 was reported by Rajput *et al.*[119] for selective deprotection of di-*O*-isopropylidene derivatives of D-glucose, D-mannose, D-fructose, and L-sorbose, leading to the corresponding mono-*O*-isopropylidene products in good to excellent yield. However, *p*-methoxyphenyl 2,3;4,6-di-*O*-isopropylidene-α-D-mannopyranoside gave only 70% of the corresponding mono-*O*-isopropylidene derivative along with the fully deprotected *p*-methoxyphenyl α-D-mannopyranoside, isolated in 15% yield.

The silica-supported Lewis acid $FeCl_3$–SiO_2 was also evaluated for its ability to promote the hydrolysis of terminal *O*-isopropylidene groups in carbohydrates.[120] Experiments were conducted starting from 1,2:5,6-di-*O*-isopropylidene-α-D-glucofuranose derivatives with OH-3 protected by various groups, including acetyl, benzyl, and TBDMS, and also starting from methyl 5,6-*O*-isopropylidene-α-D-glucopyranoside. Selective deprotection of the terminal acetal group was achieved in good yield within 1–10 h, using chloroform as solvent. When acetone was used as solvent, the acetal groups remained intact, and instead the acid-labile TBDMS group was cleaved, because of a shift in the equilibrium of hydrolysis in favor of the isopropylidene group imposed by acetone. Other Lewis acids, such as aluminum chloride and zinc chloride adsorbed on silica, were also tried for the hydrolysis of acetals and ketals, but they were less efficient than $FeCl_3$–silica gel. PMA supported on silica gel is also an active promoter for the hydrolysis of terminal isopropylidene groups[121] PMA belongs to the class of cheap and environmentally friendly heteropoly acids, whose activity is reported to be greater than that of H_2SO_4, TsOH, and $BF_3 \cdot Et_2O$.[122] The reaction proceeded in wet acetonitrile at room temperature using 1 mol% of PMA–SiO_2 without affecting the remaining hydroxyl protecting groups, such as PMB, THP, MOM, TBS, TBDPS, allyl, propargyl, Bn, Ac, Me, and Ts groups, and such sensitive functional groups as alkenes. The corresponding 1,2- and 1,3-diols were obtained in a very efficient, clean and rapid manner, and yields higher than 89% were attained within 5–7 min.

3. Hydrolysis/Isomerization of Saccharides and Glycosides

Heterogeneous catalysts, particularly zeolites, have been found suitable for performing transformations of biomass carbohydrates for the production of fine and specialty chemicals.[123] From these catalytic routes, the hydrolysis of abundant biomass saccharides, such as cellulose or sucrose, is of particular interest. The latter disaccharide constitutes one of the main renewable raw materials employed for the production of biobased products, notably food additives and pharmaceuticals.[124] Hydrolysis of sucrose leads to a 1:1 mixture of glucose and fructose, termed "invert sugar" and, depending on the reaction conditions, the subsequent formation of 5-hydroxymethylfurfural (HMF) as a by-product resulting from dehydration of fructose. HMF is a versatile intermediate used in industry, and can be derivatized to yield a number of polymerizable furanoid monomers. In particular, HMF has been used in the manufacture of special phenolic resins.[125]

a. Zeolites.—Buttersack and Laketic[126] have employed HY zeolites as catalysts for the hydrolysis of sucrose, demonstrating that dealumination contributed to an increase of the reaction rate. Dealuminated HY zeolites were also employed by Moreau *et al.*[127] to catalyze hydrolysis of the latter disaccharide and other glucose precursors, including inulin, maltose, cellobiose, and starch. Reactions were carried out at high temperature, ranging from 80 to 150 °C in the presence of a zeolite having an Si/Al ratio of 15:1. In each instance, high yields ($>90\%$) of the desired product, namely fructose from inulin, glucose, and fructose from sucrose, glucose from cellobiose, maltose from starch, were obtained. Moreover, the HMF by-product was formed in low yield in the hydrolysis of sucrose, and was not detected after hydrolysis of inulin. Hydrolysis of sucrose was also examined in the presence of various H-form dealuminated zeolites, namely HBEA, HMFI, HMOR, and HY-FAU.[128] For all catalysts, the reaction was selective to the formation of the invert sugar, in accordance with previous results.[127] In this process, zeolites were presumed to act also as specific adsorbents of colored by-products (such as HMF) formed. In particular, HY-FAU was the one for which a better balance between activity, selectivity, and by-product amount was observed.

An early report from Shukla *et al.*[129] showed efficient hydrolysis and isomerization reactions of disaccharides, including cellobiose, maltose, and lactose, over zeolites type A, X, and Y. Abbadi *et al.*[130] studied the hydrolysis of maltose, amylose, and starch over the zeolitic materials H-mordenite, H-beta, and mesoporous MCM-41. The effect of temperature and pressure, as well as that of the Si/Al ratio of H-mordenite and H-beta zeolites, on their catalytic activity was investigated for the hydrolysis of maltose. With respect to the latter reaction, all zeolites provided selectivity toward D-glucose, which was higher than 90%, and the conversion increased using dealuminated samples. This effect was more pronounced for reactions catalyzed by H-beta zeolites, with an increase of conversion from 45% when applying H-beta-25 (Si/Al ratio 12.5) to 85% with H-beta-100 (Si/Al ratio 50), after 24 h of reaction time. The MCM-41-catalyzed hydrolysis of maltose led essentially to the same results as those obtained with H-mordenite-12 (a mordenite with Si/Al ratio 12). With respect to amylose, the conversion and selectivity employing H-mordenite-12 was also similar to that obtained for maltose. On the other hand, hydrolysis of starch in the presence of the same zeolite yielded only about 18% of D-glucose. The hydrolytic activity present in solution was due to the low pH (3.9), and in view of the dimensions of the reactants, only the outer surface of the zeolite was accessible. Subsequently, the selective hydrolysis of cellulose to glucose by means of H-form zeolite catalysis was reported by Onda *et al.*[131]

Besides di- and poly-saccharides, zeolites have been applied for hydrolysis of simple glycosides as described by Le Strat and Morreau.[132] Methyl α- and β-D-glucopyranosides were treated with water in the presence of dealuminated HY faujasite with an Si/Al ratio of 15, at temperatures ranging between 100 and 150 °C. It was observed that the reaction rate for the β glycoside was about 5–6 times higher than that for the α anomer, a result that might arise from the shape-selective properties of the zeolite and stereoelectronic effects on the surface of the solid.

b. Silica Gel.—Hydrolysis of a diverse set of alkyl and aryl α- and β-1-thioglycosides mediated by NIS/H_2SO_4–SiO_2 was investigated by Dasgupta *et al.*[133] Experiments were carried out in wet dichloromethane at 0 °C, and the promoter was highly chemoselective so that the acid-labile protecting groups benzylidene, isopropylidene, *p*-methoxybenzyl, chloroacetyl, and silyl remained intact. Moreover, selective thioglycoside hydrolysis was accomplished in the presence of α- and β-*O*-glycosidic linkages in di- and tri-saccharide substrates (Scheme 28). The nature of the parent glycoside did not seem to have an effect on the results, as almost all reactions with D-*gluco*, D-*galacto*, D-*manno*, and 6-deoxy-L-*manno* derivatives were completed

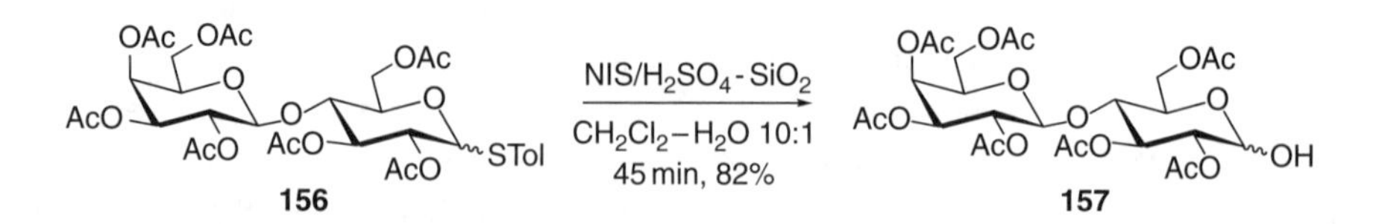

SCHEME 28. Selective thioglycoside hydrolysis mediated by NIS/H_2SO_4-SiO_2.

within 30 min in yields ranging from 75% to 95%, and the results were reproducible in large-scale preparations.

The silica-supported heteropoly acid $H_3PW_{12}O_{40}$–SiO_2 was investigated as regards reaction kinetics for the hydrolysis of sucrose to invert sugar. The authors suggested that the catalyst should not exhibit a high magnitude of ΔH_{ads} for H_2O. Hence, instead of Al_2O_3 and Nb_2O_5, which are hydrophilic catalysts, $H_3PW_{12}O_{40}$–SiO_2 seems to be a more suitable catalyst for the synthesis of invert sugar as it has a lower tendency for the absorption of water.[134]

4. Dehydration

a. Zeolites.—Dehydration of sugars can be achieved over zeolites, which are solid acid catalysts. Most of the literature for dehydration of carbohydrates employing zeolites focuses on fructose and xylose as raw materials, enabling the synthesis of 5-hydroxymethylfurfural and furfural, respectively (Scheme 29). Furfural is produced on an industrial scale, and constitutes a key intermediate for such important chemicals as furfuryl alcohol, which is used for the manufacture of synthetic fibers, rubbers, resins, or for the synthesis of fragrances, to name a few of its applications.[135] Most industrial processes for furfural production are based on the utilization of such mineral acids as sulfuric acid. Improving the chemical technology for the production of furfural and HMF by means of "green" chemistry and sustainability, remains therefore of great importance.

(i) Medium and Large-Pore Zeolites. Moreau *et al.*[136,137] reported that H-form zeolites are capable of catalyzing dehydration of fructose to HMF at high temperature (165 °C) in a solvent mixture consisting of water and methyl isobutyl ketone

CHO
H—OH
HO—H
H—OH
CH_2OH
158
D-xylose

H^+

R (furan ring, O) CHO

161 R = H, furfural
162 R = CH_2OH, HMF

H^+

CH_2OH
=O
HO—H
H—OH
H—OH
CH_2OH
159
D-fructose

SCHEME 29. Dehydration of fructose and xylose to HMF and furfural, respectively.

(1:5, v/v). Fructose conversion and selectivity for the desired product were shown to be influenced by the acidity and structural properties of the catalysts used. Thus HY faujasite and H-mordenite, with Si/Al ratios of 15 and 11, respectively, achieved a maximum in the rate of fructose conversion. The latter zeolite provided high selectivity (91–93%) to HMF, with a fructose conversion of 76%, which may be a result of its shape-selective properties and the absence of cavities within the porous structure, preventing the formation of secondary products. The higher efficiency of H-mordenites as catalysts in this referred transformation, as compared with such other zeolites as HY faujasite, H-beta, and HZSM-5, was also observed by Rivalier *et al.*[138] when developing a pilot plant using a new solid–liquid–liquid reactor.

Dehydration of xylose to furfural over zeolites was also performed by Moreau *et al.*[139] using the methodology previously employed for dehydration of fructose, with the exception of the cosolvent, which was toluene instead of methyl isobutyl ketone. Dealuminated HY faujasites and H-mordenites with different Si/Al ratios were used, and the most effective catalysts, balancing xylose conversion and product selectivity, were HY faujasite (with an Si/Al ratio of 15) and H-mordenite (with an Si/Al ratio of 11). The latter zeolite afforded the highest selectivity (90–95%) for furfural, as expected from the bidimensional structure of this solid, which allows the rapid diffusion of the furfural formed and avoids its rearrangement or degradation to by-products of higher or lower molecular weight.

A delaminated zeolite with an Si/Al ratio of 29, derived from the layered zeolite Nu-6(1), was employed as catalyst for dehydration of xylose at 170 °C, using a water–toluene biphasic reactor-system.[140] This material, designated del-Nu-6(1), proved to be efficient for this transformation, giving 47% selectivity to furfural at ~90% xylose conversion.

Dehydration of glucose has also been accomplished over zeolites. Lourvanij and Rorrer[141] studied this reaction and proposed a kinetic model for it, when performed in the presence of HY zeolite, aluminum-pillared montmorillonite (APM) and such mesoporous MSs as MCM-20 and MCM-41. The four main reaction processes, including dehydration of glucose to HMF, isomerization of glucose to fructose, fructose dehydration to HMF, rehydration and cleavage of HMF to formic acid and 4-oxopentanoic acid were considered in the study. It was observed that the rate constants for dehydration of glucose to HMF and the rehydration and cleavage of HMF to formic acid and 4-oxopentanoic acid were maximized with catalysts having pore diameters of 10–30 Å, such as APM and MCM-20 MSs.

(ii) Small-Pore Zeolites. Alditols, monoanhydroalditols, dianydroalditols, and their derivatives are commonly prepared from sugars, with alditol cyclodehydration being the

most common method for synthesis of anhydroalditols. Nevertheless this approach requires treatment with strong mineral acids, such as hydrogen fluoride in the presence of catalytic amounts of formic or acetic acid, or ion-exchange resins. The efficiency of 3Å MS to promote thermal cyclodehydration has been reported for such pentitols as D-arabinniol, ribitol, and xylitol, and hexitols such as galactitol, D-glucitol, and D-mannitol. The reactions were performed in solvent-free conditions to furnish mono-anhydroalditols and dianhydroalditols, with retention or inversion of configuration at the asymmetric carbons.[142,143] In the dehydration of galactitol (**162**, Scheme 30), three monoanhydroalditols (**163–165**) are formed in a first step, but they subsequently undergo dehydration to afford dianhydroalditols **166–169**. The unfavorable arrangement of appropriate hydroxyl groups prevented compound **168** from forming a dianhydroalditol.[142] This was the first time that formation of 1,4:3,6-dianhydro-D- and -L-iditol (**169**, racemic), well-known dehydration products of D- or L-iditol, has been reported from galactitol, a much cheaper starting material than D- or L-iditol.

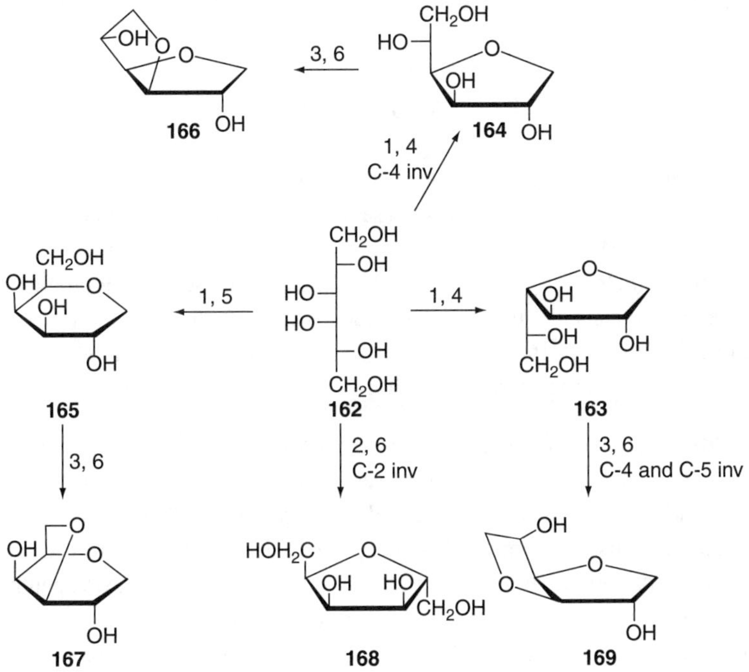

SCHEME 30. Termal cyclodehydration of galactitol promoted by 3Å MS.

b. Montmorillonite and Other Clays.—Some examples employing montmorillonite and other clays as catalysts for dehydration of glucose have appeared in the literature. Lourvanij and Rorrer[144] described the use of various clays as catalysts for this process and tested iron-, chromium-, and aluminium-pillared montmorillonites in order to evaluate the influence of the porous structure and dimensions of the material on the reaction selectivity. All of the montmorillonites catalyzed the four main steps involved, and the highest glucose conversion (100%) was achieved using the Fe-pillared montmorillonite. The latter led to the lowest selectivity to formation of HMF and to the highest selectivity to formic acid. It was concluded that catalysts possessing pore dimensions within the range 10–25 Å would preferentially direct the reaction toward the organic acid products, allowing glucose to diffuse through the microporous framework and trapping the bulkier HMF molecule inside.

Gonzales and Laird[145] have shown that smectites abiotically catalyze dehydration of glucose to form furfural under conditions similar to those found in soils. Four smectite clay minerals were used (saturated with Na, Ca, Fe, or Al), and the formation of HMF and furfural was detected by high-pressure liquid chromatography. The polymerization of furfural may thus be a pathway to the formation of new humic materials in soils.

5. Oxidation

Oxidation is a widely used procedure in carbohydrate chemistry, mainly to access sugars that contain a carbonyl function to serve as valuable intermediates for a variety of derivatizations. Many procedures have been developed, employing either chemical or biochemical methodologies.[146–148] While most of these methodologies rely on homogeneous catalysis, the use of heterogeneous catalysts has proved to be a feasible alternative.[123c] However, the utilization of catalysts based on silicon porous materials for the oxidation of carbohydrates is still a field to be further explored.

a. Zeolites.—*(i) Medium and Large-Pore Zeolites*. The Ruff oxidative degradation is a useful route to pentoses from hexoses, in particular starting from the readily accessible and cheap sugar glucose. Hourdin[149] reported the Ruff oxidative degradation of aldonic acids by titanium(IV)- and copper(II)-containing zeolites.[149,150] The oxidative degradation of calcium D-gluconate to D-arabinose with hydrogen peroxide was performed over titanium-containing zeolites, such as tEuroTS-1 (a small-pore titanium-silicalite), Ti-BEA, and Ti-FA.[149] The desired product was recovered in low yield (12–17%) and only the large-pore zeolites were effective. It was demonstrated that titanium species present in solution were responsible for the catalytic activity.

The yield of arabinose was increased when the reaction was carried out in the presence of different copper(II)-containing zeolites, including mordenite, LTL, BEA, MFI, and faujasite. The lowest conversion (37%) was observed when using the copper(II)-containing mordenite zeolite. This might result from the monodirectional porosity of this solid, limiting the diffusion rate of the gluconate. The large-pore zeolite BEA and the small-pore zeolite MFI proved to be very effective, although with the latter zeolite the reaction efficiency seemed to be more dependent on the presence of the copper(II) cations in solution than on the porosity constraints of the zeolite. The best results were obtained when copper(II)-exchanged faujasite was used as the catalyst, affording arabinose in 63% yield. The efficiency of this reaction arises from dissolution of Cu(II) cations from the zeolite.

(ii) Small-Pore Zeolites. MSs have achieved great importance as additives for oxidation reactions of sugars and nucleosides with such chromium reagents as PCC or PDC.[151,152] In the early 1980s, the Antonakis group investigated these reagent systems for carbohydrate-derived oxidations.[151] They showed that the oxidation rate of various sugars and the sugar moiety of nucleosides with PCC and PDC increases in the presence of MSs in the order 5 Å < 10 Å < 4 Å < 3 Å. The latter MS was particularly efficient and the reaction could be performed with a lower weight of MSs and smaller volume of solvent. Isotope-labeling experiments showed that the MS works by favoring the rate-determining step of the oxidation reaction, wherein which cleavage of the CH bond from the alcoholic carbon occurs.[153] It was also found that oxidation of alcohols takes place at very specific sites on the surface of the MS. Hence, in the presence of water, which takes up a proportion of the active sites of the MS, the amount of MS should be increased to provide the same number of free sites as in the dry MS. Oxidation of some sugar derivatives, previously very sluggish or even incomplete after several days of reaction, achieved completion within 60–180 min using 3–4 mol equiv of PCC and 0.5 g/mmol of alcohol, or within 60–90 min upon treatment with 1.5 mol equiv of PDC and 1.0 g MS 3 Å/mmol of alcohol in the presence of dichloroacetic acid.[153]

The system PCC/CH_2Cl_2 and 3 or 4 Å MS proved very efficient for the oxidation of secondary alcohols of furanose systems and of five-membered rings fused to furanoses, and pyranoses as well as for the regioselective oxidation of sugar diols, the latter with PCC in the presence of sodium acetate, as demonstrated by our group for the synthesis of the key intermediates **170–174** (Scheme 31).[154–157] This system permitted a high-yield synthesis of a sugar-derived α-keto amide **172**[154] by oxidation of the corresponding α-hydroxy amide. This was a promising result, considering that this type of compounds suffered degradation in the presence of other oxidation conditions. Vauzeilles and Sinaÿ[158] oxidized the anomeric position by means of

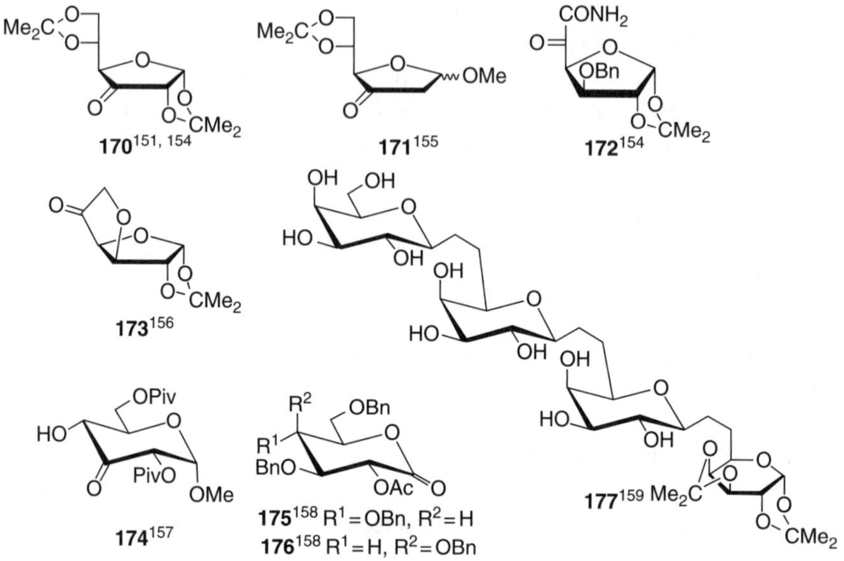

SCHEME 31. Monosaccharide derivatives obtained by oxidation at secondary hydroxyl groups using PCC in the presence of 3Å or 4Å MS.

PCC in the presence of 4 Å MS to afford the corresponding keto sugars (**175**, **176**), which after Tebbe methylenation led to the corresponding exoglycals, valuable intermediates for the synthesis of C-disaccharides. Dondoni *et al.*[159] made use of PCC in the presence of powdered 4 Å MS to introduce suitable formyl groups in the carbohydrate, and allowing the synthesis of *β*-D-*C*-(1→6)-linked oligosaccharides, such as **177**, by iterative Wittig olefination with sacharidic phosphorus ylides to afford methylene bridges, after reduction of the olefinic function.

6. Other Synthetic Applications

This section presents selected examples of the use of zeolites and related porous materials for transformations of carbohydrates, that fall beyond the scope of the previous paragraphs. They include the use of zeolites in "click" chemistry and in a variety of reactions, including the synthesis of anhydro sugars, cyclization, elimination, and addition reactions to the carbonyl group.

a. Zeolites.—These heterogeneous catalysts have been applied in the pyrolysis of cellulose, leading to a diversity of compounds, anhydro sugars among them.[160,161] The process was performed in both the absence and in the presence of the acid zeolite, among other catalysts. With respect to noncatalyzed transformation, four anhydro sugars were obtained, namely levoglucosan (1,6-anhydro-D-glucopyranose, **178**) (11% yield), and levoglucosenone (1,6-anhydro-3,4-dideoxy-β-D-*glycero*-hex-3-enopyranos-2-one, **179**) (4% yield), which have been widely explored as versatile chiral intermediates in organic synthesis (Scheme 32).[162] 1,5:3,6-Dianhydro-β-D-glucofuranose (**180**) and (1*R*)-1-hydroxy-3,6-dioxabicyclo[3.2.1]octan-2-one (**181**) were also formed in 3% and 2% yields, respectively (Scheme 31). When the acid zeolites NH_4-Y, HY, and NH_4–ZSM-5 were used, the yields were lower and the highest overall yield of anhydro sugars was obtained for the NH_4-Y zeolite-catalyzed pyrolysis, although only in 9% total yield. The latter catalyst provided selectivity toward levoglucosan **178** (8%). The pyrolytic pattern obtained using the other zeolites changed in favor of levoglucosenone, **179**, obtained in yields of 1% and 2% for HY and NH_4–ZSM-5, respectively. The low efficiency of zeolites as catalysts for cellulose pyrolysis was ascribed to be due to the deoxygenating activity promoted by these acidic solids.

"Click" chemistry, an old concept in organic synthesis recently revived, means the use of reliable and selective chemical transformations, through simple procedures and giving consistently high yields.[163] This approach has enjoyed a growing impact on drug discovery, facilitating the synthesis and assembly of specially designed scaffolds. The copper(I)-catalyzed chemoselective formation of 1,2,3-triazoles from azides and terminal alkynes is a powerful coupling reaction that has found many

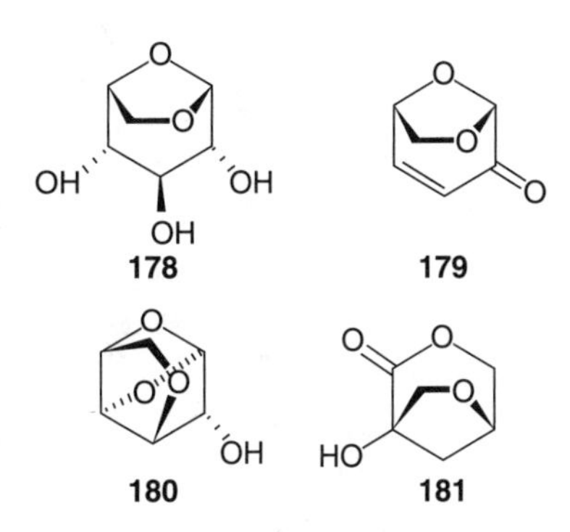

SCHEME 32. Anhydro sugars obtained by pyrolysis of cellulose.

applications.[164] This reaction in particular has been termed "click chemistry" owing to its convenience, rapidity, and quantitative yield.

Allying click chemistry with the use of environmentally benign catalysts, such as zeolites, may constitute a valuable synthetic tool, representing a true "green" and sustainable approach.

Numerous examples of oligosaccharide and glycoconjugate synthesis employing "click" chemistry have been reported[165,166], and Alix *et al.*[167] showed that zeolites modified with Cu^{I} ions are effective catalysts for "click" reactions of carbohydrates and amino acid derivatives. The H-form zeolites H-USI, HY, H-beta, H-ZSM-5, and H-MOR were mixed with CuCl at 350 °C to prepare Cu-modified materials. In initial experiments, all of the catalysts were tested for the coupling of 2-azidocyclohexanol (**182**) and methyl 6-azido-3,4-di-*O*-benzoyl-6-deoxy-2-*O*-tosyl-α-D-glucopyranoside (**184**) with ethyl propiolate (Scheme 33). The most efficient catalyst for the coupling reactions was Cu-USI, which affords the desired triazole derivatives **183** and **185** in 70% and 89% yields, respectively. Further study of the effect of the material structure and Si/Al ratio on the activity of the catalyst revealed that, among the two cage-containing zeolites (namely Cu-Y and Cu-USY), the most active one was Cu-USY, which has a higher Si/Al ratio. In channel-type zeolites (namely Cu-MOR, Cu-ZSM-5, and Cu-beta), the same relationship was observed, with the Cu-beta zeolite being the most effective. The efficiency of the coupling reactions correlates with the cross-sections of the zeolites, and the yield increased with this parameter, suggesting that the reaction takes place in the zeolite cages or channels, rather that at their external surfaces. Cu-USY was thus the

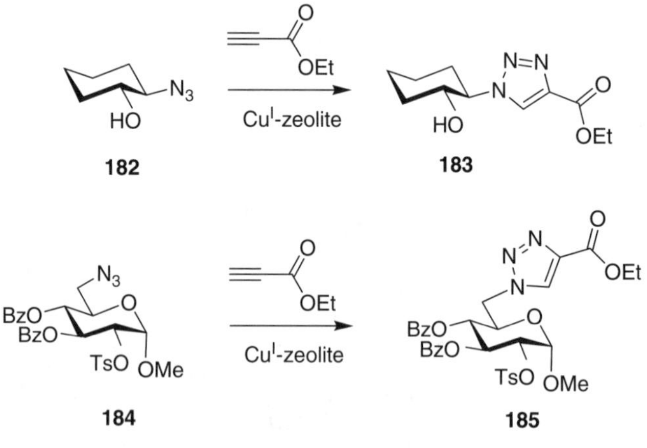

SCHEME 33. "Click" chemistry in carbohydrate derivatives using a heterogeneous Cu(I)-zeolite catalytic system.

catalyst selected for other coupling reaction of simple azides or carbohydrate azides with simple alkynes or carbohydrate- and amino acid-derived alkynes. The carbohydrate derivatives chosen included protecting groups of increasing size and volumes, such as butanediacetals (Ley acetals) or triisopropylsilyl ethers. The desired glycopeptides, oligosaccharides, and multivalent carbohydrate systems, namely glucosyl ditriazoles, disaccharide triazoles, and glucosylated triazolylamino acids, were obtained in moderate to excellent yields (47–97%), despite the large volumes of the molecules involved. These results suggested that, contrary to expectation, pore sizes, and internal shapes within the zeolite were not a limitation for its efficiency, although bulkier molecules required longer reaction times to be fully converted.

b. Montmorillonite.—In addition to the catalytic routes employing montmorillonite discussed in the previous sections, other useful and practical protocols based on these clays have been reported. The approaches disclosed here demonstrate the ability of montmorillonite to promote inter- or intra-molecular cyclization reactions that involve carbohydrate scaffolds and lead to novel derivatives.

Glycals can be transformed into 1,6-anhydro sugar derivatives by intramolecular cyclization in the presence of Lewis and Brønsted acids, a reaction that has been termed the intramolecular Ferrier glycosylation.[168] Sharma *et al.*[169] showed that a montmorillonite clay-supported silver reagent can be an efficient catalyst for this transformation. The 1,6-anhydro-2,3-dehydro sugars obtained were then selectively dihydroxylated to furnish 1,6-anhydro saccharides.

Yadav *et al.*[170] employed montmorillonite KSF as a catalyst for the reaction of arylamines with 2-deoxy-D-*erythro*-pentose. The reaction took place on the surface of the catalyst and proceeded with cyclization to afford sugar-derived chiral tetrahydroquinolines in high yields and moderate diastereoselectivity.

A new synthetic route for functionalized polyhydroxyalkyl-pyrimidines starting from unprotected aldoses and based on montmorillonite K-10 catalysis and solvent-free microwave irradiation conditions, has been reported by Yadav *et al.*[171] Thus, reaction of D-glucose and D-xylose with semicarbazide or thiosemicarbazide (**186**) in the presence of montmorillonite K-10, under microwave irradiation, proceeded via domino cycloisomerization, dehydrazination, and dehydration of the intermediate semi- or thiosemicarbazones (**187**) to afford 1,3-oxazin-2-ones or 1,3-oxazine-2-thiones (**188**) in one single step and in yields between 79% and 85% (Scheme 34). Other mineral catalysts tested, such as silica gel and basic alumina, were far less effective for this transformation and only silica gel was active at all, giving low yields (15–28%) of compounds **188a–d**. The 1,3-oxazin-2-ones(thiones) thus synthesized were subsequently converted into the target pyrimidines by reaction with aromatic

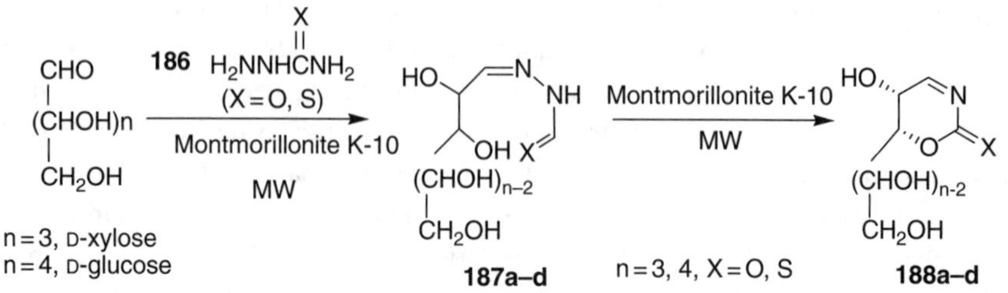

SCHEME 34. Synthesis of 1,3-oxazin-2-ones and 1,3-oxazin-2-thiones from unprotected aldoses employing Montmorillonite K-10 catalysis and MW irradiation.

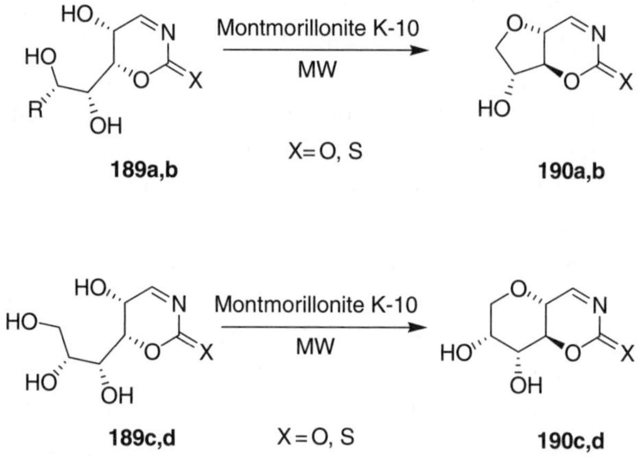

SCHEME 35. Montmorillonite K-10-catalyzed cyclodehydration of the 1,3-oxazin-2-one and 1,3-oxazin-2-thione derivatives **189** into bicyclic compounds **190**.

amines, via amine-driven dehydrative ring transformations also catalyzed by montmorillonite.

The 1,3-oxazin-2-ones and 1,3-oxazine-2-thiones previously synthesized were used to prepare various *N*- and *O*-heterocyclic systems fused with 1,3-oxazine rings.[172] For example, furan-1,3-oxazin-2-one or furan-1,3-oxazine-2-thiones (**190a,b**) and pyran-1,3-oxazin-2-ones or pyran-1,3-oxazine-2-thiones (**190c,d**) were prepared in very good yields, ranging from 83% to 90%, by montmorillonite K-10 clay-catalyzed cyclodehydration of **189a,b** and **189c,d**, respectively (Scheme 35).

c. Silica Gel.—Silica-promoted elimination of sulfinic acid from 2-deoxy-2-*C*-*p*-tolylsulfonyl-*β*-D-*arabino*-hexopyranosyl *p*-tolyl sulfones has been reported by Sakakibara *et al.*[173] These sugar derivatives assume a nonchair conformation, and

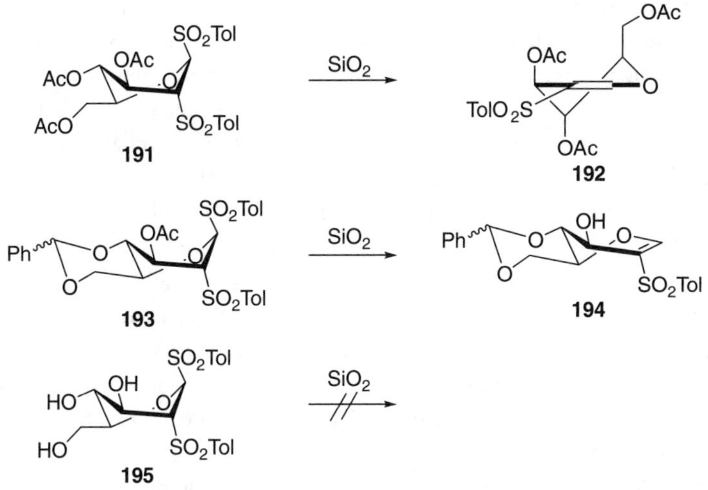

SCHEME 36. Elimination of sulfinic acid from 2-deoxy-2-*C*-*p*-tolylsulfonyl-β-D-*arabino*-Hexopyranosyl *p*-tolyl sulfones promoted by silica.

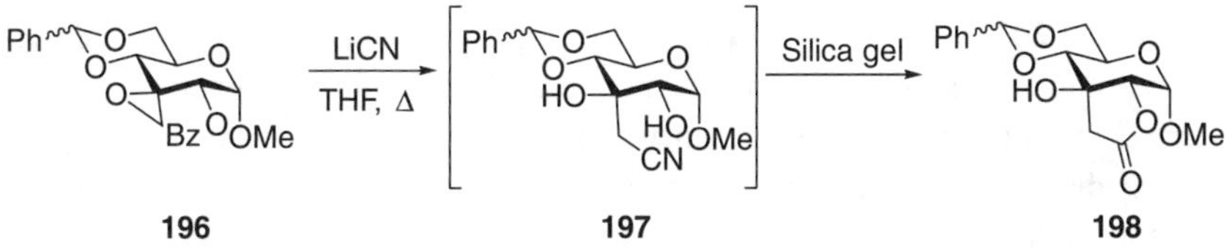

SCHEME 37. Synthesis of an Analogue of the Miharamycin Sugar Moiety Involving Silica Gel-Mediated Hydrolysis of the Intermediate Nitrile **197** and Lactonization.

the elimination of sulfinic acid from the anomeric position is conditioned by the protection of the OH-6 group (Scheme 36).

When *p*-tolylsulfonyl derivatives **191** and **193** were treated with silica gel (Merck silica gel 60, mesh), either by column chromatography or simply by stirring in a flask containing silica in DMF for 24 h, partial or complete elimination of sulfinic acid at the anomeric center occurred. In contrast, elimination did not take place when using neutral silica gel (Kanto, silica gel 60 N, 40–50 mm). Furthermore, compound **195** was completely inert to the presence of acidic silica, probably because of hydrogen bonding between the 6-hydroxyl group and the ring oxygen atom (O-5).

A paper concerning the synthesis of miharamycin analogues modified in the sugar moiety demonstrated the hydrolysis of a nitrile upon treatment with silica gel in the presence of water (Scheme 37).[174] The cyanide **197** resulting from the opening of epoxide

196 by treatment with LiCN in DMF/THF, was hydrolyzed in the presence of silica gel, causing spontaneous cyclization to the lactone **198** in 82% yield. Several attempts had been made to obtain this fused five-membered ring lactone under basic conditions, but the use of silica gel proved to be the best approach for obtaining lactone **198** from nitrile **197**.

Carbohydrate-derived oximes have been successfully cyclized to isoxazolines by intramolecular nitrile oxide–alkene cycloaddition (INOC) using Chloramine-T and silica gel,[175] without the normally required protection of the hydroxyl group of the oxime. The acidic environment of the silica gel allowed formation of the corresponding five- or six-membered ring isoxazolines in good yield, instead of the undesired oximolactones. The formation of oximolactones is favored under the basic conditions needed to transform oximes into nitrile oxides, using either NaOCl, NaOCl with Et_3N, or *N*-chlorosuccinimide with Et_3N for deprotonation of the hydroxyl function. This new methodology thus obviates protection–deprotection steps, and simplifies the entire synthetic scheme.

The nitroaldol reaction of methyl nitroacetate (**199,** Scheme 38) with 1,2:3,4-di-*O*-isopropylidene-α-D-*galacto*-hexodialdo-1,5-pyranose (**200**) and 2,3-*O*-isopropylidene-D-glyceraldehyde (**202**) catalyzed by silica gel proceeded in almost quantitative yield, with high selectivity for attack on the aldehyde carbonyl group, giving derivatives **201** and **203**, respectively. Two of the four possible diastereomers were detected as main products, and were obtained as a mixture. For the nitroaldol reaction with **200** gave similar results in either the presence or absence of silica gel, whereas the reaction with **202** did not proceed in its absence, showing that catalytic action of silica is mandatory in this case.[176]

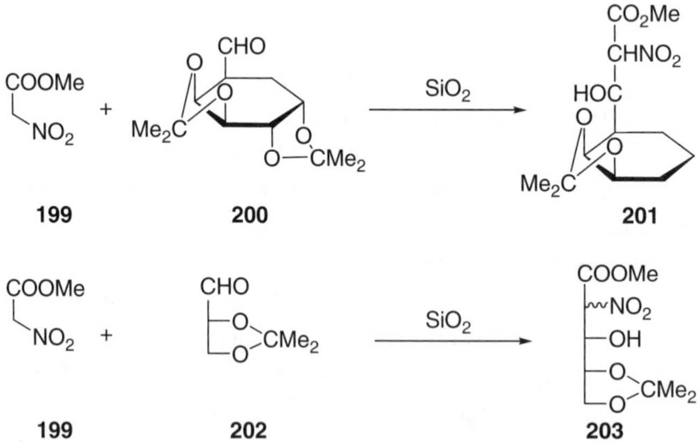

SCHEME 38. Nitroaldol reaction of methyl nitroacetate with hexodialdo-1,5-pyranose **200** and 2,3-*O*-isopropylidene-D-glyceraldehyde (**202**) catalyzed by silica gel.

IV. Conclusion

In the past few years, a significant number of contributions related to the use of zeolites and related porous silicon-based catalysts in carbohydrate chemistry have been reported. Their regular porous structure and consequent shape-selectivity properties allow stereo- and regiocontrolled reactions, which are frequently desirable attributes in synthetic strategies with carbohydrates. They often prove to be suitable alternatives to conventional catalysts, particularly for reactions that require acidic conditions, and they provide environmental friendly methodologies. The present chapter highlights newly established methods that employ these heterogeneous catalysts in transformations routinely used in carbohydrate chemistry, including glycosylation reactions, protection–deprotection of hydroxyl groups, and hydrolysis and dehydration reactions. The utility of these catalysts in "click" chemistry is emphasized. The unique properties of silicon-based materials contribute to a stimulating interplay of chemistry, processes, and products, and these catalysts have demonstrated feasibility for large-scale production of fine chemicals from carbohydrate precursors. The promising results presented herein may encourage carbohydrate chemists to explore further the potential of these materials as catalysts for new, selective, and efficient methodologies.

References

1. D. W. Breck, W. G. Eversole, R. M. Milton, T. B. Reed, and T. L. Thomas, Crystalline zeolites. I. The properties of a new synthetic zeolite, type A, *J. Am. Chem. Soc.*, 78 (1956) 5963–5971, and references cited therein.
2. (a) For books on structure and properties of zeolites, see: in M. Guisnet and J.-P. Gilson, (Eds.) *Zeolites for Cleaner Technologies, in Catalytic Science Series*, Vol. 3, Imperial College Press, London, 2002; (b) M. Guisnet and F. Ramoa-Ribeiro, *Zeólitos, um nanomundo ao serviço da catálise,* Fundação Calouste Gulbenkian, Lisbon, 2004; (c) M. Guisnet and F. Ramôa Ribeiro, *Les zéolithes, un nanomonde au service de la catalyse,* EDP Sciences, France, 2006; (d) R. Xu, W. Pang, J. Yu, Q. Huo, and J. Chen, *Chemistry of Zeolites and Related Porous Materials: Synthesis and Structure,* John Wiley & Sons, Singapore, 2007.
3. (a) For reviews and books concerning the use of zeolites as catalysts in organic synthesis, see: M. A. Martinluengo and M. Yates, Zeolitic materials as catalysts for organic syntheses, *J. Mater. Sci.*, 30 (1995) 4483–4491; (b) W. F. Hölderich, Organic reactions in zeolites, in G. Alberti and T. Bein, (Eds.),

Comprehensive Supramolecular Chemistry, Pergamon, Oxford, 1996, pp. 671–692; (c) A. Corma and H. García, Organic reactions catalyzed over solid acids, *Catal. Today*, 38 (1997) 257–308; (d) J. C. van der Waal and H. van Bekkum, *J. Por. Mater.*, 5 (1998) 289–303; (e) S. E. Sen, S. M. Smith, and K. A. Sullivan, Organic transformations using zeolites and zeotype materials, *Tetrahedron*, 55 (1999) 12657–12698; (f) K. A. Tanaka, Solvent-free organic synthesis, Wiley-VCH, Weinheim, 2003; (g) R. A. Sheldon and H. van Bekkum, *Fine Chemicals Through Heterogeneous Catalysis,* Wiley-VCH, Weinheim, 2001; (h) D. Mravec, J. Hudec, and I. Janotka, Some possibilities of catalytic and noncatalytic utilization of zeolites, *Chem. Pap.*, 59 (2005) 62–69.

4. A. F. Crønstedt, Omeno bekant bargart, som kallas zeolites, *Akad. Handl., Stockholm*, 18 (1756) 120.
5. M. Bejblová, N. Žilková, and J. Čejka, Transformations of aromatic hydrocarbons over zeolites, *Res. Chem. Intermed.*, 34 (2008) 439–454.
6. (a) For reviews, see: G. Sartori and R. Maggi, Use of solid catalysts in Friedel–Crafts acylation reactions, *Chem. Rev.*, 106 (2006) 1077–1104; (b) K. Smith and G. A. El-Hiti, Regioselective electrophilic aromatic substitution reactions over reusable zeolites, *Curr. Org. Chem.*, 10 (2006) 1603–1625.
7. S. Bursna, J. S. J. Hargreavesa, M. Stockenhuber, and R. P. K. Wells, On the mechanism of aromatic acylation over zeolites, *Micropor. Mesopor. Mater.*, 104 (2007) 217–224.
8. N. Y. Chen, W. E. Garwood, and F. G. Dwyer, *Shape Selective Catalysis in Industrial Applications*, Marcel Dekker, 1989, p. 303.
9. (a) N. Y. Chen, W. W. Keading, and G. G. Dwyer, Para-directed aromatic reactions over shape-selective molecular sieve zeolite catalysts, *J. Am. Chem. Soc.*, 101 (1979) 6783–6784; (b) W. W. Keading, C. Chu, L. B. Young, B. Weinstein, and S. A. Butter, Selective alkylation of toluene with methanol to produce para-xylene, *J. Catal.*, 67 (1981) 159–174.
10. (a) A. Botella, J. Corma, M. López-Nieto, S. Valencia, and R. Jacquot, *J. Catal.*, 195 (2000) 161; (b) E. G. Derouane, G. Crehan, C. J. Dillon, D. Bethell, H. He, and S. B. Derouane-Abd Hamid, Zeolite catalysts as solid solvents in fine chemicals synthesis, 2. Competitive adsorption of the reactants and products in the Friedel–Crafts acetylations of anisole and toluene, *J. Catal.*, 194 (2000) 410–423.
11. (a) P. M. Bhaskar and D. Loganathan, Acetylation of alcohols, phenols, thiols and amines catalysed by H-b zeolite, *Indian J. Chem., Sect. B*, 43 (2004) 892–894; (b) S. R. Kirumakki, N. Nagaraju, and S. Narayanan, A comparative

esterification of benzyl alcohol with acetic acid over zeolites Hβ, HY and HZSM5, *Appl. Catal. A Gen.*, 273 (2004) 1–9.

12. A. Corma, M. J. Climenti, H. Garcia, and J. Primo, Formation and hydrolysis of acetals catalysed by acid faujasites, *Appl. Catal.*, 59 (1990) 333–340.
13. C. Moreau, J. Lecomte, S. Mseddi, and N. Zmimita, Stereoelectronic effects in hydrolysis and hydrogenolysis of acetals and thioacetals in the presence of heterogeneous catalysts, *J. Mol. Catal. A Chem.*, 125 (1997) 143–149.
14. T. Ookoshi and M. Onaka, Zeolite-catalyzed macrolactonization of ω-hydroxyalkanoic acids in a highly concentrated solution, *Tetrahedron Lett.*, 39 (1998) 293–296.
15. P. O'Sullivan, L. Forni, and B. K. Hodnett, The role of acid site strength in the Beckmann rearrangement, *Ind. Eng. Chem. Res.*, 40 (2001) 1471–1475 and references cited therein.
16. E. Baburek and J. Nováková, Isomerization of *n*-butane over acid zeolites, role of Brønsted and Lewis acid sites, *Appl. Catal. A Gen.*, 185 (1999) 123–130.
17. (a) H. Wang and Y. Zou, Modified beta zeolite as catalyst for Fries rearrangement reaction, *Catal. Lett.*, 86 (2003) 163–167; (b) P. Adryan, R. A. Isnail, and F. Roessner, On the equivalence of acidic centers in H-beta zeolites tested by Fries rearrangement of phenyl acetate, *Kinet. Catal.*, 49 (2008) 587–593.
18. K. Smith, G. A. El-Hiti, and M. Al-Shamalia, Rearrangement of epoxides to carbonyl compounds in the presence of reusable acidic zeolite catalysts under mild conditions, *Catal. Lett.*, 109 (2006) 77–82.
19. (a) For reviews on structure, properties of clays and their use in organic synthesis, see:M. Balogh and P. Laszlo, *Organic Chemistry Using Clays,* Springer, Berlin, 1993; (b) G. Nagendrappa, Organic synthesis using clay catalysts, clays for green chemistry, *Resonance* (2002) 64–77; (c) R. S. Varma, Clay and clay-supported reagents in organic synthesis, *Tetrahedron*, 58 (2002) 1235–1255; (d) S. Dasgupta and B. Torok, Application of clay catalysts in organic synthesis. A review, *Org. Prep. Proced. Int.*, 40 (2008) 1–65; (e) J. A. Kappel, Montmorillonite K-10: Recyclable and useful catalyst, *Res. J. Chem. Environ.*, 12 (2008) 102–103.
20. G. Sharma, R. Kumar, and A. K. Chakraborti, A novel environmentally friendly process for carbon–sulfur bond formation catalyzed by montmorillonite clays, *J. Mol. Cat. A Chem.*, 263 (2007) 143–148.
21. K. Motokura, N. Nakagiri, T. Mizugaki, K. Ebitani, and K. Kaneda, Nucleophilic substitution reactions of alcohols with use of montmorillonite catalysts as solid Brønsted acids, *J. Org. Chem.*, 72 (2007) 6006–6015.

22. B. Thomas, S. Prathapan, and S. Sugunan, Effect of pore size on the catalytic activities of K-10 clay and H-zeolites for the acetalization of ketones with methanol, *Appl. Cat. A Gen.*, 277 (2004) 247–252.
23. E. C. L. Gautier, A. E. Graham, A. McKillop, S. P. Standen, and R. J. K. Taylor, Acetal and ketal deprotection using montmorillonite K10: The first synthesis of syn-4,8-dioxatricyclo[5.1.0.0^{3,5}]-2,6-octanedione, *Tetrahedron Lett.*, 38 (1997) 1881–1884.
24. (a) J.-M. Lu and M. Shi, Montmorillonite K-10-catalyzed intramolecular rearrangement of vinylidenecyclopropanes, *Tetrahedron*, 63 (2007) 7545–7549; (b) P. Shanmugam and P. Rajasingh, Studies on montmorillonite K10-microwave assisted isomerisation of Baylis–Hillman adduct. Synthesis of E-trisubstituted alkenes and synthetic application to lignan core structures by vinyl radical cyclization, *Tetrahedron*, 60 (2004) 9283–9295; (c) P. Shanmugam and P. Rajasingh, Montmorillonite K10 clay catalyzed mild, clean, solvent free one-pot protection-isomerisation of the Baylis-Hillman adducts with alcohols, *Chem. Lett.* (2002) 1212–1213.
25. D. A. Young and J. W. Koepke, Chemical reaction promoted by catalytically active amorphous silica, US Patent 4501925, February 1985.
26. E. G. Rochow, *The Chemistry of Silicon,* Pergamon Press, Oxford, 1973.
27. R. K. Bansal, *Synthetic Approaches in Organic Chemistry,* Jones & Bartlett Publishers, Sudbury, Massachusetts, 1998, pp. 386–387.
28. (a) A. K. Banerjee, M. S. Laya, and W. J. Vera, Silica gel in organic synthesis, *Russ. Chem. Rev.*, 70 (2001) 971–990; (b) D. H. Drewry, D. M. Coe, and S. Poon, Solid-supported reagents in organic synthesis, *Med. Res. Chem.*, 19 (1999) 97–148.
29. J. Tercio, B. Ferreira, W. O. Cruz, P. C. Vieira, and M. Yonashiro, Carbon–carbon double bond cleavage using solid-supported potassium permanaganate on silica gel, *J. Org. Chem.*, 52 (1987) 3698–3699.
30. H. Kotsuki, T. Shimanouchi, R. Ohshima, and S. Fujiwara, Solvent-free organic reactions on silica gel supports. Facile transformation of epoxides to β-halohydrins with lithium halides, *Tetrahedron*, 54 (1998) 2709–2722.
31. G. D. Kishore Kumar and S. Baskaran, A facile, catalytic, and environmentally benign method for selective deprotection of *tert*-butyldimethylsilyl ether mediated by phosphomolybdic acid supported on silica gel, *J. Org. Chem.*, 70 (2005) 4520–4523.
32. J.-Q. Wang, D.-L. Kong, J.-Y. Chen, F. Cai, and L.-N. He, Synthesis of cyclic carbonates from epoxides and carbon dioxide over silica-supported quaternary ammonium salts under supercritical conditions, *J. Mol. Cat. A Chem.*, 249 (2006) 143–148.

33. E. Fischer, Ueber die Glucoside der Alkohole, *Ber. Deutsch. Chem. Ges.*, 26 (1893) 2400–2412.
34. (a) K. Toshima and K. Tatsuta, Recent progress in O-glycosylation methods and its application to natural products synthesis, *Chem. Rev.*, 93 (1993) 1503–1531; (b) S. Hanessian, Stereocontrolled glycosyl transfer reactions with unprotected glycosyl donors, *Chem. Rev.*, 100 (2000) 4443–4463; (c) K. J. Jensen, O-Glycosylations under neutral or basic conditions, *J. Chem. Soc., Perkin Trans.*, 1 (2002) 2219–2233; (d) K. Toshima, Novel glycosylation methods and their application to natural products synthesis, *Carbohydr. Res.*, 341 (2006) 1282–1297; (e) K. Toshima and K. Sasaki, O-glycosidation methods, in J. P. Kamerling, G.-J. Boons, Y. C. Lee, A. Suzuki, N. Taniguchi, and A. G. J. Voragen, (Eds.), *Comprehensive Glycoscience, Vol. 1,* Elsevier, Amsterdam, 2007, pp. 261–311; (f) Chemical glycosylation reactions, in B. O. Fraser-Reid, K. Tatsuta, and J. Thiem, (Eds.), *Glycoscience, Chemistry and Chemical Biology*, Springer Verlag, Heidelberg, 2008, pp. 427–811.
35. A. Corma, S. Iborra, S. Miquel, and J. Primo, Preparation of environmentally friendly alkylglucoside surfactants using zeolites as catalysts, *J. Catal.*, 161 (1996) 713–719.
36. M. A. Camblor, A. Corma, S. Iborra, S. Miquel, J. Primo, and S. Valencia, Beta zelite as a catalyst for the preparation of alkyl glucoside surfactants: The role of crystal size and hydrophobicity, *J. Catal.*, 172 (1997) 76–84.
37. A. Corma, S. Iborra, S. Miquel, and J. Primo, Preparation of long-chain alkyl glucoside surfactants by one-step direct Fischer glucosidation, and by transacetalation of butyl glucosides, on beta zeolite catalysts, *J. Catal.*, 180 (1998) 218–224.
38. J.-F. Chapat, A. Finiels, J. Joffre, and C. Moreau, Synthesis of butyl-α and β-D-glucopyranosides in the presence of dealuminated H-Y faujasites: Kinetic study, mechanism, stereoelectronic effects, and microreversibility principle, *J. Catal.*, 185 (1999) 445–453.
39. A. P. Rauter, T. Almeida, N. M. Xavier, F. Siopa, A. I. Vicente, S. D. Lucas, J. P. Marques, F. Ramôa Ribeiro, M. Guisnet, and M. J. Ferreira, Acid zeolites as efficient catalysts for *O*- and *S*-glycosylation, *J. Mol. Cat. A: Chem.*, 275 (2007) 206–213.
40. B. Helferich and E. Shimitz-Hillebrecht, Eine neue Methode zur Synthese von Glykosiden der Phenole, *Chem. Ber.*, 66 (1933) 378–383.
41. K. S. Kim and H. B. Jeon, Further anomeric esters, in B. O. Fraser-Reid, K. Tatsuta, and J. Thiem, (Eds.), *Glycoscience, Chemistry and Chemical Biology*, Springer Verlag, Heidelberg, 2008, pp. 525–564.

42. U. Aich and D. Loganathan, Stereoselective single-step synthesis and X-ray crystallographic investigation of acetylated aryl 1, 2-*trans* glycopyranosides and aryl 1, 2-*cis* C2-hydroxy-glycopyranosides, *Carbohydr. Res.*, 341 (2006) 19–28.
43. M. Bols, Stereocontrolled synthesis of α glucosides by intramolecular glycosidation, *J. Chem. Soc., Chem. Commun.*, 12 (1992) 913–914.
44. U. Aich and D. Loganathan, Zeolite-catalyzed Helferich-type glycosylation of long-chain alcohols. Synthesis of acetylated alkyl 1, 2-*trans* glycopyranosides and alkyl 1, 2-*cis* C2-hydroxy-glycopyranosides, *Carbohydr. Res.*, 342 (2007) 704–709.
45. (a) R. L. Halcomb and S. J. Danishefsky, On the direct epoxidation of glycals: Application of a reiterative strategy for the synthesis of β-linked oligosaccharides, *J. Am. Chem. Soc.*, 111 (1989) 6661–6666; (b) K. Toshima and K. Tatsuta, Recent progress in O-glycosylation methods and its application to natural products synthesis, *Chem. Rev.*, 93 (1993) 1503–1531and references cited herein; (c) S. Jarosz and M. Nowogródski, Anhydro sugars, in B. O. Fraser-Reid, K. Tatsuta, and J. Thiem, (Eds.), *Glycoscience, Chemistry and Chemical Biology,* Springer Verlag, Heidelberg, 2008, pp. 280–283.
46. Y.-i. Matsushita, K. Sugamoto, Y. Kita, and T. Matsui, Silica gel-catalyzed β-O-glucosylation of alcohols with 1,2-anhydro-3,4,6-tri-O-pivaloyl-α-d-glucopyranose, *Tetrahedron Lett.*, 38 (1997) 8709–8712.
47. R. M. de Lederkremer and C. Marino, Deoxy sugars: occurrence and synthesis, *Adv. Carbohydr. Chem Biochem.*, 61 (2007) 148–151.
48. (a) R. J. Ferrier, Substitution-with-allylic-rearrangement reactions of glycal derivatives, *Top. Curr. Chem.*, 215 (2001) 153–175; (b) R. J. Ferrier and O. A. Zubkov, Transformation of glycals into 2,3-unsaturated glycosyl derivatives, *Org. React.*, 62 (2003) 569–736; (c) S. Jarosz and M. Nowogródski, C=C bond formation, in B. O. Fraser-Reid, K. Tatsuta, and J. Thiem, (Eds.), *Glycoscience, Chemistry and Chemical Biology,* Springer Verlag, Heidelberg, 2008, pp. 354–357(See also p. 350); (d) W. Priebe, I. Fokt, and G. Grynkiewicz, Glycal derivatives, in B. O. Fraser-Reid, K. Tatsuta, and J. Thiem, (Eds.), *Glycoscience, Chemistry and Chemical Biology,* Springer Verlag, Heidelberg, 2008, pp. 716–722.
49. L. V. Dunkerton, N. K. Adair, J. M. Euske, K. T. Brady, and P. D. Robinson, Regioselective synthesis of substituted 1-thiohex-2-enopyranosides, *J. Org. Chem.*, 53 (1988) 845–850.
50. E. Wieczorek and J. Thiem, Preparation of modified glycosyl glycerol derivatives by glycal rearrangement, *Carbohydr. Res.*, 307 (1998) 263–270.

51. M. A. Probert, J. Zhang, and D. R. Bundle, Synthesis of α- and β-linked tyvelose epitopes of the *Trichinella spiralis* glycan: 2-Acetamido-2-deoxy-3-*O*-(3, 6-dideoxy-D-*arabino*-hexopyranosyl)-β-D-galactopyranosides, *Carbohydr. Res.*, 296 (1996) 149–170.
52. G. Posner and D. S. Bull, Molecular sieves promote stereocontrolled α,α-disaccharide formation via direct dimerization of free sugars, *Tetrahedron Lett.*, 37 (1996) 6279–6282.
53. M. Thomas, J.-P. Gesson, and S. Papot, First O-glycosylation of hydroxamic acids, *J. Org. Chem.*, 72 (2007) 4262–4264.
54. J.-I. Matsuo, T. Shirahata, and S. Omura, Catalytic and stereoselective glycosylation with glycosyl *N*-trichloroacetylcarbamate, *Tetrahedron Lett.*, 47 (2006) 267–271 and references cited therein.
55. M. Wilstermann, J. Balogh, and G. J. Magnusson, Synthesis of XylβCer, Galβ1-4XylβCer, NeuAcβ2-3Galβ1-4XylβCer and the corresponding lactone and lactam trisaccharides, *J. Org. Chem.*, 62 (1997) 7961–7971.
56. M. Wilstermann, J. Balogh, and G. J. Magnusson, Restriction of conformation in galabiosides via an *O*-*O*′-methylene bridge, *J. Org. Chem.*, 62 (1997) 3659–3665.
57. M. Wilstermann and G. Magnusson, Synthesis of disaccharide glycosyl donors suitable for introduction of the β-D-Gal*p*-(1 → 3)-α- and -β-D-Gal*p*NAc groups, *Carbohydr. Res.*, 272 (1995) 1–7.
58. H. P. Wessel and N. Ruiz, α-Glucosylation reactions with 2,3,4,6-tetra-*O*-benzyl-β-D-glucopyranosyl fluoride and triflic anhydride as promoter, *J. Carbohydr. Chem.*, 10 (1991) 901–910.
59. H. Nagai, S. Matsumura, and K. Toshima, A novel promoter, heteropoly acid, mediated chemo- and stereoselective sulfoxide glycosidation reactions, *Tetrahedron Lett.*, 41 (2000) 10233–10237.
60. A. Ben, T. Yamauchi, T. Matsumotu, and K. Suzuki, $Sc(OTf)_3$ as efficient catalyst for aryl *C*-glycoside synthesis, *Synlett*, 2 (2004) 225–230.
61. S. Komba and Y. Ito, Novel substrates for efficient enzymatic transglycosylation by *Bacillus circulans*, *Can. J. Chem.*, 80 (2002) 1174–1185.
62. M. Adinolfi, B. Gaspare, A. Iadonisi, and M. Schiattarella, Iodine/triethylsilane as a convenient promoter system for the activation of disarmed glycosyl trichloro- and *N*-(phenyl)trifluoroacetimidates, *Synlett*, 2 (2002) 269–270.
63. M. Adinolfi, G. Barone, A. Iadonisi, and M. Schiattarella, Activation of glycoyl trihaloacetimidates with acid-washed molecular sieves in the glycosidation reaction, *Org Lett.*, 5 (2003) 987–989.

64. M. Adinolfi, A. Iadonisi, A. Pezzella, and A. Ravidà, Regioselective phenol or carbinol glycosidation of 17p-estradiol and derivatives thereof, *Synlett*, 12 (2005) 1848–1852.
65. M. Adinolfi, A. Iadonisi, A. Ravidà, and M. Schiattarella, Versatile use of ytterbium(III) triflate and acid washed molecular sieves in the activation of glycosyl trifluoroacetimidate donors. Assemblage of a biologically relevant tetrasaccharide sequence of Globo H, *J. Org. Chem.*, 70 (2005) 5316–5319.
66. T. Chandra and K. L. Brown, Direct glycosylation: Synthesis of α-indoline ribonucleosides, *Tetrahedron Lett.*, 46 (2005) 2071–2074.
67. T. Chandra, X. Zou, and K. L. Brown, Low temperature dehydrogenation of α-indoline nucleosides, *Tetrahedron Lett.*, 45 (2004) 7783–7786.
68. J.-C. Florent and C. Monneret, Stereocontrolled route to 3-amino-2,3,6-trideoxy-hexopyranoses. K-10 Montmorillonite as a glycosidation reagent for acosaminide synthesis, *J. Chem. Soc., Chem. Commun.* (1987) 1171–1172.
69. S. Brochette, G. Descotes, A. Bouchu, Y. Queneau, N. Monnier, and C. Pétrier, Effect of ultrasound on KSF/O mediated glycosylations, *J. Mol. Cat. A Chem.*, 123 (1997) 123–130.
70. R. M. de Lederkremer and C. Marino, Deoxy sugars: Occurrence and synthesis, *Adv. Carbohydr. Chem. Biochem.*, 61 (2007) 163–167.
71. T. Jyojima, N. Miyamoto, Y. Ogawa, S. Matsumura, and K. Toshima, Novel stereocontrolled glycosidations of olivoses using montmorillonite K-10 as an environmentally benign catalyst, *Tetrahedron Lett.*, 40 (1999) 5023–5026.
72. K. Toshima, K. Uehara, H. Nagai, and S. Matsumura, A highly stereoselective synthesis of β-olivosides by glycosidations of 2-iodo-olivosyl fluoride using montmorillonite K-10 as an environmentally benign solid acid, *Green Chem.*, 4 (2002) 27–29.
73. K. Toshima, T. Ishizuka, G. Matsuo, and M. Nakata, Practical glycosidation method of glycals using montmorillonite K-10 as an environmentally acceptable and inexpensive industrial catalyst, *Synlett* (1995) 306–308.
74. S. Sowmya and K. K. Balasubramanian, Microwave induced Ferrier rearrangement, *Synthetic Commun.*, 24 (1994) 2097–2101.
75. B. Shanmugasundaram, A. K. Boseb, and K. K. Balasubramanian, Microwave-induced, montmorillonite K10-catalyzed Ferrier rearrangement of tri-*O*-acetyl-D-galactal: Mild, eco-friendly, rapid glycosidation with allylic rearrangement, *Tetrahedron Lett.*, 43 (2002) 6795–6798.
76. K. Ikeda, Y. Ueno, S. Kitani, R. Nishino, and M. Sato, Ferrier glycosylation reaction catalyzed by Bi(OTf)3-Montmorillonite K-10: Efficient synthesis of 3,4-unsaturated sialic acid derivatives, *Synlett* (2008) 1027–1030.

77. (a) For reviews on C-Glycosylation, see:Y. Du and R. J. Linhardt, Recent advances in stereoselective *C*-glycoside synthesis, *Tetrahedron*, 54 (1998) 9913–9959; (b) T. Nishikawa, M. Adachi, and M. Isobe, *C*-Glycosylation, in B. O. Fraser-Reid, K. Tatsuta, and J. Thiem, (Eds.), *Glycoscience, Chemistry and Chemical Biology,* Springer Verlag, Heidelberg, 2008, pp. 755–811.
78. K. Toshima, N. Miyamoto, G. Matsuo, M. Nakata, and S. Matsumura, Environmentally compatible *C*-glycosidation of glycals using montmorillonite K-10, *Chem. Commun.*, 11 (1996) 1379–1380.
79. (a) For reviews concerning the synthesis and biological profile of C-glycosyl aromatic compounds, see: F. Nicotra, Synthesis of *C*-glycosides of biological interest, *Top. Curr. Chem.*, 187 (1997) 55–83; (b) T. Bililign, B. R. Griffith, and J. S. Thorson, Structure, activity, synthesis and biosynthesis of aryl-C-glycosides, *Nat. Prod. Rep.*, 22 (2005) 742–760; (c) D. Y. W. Lee and M. He, Recent advances in aryl C-glycoside synthesis, *Curr. Top. Med. Chem.*, 5 (2005) 1333–1350; (d) A. P. Rauter, R. G. Lopes, and A. Martins, C-Glycosylflavonoids: Identification, bioactivity and synthesis, *Nat. Prod. Commun.*, 2 (2007) 1175–1196.
80. K. Toshima, Y. Ushiki, G. Matsuo, and S. Matsumura, Environmentally benign aryl *C*-glycosidations of unprotected sugars using montmorillonite K-10 as a solid acid, *Tetrahedron Lett.*, 38 (1997) 7375–7378.
81. A. K. Chakraborti and R. Gulhane, Perchloric acid adsorbed on silica gel as a new, highly efficient, and versatile catalyst for acetylation of phenols, thiols, alcohols, and amines, *Chem. Commun.*, 15 (2003) 1896–1897.
82. A. Agarwal, S. Rani, and Y. D. Vankar, Protic acid ($HClO_4$ supported on silica gel)-mediated synthesis of 2, 3-unsaturated-O-glucosides and a chiral furan diol from 2,3-glycals, *J. Org. Chem.*, 69 (2004) 6137–6140.
83. A. K. Misra, P. Tiwari, and G. Agnihotri, Ferrier rearrangement catalyzed by $HClO_4$–SiO_2: Synthesis of 2,3-unsaturated glycopyranosides, *Synthesis*, 2 (2005) 260–266.
84. P. Tiwari, G. Agnihotriand, and A. K. Misra, Synthesis of 2, 3-unsaturated C-glycosides by $HClO_4$–SiO_2 catalyzed Ferrier rearrangement of glycals, *Carbohydr. Res.*, 340 (2005) 749–752.
85. B. Mukhopadhyay, B. Collet, and R. A. Field, Glycosylation reactions with 'disarmed' thioglycoside donors promoted by *N*-iodosuccinimide and $HClO_4$–silica, *Tetrahedron Lett.*, 46 (2005) 5923–5925.
86. B. Mukhopadhyay, S. V. Maurer, N. Rudolph, R. M. van Well, D. A. Russell, and R. A. Field, From solution phase to, "on-column" chemistry:

Trichloroacetimidate-based glycosylation promoted by perchloric acid-silica, *J. Org. Chem.*, 70(22), (2005) 9059–9062.

87. Y. Du, G. Wei, S. Cheng, Y. Hua, and R. J. Linhardt, $HClO_4$–SiO_2 catalyzed glycosylation using sugar trichloroacetimidates as glycosyl donors, *Tetrahedron Lett.*, 47 (2006) 307–310.
88. G. Agnihotri and A. K. Misra, Synthesis of a di- and a trisaccharide related to the *O*-antigen of *Escherichia coli* O83:K24:H31, *Carbohydr. Res.*, 341 (2006) 2420–2425.
89. B. Roy and B. Mukhopadhyay, Sulfuric acid immobilized on silica: An excellent catalyst for Fischer type glycosylation, *Tetrahedron Lett.*, 48 (2007) 3783–3787.
90. S. Dasgupta, K. Pramanik, and B. Mukhopadhyay, Oligosaccharides through reactivity tuning: Convergent synthesis of the trisaccharides of the steroid glycoside Sokodoside B isolated from marine sponge *Erylus placenta*, *Tetrahedron*, 63 (2007) 12310–12316.
91. S. Mandaland and B. Mukhopadhyay, Concise synthesis of two trisaccharides related to the saponin isolated from *Centratherum anthelminticum*, *Tetrahedron*, 63 (2007) 11363–11370.
92. V. K. Rajput and B. Mukhopadhyay, Concise synthesis of a pentasaccharide related to the anti-leishmanial triterpenoid saponin isolated from *Maesa balansae*, *J. Org. Chem.*, 73 (2008) 6924–6927.
93. B. Roy, K. Pramanik, and B. Mukhopadhyay, Synthesis of a tetra- and a trisaccharide related to an anti-tumor saponin, "Julibroside J28" from *Albizia julibrissin*, *Glycoconj. J.*, 25 (2008) 157–166.
94. Y. Wu, J. G. Yu, X. F. Ma, and J. S. Zhang, An efficient and facile procedure for synthesis of octyl polyglucoside, *Chinese Chem. Lett.*, 18 (2007) 1173–1175.
95. For a recent review, see:A. M. Gómez, Reactions at oxygen atoms, in B. O. Fraser-Reid, K. Tatsuta, and J. Thiem, (Eds.), *Glycoscience, Chemistry and Chemical Biology,* Springer Verlag, Heidelberg, 2008, pp. 129–133.
96. P. M. Bhaskar and D. Loganathan, H-beta zeolite as an efficient catalyst for per-*O*-acetylation of mono- and disaccharides, *Synlett* (1999) 129–131.
97. V. L. C. Gonçalves, B. P. Pinto, J. C. Silva, and C. J. A. Mota, Acetylation of glycerol catalyzed by different solid acids, *Catal. Today*, 133–135 (2008) 673–677.
98. (a) For reviews, see: P. Calinaud and J. Gelas, in S. Hanessian, (Ed.), *Preparative Carbohydrate Chemistry, Synthesis of Isopropylidene, Benzylidene and Related Acetals,* Marcel Dekker, New York, 1997, pp. 3–33; (b) A. M. Gómez, Reactions at oxygen atoms, in B. O. Fraser-Reid, K. Tatsuta, and J. Thiem, (Eds.),

Glycoscience, Chemistry and Chemical Biology, Springer Verlag, Heidelberg, 2008, pp. 121–122.

99. A. P. Rauter, F. Ramôa-Ribeiro, A. C. Fernandes, and J. A. Figueiredo, A new method of acetonation with the zeolite HY as catalyst. Synthesis of O-isopropylidene sugar derivatives, *Tetrahedron*, 51 (1995) 6529–6540.
100. (a) A. H. Haines, The selective removal of protecting groups in carbohydrate chemistry, *Adv. Carbohydr. Chem. Biochem.*, 39 (1981) 13–70; (b) A. M. Gómez, Reactions at oxygen atoms, in B. O. Fraser-Reid, K. Tatsuta, and J. Thiem, (Eds.), *Glycoscience, Chemistry and Chemical Biology,* Springer Verlag, Heidelberg, 2008, pp. 122–123.
101. P. M. Bhaskar, M. Mathiselvam, and D. Loganathan, Zeolite-catalyzed selective deprotection of di- and tri-*O*-isopropylidene sugar acetals, *Carbohydr. Res.*, 343 (2008) 1801–1807.
102. M. Adinolfi, G. Barone, A. Iadonisi, and M. Schiattarella, An easy approach for the acetylation of saccharidic alcohols. Applicability for regioselective protections, *Tetrahedron Lett.*, 44 (2003) 4661–4663.
103. M. M. Sá and L. Meier, Pyridine-free and solvent-free acetylation of nucleosides promoted by molecular sieves, *Synlett*, 20 (2006) 3474–3478.
104. M. Adinolfi, G. Barone, A. Iadonisi, and M. Schiattarella, Mild benzhydrylation and tritylation of saccharidic hydroxyls promoted by acid washed molecular sieves, *Tetrahedron Lett.*, 44 (2003) 3733–3735.
105. M. Mizuno, K. Kobayashi, H. Nakajima, M. Koya, and T. Inazu, Unexpected reaction using methanol dried over molecular sieves, *Synth. Commun.*, 32 (2002) 1665–1670.
106. K. P. R. Kartha, B. Mukhopadhyay, and R. A. Field, Practical de-*O*-acylation reactions promoted by molecular sieves, *Carbohydr. Res.*, 339 (2004) 729–732.
107. P. M. Bhaskar and D. Loganathan, Per-*O*-acetylation of sugars catalysed by montmorillonite K-10, *Tetrahedron Lett.*, 39 (1998) 2215–2218.
108. J. Asakura, Y. Matsubara, and M. Yoshihara, Clay catalyzed acetonation: A simple method for the preparation of isopropylidene carbohydrates, *J. Carbohydr. Chem.*, 15 (1996) 231–239.
109. J. I. Asakura, M. J. Robins, Y. Asaka, and T. H. Kim, Removal of acetal, silyl, and 4, 4′-dimethoxytrityl protecting groups from hydroxyl functions of carbohydrates and nucleosides with clay in aqueous methanol, *J. Org. Chem.*, 61 (1996) 9026–9027.
110. A. K. Misra, P. Tiwari, and S. K. Madhusudan, $HClO_4$–SiO_2 catalyzed per-*O*-acetylation of carbohydrates, *Carbohydr. Res.*, 340 (2005) 325–329.

111. A. Bianco, M. Brufani, C. Melchioni, and P. Romagnoli, A new method of regioselective protection of primary alcoholic function with rare earths salts, *Tetrahedron Lett.*, 38 (1997) 651–652.
112. B. Mukhopadhyay, D. A. Russell, and R. A. Field, One-pot acetalation–acetylation of sugar derivatives employing perchloric acid immobilised on silica, *Carbohydr. Res.*, 340 (2005) 1075–1080.
113. B. Mukhopadhyay, Sulfuric acid immobilized on silica: An efficient promoter for one-pot acetalation–acetylation of sugar derivatives, *Tetrahedron Lett.*, 47 (2006) 4337–4341.
114. V. K. Rajput and B. Mukhopadhyay, Sulfuric acid immobilized on silica: An effcient reusable catalyst for the synthesis of O-isopropylidene sugar derivatives, *Tetrahedron Lett.*, 47 (2006) 5939–5941.
115. Y. Niu, N. Wang, X. Cao, and X. Ye, Efficient formation and cleavage of benzylidene acetals by sodium hydrogen sulfate supported on silica gel, *Synlett*, 13 (2007) 2116–2120.
116. B. Das, G. Mahender, V. S. Kumar, and N. Chowdhury, Chemoselective deprotection of trityl ethers using silica-supported sodium hydrogen sulfate, *Tetrahedron Lett.*, 45 (2004) 6709–6711.
117. M. R. C. Couri, E. A. Evangelista, R. B. Alves, M. A. F. Prado, R. P. F. Gil, M. V. De Almeida, and D. S. Raslan, Microwave-assisted rapid deacetalation of carbohydrates, *Synth. Comm.*, 35 (2005) 2025–2031.
118. A. Agarwal and Y. D. Vankar, Selective deprotection of terminal isopropylidene acetals and trityl ethers using $HClO_4$ supported on silica gel, *Carbohydr. Res.*, 340 (2005) 1661–1667.
119. V. K. Rajput, B. Roy, and B. Mukhopadhyay, Sulfuric acid immobilized on silica: An efficient reusable catalyst for selective hydrolysis of the terminal *O*-isopropylidene group of sugar derivatives, *Tetrahedron Lett.*, 47 (2006) 6987–6991.
120. K. S. Kim, Y. H. Song, B. H. Lee, and C. S. Hahn, Efficient and selective cleavage of acetals and ketals using ferric chloride adsorbed on silica gel, *J. Org. Chem.*, 51 (1986) 404–407.
121. J. S. Yadav, S. Raghavendra, M. Satyanarayana, and E. Balanarsaiah, Phosphomolybdic acid supported on silica gel: An efficient, mild and reusable catalyst for the chemoselective hydrolysis of acetonides, *Synlett*, 16 (2005) 2461–2464.
122. Y. Izumi, R. Hasebe, and K. Urabe, Catalysis by heterogeneous supported heteropoly acid, *J. Catal.*, 84 (1983) 402–409.
123. (a) A. Corma, S. Iborra, and A. Velty, Chemical routes for the transformation of biomass into chemicals, *Chem. Rev.*, 107 (2007) 2411–2502; (b) J. N. Chheda

and J. A. Dumesic, Catalytic routes from renewables to fine chemicals, *Catal. Today*, 123 (2007) 59–70; (c) P. Mäki-Arvela, B. Holmbom, T. Salmi, and D. Yu. Murzin, Recent progress in synthesis of fine and specialty chemicals from wood and other biomass by heterogeneous catalytic processes, *Catal. Rev.*, 49 (2007) 197–340.

124. Y. Queneau, S. Jarosz, B. Lewandowski, and J. Fitremann, Sucrose chemistry and applications of sucrochemicals, *Adv. Carbohydr. Chem. Biochem.*, 61 (2007) 217–292.
125. C. Moreau, M. N. Belgacem, and A. Gandini, Recent catalytic advances in the chemistry of substituted furans from carbohydrates and in the ensuing polymers, *Top. Catal.*, 27 (2004) 11–30.
126. C. Buttersack and D. Laketic, Hydrolysis of sucrose by dealuminated Y-zeolites, *J. Mol. Cat.*, 94 (1994) L283–L290.
127. C. Moreau, R. Durand, J. Duhamet, and P. Rivalier, Hydrolysis of fructose and glucose precursors in the presence of H-form zeolites, *J. Carbohydr. Chem.*, 16 (1997) 709–714.
128. C. Moreau, R. Durand, F. Aliès, M. Cotillon, M.-A. Frutz, and T. Théoleyre, Hydrolysis of sucrose in the presence of H-form zeolites, *Ind. Crops Prod.*, 11 (2000) 237–242.
129. R. Shukla, X. E. Verykios, and R. Mutharasan, Isomerization and hydrolysis reactions of important disaccharides over inorganic heterogeneous catalysts, *Carbohydr. Res.*, 143 (1985) 97–106.
130. A. Abbadi, K. F. Gotlieb, and H. van Bekkum, Study on solid acid catalyzed hydrolysis of maltose and related polysaccharides, *Starch*, 50 (1998) 23–28.
131. A. Onda, T. Ochi, and K. Yanagisawa, Selective hydrolysis of cellulose into glucose over solid acid catalysts, *Green Chem.*, 10 (2008) 1033–1037.
132. V. Le Strat and C. Moreau, Hydrolysis of methyl α- and β-D-glucopyranosides in the presence of a dealuminated H-Y faujasite, *Catal. Lett.*, 51 (1998) 219–222.
133. S. Dasgupta, B. Roy, and B. Mukhopadhyay, NIS/H_2SO_4–Silica: A mild and efficient reagent system for the hydrolysis of thioglycosides, *Carbohydr. Res.*, 341 (2006) 2708–2713.
134. H. Iloukhani, S. Azizian, and N. Samadani, Hydrolysis of sucrose by heterogeneous catalysis, *Phys. Chem. Liq.*, 40 (2002) 159–165.
135. K. J. Zeitsch, *The chemistry and technology of furfural and its many by-products,* 1st ed. Elsevier, Amsterdam, Vol. 13, 2000.
136. C. Moreau, R. Durand, C. Pourcheron, and S. Razigade, Preparation of 5-hydroxymethylfurfural from fructose and precursors over H-form zeolites, *Ind. Crops Prod.*, 3 (1994) 85–90.

137. C. Moreau, R. Durand, S. Razigade, J. Duhamet, P. Faugeras, P. Rivalier, P. Ros, and G. Avignon, Dehydration of fructose to 5-hydroxymethylfurfural over H-mordenites, *Appl. Catal. A Gen.*, 145 (1996) 211–224.
138. P. Rivalier, J. Duhamet, C. Moreau, and R. Durand, Development of a continuous catalytic heterogeneous column reactor with simultaneous extraction of an intermediate product by an organic solvent circulating in countercurrent manner with the aqueous phase, *Catal. Today*, 24 (1995) 165–171.
139. C. Moreau, R. Durand, D. Peyron, J. Duhamet, and P. Rivalier, Selective preparation of furfural from xylose over microporous solid acid catalysts, *Ind. Crops Prod.*, 7 (1998) 95–99.
140. S. Lima, M. Pillinger, and A. A. Valente, Dehydration of D-xylose into furfural catalysed by solid acids derived from the layered zeolite Nu-6(1), *Catal. Commun.*, 9 (2008) 2144–2148.
141. K. Lourvanij and G. L. Rorrer, Reaction rates for the partial dehydration of glucose to organic acids in solid-acid, molecular-sieving catalyst powders, *J. Chem. Tech. Biotechnol.*, 69 (1997) 35–44.
142. M. Kurszewska, E. Skorupowa, J. Madaj, A. Konitz, W. Wojnowski, and A. Wiśniewski, The solvent-free thermal dehydration of hexitols on zeolites, *Carbohydr. Res.*, 337 (2002) 1261–1268.
143. M. Kurszewska, E. Skorupowa, J. Madaj, and A. Wiśniewski, Solvent-free thermal dehydration of pentitols on zeolites, *J. Carbohydr. Chem.*, 23 (2004) 169–177.
144. K. Lourvanij and G. L. Rorrer, Dehydration of glucose to organic acids in microporous pillared clay catalysts, *Appl. Catal. A Gen.*, 109 (1994) 147–165.
145. J. M. Gonzales and D. A. Laird, Smectite-catalysed dehydration of glucose, *Clays Clay Min.*, 54 (2006) 38–44.
146. P. M. Collins and R. J. Ferrier, *Monosaccharides: Their Chemistry and Their Role in Natural Products,* John Wiley & Sons, New York, 1995, p. 115.
147. R. Madsen, Oxidation, reduction and deoxygenation, in B. O. Fraser-Reid, K. Tatsuta, and J. Thiem, (Eds.), *Glycoscience, Chemistry and Chemical Biology,* Springer Verlag, Heidelberg, 2008, pp. 179–225.
148. S. Ramachandran, P. Fontanille, A. Pandey, and C. Larroche, Gluconic acid: Properties, applications and microbial production, *Food. Technol. Biotechnol.*, 44 (2006) 185–195.
149. G. Hourdin, The catalysis of the Ruff oxidative degradation of aldonic acids by titanium-containing zeolites, *Catal. Lett.*, 69 (2000) 241–244.
150. G. Hourdin, A. Germain, C. Moreau, and F. Fajula, The catalysis of the Ruff oxidative degradation of aldonic acids by copper(II)-containing solids, *J. Catal.*, 209 (2002) 217–224.

151. J. Herscovici and K. Antonakis, Molecular sieve-assisted oxidations: New methods for carbohydrate derivative oxidations, *J. Chem. Soc. Chem. Commun.* (1980) 561–562.
152. G. Piancatelli, A. Scettri, and M. D'Auria, Pyridinium chlorochromate: A versatile oxidant in organic synthesis, *Synthesis* (1982) 245–258.
153. J. Herscovici, M.-J. Egron, and K. Antonakis, New oxidative systems for alcohols: Molecular sieves with chromium-(VI) reagents, *J. Chem. Soc. Perkin Trans. I* (1982) 1967–1973.
154. A. P. Rauter, J. A. Figueiredo, I. Ismael, and M. S. Pais, Synthesis of α-methylene-γ-lactones in furanosidic systems, *J. Carbohydr. Chem.*, 6 (1987) 259–272.
155. A. P. Rauter, J. A. Figueiredo, I. Ismael, T. Canda, J. Font, and M. Figueiredo, Efficient synthesis of α,β-unsaturated γ-lactones linked to sugars, *Tetrahedron Asym.*, 12 (2001) 1131–1146.
156. A. P. Rauter and H. Weidmann, Versuche zur Verzweigung von Hexofuranurono-6,3-lactonen; Synthese von 5-Desoxy-1,2-*O*-isopropyliden-5-*C*-methylen-α-D-*xylo*-hexofuranurono-6,3-lacton, *Liebigs Ann. Chem.* (1982) 2231–2237.
157. A. P. Rauter, F. Piedade, T. Almeida, R. Ramalho, M. J. Ferreira, R. Resende, J. Amado, H. Pereira, J. Justino, A. Neves, F. V. M. Silva, and T. Canda, Sugar bislactones by one-step oxidative dimerisation with pyridinium chlorochromate versus regioselective oxidation of vicinal diols, *Carbohydr. Res.*, 339 (2004) 1889–1897.
158. B. Vauzeilles and P. Sinaÿ, Selective radical synthesis of β-*C*-disaccharides, *Tetrahedron Lett.*, 42 (2001) 7269–7272.
159. A. Dondoni, A. Marra, M. Mizuno, and P. P. Giovannini, Linear total synthetic routes to β-D-C-(1, 6)-linked oligoglucoses and oligogalactoses up to pentaoses by iterative Wittig olefination assembly, *J. Org. Chem.*, 67 (2002) 4186–4199.
160. J. Piskorz, P. Majerski, D. Radlein, A. Vladars-Usas, and D. S. Scott, Flash pyrolysis of cellulose for production of anhydro-oligomers, *J. Anal. Appl. Pyrol.*, 56 (2000) 145–166.
161. D. Fabbri, C. Torri, and V. Baravelli, Effect of zeolites and nanopowder metal oxides on the distribution of chiral anhydro sugars evolved from pyrolysis of cellulose: An analytical study, *J. Anal. Appl. Pyrolysis*, 80 (2007) 24–29.
162. (a) Z. J. Witczak, in Z. J. Witczak, (Ed.), *Levoglucosenone and Levoglucosans: Chemistry and Applications,* ATL Press, Mount Prospect, 1994; (b) A. V. Samet, A. M. Shestopalov, D. N. Lutov, L. A. Rodinovskaya, A. A. Shestopalov, and V. V. Semenov, *Tetrahedron Asymm.*, 18 (2007) 1986–1989, and references cited herein.

163. (a) For reviews on click chemistry, see: (a) H. C. Kolb, M. G. Finn, and K. B. Sharpless, Click chemistry: Diverse chemical function from a few good reaction, *Angew. Chem. Int. Ed.*, 40 (2001) 2004–2021; (b) H. C. Kolb and K. B. Sharpless, The growing impact of click chemistry on drug discovery, *Drug Discov. Today*, 8 (2003) 1128–1137; (c) J. E. Moses and A. D. Moorhouse, The growing applications of click chemistry, *Chem. Soc. Rev.*, 36 (2007) 1249–1262.
164. V. V. Rostovtsev, L. G. Green, V. V. Fokin, and K. B. Sharpless, A stepwise Huisgen cycloaddition process: Copper(I)-catalysed regioselective ligation of azides and terminal alkynes, *Angew. Chem. Int. Ed.*, 41 (2002) 2596–2599.
165. T. Hasegawa, M. Umeda, M. Numata, C. Li, A.-H. Bae, T. Fujisawa, S. Haraguchi, K. Sakuraib, and S. Shinkai, 'Click chemistry' on polysaccharides: A convenient, general, and monitorable approach to develop (1→3)-β-D-glucans with various functional appendages, *Carbohydr. Res.*, 341 (2006) 35–40.
166. S. G. Gouin, L. Bultel, C. Falentin, and J. Kovensky, Simple procedure for connecting two carbohydrate moieties by click chemistry techniques, *Eur. J. Org. Chem.* (2007) 1160–1167.
167. A. Alix, S. Chassaing, P. Pale, and J. Sommer, 'Click chemistry' in Cu^{I}-zeolites: A convenient access to glycoconjugates, *Tetrahedron*, 64 (2008) 8922–8929.
168. (a) R. Haeckel, G. Lauer, and F. Oberdorfer, Facile synthesis of 1,6-anhydrohalosugars via a novel rearrangement of galactal, *Synlett* (1996) 21–23; (b) M. Černý, Chemistry of anhydro sugars, *Adv. Carbohydr. Chem. Biochem.*, 58 (2003) 121–198; (c) S. S. Kulkarni, J.-C. Lee, and S.-C. Hung, Recent advances in the applications of D- and L-form 1,6-anhydrohexopyranoses for the synthesis of oligosaccharides and nature products, *Curr. Org. Chem.*, 8 (2004) 475–509.
169. G. V. M. Sharma, K. C. V. Ramanaiah, and K. Krishnudu, Clay montmorillonite in carbohydrates use of claysil as an efficient heterogenous catalyst for the intramolecular Ferrier reactions leading to 1,6-anhydro rare saccharides, *Tetrahedron Asymm.*, 5 (1994) 1905–1908.
170. J. S. Yadav, B. V. S. Reddy, S. Meraj, P. Vishnumurthy, K. Narsimulu, and A. C. Kunwar, Montmorillonite clay catalyzed synthesis of enantiomerically pure 1, 2, 3, 4-tetrahydroquinolines, *Synthesis*, 17 (2006) 2923–2926.
171. L. D. S. Yadav, C. Awasthi, V. K. Rai, and A. Rai, A route to functionalized pyrimidines from carbohydrates via amine-driven dehydrative ring transformations, *Tetrahedron Lett.*, 49 (2008) 2377–2380.

172. L. D. S. Yadav, V. P. Srivastava, V. K. Rai, and R. Patel, Diversity oriented synthesis of fused-ring 1, 3-oxazines from carbohydrates as biorenewable feedstocks, *Tetrahedron*, 64 (2008) 4246–4253.
173. T. Sakakibara, T. Suganuma, and Y. Kajihara, Preparation of 2-deoxy-2-*C*-*p*-tolylsulfonyl-*β*-D-glucopyranosyl *p*-tolyl sulfones having non-chair conformation and their elimination reactions, *Chem. Commun.*, 34 (2007) 3568–3570.
174. A. Rauter, M. Ferreira, C. Borges, T. Duarte, F. Piedade, M. Silva, and H. Santos, Construction of a branched chain at C-3 of a hexopyranoside. Synthesis of miharamycin sugar moiety analogues, *Carbohydr. Res.*, 325 (2000) 1–15.
175. T. K. M. Shing, W. F. Wong, H. M. Cheng, W. S. Kwok, and K. H. So, Intramolecular nitrile oxide–alkene cycloaddition of sugar derivatives with unmasked hydroxyl group(s), *Org. Lett.*, 9 (2007) 753–756.
176. P. Borrachero, M. J. Diánez, M. D. Estrada, M. Gómez-Guillén, A. Gómez-Sánchez, A. López-Castro, and S. Pérez-Garrido, Silica gel-catalysed addition of methyl nitroacetate to 1, 2:3, 4-di-*O*-isopropylidene-α-D-*galacto*-hexodialdo-1, 5-pyranose and 2, 3-*O*-isopropylidene-D-glyceraldehyde. Crystal structure of methyl 7-acetamido-7-deoxy-1,2:3,4-di-*O*-isopropylidene-L-*threo*-α-D-*galacto*-octopyranuronate, *Carbohydr. Res.*, 271 (1995) 79–99.

TOOLS IN OLIGOSACCHARIDE SYNTHESIS: CURRENT RESEARCH AND APPLICATION

BY JÜRGEN SEIBEL[a] and KLAUS BUCHHOLZ[b]

[a]Institute of Organic Chemistry, University of Würzburg, Am Hubland, D-97074, Würzburg, Germany
[b]Department of Technical Chemistry, Technical University, Braunschweig, D-38106, Germany

ABBREVIATIONS

CD, cyclodextrin; DP, degree of polymerization; DS, dextransucrase; FTF, fructosyl transferase; FOS, fructosyl-oligosaccharide; Fru, fructose; FS, fructansucrase; GH, glycoside hydrolase; Glc, glucose; GOS, glucosyl-oligosaccharide; GTF, glycosyl-transferase; IM, isomaltulose; IMOS, isomalto-oligosaccharides; IS, inulosucrase; LS, levansucrase; SI, sucrose isomerase; WT, wild-type; XOS, different oligosaccharide

ISBN: 978-0-12-380856-1
DOI: 10.1016/S0065-2318(10)63004-1.

I. Introduction

Manifold applications of oligosaccharides have been established during recent decades. Food, sweeteners, and food ingredients comprise important sectors where oligosaccharides are used in major amounts worldwide. Increasing attention has been devoted to the sophisticated roles of oligosaccharides and glycosylated compounds at cell or membrane surfaces. Oligosaccharides found on cell surfaces as glycoprotein or glycolipid conjugates play important structural and functional roles in numerous biological recognition processes. These processes include viral and bacterial infection, cancer metastasis, the inflammatory response, innate and adaptive immunity, and many other receptor-mediated signaling processes.[1,2] The inhibition of cell or bacterial adhesion processes, along with vaccine formulation, have therefore become key topics of pharmaceutical research. Specific oligosaccharides have been reported to stimulate the immune system.[3] Prospects may be anticipated for semisynthetic glycan vaccines protecting against candida, meningitis, and leishmania infections. Expectations for a vaccine against malaria are related to the synthesis of an oligosaccharide identified on the surface of the pathogen *Plasmodium falciparum*.[4]

The challenge for developing methods for synthesizing oligosaccharides, notably for complex oligosaccharides, that can be scaled up to an industrial level is obvious, based on the use of substrates and enzymes that are available at reasonable cost and purity.

Strategies for oligosaccharides synthesis include:

- Chemical synthesis, an approach limited by complex synthetic sequences using protection of hydroxyl groups to achieve regio- and stereo-selectivity
- Highly selective enzymatic synthesis using glycosyltransferases (GTFs), an approach restricted by the limited availability of Leloir-type enzymes, and expensive nucleotide-activated substrates
- Sucrase-type (non-Leloir-type) enzymes that operate both regio- and stereo-selectively, using sucrose as a cheap substrate, or, in some cases (such as cyclodextrin (CD) transferases) starch; these enzymes are, however, limited to the transfer of only glucose or fructose

(For an overview on activated sugars in synthesis see ref. 5 and 24).

Several enzymatic routes with sucrase-type enzymes have been established for oligosaccharide synthesis, some of which are applied industrially for food and feed ingredients, and important examples are compiled.[5] Convenient routes for new oligosaccharides, however, are rarely available.

This report surveys new routes for the synthesis of oligosaccharides, with emphasis on enzymatic reactions, since they offer unique properties, proceeding highly regio- and stereo-selectively in water solution, and generally afford high yields. Summarized are approaches with glycosynthases and GTFs. Also discussed is the use of new or modified substrates, and both wild-type (WT) and genetically modified GTFs of the non-Leloir (sucrase) type, both being glucosyl- and fructosyltransferases that synthesize oligosaccharides and polysaccharides. Those selected are the ones that may be scaled up and adapted for industrial manufacturing. Nucleotide-activated sugars are therefore not included.

One of the aims was to combine structural features of glucosyl-oligosaccharides (GOSs) and other (XOSs) with fructosyl-oligosaccharides (FOSs) to design a new class of FOSs. The different glycopyranosyl residues may be recognized by carbohydrate-binding cell receptors, and function as inhibitors of bacterial adhesion.[6]

Finally, selected industrial processes and commercial oligosaccharides are summarized.

II. Modified Enzymes and Substrates

1. Glycosidases, Glycosynthases, Thioglycosidases

Glycosidases are a broad family of enzymes (EC 3.2.1), which hydrolytically cleave glycosyl residues of glycoconjugates. Most common glycosidases hydrolyze their substrate by releasing a specific monosaccharide from the nonreducing end of a sugar chain (exo-glycosidases), or by cleaving a bond within the carbohydrate polymer chain (endoglycosidases). They are highly stereospecific, and according to the stereochemical outcome of the reaction, they are classified as inverting or retaining enzymes (see also CAZY, ref. 85).[7,8] Because this hydrolytic reaction is an equilibrium, it can be used for an endergonic glycosylation termed reverse hydrolysis[9,10] or reversion. In this situation, the concentration of free monosaccharides needs to be high, but the yields are low in most instances. One strategy to enhance the yield of a transglycosylation reaction is to lower the water concentration, but this in turn is problematic, as enzyme activity decreases when solvents other than water are used. The immobilization of glycosidases has been examined to control the water level, but reverse hydrolysis is still not capable of productive synthesis.[11] Interestingly, glycosidases can use different substrates such as *p*-nitrophenyl glycosides[12] (pNP-sugars) or sugar fluorides,[13] which can be considered as activated sugars because of their linkage energy useable in an exergonic transglycosylation reaction. This method still suffers from similar problems as reverse hydrolysis, in trying to avoid the hydrolysis of the products formed.

Transglycosylation with activated substrates and glycosidases has been reported in several applications on a 100-mg scale, but the yields achieved with these methods are low, and the need for solvent systems to dissolve pNP-sugars limits their utility.[9] To avoid the disadvantage that glycosidases hydrolyze their glycosylation products again, efforts are being made to inhibit hydrolysis via mutagenesis, constructing the so-called glycosynthases (Fig. 1).[14] In such developments one of the key initial steps was the replacement of the active-site carboxylate-Glu358Ala mutant of the *Agrobacterium* sp. β-glucosidase (AbgGlu358Ala) nucleophile with a nonnucleophilic amino acid side chain, resulting in a correctly folded enzyme showing no detectable hydrolytic activity against natural substrates.[15] The α-glycosyl fluorides were chosen as activated glycosyl donors, as these mimic the enzyme–substrate intermediate, and

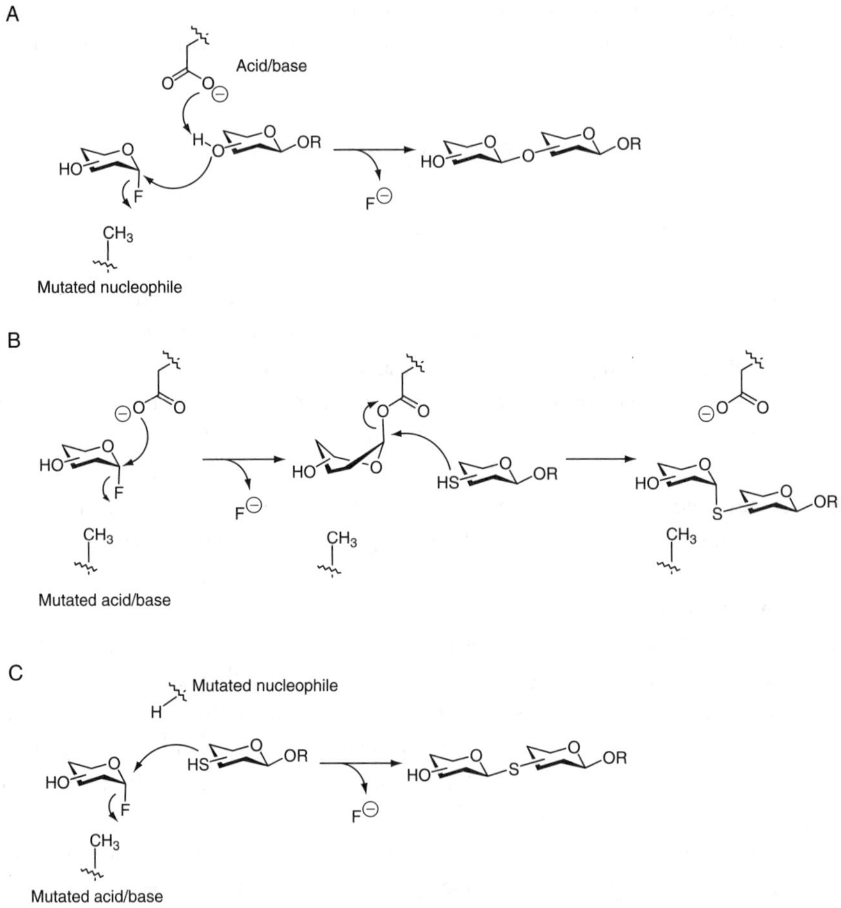
A
Acid/base
OR
HO
F
CH3
Mutated nucleophile
B
HO
F
HS
OR
S
CH3
Mutated acid/base
C
Mutated nucleophile
H
HS
OR
HO
F
S
CH3
Mutated acid/base

FIG. 1. Scheme of mechanism-based principle of transglycosyldation with (A) an inverting β-glycosidase, (B) a retaining thioglucosidase, and (C) an inverting thioglycosynthase.

aryl glycosides as acceptors, leading to transglycosylation with inverted configuration. The retaining β-glucosidase thus has been changed into an inverting glycosynthase.[15] Those findings led to the development of further endo- and exoglycosynthases, which have been summarized in various reviews.[16,17]

An equivalent approach has been successfully accomplished using thioglycosides as nucleophilic acceptors for ''thioglycoligation.'' The development of thioglycoligases succeeded through changing the acid–base catalytic groups of retaining β-glycosidases to the nonnucleophilic residue alanine.[18] Activated 2,4-dinitrophenyl β-glycosides and α-glycosyl fluorides were accepted as donor substrates by these variants and thio sugar as acceptor substrates because of their nucleophilic thiol group.[19] The resulting disaccharides contain thioglycosidic linkages with retained configuration. It has been further demonstrated that the double mutant Glu171Ala, Glu358Gly of the β-glycosidase from *Agrobacterium* yielded, in the presence of α-glycosyl fluorides and thio sugars, oligosaccharides having inverted configuration. Those variants may be termed thioglycosynthases.[20]

2. Glucansucrases

For the synthesis of oligosaccharides, the concept of using the so-called glycosynthases is very promising. Here, the hydrolytic activity of glycosidases has been suppressed via random and rational site-directed mutagenesis, allowing transglycosylation reactions to occur instead.[14,17,21–23]

Glucansucrases (non-Leloir glycosyltransferases), in contrast, normally act on substrates, sucrose in most cases, with the transfer of one glycosyl group, either glucose or fructose, to growing polysaccharide chains, or to acceptors, producing oligosaccharides elongated by one or more glycosyl groups. The reactions proceed very efficiently, with high yields, regio- and stereoselecivity, and in water as the solvent. The background of this convenient synthetic pathway is the high energy of the glycosidic bond of sucrose, which is similar to that of nucleotide-activated sugars.[24] The GTFs are bacterial enzymes expressed extracellularly by various bacterial species of *Leuconostoc*, *Lactobacillus*, and *Streptococcus*.

a. Kinetics.—Among the GTFs, dextransucrase (EC 2.4.1.5) has attracted much attention in research, since it elaborates two kinds of reactions: (1) the primary reaction (substrate reaction), the synthesis of dextran from sucrose and (2) the secondary reaction (acceptor reaction), the transfer of D-glucose from sucrose to carbohydrate acceptors that are added to the reaction solution, thus producing a range of oligosaccharides.

Dextran is a D-glucopyranosyl glucan containing mostly α-(1→6) glycosidic bonds, with branches linked α-(1→2) or α-(1→3) to the main chain, and with a molecular weight ranging from 0.5 to 6.10^6 kDa. [25] The most widely used dextran is produced by the dextransucrase of the strain *Leuconostoc mesenteroides* NRRL B-512F, which synthesizes a highly linear polysaccharide with about 95% α-(1→6) bonds.[25] The kinetics of dextran synthesis were investigated early by Ebert, Patat, and Schenk.[26,27] With respect to the acceptor reaction, the glucosyl moiety is transferred from sucrose to the acceptor molecule instead of the growing dextran chain, to give the acceptor molecule elongated by one single glucosyl group as the primary product. The new bond formed is usually an α-(1→6)-glucosidic bond. In most instances, the product can itself serve as an acceptor, so that a homologous series of glycosylated oligosaccharides is formed.[28,29]

In the presence of sucrose alone as the single substrate, initial reaction rates follow Michaelis–Menten kinetics up to 200 mM sucrose concentration, but the enzyme is inhibited by higher concentrations of substrate.[30] The inhibitor constant for sucrose is 730 mM. This inhibition can be overcome by the addition of acceptors.[31,32] The enzyme activity is significantly enhanced, and stabilized, by the presence of dextran, and by calcium ions.

For the production of α-gluco-oligosaccharides, the kinetic behavior of dextransucrase has been investigated in order to optimize the synthesis. In fact, with maltose as an acceptor, the dextransucrase of *L. mesenteroides* NRRL B-1299 produces three families of oligosaccharides of increasing degree of polymerization.[33] These respectively contain, besides a maltose residue at the reducing end (Fig. 2): only α-(1→6) glucosidic bonds; α-(1→6) glucosidic bonds, and one α-(1→2)-glucosidic bond at the nonreducing end; α-(1→6) glucosidic bonds, and one α-(1→2)-glucosidic bond on the penultimate D-glucosyl residue (for commercial utilization see Section III).

It must be pointed out that Michaelis kinetics do not apply in the presence of acceptors. Such constants as V_{max}, K_m, and K_I are dependent on acceptor concentrations. To understand the reactions occurring in the presence of different types of acceptors, and to have an appropriate background for reaction engineering in industrial processes, the kinetics of the different reaction pathways have been analyzed in detail, and a mathematical model has been established. These factors take into account all of the different reactions occurring: enzyme complexes with substrate, a covalent glucosyl–enzyme intermediate, oligosaccharides formed, and the growing chain of dextran [with α-(1→6) glucosidic bonds mainly]. Kinetics and modeling have been based on a wide range of experimental investigations.[34,35]

With good acceptors, apparent V_{max} values increase with increasing acceptor concentration by a factor of about three at 600 mM maltose concentration.[36] They decrease, as do initial reaction rates, with increasing concentrations of such weak

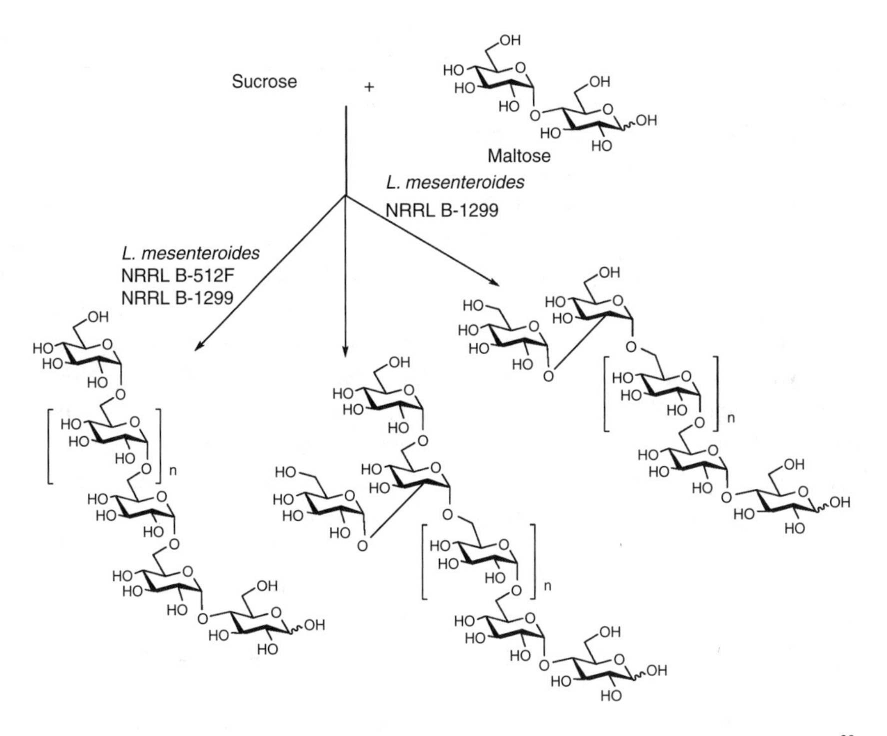

FIG. 2. Types of oligosaccharides formed by dextransucrase of *L. mesenteroides* NRRL B-1299.[33]

acceptors as fructose, again by a factor of three, at 2.75 M fructose concentration.[34] K_m values ranging from 12 to 27 mM were obtained in the absence of acceptors, whereas apparent K_m values increase with increasing acceptor concentration, up to 163 mM at 600 mM maltose concentration. At a fructose concentration of 1.39 M this value was 40 mM.[34] Dextran formation may be almost suppressed at high acceptor concentration, which is important for oligosaccharide production.

The final product concentration depends significantly on the ratio of substrate and acceptor. The first acceptor product (panose in the presence of maltose) is favored at high maltose in excess, further tetra- and pentasaccharides are formed in significant amounts.[37] Thus isomalto-oligosaccharides (IMOS) can be produced in good yield under appropriate reaction conditions (see Section III).

In the presence of such weak acceptors as fructose, initial overall reaction rates decrease significantly with increasing concentration of acceptor, whereas they increase with the substrate concentration. Initial rates of acceptor-product formation increase with both substrate and acceptor formation, the formation of dextran being suppressed to a major extend only at very high acceptor concentration (at 3 M fructose concentration).[34,38] The acceptor product of fructose, leucrose, does not act as an acceptor, so that high yields can be obtained, and this is of interest for industrial application. Leucrose production has been developed up to the pilot scale for application as an alternative sweetener.[39,40]

The quantitative description of all reactions occurring—in parallel or in sequence—requires a model that incorporates all of the relevant parameters. Such a model has been developed to predict the optimal reaction conditions and reactor configurations for different acceptors, most notably for the production of leucrose and IMOS. The kinetic parameters of the model identified are based on a wide range of experimental data.[38] The enzyme is assumed to catalyze glucosyl transfer via a covalent glucosyl–enzyme intermediate, and binding sites for the acceptor and the growing dextran chain, respectively.[38,41–44] For leucrose synthesis Böker and coworkers presented approximate kinetic correlations that permitted optimization of the yield.[34] With initial concentrations of 0.6 M sucrose and 2.2 M fructose, the yield of leucrose was in the range of 64–72% depending on the sucrose conversion (95–55%). Byproducts are isomaltulose (IM) and trehalose, in the range of 4–5% each. Immobilization of the enzyme has been achieved efficiently by its inclusion in alginate beads.[45] The productivity of the immobilized enzyme was found optimal in a continuous tubular reactor.[35]

Summarizing the results of many investigations, monosaccharides and such derivatives as D-mannitol and D-glucitol are rather weak acceptors. Disaccharides, including such acceptor products as isomaltose, are much better acceptors, except for certain molecules, for instance leucrose, which is not an acceptor.[29,46,47] The decrease of enzyme activity with time has been described in terms of a first-order reaction. The inactivation parameters have been calculated for the immobilized enzyme. The inactivation constants k_d were 0.0135 (1/d) when maltose was the acceptor (stabilizing), and 0.029 (1/d) when fructose was the acceptor.[38]

b. Structure and Enzyme Engineering.—Glucansucrases (non-Leloir glycosyltransferases) are structurally related to some glycosidases, their main reaction being the transfer reaction yielding large glucan polymers synthesized from sucrose, which provides the energy for this exergonic glycosylation.[48] Glucansucrases are expressed by lactic acid bacteria for the production of exopolysaccharides that form a kind of extracellular matrix for these bacteria. Glucansucrases are classified as glycoside hydrolase (GH) enzymes of Henrissat's family 70.[49] In terms of the EC-classification system, glucansucrases are designated as EC 2.4.1.5 (dextransucrase, but mutansucrase and reuteransucrase are still currently classified here) and EC 2.4.1.140 (alternansucrase). Over 40 glucansucrase primary structures are now known, all showing a high similarity in organization. In all glucansucrases, an N-terminal, variable region of different length can be found, followed by a conserved catalytic domain and a C-terminal, so-called ''glucan binding'' domain.[50] Unfortunately, no crystal structure of a lactic acid glucansucrase has yet been published. Structural comparison with GH-13 (alpha-amylase) enzymes predicts that the catalytic domain is organized as a $(\beta/\alpha)_8$-barrel, but circularly permutated.[51] This prediction has been confirmed by Dijkstra *et al.*, presenting a first, as yet unpublished crystal structure of GTF180, a glucansucrase of *L. mesenteroides* 180.[52] Interestingly, the enzyme showed not only a circular permutation, but also a huge U-shaped folding of the whole enzyme, forming distinct domains out of severed sequence-sections (Personal communication, see Fig. 8 by A. Vujicic *et al.*).[53]

One industrial process is the fermentative production of the polysaccharide dextran (α-glucan) by *L. mesenteroides*. Such polymers as dextran are used in food industry, as additives for dyes, and in health care products.[54,55] Variants and differences in glycosidic linkage-type, degree and type of branching, and molecular mass of glucans yield products having different structural and functional properties. For example, the glucans can be dextrans of varied structural identity [α-(1$\rightarrow$6)-linked glucose (Glc) backbone, with α-(1$\rightarrow$2), α-(1$\rightarrow$3) side chains],[42,56] mutan (α-(1$\rightarrow$3)-linked Glc residues),[44] alternan [α-(1$\rightarrow$6)- and α-(1$\rightarrow$3)-linked Glc residues],[43] amylose α-(1$\rightarrow$4)-linked, and reuteran [α-(1$\rightarrow$4)- and α-(1$\rightarrow$6)-linked Glc residues].[57]

WT enzymes among the glucansucrases (E.C. 2.4.1.5) have been demonstrated to control the regiospecificity and length of the polymer synthesis. However, how do they control oligosaccharide versus polysaccharide synthesis, and direct their glycosidic linkage specificity? This important question was very recently addressed for the design of enzymes that should enable production of tailor-made saccharides. In an enzyme-engineering approach, Hellmuth *et al.*[58] chose the most conserved motif around the transition-state stabilizer in glucansucrases for a random mutagenesis of the glucansucrase glycosyltransferase R (GTFR) of *Streptococcus oralis*. Modifications at position S628 achieved by saturation mutagenesis guided the reaction toward the synthesis of short-chain oligosaccharides with a drastically increased yield of the prebiotic isomaltose (47%) or alternatively leucrose (64%), a potential sweetener.[58] They also found that GTFR variant R624G:V630I:D717A showed a drastic switch in regioselectivity from a dextran type linked mainly α-(1$\rightarrow$6) to a mutan-type polymer with predominantly α-(1$\rightarrow$3)-glucosidic linkages. Targeted modifications demonstrated that both mutations near the transition-state stabilizer, R624G and V630I, are contributing to this alteration. It was shown that mutagenesis can guide the transglycosylation reaction of glucansucrase enzymes toward synthesis of (a) various short-chain oligosaccharides (see Fig. 3) or (b) novel polymers having completely altered linkages, without compromising its high transglycosylation activity and efficiency (Table I).[58]

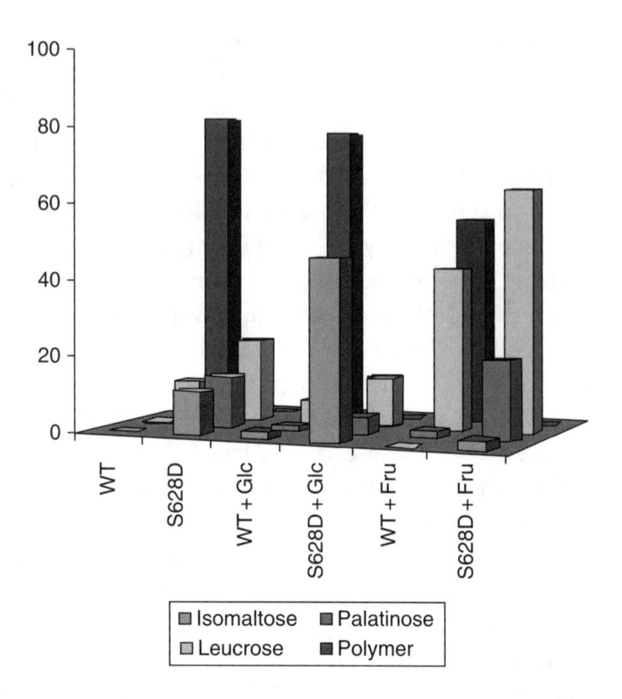

FIG. 3. Product spectra of the wild-type (WT) GTFR and mutant S628D enzymes (200 U/L) incubated (7 days at 30 °C) with sucrose (146 mM) and different acceptor substrates (292 mM, Glc: glucose; Fru: fructose). Yields are given in percentage (mol/mol Glc). Yields of higher oligosaccharides (DP > 5) and hydrolysis products are not shown.[58] (See Color Plate 1.)

Other important regions with relevance for acceptor specificity were identified in the dextransucrase DSRS. Kralj *et al.*[62] suggested that amino acids C-terminal to the invariant His626Asp627 motif of glucansucrases (GTFR numbering) influence the structure of the oligo- and poly-saccharides formed. Decrease or absence of polymer formation by rational mutations at this position is consistent with the findings of Kralj *et al.*[62,64] and Moulis *et al.*,[65] but in the latter studies, with other glucansucrases, polymer formation was slightly attenuated[62,64] or associated with a great loss of enzyme activity.[65] In addition to defined polymers, oligosaccharides having specific structures are also urgently needed. Oligosaccharides currently produced for commercial markets,[66] including IMOS,[67] leucrose,[68] and palatinose,[69] are of interest in the fields of food, pharmaceuticals, and cosmetics because of their ability to prevent and treat diseases of various biological origins. Isomaltose, for instance, enhances production of cytokine IL-12 by macrophages stimulated with *Lactobacillus gasseri in vitro*, and dietary IMOS significantly increased numbers of lactobacilli in the intestinal microflora.[70]

TABLE I

Polymer Linkage-Type and Alignment of Amino Acid Sequences of Various (Mutant) Glucansucrase Enzymes with GTFR Wild-Type and Mutant Variants[58]

Wild-type and mutant enzymes	Main α-linkage in glucan products	Amino acid sequence around transition-state stabilizer D627 in GTFR
DSRS	(1→6) PS	655 YSFVRAH**D**SEVQTVI
GTFI	(1→3) PS	557 YSFIRAH**D**SEVQDLI
GTFA	(1→4) PS	1126 YSFVRAH**D**NNSQDQI
GTFA′	(1→6) PS	1126 YSFVRAH**D**SEVQDQI
GTFR	(1→6) PS	620 YIFVRAH**D**SEVQTVI 713 SPYHDAIDA
S628D	OS	620 YIFVRAH**D**DEVQTVI
S628R	OS, PS	620 YIFVRAH**D**REVQTVI
R624G/V630I/D717A	(1→3, 1→6) PS	620 YIFVGAH**D**SEIQTVI 713 SPYHAAIDA
R624G		620 YIFVGAH**D**SEVQTVI
V630I	(1→6) PS	620 YIFVRAH**D**SEIQTVI
R624G/V630I	(1→3, 1→6) PS	620 YIFVGAH**D**SEIQTVI

DSRS, *Leuconostoc mesenteroides* NRRL B-1355[59]; GTFI, *Streptococcus downei* Mfe28[60]; GTFA, *Lactobacillus reuteri* 121[61]; GTFA′, mutant GTFA[62]; GTFR, *Streptococcus oralis*[63]; and GTFR variants,[63] mutants constructed in this study.
Mutations are shown in italics; the putative transition-state stabilizer is shown in bold.
PS, polysaccharide; OS, oligosaccharide.

c. Donor and Acceptor Substrate Engineering.—In the presence of suitable (mostly saccharide) acceptor substrates, glucan synthesis is directed toward oligosaccharide products. Thus the acceptor reactions of dextransucrase offer the potential for a targeted synthesis of a wide range of di-, tri-, and higher oligosaccharides by the transfer of a glucosyl group from sucrose to the acceptor, and a broad range of acceptor reactions with small acceptor molecules has been reported and characterized.[5,26–29,31,40] However, the synthetic potential of this enzyme is not restricted to ''normal'' saccharides: such modified acceptors, as additionally functionalized saccharides, alditols, aldosuloses, aldonic acids, alkyl glycosides, and glycals, along with rather unconventional saccharides, for instance a fructose dianhydride, may act as acceptors.

Some of these acceptors even turned out to be relatively efficient: α-D-glucopyranosyl-(1→5)-D-arabinonic acid, α-D-glucopyranosyl-(1→4)-D-glucitol, α-D-glucopyranosyl-(1→6)-D-glucitol, α-D-glucopyranosyl-(1→6)-D-mannitol, α-D-fructofuranosyl-β-D-fructofuranosyl-(1,2′:2,3′)-dianhydride, 1,5-anhydro-2-deoxy-D-*arabino*-hex-1-enitol (''D-glucal''). These may therefore be of interest for future applications of the dextransucrase acceptor reaction (Fig. 4).[71]

In a very recent approach Seibel *et al.* used a microarray approach on microtiter plates to identify new acceptor specificities of the non-Leloir GTFR from *S. oralis*,

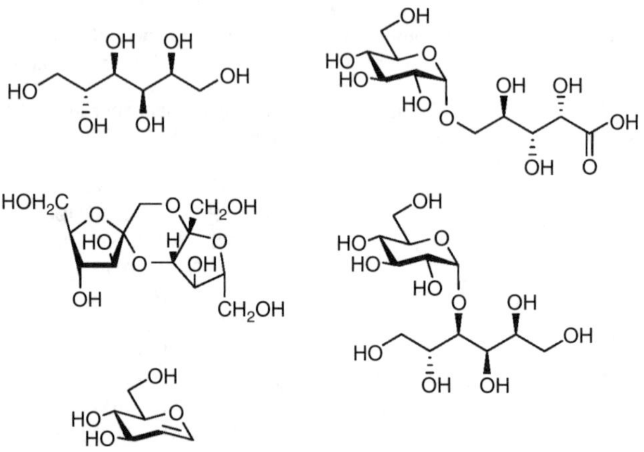

FIG. 4. Sugar derivatives serving as acceptors for glycosylation by dextransucrase: alditols, aldosuloses, aldonic acids, a fructose dianhydride, and a glycal.[71]

using sucrose as a glucose donor. Interestingly, it was found that the GTFR was able to glycosylate not only maltose, but also immobilized primary alcohols to yield the corresponding glucosides.[72] This format and technique enables fast and efficient screening of enzyme activities.

In a new strategy, acceptor substrates were designed that allow control of the linkage specificity by the enzyme and further chemical reactions to enhance glycodiversity. The blockage of the favored transglycosylation site by α-6-*O*-tosyl-group switched GTFR glycosylation activity from the major α-(1→6) to the minor α-(1→3) activity (Fig. 5).[73] This principle was expanded and applied in a two-step chemoenzymatic approach where the chemoselectivity of the glucansucrases employed can be guided from α-(1→6)- to α-(1→2)-, α-(1→3)-, or α-(1→4)-linked glucose by the acceptor, allowed the successful construction of various complex glycoconjugates containing thioglycosidic linkages with the glycopyranosides (galactose, glucose, neuraminic acid) of choice, giving access to a ''toolkit'' providing different linkages and offering another way for the synthesis of complex glycostructures (Figs. 5 and 6).[73]

Pijning *et al.* have reported a successful approach for crystallizing a glucansucrase GTFA from *Lactobacillus reuteri*[74]; an awaited structure may help in future in interpreting the data obtained.

Here we have demonstrated that nature of the oligosaccharide product and the glycosidic linkage specificity of glucansucrases can be modified by site-directed/random

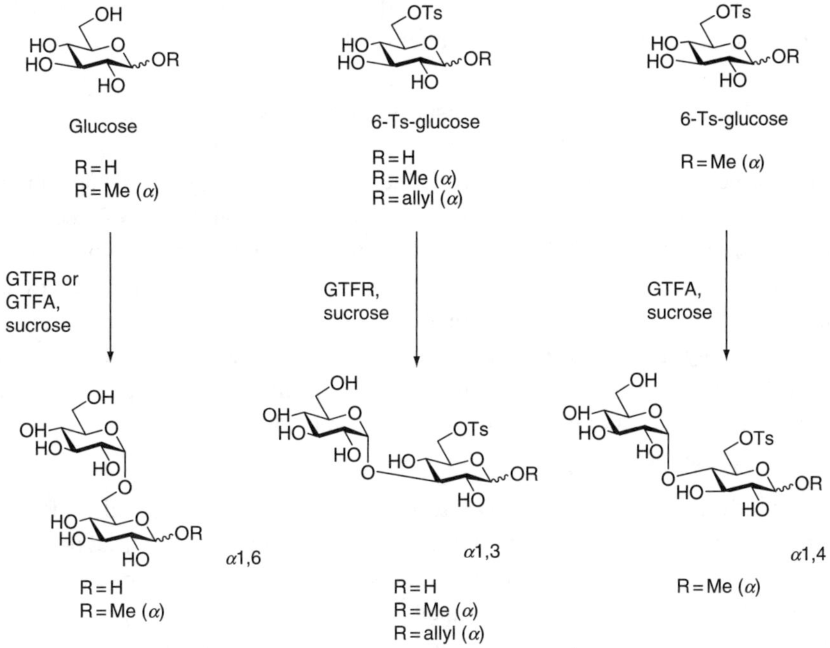

FIG. 5. Acceptor–substrate-directed synthesis by GTFR and GTFA enzymes.[73]

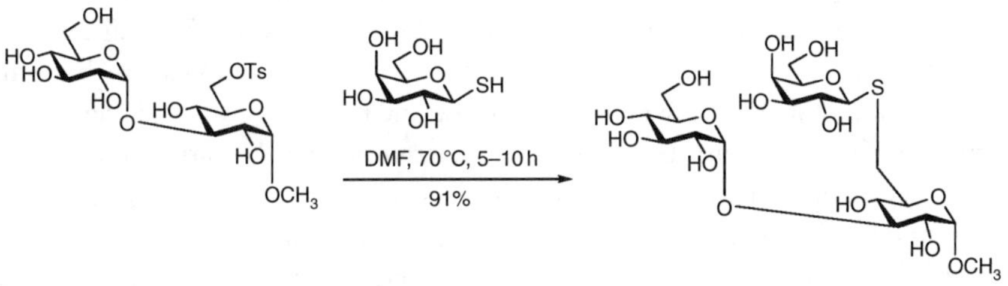

FIG. 6. Chemical synthesis of thio sugars.[73]

mutagenesis and substrate engineering, while maintaining a high transglycosylation efficiency, resulting in excellent product yields. Other glucansucrase and fructansucrase, and also glycosyltransfer enzymes, known for efficiently producing homopolysaccharides and oligosaccharides of different monosaccharide and linkage compositions,[50] might be subjected to similar enzyme- and substrate-engineering approaches. They provide very interesting enzyme-engineering targets directed toward the development of modified and new biocatalysts to produce tailor-made oligo- and poly-saccharides.

3. Fructansucrase Enzymes

Fructansucrase enzymes (FSs) of lactic acid bacteria employ the cheap substrate sucrose (Glc–Fru) to synthesize a variety of polyfructoside products.[75] These include two fructans,[76] levan with β-(2$\rightarrow$6)-linked fructose (Fru) residues[77,78] synthesized by levansucrases (LSs), and inulin [(β-(2$\rightarrow$1)-linked Fru residues)] elaborated by inulosucrases (ISs).[50,79] Besides this polymerization reaction, the enzymes can also perform hydrolysis, and, in the presence of suitable acceptors, oligosaccharides are formed.[78,79] LSs are present in Gram-positive as well as Gram-negative bacteria. Some LS and IS enzymes carry additional (glucan binding) domains, which are also present in glucansucrases.[81,82]

a. Kinetics.—The kinetics of FTF reactions have been investigated in detail by Chambert *et al.*[78,83] and by Ouarne and Guibert[84] with respect to industrial production of fructo-oligosaccharides. Both groups also discussed the reaction mechanism, notably the fructosyl transfer reaction that yields fructo-oligosaccharides. The second group also developed a mathematical model describing the experimental results with good accuracy.

b. Structure–Function Relationship.—LS and ISs belong to GH family 68 (GH68) and together with GH32 (such as invertase, inulinase) they comprise clan GH-J, according to the CAZy system,[85] (http://www.cazy.org). Two different LS X-ray structures have been elucidated, the first from the Gram-positive bacterium *Bacillus subtilis* (SacB, also with bound sucrose and raffinose) [86,87] and one from the Gram-negative bacterium *Gluconobacter diazotrophicus*.[88] Furthermore, GH32 three-dimensional structures of invertase from *Thermotoga maritima* (also with the trisaccharide raffinose in the active site), the fructan 1-exohydrolase from *Cichorium intybus*, the exoinulinase from *Aspergillus niger*, and an E203Q *Arabidopsis* invertase mutant[89–93] have been solved. Interestingly all protein structures contain a five-bladed β-propeller fold with a deep, negatively charged, central cavity, with the active site at the end of this cavity. The substrate-binding mode of sucrose can be seen in the GH68 *B. subtilis* SacB X-ray structure, where sucrose is situated in subsites -1 (fructosyl residue) and $+1$ (glucosyl residue).[86]

Amino acid-sequence alignments of various LSs and ISs, in combination with site-directed mutagenesis experiments and structural data of family members, have highlighted functional amino acid residues, and has provided first insights into structure–function relationships.[86,88,94,95] In the LSs and ISs from various bacteria the residues Asp[95], Glu[352], and Asp[257] (*Bacillus megaterium* numbering) constitute the catalytic triad.[95,96] Thus far the mechanism may be summarized as follows: the - acid–base catalyst Glu[352] protonates the glycosidic bond of sucrose, while the Asp[95] residue acts as a nucleophile,[95] attacking the glucopyranosyl residue of the sucrose, most probably forming an enzyme–fructosyl intermediate with inversion of the glycosidic bond in the intermediate state. The Asp[257] residue may assist in coordinating the hydrogen bonds of the fructofuranoside at positions 3-OH and 4-OH.

Apart from the activation of sucrose, the question of the fructosyl transfer for polymer versus oligosaccharide synthesis is still open. The ratio between polysaccharide and oligosaccharide synthesis activities differs significantly, depending on the enzyme source. While LSs of *B. subtilis*, *L. reuteri* 121,[97] and *B. megaterium*[95] synthesize high-molecular-mass levan without transient accumulation of oligofructan molecules,[6,78,98] the LS of *G. diazotrophicus*, *Zymomonas mobilis*, and *Lactobacillus sanfranciscensis*, and the IS of *L. reuteri* 121 synthesize short fructo-oligosaccharides (kestose and nystose) from sucrose.[97–101] In all FSs, subsite −1 is highly specific for binding fructose residues,[78,98,102] whereas subsite +1 is more flexible, exhibiting affinity for different glycopyranosides (mannose, galactose, fucose, and xylose) and disaccharides (maltose, lactose, and melibiose).[80,103] Novel insights regarding the products formed have been obtained. Key amino acid residues in the LSs important for oligosaccharide formation have been identified. For instance, changes in Arg[360] from *B. subtilis* SacB[104] at subsite −1, and the similar amino acid residue Arg[370] of *B. megaterium* SacB, resulted in time-dependent accumulation of different oligosaccharides during their catalysis, and accumulation of neokestose [β-Fru*f*-(2→6)-α-Glc*p*-(1→2)-β-Fru*f*] and blastose [β-Fru*f*-(2→6)-α-Glc*p*].[95] Also Arg[423] (subsites −1 and +1) in IS is part of the highly conserved ''RDP motif'' in clan GH-J.[86] Changes in *L. reuteri* 121 IS residues at acceptor subsite −1, R423K (also interacting with a glucosyl residue at subsite +1; Arg[246] residue in *B. subtilis* SacB), or W271N (Trp[85] residue in *B. subtilis* SacB), caused altered fructo-oligosaccharide product patterns with sucrose, yielding much less oligosaccharides and significantly more polysaccharides.[94] This observation suggests that modification of amino acid residues, or the glycosidic binding mode, at subsite −1 is important in determining the product profile, namely polysaccharide versus oligosaccharide synthesis. It was shown that the corresponding arginine residue (R188) in exoinulinase of family GH32 participates in substrate binding, is important for recognition of the sugar ring, and might be responsible for specificity of the enzyme toward the fructofuranosyl residue.[91]

Interestingly, W271N synthesized larger fructan products as compared to the WT enzyme.[97] *B. megaterium* LS residue Asn[252] (subsite +2) clearly plays an important role in transfructosylation.[95] Substitution of Asn^{252} by Ala or Gly completely abolished polysaccharide production without significantly affecting K_m and k_{cat} values, but causing these variants to switch from mainly polysaccharide synthesis to hydrolysis. In contrast, mutation N252D diminished polysaccharide synthesis without changing K_m and k_{cat} values significantly.

The X-ray structure of the *B. subtilis* SacB (74% amino acid identity with *B. megaterium* LS) provides insights into the function of this Asn^{252} residue. Its position at subsite +2 may allow it to stabilize the third fructosyl residue of the growing oligosaccharide chain and direct it as an acceptor substrate into the optimal position for further transfructosylation.[95] The mutation N252D in *B. megaterium* LS clearly weakened the coordination of a fructosyl unit at subsite +2; however, it still occurred, as indicated by the residual formation of polysaccharide. Residues Arg^{370} and Asn^{252}, which appear to be crucial for fructan synthesis, are conserved in LSs from Gram-positive bacteria. In contrast, the endophytic Gram-negative bacterium *G. diazotrophicus* SRT4 secretes a constitutively expressed LS (LsdA, EC 2.4.1.10), which mainly converts sucrose into fructo-oligosaccharides. It contains His^{419} instead of Arg^{370} at the equivalent position.[88] This His residue is strictly conserved in Gram-negative LSs. Furthermore, Asn^{252} is strictly conserved in the LSs of Gram-positive bacteria, whereas Gram-negative bacteria possess a conserved Arg residue in the equivalent position.[95] This section of the 3D structure of LsdA is almost perfectly superposable upon the equivalent residues of *B. subtilis* SacB in the active site. In contrast, the region constituting subsite +2 shows clear differences.

4. Sucrose Analogues

Sucrose analogues are nonnatural activated substrates, intended to serve in the synthesis of new oligo- and poly-saccharides. They provide a convenient and rather low-cost access to synthesis and glycosidation.[105]The enzymatic synthesis of these sucrose analogues has been established for a variety of structures. The exofructosyltransferase (EC 2.4.1.162) from *B. subtilis* NCIMB 11871 transfers the fructosyl residue of the substrate sucrose to a series of monosaccharide D-glucopyranosyl acceptors (D-mannose, D-galactose, 2-deoxy-D-*arabino*-hexose, D-fucose, and D-xylose) to yield the β-D-fructofuranosyl α-D-glycopyranosides (D-Man-Fru, D-Gal-Fru, 2-deoxy-D-*arabino*-Hex-Fru, D-Fuc-Fru, D-Xyl-Fru).[80,106] A range of these has been shown to be formed in good, or high yields, with respect to the quasi-equilibrium, under kinetic control and optimum conditions.

Using L-glycopyranosides as acceptors led to the formation of β-D-fructofuranosyl-β-L-glycopyranosides (L-Glc-Fru, L-Gal-Fru, L-Fuc-Fru, L-Xyl-Fru, Rha-Fru), sucrose analogues having a β-(1 $\rightarrow$ 2)-glycosidic linkage.[80] In the chemoenzymatic approach, a range of new sucrose derivatives can be obtained from "3-ketosucrose" in aqueous medium with few reaction steps. *Agrobacterium tumefaciens* elaborates a dehydrogenase that oxidizes sucrose specifically at the 3-position of Glc, yielding α-D-*ribo*-hex-3-ulopyranosyl-β-D-fructofuranoside ("3-ketosucrose") (Fig. 7).[107] Subsequent chemical steps provide access to a range of products, including α-D-allopyranosyl β-D-fructofuranoside (allosucrose).[108,109]

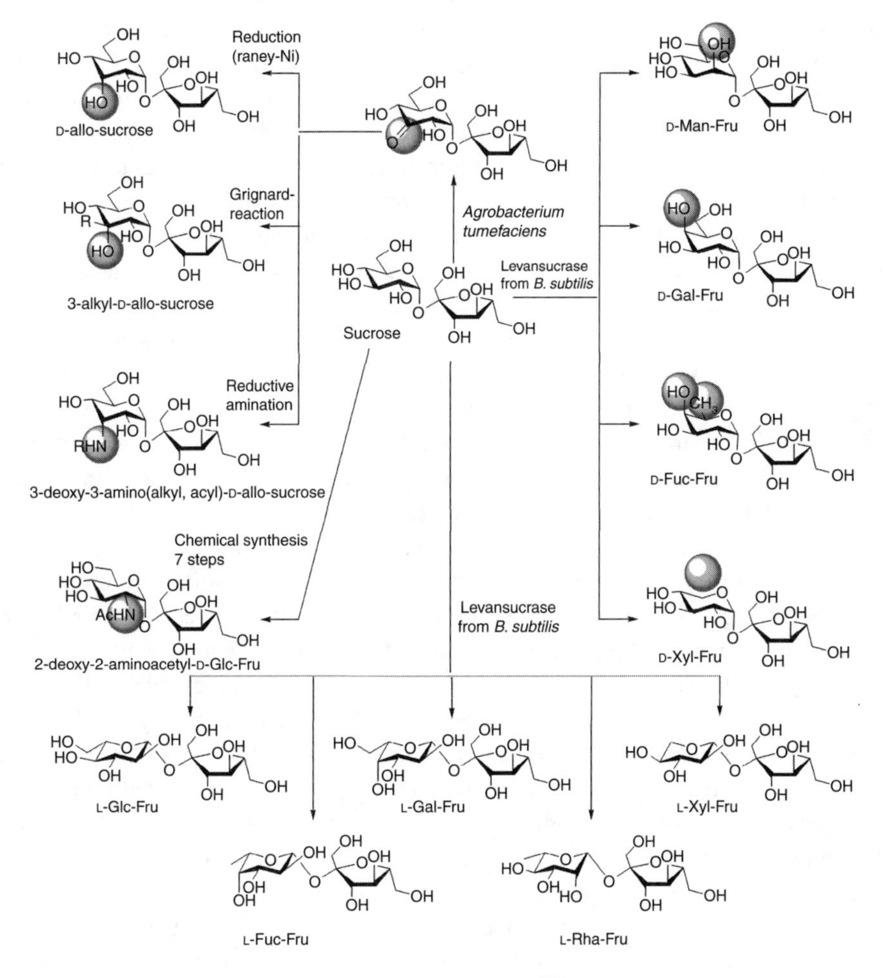

FIG. 7. Different routes to sucrose analogues.[105] (See Color Plate 2.)

5. Oligo- and Poly-saccharide Synthesis with Sucrose Analogues

Previous sections of this chapter have demonstrated that acceptor substrates can direct the regioselectivity of a glucansucrase for acceptor reactions without mutating the enzyme. This approach has been expanded to sucrose analogues, which should function as alternative donor substrates. Sucrose is a highly activated substrate for GTFs ($\Delta G_A^0 = -26.5\,\text{kJ mol}^{-1}$)[110,111] as compared to other disaccharides, such as isomaltose, lactose, and maltose ($\Delta G_A^0 = -7,\ -8.8,\ -15.5\,\text{kJ mol}^{-1}$).[112] This difference in the Gibbs energy change is available to enable synthesis of oligo- and poly-saccharides by sucrose-type enzymes that use sucrose as a substrate for the transfer of glucose or fructose.

Sucrose analogues (β-D-fructofuranosyl α-D-glycopyranosides) open a gate for the transfer of a wide repertoire of monosaccharides, and may be used for oligosaccharide and polysaccharide synthesis (a) using fructansucrases for the synthesis of new glycopyranosyloligofructosides or (b) alternatively by using modified glucansucrases for the transfer of the glucopyranoside.[24,113] Sucrose analogues have been shown to serve as efficiently for synthesis (transfer of monosaccharides) as sucrose. When the glucose part of sucrose is modified, as, for example, in galactose-, mannose-, or xylose-fructosides, the transfer of such monosaccharides (galactose, xylose, and others) is under investigation, and is a topic in promising recent studies in our laboratory. However, the transfer of either fructose or the glycopyranoside to other sugars or different natural products as acceptors is a challenging prospect. Advantages of this system, apart from the currently employed enzymes such as glycosynthases and GTFs[17,23] are that industrial established glucansucrases and fructansucrases may also serve for a extended range of substrates and products.

Not surprisingly, fructosyltransferases can use most of the sucrose analogues (with a remarkable exception, allosucrose) producing new oligosaccharides or polysaccharides. Such oligofructans as kestose and nystose are well regarded for their health benefits in human nutrition. The substitution of a glucose residue by another glycopyranosyl residue such as xylose can be expected to affect the structural and biochemical properties of the new molecule. In contrast to short-chain fructo-oligosaccharides, the glucopyranosyl residue in polysaccharides may not influence its structure and properties significantly. Thus in a recent approach, a LS was generated by random mutagenesis that does not produce polysaccharides but instead yields short-chain fructo-oligosaccharides.[6]

Sequence analysis of the variant A5 revealed the presence of two mutations in the gene, V106A and N242H. These novel insights were used for the functional characterization of a novel LS from *B. megaterium* (74% identity with SacB from

B. subtilis).[95] There it was found that V115A (Val106 *B. subtilis* numbering) mutation has no effect on LSs.[95] It was also demonstrated that sucrose analogues indeed provide a powerful new tool for the highly efficient and inexpensive preparative synthesis of tailor-made saccharides. In early studies, the novel substrates have been converted by LS enzymes into unique oligo- and polyfructans[24,113]; and their kinetics[6] have been investigated (Table II). These results, combined with Xyl-Fru as substrate and LS variant A5, resulted in the production of one main product, a 6-kestose analogue [α-Xyl-(1→2)-β-Fru-(2→6)-β-Fru].[6] As 6-kestose and XOFs have strong prebiotic activity, the impact of this novel product on the prebiotic properties should be tested in the mammalian gut system.

Moreover, in further studies, a genetically optimized strain of *A. niger* with an overexpressed β-fructofuranosidase transformed sucrose analogues efficiently and with high yield into the 1-kestose and 1-nystose analogues functionalized with different monosaccharides of potential interest (Fig. 8).[115]

6. Sucrose Isomerase

Sucrose isomerase (SI, also termed sucrose mutase, or IM synthase, EC 5.4.99.11) catalyzes the isomerization of sucrose to isomaltulose [IM, α-D-glucopyranosyl-(1→6)-D-fructose], a rearrangement of the α-(1→2)-bond to an α-(1→6)-bond, and, to a minor extent, to trehalulose [α-D-glucopyranosyl-(1→1)-D-fructose].

TABLE II

Kinetic Parameters for Sucrose Analogues as Substrates for Levansucrases (Wild-Type and Variant A5)

Substrate	K_m (mM)[a,b]	k_{cat} (s^{-1})[a,b]	k_{cat}/K_m ($mM^{-1}s^{-1}$)	$\Delta\Delta G$ (kcal mol^{-1})
Sucrose (wild-type)	14[c]	85.5 (3.2)	6.1	0
Man-Fru (wild-type)	30.8 (7.9)	0.4 (0.1)	0.01	3.8
All-Fru (wild-type)	–	–	–	–
Gal-Fru (wild-type)	20.3 (4.1)	36.7 (2.8)	1.8	0.7
D-Fuc-Fru (wild-type)	40.5 (5.7)	70.6 (4.9)	1.7	0.8
Xyl-Fru (wild-type)	28.5 (4.3)	72.6 (3.1)	2.6	0.5
Sucrose (variant A5)	222 (44.5)	40.4 (4.2)	0.2	–

[a] Parameters determined as described in the experimental section. Reaction times were 60 min, and enzyme concentrations used for each substrate were as follows: Gal-Fru 0.2–5 U mL^{-1}, Man-Fru 7.4–148 U mL^{-1}, Xyl-Fru 0.1–1.0 U mL^{-1}, and D-Fuc-Fru 0.3–0.6 U mL^{-1}. Enzyme concentrations used for each substrate were added as estimated to convert about 10% of the different initial sugar concentrations within 60 min.
[b] Values in parentheses represent the error limits in the reported number.[6]
[c] Data correspond to literature data.[114]

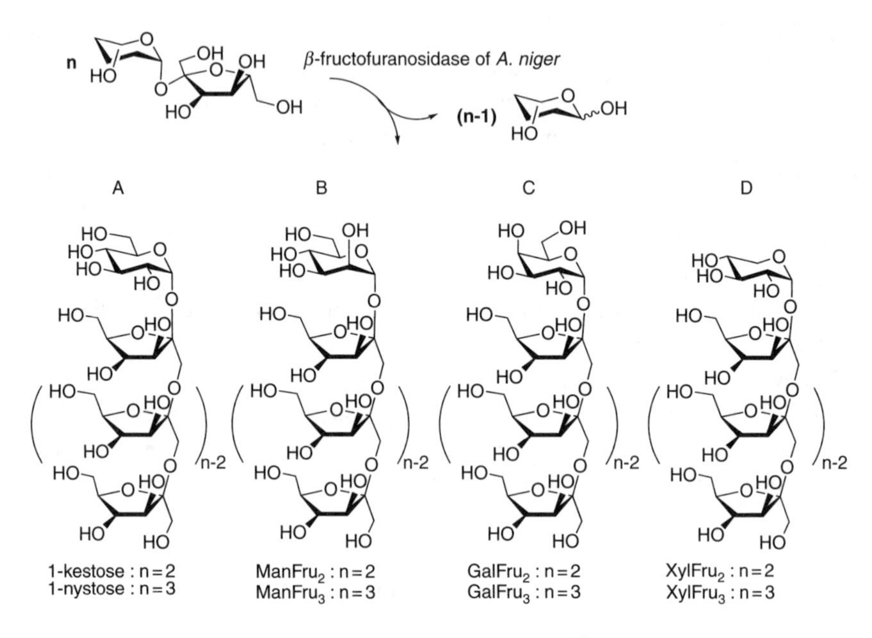

FIG. 8. Different products with sucrose analogues as substrates.[115] Enzymatic synthesis of 1-kestose, 1-nystose, and their analogues by β-fructofuranosidase of *A. niger*. Structures of fructo-oligosaccharides: (A) commercial products, (B) mannose- (C) galactose-, and (D) xylose-substituted analogues.

The ratio of the enzyme products varies from mainly IM (up to 85%) (used in industrial processes) to predominantly trehalulose (up to 90%), depending on the bacterial strain that synthesizes the enzyme. SI is produced by several microorganisms, *Klebsiella planticola*, *Pantoea dispersa*, and *Protaminobacter rubrum*.[116–120] The enzyme kinetics from the three species mentioned have been investigated in detail. The three different SI produce IM with high yield, around 85%, with trehalulose as a by-product, and low hydrolytic activity. The yield of IM increased with temperature, up to 50 °C, for the *Klebsiella* enzyme. A remarkably high activity, 638 U/mg, was found for the *P. dispersa* enzyme, while those for different *Klebsiella* enzymes were in the range from 330 to 420 U/mg; K_m values were in the range from 40 to 120 mM. Competitive inhibition by glucose and fructose, most pronounced by a mixture of both (at a concentration of 277 mM for each), was observed for the enzymes from *Pantoea* and *Klebsiella*. The SI from *Pseudomonas mesoacidophila* showed a high synthetic potential for trehalulose, with 90% yield.[117,119,121]

Formation of IM attains a maximum yield, with subsequent formation of trehalulose as a by-product, at extended reaction (residence) time, thus kinetic control is essential for good yields. For industrial production of SI, and IM, cells of *P. rubrum* have been entrapped in alginate beads and cross-linked with hexanedial (SI remaining active, whereas the cells are inactivated). Immobilization by entrapment in calcium alginate proved by far the most efficient method, as compared to entrapment in other gels, or adsorption on carriers; these latter resulted in rather poor stability. The activity after immobilization was about 50% of free cells. However, the half life was increased by more than two orders of magnitude. Thus the operational stability of the alginate-immobilized biocatalyst exhibited, when operating at high sugar concentration (1.6 M), a half-life time of over 8000 h.[122] The thermostability has further been improved by engineering the enzyme from *Klebsiella* through replacing selected amino acid residues by proline, resulting in an 11-fold increase in half life at 50 °C.[121] IM, in contrast to sucrose, is a reducing sugar, since the carbonyl group of fructose is not blocked by the glycosidic bond. Thus it can be hydrogenated by classical catalysis to yield two alditols, α-D-glucosyl-D-mannitol and α-D-glucosyl-D-glucitol. These are the main components of a commercial product, widely used as alternative sweetener, mainly because they are noncariogenic (see Section III).[116]

To understand the mechanism of sucrose isomerization at the molecular level and to identify the key amino acids involved in the enzyme reaction, enzymes of *Klebsiella* sp. and *P. rubrum* have been cloned and crystallized. The three-dimensional structure of the *Klebsiella* enzyme has been revealed by Zhang *et al.*[118] It consist of three domains, an N-terminal catalytic $(\beta/\alpha)_8$ domain, a subdomain between Nβ3 and Nα3, and a C-terminal domain having seven β-strands. The active-site architecture is similar to that of the 13-member GH family. However, a unique RLDRD motif in the proximity of the active site has been identified and shown biochemically to be responsible for sucrose isomerization. The catalytic triad (Asp^{241}, Glu^{295}, and Asp^{369}) and two histidine residues (His^{145} and His^{368}) in SI are highly conserved in alpha-amylase and GTFs, for instance in amylosucrase.[123]

The active-site cleft is surrounded by a loop, forming a pocket with the dimensions $2 \times 2 \times 2.5$ nm^3, large enough to accommodate a sucrose molecule. A mechanism proposed, by analogy to the other enzymes mentioned, suggests that the glycosidic bond is protonated by a proton donor, and the anomeric carbon of the glucose moiety is attacked by the nucleophile leading to the formation of the covalently linked enzyme–glucosyl intermediate. In the isomerization step, the fructose substitutes the glucosyl group of the intermediate to form the sucrose isomers, either IM or trehalulose, depending on the mode of binding of fructose in the active center.[118] Further insight into the mechanism of sucrose isomerization has been provided by high-resolution three-dimensional structures of native and mutant complexes of a trehalulose synthase from *P. mesoacidophila* that mimics successive states of the enzyme reaction.[124]

In addition to the $(\beta/\alpha)_8$ domain and amino acids highly conserved as mentioned before (the catalytic triad, histidine residues) an aromatic clamp of two phenylalanine residues has been identified that play an essential role in substrate recognition and in controlling the reaction specificity. These aromatic residues are presumed to be involved in hydrophobic interaction with the fructose moiety of the substrate, and thus in substrate binding and positioning, in addition to three hydrogen bonds. These interactions are assumed to prevent release of the fructosyl unit, and are thus responsible for an intramolecular glucosyl transfer and rearrangement of the glycosidic bond. It also prevents access of water to a large extent, being thus responsible for a high ratio of glucosyl transfer versus hydrolysis—an essential condition for industrial application. Binding of the glucose moiety is stronger, mediated by 14 direct hydrogen bonds and by stacking to tyrosine and phenylalanine residues (for the technical process see Section III).[124]

III. Commercial Products

The preceding sections should illustrate how far our insight and understanding of GTF has progressed, in structure, mechanism, and kinetics, and how the design of GTF has been achieved to optimize properties. This last section here documents how many and diverse processes have been established industrially, and to what impressive volume and commercial value products are being manufactured using GTF. Commercial products comprise IM, and derivatives, IMOS, CDs, FOSs, galactosyl-oligosaccharides (GalOS), and other special products. They are produced based on sucrose or starch, which are available in high quality and at low cost as substrates, along with acceptor saccharides, for instance galactose, when part of the product. They are generally produced using immobilized GTF (glycosyltransferases of the sucrase type). Table III gives some examples, with current figures (2004–2009) on market estimates. For overviews on enzymes and oligosaccharides used in food manufacture see Whitaker *et al.* and Eggleston and Côté.[125,126]

GTFs are applied as immobilized biocatalysts in a range of processes for manufacturing such oligo- and poly-saccharides as IM, malto-, and isomalto-oligosacchrides, galacto- and fructo-oligosaccharides, and CDs, along with dextran, mostly in the range of 3000–15,000 tonnes/annum.[127–129] The price range for oligosaccharides in the food sector is mostly in the range of 2–5 euros/kg; and thus these products (except CDs) constitute rather low-price commodities requiring inexpensive raw materials (sucrose, starch, inulin), and economic processing with low cost for the biocatalyst, which must be immobilized and exhibit high productivity.[125,128]

TABLE III
Application of Immobilized Glycosyltransferases[5,66,84,127,130,131]

Enzyme	Product	Market (estimated, tons/annum)	Company
Sucrose isomerase (Sucrose mutase)	Isomaltulose	100,000	Südzucker, Cerestar (Germany) Mitsui Seito Co (Japan)
α-Glucosidase and Glucosyltransferase	Isomalto-oligosaccharides	15,000	Hayashibara (Japan) and others
α-Glucosidases	Malto-oligosaccharides	15,000	
Glucosyltransferase and α-Glucosidase	Trehalose	30,000	Hayashibara (Japan)
Dextransucrase	Isomalto-oligosaccharides		BioEurope (France)
Dextransucrase[a]	Dextran	200	Pharmacia (Sweden)
β-Galactosidase	Galacto-oligosaccharides	7000–15,000	Yacult (Japan), Milupa/Numico (Germany)
Fructosyltransferase	Neosugar Fructo-oligosaccharides	4000 3500	Beghin-Meji Industries (France, Japan)
Cyclodextrin transferase	β-Cyclodextrin and other cyclodextrins	3000–4000	Wacker-Chemie (Germany)
Total, approx.		180,000	

[a] Single example where the product is not made by an immobilized enzyme; Pharmacia is now a Pfizer company.

Oligosaccharides manifest a broad range of different properties, including sweetness, bitterness, hygroscopicity, specific water activity, stabilization of active substances, including proteins, and flavor and color components. Syrups of branched oligosaccharides (such as isomaltose and panose) are mildly sweet and exhibit relatively low viscosity and lower the physicochemical activity of water in solution, as well as high moisture-retaining properties. In addition, such factors as digestibility, or nondigestibility, noncariogenicity, or anticariogenicity, and immunopotentiating activity may constitute significant roles for practical applications.[129] The worldwide production of oligosaccharides may be estimated at some 175,000 tonnes/annum, IM and its derivatives constituting the largest single type of oligosaccharide. In Japan some 90,000 tonnes of oligosaccharides were produced in the late 1990s, worth about 650 million euros. The production of nondigestible oligosaccharides worldwide was estimated at 85,000 tonnes in the late 1990s, more than half of this in Japan. In Europe, sales of oligosaccharides may be estimated in the range of about 300 million euros (Table III).

Manufacture of IM [palatinose, α-D-glucosyl-(1 → 6)-D-fructose] is a major process that uses a glucosyltransfer activity of a sucrose mutase (sucrose isomerase) from

P. rubrum. Sucrose is the substrate, and the yield from the isomerization is 80% to 85% of IM. The catalyst consists of immobilized cells of *P. rubrum* which are cross-linked and entrapped in alginate beads (the sucrose mutase remaining active). Most remarkable is the greatly increased stability of the biocatalyst at high sugar concentrations, being further improved under operational conditions in sucrose solution (1.6 M; half-life time of over 8000 h).[122] The yield of IM depends critically on kinetic control of the reaction, with increasing formation of trehalulose as a by-product at extended reaction (residence) times. Figure 9 presents a scheme of the technical process.[5,116,132]

IM is mostly produced by Südzucker AG, Germany, and by Cerestar, Germany, and Mitsui Seito Co, Japan, with a total of about 100,000 tonnes/annum. Most of it is hydrogenated by classical Raney nickel catalysis to give Isomalt®, a mixture of two isomeric alditols, 6-*O*-α-D-glucosyl-D-glucitol and -D-mannitol. The latter is used in the food sector, suited for diabetics and especially as an alternative sweetener with noncariogenic properties. It is largely employed in caramels, chewing gum, tablets, and so on.[116]

IMOS are produced by companies in Japan and Europe for use in food, animal feed, and in dermocosmetics.[42] Raw materials are starch or sucrose along with an acceptor, such as glucose or maltose, to which a glucosyl group from sucrose is transferred. Such nondigestible IMOS are manufactured by using the dextransucrase from the soil bacterium *L. mesenteroides* NRRL B-1299, which is known to catalyze the synthesis of dextran polymers containing α-(1→2)-linked chain branches. When this specific glucosyltransferase is used in the presence of maltose as acceptor and of sucrose as the

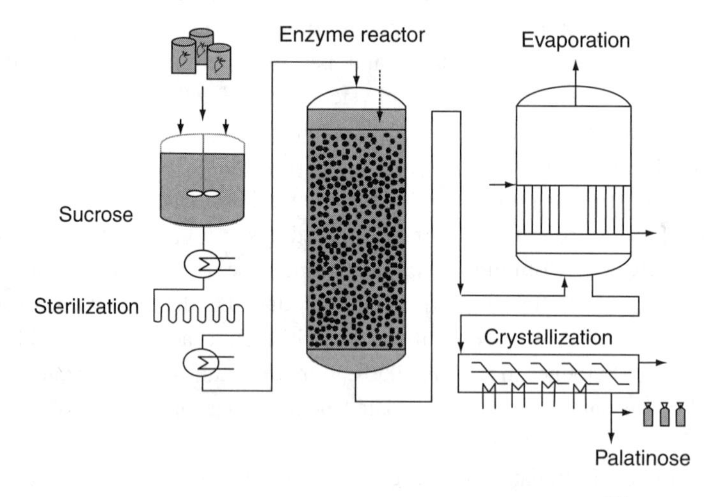

FIG. 9. Scheme of the technical production of isomaltulose.[116]

D-glucosyl donor, α-gluco-oligosaccharides are obtained that contain α-(1→2) glucosidic bonds at their nonreducing end and a maltose residue at the reducing end.[42] The presence of these α-(1→2) linkages endows a very high resistance of these oligosaccharides to attack by the digestive enzymes of humans and animals.[133] Such α-gluco-oligosaccharides cannot be metabolized, as demonstrated in studies with germ-free rats. That is the reason why such α-gluco-oligosaccharides were initially developed as low-calorie bulking agents, to be used in food formulations to attenuate the effect of intense artificial sweeteners.[134] These α-gluco-oligosaccharides are, however, metabolized by the intestinal bacterial flora. In contrast to fructo-oligosaccharides and galacto-oligosaccharides, these α-gluco-oligosaccharides are not bifidogenic, but they do promote growth of the cellulolytic intestinal flora. In addition, they induce a broader range of glycolytic enzymes than fructo-oligosaccharides and galacto-oligosaccharides, without any significant side-production of gasses and thus of any detrimental effect.[135]

These oligosaccharides are presently marketed for human nutritional application as food complements, in combination with specific microbial flora and vitamins. The prebiotic effect of such α-gluco-oligosaccharides has also been demonstrated in relation to skin microbial flora, wherein lactic bacteria also play a key protective role. These studies have resulted in the development of dermocosmetic applications for the α-gluco-oligosaccharides, under the trade name BioEcolia.[136] Fructo-oligosaccharides have gained major attention as functional food ingredients and medical foods during recent years.[137,138] They are manufactured by partial hydrolysis of inulin. An alternative method for production of fructo-oligosaccharides is by fructosyltransferases acting on sucrose, with the transfer a fructosyl group from sucrose to another sucrose molecule linking the fructosyl groups by β-(2→1) bonds to the fructosyl moiety of sucrose.[138–140] Fructosyltransferases, mainly from lactic acid bacteria, may be of both the IS type, forming β-(2→1) bonds, and of the LS type, forming β-(2→6) bonds.[77,79] The product mixture obtained with ISs is composed mainly of glucose, sucrose and tri- to pentasaccharides having 1 glucose and 2–4 fructosyl residues linked by β-(2→1) bonds (kestose, nystose, and others) exhibiting prebiotic properties. Physiological attributes claimed for fructo-oligosaccharides are prevention of dental caries, provision of dietary fiber, stimulation of bifidobacteria, prevention of diarrhea and constipation, and reduction of serum cholesterol. They also provide texture, mouth feel, and taste improvement by replacing fat.[137–139]

Galacto-oligosaccharides are produced by the action of β-galactosidase (a lactose-hydrolyzing enzyme) on lactose under appropriate conditions, at high substrate concentrations. They comprise a mixture of β-glycosidically bound galactose moieties and with a glucose residue from the substrate (lactose). They are commercially available and used in human nutrition, as in milk products. One application is in a mixture of 90% short-chain galacto-oligosaccharides and 10% long-chain (DP greater than 10) fructo-oligosaccharides for infant nutrition. This mixture promotes the growth of beneficial intestinal bacteria, in particular *Bifidobacteria* and *Lactobacilli*, in a synergistic way. In Japan, over 90% of infant formulas have been supplemented with nondigestible oligosaccharides. Data of experimental research and clinical studies have been reported in 2005 by Boehm *et al.*[142]

Japan constitutes the most dynamic market for oligosaccharide development and application, having available a wide diversity of research approaches, technical development, and commercial products. Among these are specific starch-derived oligosaccharides, such as β-(1$\rightarrow$6)-linked gentio-oligosaccharide, α,α-(1$\rightarrow$1)-linked trehalose, α-(1$\rightarrow$3)-linked nigero-oligosaccharide, α-(1$\rightarrow$2)-linked koji-oligosaccharide, branched CDs having glucosyl and maltosyl branches, and macrocyclic dextrins. Additional oligosaccharides are produced from sucrose or mixed substrates, yielding IM, fructo-oligosaccharides, and such glycosylsucroses, as lactosucrose, xylosucrose, or trehalose.[129] Some more details are also summarized here.

A product manufactured in Japan as ''Alo mixture'' (Anomalously Linked Oligosaccharides) contains mainly a range of IMOS (glucose, isomaltose, and panose as principal constituents). It is produced from starch as substrate by the action of alpha-amylase, beta-amylase, and a transglucosidase. It is claimed to have favorable properties for application in the food industry.[143]

Trehalose [α-D-glucosyl-(1$\rightarrow$1)-α-D-glucose] is manufactured from starch, mainly by the Japanese company Hayashibara.[144] It has found wide application as a result of its organoleptic and physicochemical properties, among them sweetness, but a key characteristic is that it increases the shelf life of many food products, and exhibits protecting properties in frozen food products. The enzymes used are malto-oligosyl trehalose synthase, a glucosyltransferase, and malto-oligosyl trehalose trehalohydrolase, an alpha-amylase from *Arthrobacter* sp.

Gentio-oligosaccharides exhibit a bitter taste and are used in beverages. Syrups containing nigero-oligosaccharides have recently been used for various kinds of foods and beverages as taste-improving and color-stabilizing additives.[129,144] Large-ring, macrocyclic dextrins that display specific complexation properties are produced by Ezaki Glico Co.[145] ''Lactosucrose'' is produced by fructosyltransferase that transfers the galactosyl group of lactose to sucrose, and is marketed as a prebiotic in Japan. It is further claimed to enhance indirectly the immune function in the gut, where increased amounts of IgA have been found.[146]

CDs are torus-shaped cyclic molecules of α-(1$\rightarrow$4)-linked glucose, with 6, 7, or 8 glucose residues in the ring (α-,β-, or γ-cyclodextrins, respectively). They are products of major importance in the fields of food (aroma complexation and slow release, stabilization of flavors), pharmaceuticals (drug protection, slow release), and in commodities, for example, uses in textile drying as a perfume carrier. The enzymes used for their manufacture from starch or dextrins are CD GTFs, as from *Bacillus macerans*, *B. stereothermophilus*, and *B. coagulans*. Several have been crystallized, and their structures elucidated and characterized in detail, including the reaction mechanism.[141]

The Wacker company (Germany) has optimized, via genetic tools, the production of the enzyme, including improved selectivity to increase the yield of specific CDs, α, or β, or γ, respectively, that correspond best to the particular application.[131]

Acknowledgment

This project was supported by the German Research Foundation via the grant SFB 578 ''From Gene to Product.''

References

1. A. Varki, Biological roles of oligosaccharides: All of the theories are correct, *Glycobiology*, 3 (1993) 97–130.
2. C. H. Wong, Protein glycosylation: New challenges and opportunities, *J. Org. Chem.*, 70 (2005) 4219–4225.
3. J. G. Joyce, I. J. Krauss, H. C. Song, D. W. Opalka, K. M. Grimm, D. D. Nahas, M. T. Esser, R. Hrin, M. Feng, V. Y. Dudkin, M. Chastain, J. W. Shiver, *et al.*, An oligosaccharide-based HIV-1 2G12 mimotope vaccine induces carbohydrate-specific antibodies that fail to neutralize HIV-1 virions, *Proc. Natl. Acad. Sci. USA*, 105 (2008) 15684–15689.
4. F. Kamena, M. Tamborrini, X. Liu, Y. U. Kwon, F. Thompson, G. Pluschke, and P. H. Seeberger, Synthetic GPI array to study antitoxic malaria response, *Nat. Chem. Biol.*, 4 (2008) 238–240.
5. K. Buchholz and J. Seibel, Industrial carbohydrate biotransformations, *Carbohydr. Res.*, 343 (2008) 1966–1979.
6. R. Beine, R. Moraru, M. Nimtz, S. Na'amnieh, A. Pawlowski, K. Buchholz, and J. Seibel, Synthesis of novel fructooligosaccharides by substrate and enzyme engineering, *J. Biotechnol.*, 138 (2008) 33–41.

7. E. J. Hehre, Glycosyl transfer: A history of the concept's development and view of its major contributions to biochemistry, *Carbohydr. Res.*, 331 (2001) 347–368.
8. S. Kitahata, C. F. Brewer, D. S. Genghof, T. Sawai, and E. J. Hehre, Scope and mechanism of carbohydrase action. Stereocomplementary hydrolytic and glucosyl-transferring actions of glucoamylase and glucodextranase with alpha- and beta-D-glucosyl fluoride, *J. Biol. Chem.*, 256 (1981) 6017–6026.
9. C. J. Hamilton, Enzymes in preparative mono- and oligo-saccharide synthesis, *Nat. Prod. Rep.*, 21 (2004) 365–385.
10. S. Singh, M. Scigelova, and D. H. G. Crout, Glycosidase-catalysed synthesis of mannobioses by the reverse hydrolysis activity of alpha-mannosidase: Partial purification of alpha-mannosidases from almond meal, limpets and *Aspergillus niger*, *Tetrahedron Asymmetry*, 11 (2000) 223–229.
11. A. Basso, A. Ducret, L. Gardossi, and R. Lortie, Synthesis of octyl glucopyranoside by almond β-glucosidase adsorbed onto Celite R-640®, *Tetrahedron Lett.*, 43 (2002) 2005–2008.
12. T. P. Binder and J. F. Robyt, *p*-Nitrophenyl α-D-glycopyranoside, a new substrate for glucansucrases, *Carbohydr. Res.*, 124 (1983) 287–299.
13. W. R. Figures and J. R. Edwards, α-D-Glucopyranosyl fluoride as a D-glucopyranosyl donor for a glycosyltransferase complex from *Streptococcus mutans* FA1, *Carbohydr. Res.*, 48 (1976) 245–253.
14. H. D. Ly and S. G. Withers, Mutagenesis of glycosidases, *Annu. Rev. Biochem.*, 68 (1999) 487–522.
15. L.F Mackenzie, Glycosynthases: Mutant glycosidases for oligosaccharide synthesis, *J. Am. Chem. Soc.*, 120 (1998) 5583–5584.
16. M. Jahn, D. Stoll, R. A. Warren, L. Szabo, P. Singh, H. J. Gilbert, V. M. Ducros, G. J. Davies, and S. G. Withers, Expansion of the glycosynthase repertoire to produce defined manno-oligosaccharides, *Chem. Commun. (Camb.)*, 12 (2003) 1327–1329.
17. S. M. Hancock, M. D. Vaughan, and S. G. Withers, Engineering of glycosidases and glycosyltransferases, *Curr. Opin. Chem. Biol.*, 10 (2006) 509–519.
18. M. Jahn, J. Marles, R. A. Warren, and S. G. Withers, Thioglycoligases: Mutant glycosidases for thioglycoside synthesis, *Angew. Chem. Int. Ed. Engl.*, 42 (2003) 352–354.
19. Y. W. Kim, A. L. Lovering, H. Chen, T. Kantner, L. P. McIntosh, N. C. Strynadka, and S. G. Withers, Expanding the thioglycoligase strategy to the synthesis of alpha-linked thioglycosides allows structural investigation of

the parent enzyme/substrate complex, *J. Am. Chem. Soc.*, 128 (2006) 2202–2203.
20. M. Jahn, H. Chen, J. Mullegger, J. Marles, R. A. Warren, and S. G. Withers, Thioglycosynthases: Double mutant glycosidases that serve as scaffolds for thioglycoside synthesis, *Chem. Commun. (Camb.)* (2004) 274–275.
21. S. G. Withers, Mechanisms of glycosyl transferases and hydrolases, *Carbohydr. Polym.*, 44 (2001) 325–337.
22. M. Faijes, X. Perez, O. Perez, and A. Planas, Glycosynthase activity of *Bacillus licheniformis* 1, 3-1, 4-beta-glucanase mutants: Specificity, kinetics, and mechanism, *Biochemistry*, 42 (2003) 13304–13318.
23. A. Trincone and A. Giordano, Glycosyl hydrolases and glycosyltransferases in the synthesis of oligosaccharides, *Curr. Org. Chem.*, 10 (2006) 1163–1193.
24. J. Seibel, K. Buchholz, and H. J. Jördening, Glycosylation with activated sugars using glycosyltransferases and transglycosidases, *Biocatal. Biotransformation*, 24 (2006) 311–342.
25. K. H. Ebert and G. Schenk, Mechanism of biopolymer growth: The formation of dextran and levan, *Adv. Enzymol.*, 30 (1968) 179–210.
26. K. H. Ebert, G. Schenk, and H. Stricker, Über den Mechanismus des Aufbauschrtittes der enzymatischen Dextran- und Laevanbildung, *Ber. Bunsenges.*, 68 (1964) 765–767.
27. K. H. Ebert and F. Patat, Kinetische Betrachtungen ueber die enzymatische Dextransynthese, *Z. Naturforsch. B*, 17 (1962) 738–748.
28. J. F. Robyt and T. F. Walseth, The mechanism of acceptor reactions of *Leuconostoc mesenteroides* B-512F dextransucrase, *Carbohydr. Res.*, 61 (1978) 433–445.
29. J. F. Robyt and S. H. Eklund, Stereochemistry involved in the mechanism of action of dextransucrase in the synthesis of dextran and the formation of acceptor products, *Bioorg. Chem.*, 11 (1982) 115–132.
30. E. J. Hehre and D. M. Hamilton, Bacterial synthesis of an amylopectin-like polysaccharide from sucrose, *J. Biol. Chem.*, 166 (1946) 77–78.
31. R. M. Mayer, M. M. Matthews, C. L. Futerman, V. K. Parnaik, and S. M. Jung, Dextransucrase: Acceptor substrate reactions, *Arch. Biochem. Biophys.*, 208 (1981) 278–287.
32. M. Quirasco, A. Lopez-Munguia, V. Pelenc, M. Remaud, F. Paul, and P. Monsan, Enzymatic production of glucooligosaccharides containing α-$(1\rightarrow 2)$ osidic bonds. Potential application in nutrition, *Ann. NY Acad. Sci.*, 750 (1995) 317–320.

33. M. Dols, M. Remaud-Siméon, R. M. Willemot, M. Vignon, and P. Monsan, Structural characterization of the maltose acceptor products synthesized by *Leuconostoc mesenteroides* NRRL B-1299 dextransucrase, *Carbohydr. Res.*, 305 (1998) 549–559.
34. M. Böker, H. J. Jördening, and K. Buchholz, Kinetics of leucrose formation from sucrose by dextransucrase, *Biotechnol. Bioeng.*, 43 (1994) 856–864.
35. K. D. Reh, M. Noll-Borchers, and K. Buchholz, Productivity of immobilized dextransucrase for leucrose formation, *Enzyme Microb. Technol.*, 19 (1996) 518–524.
36. F. Paul, E. Oriol, D. Auriol, and P. Monsan, Acceptor reaction of a highly purified dextransucrase with maltose and oligosaccharides. Application to the synthesis of controlled molecular weight dextrans, *Carbohydr. Res.*, 149 (1986) 433–441.
37. K. Heincke, B. Demuth, H. J. Joerdening, and K. Buchholz, Kinetics of the dextransucrase acceptor reaction with maltose: Experimental results and modeling, *Enzyme Microb. Technol.*, 24 (1999) 523–534.
38. B. Demuth, H. J. Jordening, and K. Buchholz, Modelling of oligosaccharide synthesis by dextransucrase, *Biotechnol. Bioeng.*, 62 (1999) 583–592.
39. D. Schwengers, Leucrose, a ketodisaccharide of industrial design, in F. W. Lichtenthaler, (Ed.), *Carbohydrates as Organic Raw Materials,* VCH, Weinheim, 1991, pp. 183–195.
40. K. Buchholz, M. Noll-Borchers, and D. Schwengers, Production of leucrose by dextransucrase, *Starch/Stärke*, 50 (1998) 164–172.
41. A. Tanriseven and J. F. Robyt, Inhibition of dextran synthesis by acceptor reactions of dextransucrase and the demonstration of a separate acceptor binding-site, *Carbohydr. Res.*, 225 (1992) 321–329.
42. M. Remaud-Simeon, A. Lopez-Munguia, V. Pelenc, F. Paul, and P. Monsan, Production and use of glucosyltransferases from *Leuconostoc mesenteroides* NRRL B-1299 for the synthesis of oligosaccharides containing α-(1→2) linkages, *Appl. Biochem. Biotechnol.*, 44 (1994) 101–117.
43. G. L. Côté and J. F. Robyt, Isolation and partial characterization of an extracellular glucansucrase from *Leuconostoc mesenteroides* NRRL B-1355 that synthesizes an alternating (1→6), (1→3)-α-D-glucan, *Carbohydr. Res.*, 101 (1982) 57–74.
44. S. Kralj, G. H. van Geel-Schutten, M. M. Dondorff, S. Kirsanovs, M. J. van der Maarel, and L. Dijkhuizen, Glucan synthesis in the genus *Lactobacillus*: Isolation and characterization of glucansucrase genes, enzymes and glucan products from six different strains, *Microbiology*, 150 (2004) 3681–3690.

45. A. Reischwitz, K.-D. Reh, and K. Buchholz, Unconventional immobilization of dextransucrase with alginate, *Enzyme Microb. Technol.*, 17 (1995) 457–461.
46. J. F. Robyt, Mechanism in the glucansucrase synthesis of polysaccharides and oligosaccharides from sucrose, *Adv. Carbohydr. Chem. Biochem.*, 51 (1995) 133–168.
47. K. Demuth, H. J. Jördening, and K. Buchholz, Oligosaccharide synthesis with dextransucrase, in S. Bielecki, J. Tramper, and J. Polak, (Eds.), *Food Biotechnology,* Elsevier Science, Amsterdam, 2000, pp. 123–135.
48. V. Monchois, R. M. Willemot, and P. Monsan, Glucansucrases: Mechanism of action and structure–function relationships, *FEMS Microbiol. Rev.*, 23 (1999) 131–151.
49. B. Henrissat, A classification of glycosyl hydrolases based on amino acid sequence similarities, *Biochem. J.*, 280(Pt 2), (1991) 309–316.
50. S. A. van Hijum, S. Kralj, L. K. Ozimek, L. Dijkhuizen, and I. G. van Geel-Schutten, Structure–function relationships of glucansucrase and fructansucrase enzymes from lactic acid bacteria, *Microbiol. Mol. Biol. Rev.*, 70 (2006) 157–176.
51. E. A. MacGregor, H. M. Jespersen, and B. Svensson, A circularly permuted alpha-amylase-type alpha/beta-barrel structure in glucan-synthesizing glucosyltransferases, *FEBS Lett.*, 378 (1996) 263–266.
52. T. Pijning, A. Vujicic-Zagar, S. Kralj, W. Eeuwema, L. Dijkhuizen, and B. W. Dijkstra, Biochemical and crystallographic characterization of a glucansucrase from *Lactobacillus reuteri* 180, *Biocatal. Biotransformation*, 26 (2008) 12–17.
53. K. Buchholz, Introduction: The spirit of carbohydrates—Carbohydrate bioengineering, *Biocatal. Biotransformation*, 26 (2008) 3–11.
54. G. Lacaze, M. Wick, and S. Cappelle, Emerging fermentation technologies: Development of novel sourdoughs, *Food Microbiol.*, 24 (2007) 155–160.
55. T. Coviello, P. Matricardi, and F. Alhaique, Drug delivery strategies using polysaccharidic gels, *Expert Opin. Drug Deliv.*, 3 (2006) 395–404.
56. J. F. Robyt and T. F. Walseth, Production, purification, and properties of dextransucrase from *Leuconostoc mesenteroides* NRRL B-512F, *Carbohydr. Res.*, 68 (1979) 95–111.
57. S. Kralj, G. H. van Geel-Schutten, M. J. van der Maarel, and L. Dijkhuizen, Biochemical and molecular characterization of *Lactobacillus reuteri* 121 reuteransucrase, *Microbiology*, 150 (2004) 2099–2112.
58. H. Hellmuth, S. Wittrock, S. Kralj, L. Dijkhuizen, B. Hofer, and J. Seibel, Engineering the glucansucrase GTFR enzyme reaction and glycosidic bond

specificity: Toward tailor-made polymer and oligosaccharide products, *Biochemistry*, 47 (2008) 6678–6684.

59. V. Monchois, M. Remaud-Simeon, R. R. Russell, P. Monsan, and R. M. Willemot, Characterization of *Leuconostoc mesenteroides* NRRL B-512F dextransucrase (DSRS) and identification of amino-acid residues playing a key role in enzyme activity, *Appl. Microbiol. Biotechnol.*, 48 (1997) 465–472.
60. J. J. Ferretti, M. L. Gilpin, and R. R. Russell, Nucleotide sequence of a glucosyltransferase gene from *Streptococcus sobrinus* MFe28, *J. Bacteriol.*, 169 (1987) 4271–4278.
61. S. Kralj, G. H. van Geel-Schutten, H. Rahaoui, R. J. Leer, E. J. Faber, M. J. van der Maarel, and L. Dijkhuizen, Molecular characterization of a novel glucosyltransferase from *Lactobacillus reuteri* strain 121 synthesizing a unique, highly branched glucan with α-(1→4) and α-(1→6) glucosidic bonds, *Appl. Environ. Microbiol.*, 68 (2002) 4283–4291.
62. S. Kralj, I. G. van Geel-Schutten, E. J. Faber, M. J. van der Maarel, and L. Dijkhuizen, Rational transformation of *Lactobacillus reuteri* 121 reuteransucrase into a dextransucrase, *Biochemistry*, 44 (2005) 9206–9216.
63. T. Fujiwara, T. Hoshino, T. Ooshima, S. Sobue, and S. Hamada, Purification, characterization, and molecular analysis of the gene encoding glucosyltransferase from *Streptococcus oralis*, *Infect. Immun.*, 68 (2000) 2475–2483.
64. S. Kralj, W. Eeuwema, T. H. Eckhardt, and L. Dijkhuizen, Role of asparagine 1134 in glucosidic bond and transglycosylation specificity of reuteransucrase from *Lactobacillus reuteri* 121, *FEBS. J.*, 273 (2006) 3735–3742.
65. C. Moulis, G. Joucla, D. Harrison, E. Fabre, G. Potocki-Veronese, P. Monsan, and M. Remaud-Simeon, Understanding the polymerization mechanism of glycoside-hydrolase family 70 glucansucrases, *J. Biol. Chem.*, 281 (2006) 31254–31267.
66. T. Nakakuki, Present status and future of functional oligosaccharide development in Japan, *Pure Appl. Chem.*, 74 (2002) 1245–1251.
67. K. Buchholz and J. Seibel, Isomaltooligosaccharides, in G. Eggleston and G. L. Côté, (Eds.) *Oligosaccharides in Food and Agriculture,* Oxford University Press, Washington, 2003, pp. 63–65.
68. K. Buchholz, M. Noll-Borchers, and D. Schwengers, Production of leucrose by dextransucrase, *Starch*, 50 (1998) 162–164.
69. B. A. Lina, D. Jonker, and G. Kozianowski, Isomaltulose (palatinose): A review of biological and toxicological studies, *Food Chem. Toxicol.*, 40 (2002) 1375–1381.

70. H. Mizubuchi, T. Yajima, N. Aoi, T. Tomita, and Y. Yoshikai, Isomalto-oligosaccharides polarize Th1-like responses in intestinal and systemic immunity in mice, *J. Nutr.*, 135 (2005) 2857–2861.
71. K. Demuth, H. J. Jordening, and K. Buchholz, Oligosaccharide synthesis by dextransucrase: New unconventional acceptors, *Carbohydr. Res.*, 337 (2002) 1811–1820.
72. J. Seibel, H. Hellmuth, B. Hofer, A. M. Kicinska, and B. Schmalbruch, Identification of new acceptor specificities of glycosyltransferase R with the aid of substrate microarrays, *Chembiochem*, 7 (2006) 310–320.
73. H. Hellmuth, L. Hillringhaus, S. Hobbel, S. Kralj, L. Dijkhuizen, and J. Seibel, Highly efficient chemoenzymatic synthesis of novel branched thiooligosaccharides by substrate direction with glucansucrases, *Chembiochem*, 8 (2007) 273–276.
74. T. Pijning, A. Vuijičić-Žagar, S. Kralj, W. Eeuwema, L. Dijkhuizen, and B. W. Dijkstra, Biochemical and crystallographic characterization of a glucansucrase from *Lactobacillus reuteri* 180, *Biocatal. Biotransformation*, 26 (2008) 12–17.
75. J. Cerning, Exocellular polysaccharides produced by lactic acid bacteria, *FEMS Microbiol. Rev.*, 7 (1990) 113–130.
76. J. W. Yun, Fructooligosaccharides—Occurrence, preparation, and application, *Enzyme Microb. Technol.*, 19 (1996) 107–117.
77. S. A. van Hijum, K. Bonting, M. J. van der Maarel, and L. Dijkhuizen, Purification of a novel fructosyltransferase from *Lactobacillus reuteri* strain 121 and characterization of the levan produced, *FEMS Microbiol. Lett.*, 205 (2001) 323–328.
78. R. Chambert, G. Treboule, and R. Dedonder, Kinetic studies of levansucrase of *Bacillus subtilis.*, *Eur. J. Biochem.*, 41 (1974) 285–300.
79. S. A. van Hijum, G. H. van Geel-Schutten, H. Rahaoui, M. J. van der Maarel, and L. Dijkhuizen, Characterization of a novel fructosyltransferase from *Lactobacillus reuteri* that synthesizes high-molecular-weight inulin and inulin oligosaccharides, *Appl. Environ. Microbiol.*, 68 (2002) 4390–4398.
80. J. Seibel, R. Moraru, S. Gotze, K. Buchholz, S. Na'amnieh, A. Pawlowski, and H. J. Hecht, Synthesis of sucrose analogues and the mechanism of action of *Bacillus subtilis* fructosyltransferase (levansucrase), *Carbohydr. Res.*, 341 (2006) 2335–2349.
81. V. Olivares-Illana, C. Wacher-Odarte, S. Le Borgne, and A. Lopez-Munguia, Characterization of a cell-associated inulosucrase from a novel source: A

Leuconostoc citreum strain isolated from Pozol, a fermented corn beverage of Mayan origin, *J. Ind. Microbiol. Biotechnol.*, 28 (2002) 112–117.

82. S. Morales-Arrieta, M. E. Rodriguez, L. Segovia, A. Lopez-Munguia, and C. Olvera-Carranza, Identification and functional characterization of levS, a gene encoding for a levansucrase from *Leuconostoc mesenteroides* NRRL B-512 F, *Gene*, 376 (2006) 59–67.
83. R. Chambert and G. Gonzy-Treboul, Levansucrase of *Bacillus subtilis*: Kinetic and thermodynamic aspects of transfructosylation processes, *Eur. J. Biochem.*, 62 (1976) 55–64.
84. F. Ouarne and A. Guibert, Fructo-oligosaccharides: Enzymatic synthesis from sucrose, *Zuckerind*, 120 (1995) 793–798.
85. CAZy, Carbohydrate-active enzymes, (2006) online database http://www.cazy.org; 21.10.2006.
86. G. Meng and K. Fütterer, Structural framework of fructosyl transfer in *Bacillus subtilis* levansucrase, *Nat. Struct. Biol.*, 10 (2003) 935–941.
87. G. Meng and K. Futterer, Donor substrate recognition in the raffinose-bound E342A mutant of fructosyltransferase *Bacillus subtilis* levansucrase, *BMC Struct. Biol.*, 8 (2008) 16.
88. C. Martinez-Fleites, M. Ortiz-Lombardia, T. Pons, N. Tarbouriech, E. J. Taylor, J. G. Arrieta, L. Hernandez, and G. J. Davies, Crystal structure of levansucrase from the Gram-negative bacterium *Gluconacetobacter diazotrophicus*, *Biochem. J.*, 390 (2005) 19–27.
89. F. Alberto, C. Bignon, G. Sulzenbacher, B. Henrissat, and M. Czjzek, The three-dimensional structure of invertase (beta-fructosidase) from *Thermotoga maritima* reveals a bimodular arrangement and an evolutionary relationship between retaining and inverting glycosidases, *J. Biol. Chem.*, 279 (2004) 18903–18910.
90. M. Verhaest, W. V. Ende, K. L. Roy, C. J. De Ranter, A. V. Laere, and A. Rabijns, X-ray diffraction structure of a plant glycosyl hydrolase family 32 protein: Fructan 1-exohydrolase IIa of *Cichorium intybus*, *Plant J.*, 41 (2005) 400–411.
91. R. A. Nagem, A. L. Rojas, A. M. Golubev, O. S. Korneeva, E. V. Eneyskaya, A. A. Kulminskaya, K. N. Neustroev, and I. Polikarpov, Crystal structure of exo-inulinase from *Aspergillus awamori*: The enzyme fold and structural determinants of substrate recognition, *J. Mol. Biol.*, 344 (2004) 471–480.
92. F. Alberto, E. Jordi, B. Henrissat, and M. Czjzek, Crystal structure of inactivated *Thermotoga maritima* invertase in complex with the trisaccharide substrate raffinose, *Biochem. J.*, 395 (2006) 457–462.

93. J. Matrai, W. Lammens, A. Jonckheer, K. Le Roy, A. Rabijns, W. Van den Ende, and M. De Maeyer, An alternate sucrose binding mode in the E203Q *Arabidopsis* invertase mutant: An X-ray crystallography and docking study, *Proteins*, 71 (2008) 552–564.
94. L. K. Ozimek, S. Kralj, T. Kaper, M. J. van der Maarel, and L. Dijkhuizen, Single amino acid residue changes in subsite-1 of inulosucrase from *Lactobacillus reuteri* 121 strongly influence the size of products synthesized, *FEBS J.*, 273 (2006) 4104–4113.
95. A. Homann, R. Biedendieck, S. Gotze, D. Jahn, and J. Seibel, Insights into polymer versus oligosaccharide synthesis: Mutagenesis and mechanistic studies of a novel levansucrase from *Bacillus megaterium*, *Biochem. J.*, 407 (2007) 189–198.
96. L. K. Ozimek, S. A. van Hijum, G. A. van Koningsveld, M. J. van Der Maarel, G. H. van Geel-Schutten, and L. Dijkhuizen, Site-directed mutagenesis study of the three catalytic residues of the fructosyltransferases of *Lactobacillus reuteri* 121, *FEBS Lett.*, 560 (2004) 131–133.
97. L. K. Ozimek, S. Kralj, M. J. van der Maarel, and L. Dijkhuizen, The levansucrase and inulosucrase enzymes of *Lactobacillus reuteri* 121 catalyse processive and non-processive transglycosylation reactions, *Microbiology*, 152 (2006) 1187–1196.
98. L. Hernandez, J. Arrieta, C. Menendez, R. Vazquez, A. Coego, V. Suarez, G. Selman, M. F. Petit-Glatron, and R. Chambert, Isolation and enzymic properties of levansucrase secreted by *Acetobacter diazotrophicus* SRT4, a bacterium associated with sugar cane, *Biochem. J.*, 309(Pt 1), (1995) 113–118.
99. M. Korakli, A. Rossmann, M. G. Ganzle, and R. F. Vogel, Sucrose metabolism and exopolysaccharide production in wheat and rye sourdoughs by *Lactobacillus sanfranciscensis*, *J. Agric. Food Chem.*, 49 (2001) 5194–5200.
100. M. Korakli, M. Pavlovic, M. G. Ganzle, and R. F. Vogel, Exopolysaccharide and kestose production by *Lactobacillus sanfranciscensis* LTH2590, *Appl. Environ. Microbiol.*, 69 (2003) 2073–2079.
101. H. W. Doelle, L. Kirk, R. Crittenden, H. Toh, and M. B. Doelle, *Zymomonas mobilis*—Science and industrial application, *Crit. Rev. Biotechnol.*, 13 (1993) 57–98.
102. K. B. Song, J. W. Seo, and S. K. Rhee, Transcriptional analysis of levU operon encoding saccharolytic enzymes and two apparent genes involved in amino acid biosynthesis in *Zymomonas mobilis*, *Gene*, 232 (1999) 107–114.
103. I. E. Baciu, H. J. Jordening, J. Seibel, and K. Buchholz, Investigations of the transfructosylation reaction by fructosyltransferase from *B. subtilis* NCIMB 11871 for the synthesis of the sucrose analogue galactosyl-fructoside, *J. Biotechnol.*, 116 (2005) 347–357.

104. R. Chambert and M. F. Petit-Glatron, Polymerase and hydrolase activities of *Bacillus subtilis* levansucrase can be separately modulated by site-directed mutagenesis, *Biochem. J.*, 279 (1991) 35–41.
105. S. Kralj, K. Buchholz, L. Dijkhuizen, and J. Seibel, Fructansucrase enzymes and sucrose analogues: New approach for the synthesis of unique fructo-oligosaccharides, *Biocatal. Biotransformation*, 26 (2008) 32–41.
106. P. S. J. Cheetam, A. J. Hacking, and M. Vlitos, Synthesis of novel disaccharides by a newly isolated fructosyltransferase from *Bacillus subtilis*, *Enzyme Microb. Technol.*, 11 (1989) 212–219.
107. E. Stoppok, K. Matalla, and K. Buchholz, Microbial modification of. sugars as building blocks for chemicals, *Appl. Microbiol. Biotechnol.*, 36 (1992) 604–610.
108. V. Timme, R. Buczys, and K. Buchholz, Kinetic investigations on the hydrogenation of 3-ketosucrose, *Starch/Stärke*, 50 (1998) 29–32.
109. M. Pietsch, W. Walter, and K. Buchholz, Regioselective synthesis of new sucrose derivatives via 3-ketosucrose, *Carbohydr. Res.*, 254 (1994) 183–194.
110. R. Goldberg and J. Tewari, Thermodynamic and transport properties of carbohydrates and their monophosphates: The pentoses and hexoses, *J. Phys. Chem. Ref. Data*, 18 (1989) 809–822.
111. R. Goldberg, J. Tewari, and J. Ahluwalia, Thermodynamics of the hydrolysis of sucrose, *J. Biol. Chem.*, 264 (1989) 9901–9904.
112. J. Tewari and R. Goldberg, Thermodynamics of hydrolysis of disaccharides, *Biophys. Chem.*, 40 (1991) 59–67.
113. J. Seibel, R. Beine, R. Moraru, C. Behringer, and K. Buchholz, A new pathway for the synthesis of oligosaccharides by the use of non-Leloir glycosyltransferases, *Biocatal. Biotransformation*, 24 (2006) 157–165.
114. P. Mäntsälä and M. Puntala, Comparison of levansucrase from *Bacillus subtilis* and from *Bacillus amyloliquefaciens*, *FEMS Microbiol. Lett.*, 13 (1982) 395–399.
115. A. Zuccaro, S. Götze, S. Kneip, P. Dersch, and J. Seibel, Tailor-made fructooligosaccharides by a combination of substrate and genetic engineering, *Chembiochem*, 9 (2008) 143–149.
116. T. Rose and M. Kunz, Braunschweig, Bundesanstalt für Landwirtschaft, *Landbauforschung Voelkenrode, Sonderheft*, 241 (2002) 75–80.
117. T. Veronese and P. Perlot, Proposition for the biochemical mechanism occurring in the sucrose isomerase active site, *FEBS Lett.*, 441 (1998) 348–352.
118. D. Zhang, N. Li, S. M. Lok, L. H. Zhang, and K. Swaminathan, Isomaltulose synthase (PalI) of *Klebsiella* sp. LX3. Crystal structure and implication of mechanism, *J. Biol. Chem.*, 278 (2003) 35428–35434.

119. L. Wu and R. G. Birch, Characterization of the highly efficient sucrose isomerase from *Pantoea dispersa* UQ68J and cloning of the sucrose isomerase gene, *Appl. Environ. Microbiol.*, 71 (2005) 1581–1590.
120. S. Ravaud, H. Watzlawick, R. Haser, R. Mattes, and N. Aghajari, Overexpression, purification, crystallization and preliminary diffraction studies of the *Protaminobacter rubrum* sucrose isomerase SmuA, *Acta Crystallogr. Sect. F Struct. Biol. Cryst. Commun.*, 62 (2006) 74–76.
121. D. Zhang, X. Li, and L. H. Zhang, Isomaltulose synthase from *Klebsiella* sp. strain LX3: Gene cloning and characterization and engineering of thermostability, *Appl. Environ. Microbiol.*, 68 (2002) 2676–2682.
122. P. S. J. Cheetham, Production of isomaltose using immobilized microbial cells, in K. Mosbach, (Ed.), *Methods Enzymology*, Vol. 136, 1987, pp. 432–454.
123. L. K. Skov, O. Mirza, A. Henriksen, G. P. De Montalk, M. Remaud-Simeon, P. Sarcabal, R. M. Willemot, P. Monsan, and M. Gajhede, Amylosucrase, a glucan-synthesizing enzyme from the alpha-amylase family, *J. Biol. Chem.*, 276 (2001) 25273–25278.
124. S. Ravaud, X. Robert, H. Watzlawick, R. Haser, R. Mattes, and N. Aghajari, Trehalulose synthase native and carbohydrate complexed structures provide insights into sucrose isomerization, *J. Biol. Chem.*, 282 (2007) 28126–28136.
125. G. Eggleston and G. L. Côté, (Eds.), Oligosaccharides in food and agriculture, *ACS Symp. Ser.*, 849 (2003).
126. J. R. Whitaker, A. G. J. Voragen,, and D. W. S. Wong, (Eds.), *Handbook of Food Enzymology,* M. Dekker, New York, 2003, pp. 589–603.
127. S. Fuji and M. Komoto, Novel carbohydrate sweeteners in Japan, *Zuckerind,* 116 (1991) 197–200.
128. K. Buchholz, V. Kasche, and U. Bornscheuer, Biocatalysts and Enzyme Technology, (2005). Wiley-VCH, Weinheim.
129. T. Nakakuki, Present status and future prospects of functional oligosaccharide development in Japan, *J. Appl. Glycosci.*, 52 (2005) 267–271.
130. K. Buchholz and P. Monsan, Dextransucrase, in J. R. Whitaker, A. G. J. Voragen, and D. W. S. Wong, (Eds.), *Handbook of Food Enzymology,* M. Dekker, New York, 2003, pp. 589–603.
131. T. Wimmer, München, Wacker.-Chemie, (1999). Personal communication.
132. K. Buchholz and B. Ekelhof, Technologie der Kohlenhydrate—1: Zucker, in R. Dittmeyer, W. Keim, G. Kreysa, and A. Oberholz, (Eds.), *Winnacker-Küchler. Chemische Technik: Prozesse und Produkte*, Vol. 8, Wiley-VCH, Weinheim, 2005, pp. 315–354.

133. P. Valette, V. Pelenc, Z. Djouzi, C. Andrieux, F. Paul, P. Monsan, and O. Szylit, Bioavailability of new synthesized glucooligosaccharides in the intestinal tract of gnotobiotic rats, *J. Sci. Food Agric.*, 62 (1993) 121–127.
134. F. Paul, A. Lopez-Munguia, M. Remaud, V. Pelenc, and P. Monsan, Process for the enzymatic preparation of oligodextrans useful in the production of sugar substitutes, and these oligodextrans, (1992). U.S. Patent 5,141,858.
135. Z. Djouzi and C. Andrieux, Compared effects of three oligosaccharides on metabolism of intestinal microflora in rats inoculated with a human faecal flora, *Br. J. Nutr.*, 78 (1997) 313–324.
136. J. P. Lamothe, Y. Marchenay, P. Monsan, and F. Paul, Cosmetic compositions containing oligosaccharides, (1993). WO patent 9300067.
137. R. A. Rastall and J. A. T. Hotchkiss, Potential for the development of prebiotic oligosaccharides from biomass, *ACS Symp. Ser.*, 849 (2003) 44–53.
138. B. C. Tungland, Fructooligosaccharides and other fructans, in G. Eggleston and G. L. Côté, (Eds.), *Oligosaccharides in Food and Agriculture, ACS Symp. Ser.*, Vol. 849, American Chemical Society, Washington, DC.
139. J. W. Yun and S. K. Song, Enzymatic production of fructooligosaccharides from sucrose, in C. Bucke, (Ed.), *Methods in Biotechnology—Carbohydrate Biotechnology Protocols,* Humana Press, New Jersey, 1999, pp. 141–151.
140. J. W. Yun and D. H. Kim, Enzymatic production of inulooligosaccharides from inulin, in C. Bucke, (Ed.), *Methods in Biotechnology—Carbohydrate Biotechnology Protocols,* Humana Press, New Jersey, 1999, pp. 153–163.
141. L. Dijkhuizen and B. van der Veen, Cyclodextrin Glycosyltransferase, in J. R. Whitaker, A. G. J. Voragen, and D. W. S. Wong, (Eds.), *Handbook of Food Enzymology,* M. Dekker, New York, 2003, pp. 615–627.
142. G. Boehm, B. Stahl, J. Jelinek, J. Knol, V. Miniello, and G. E. Moro, Prebiotic carbohydrates in human milk and formulas, *Acta Paediatr. Suppl.*, 94 (2005) 18–21.
143. H. Takaku, Anomalously linked oligosaccharides mixture (''Alo mixture''), in T. A. R. S. O. Japan, (Ed.), *Handbook of Amylases and Related Enzymes,* Pergamon Press, Oxford, 1988, pp. 215–217.
144. Hayashibara, www.hayashibara.co.jp/english/main.html/(2007).
145. K. Tomono, A. Mugishima, T. Suzuki, H. Goto, H. Uedal, T. Nagail, and J. Watanabe, Interaction between cycloamylose and various drugs, *J. Inclusion Phenom. Macrocyclic Chem.*, 44 (2002) 267–279.
146. K. Hino, M. Kurose, T. Sakurai, S. Inoue, K. Oku, H. Chaen, and S. Fukuda, Effect of dietary lactosucrose on the intestinal immune functions in mice, *J. Appl. Glycosci.*, 54 (2007) 169–172.

MULTIVALENT LECTIN—CARBOHYDRATE INTERACTIONS: ENERGETICS AND MECHANISMS OF BINDING

BY TARUN K. DAM[a] and C. FRED BREWER[b]

[a]Department of Molecular Pharmacology, Albert Einstein College of Medicine, Bronx, New York, 10461, USA
[b]Department of Molecular Pharmacology, Microbiology and Immunology, Albert Einstein College of Medicine, Bronx, New York, 10461, USA

ISBN: 978-0-12-380856-1
DOI: 10.1016/S0065-2318(10)63005-3.

Abbreviations

38/40-mer Tn-PSM, 38/40-residue amino acid cleavage product of 81-mer Tn-PSM containing αGalNAc1-*O*-Ser/Thr residues; 81-mer Tn-PSM, 81-residue amino acid repeat of domain of porcine submaxillary mucin containing αGalNAc1-*O*-Ser/Thr residues; ASF, asialofetuin; ConA, concanavalin A; DGL, *Dioclea grandiflora* lectin; Fd-PSM, fully carbohydrate-decorated porcine submaxillary mucin; Fuc, L-fucose; Gal, D-galactose; GalNAc, *N*-acetyl-D-galactosamine; ITC, isothermal titration microcalorimetry; LacNAc, 2-Acetamido-2-deoxylactose; Neu5Gl, *N*-glycoloylneuraminic acid or sialic acid; NeuAc, *N*-acetylneuraminic acid; NMR, nuclear magnetic resonance; PSM, porcine submaxillary mucin; SBA, soybean agglutinin; Tn-antigen, αGalNAc1-*O*-Ser/Thr; Tn-PSM, porcine submaxillary mucin containing αGalNAc1-*O*-Ser/Thr residues; TF antigen, βGal1 → 3GalNAc.

I. Introduction

Lectins are carbohydrate-binding proteins that are widely conserved in Nature including in animals, plants, and microorganisms.[1] The biological activities of many animal lectins have been determined, including receptor-mediated endocytosis of glycoproteins, cellular recognition and adhesion,[2] inflammation,[3] and cell growth and metastasis.[4,5] The X-ray crystal structures of lectins demonstrate their oligomeric structures and multivalent binding properties.[6,7] As a consequence, lectin binding to multivalent glycoconjugate receptors on the surface of cells leads to their cross-linking, aggregation and subsequent signal transduction effects, including apoptosis of T cells,[8,9] regulation of the T-cell receptor,[10,11] and growth regulation of neuroblastoma cells.[12] Recently, galectin-9 has been implicated in the localization and regulation of the Glut-2 transporter on the surface of pancreatic beta cells.[13] In addition, galectin-3-mediated crosslinking of glycoprotein receptors on the surface of cells has been shown to be involved in regulating the cell cycling kinetics and activities of different cytokine receptors.[14] Hence, the binding and crosslinking activities of lectins with multivalent glycoprotein receptors on normal and transformed cells are key features of their biological activities.

While the mechanisms of formation of noncovalent crosslinked lattices of lectins with multivalent carbohydrates and glycoproteins have been well investigated,[15–17] the mechanisms associated with the enhanced affinities of lectins binding to multivalent carbohydrates and glycoproteins have been less well investigated until recently.[18,19]

It is known that the enhanced affinities of lectins for multivalent carbohydrates and glycoproteins depend on the structures and interactions that occur between the molecules.[20,21] These interactions include face-to-face binding of a lectin, which possesses multiple subsites in a common direction, to a carbohydrate or glycoprotein with clustered epitopes, as shown in Fig. 1A. The affinity enhancements in this case are large and relatively well understood. For example, binding of a trivalent carbohydrate with terminal LacNAc residues to the asialoglycoprotein receptor results in a $\sim 10^6$-fold increase in affinity relative to LacNAc.[22] In this case, the affinity of the receptor for the trivalent carbohydrate is the sum of the free energies of binding of the three subsites of the receptor, as shown in Fig. 1A. Studies of the binding of the Shiga toxin, which is a carbohydrate-binding protein, to multivalent glycoconjugates also fit

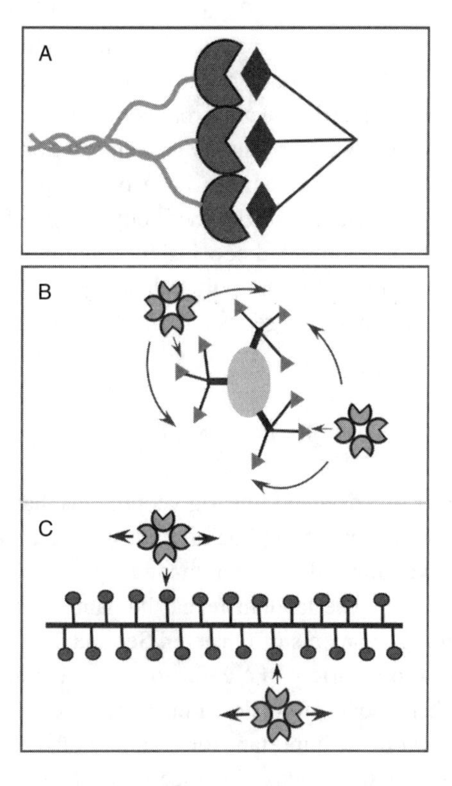

FIG. 1. Schematic representations of (A) face-to-face binding of a lectin with three subsites (green) to a trivalent carbohydrate (blue); (B) binding of a nonavalent glycoprotein (orange/black/pink) to two lectin molecules (green); (C) binding of a linear glycoprotein (black/red) to two lectin molecules (green). (See Color Plate 3.)

the face-to-face model with similar large increases in affinities.[23] In these cases, noncovalent crosslinking of the lectin with the glycoprotein does not occur.

In the second case, the interactions occur between lectins with binding sites at opposite ends of the molecule with epitopes of a multivalent carbohydrate or glycoprotein, as shown in Fig. 1B and C for globular and linear glycoproteins, respectively. The observed affinity enhancements are generally smaller than those in Fig. 1A,[24,25] and crosslinking of the complexes often occurs under appropriate stoichiometric conditions. The mechanisms of enhanced affinities in these cases have not been well understood until recently. For example, the affinities of two Man-specific lectins, concanavalin A (ConA), and *Dioclea grandiflora* lectin (DGL), for a synthetic tetra-antennary clustered glycoside are ~35–50-fold greater, respectively, than the corresponding monovalent carbohydrate.[26] The affinities of a series of galectins for asialofetuin (ASF), a nonavalent glycoprotein with terminal LacNAc residues (Fig. 1B), are ~50–80-fold greater than LacNAc.[27] However, the affinity of the GalNAc-specific soybean agglutinin (SBA) for a mucin that possesses ~2300 GalNAc residues has been recently reported to be ~10^6-fold greater than the corresponding monovalent carbohydrate (Fig. 1C).[28]

Previous comprehensive reviews of multivalent interactions in biological systems including protein–carbohydrate interactions have been reported.[18,21,29] This article focuses primarily on recent studies of the thermodynamics of binding of lectins to mucins, since the results have provided new insights into the energetics and mechanisms of binding of lectins to multi- and poly-valent glycoproteins. Furthermore, the results have suggested common energetics and mechanisms of binding of ligands to biopolymers in general.

II. Mucins: Background

Mucins are heavily O-glycosylated linear glycoproteins that are secreted by higher organisms to protect and lubricate epithelial cell surfaces. Mucin and mucin-like domains are also involved in modulating immune response, inflammation, adhesion, and tumorigenesis.[(cf. 30–32)] The tandem repeat domains of mucins and mucin-like glycoproteins contain high contents of clustered Ser and Thr residues, with many of these residues O-linked to a variety of "core" oligosaccharide structures, including linear and branched-chain blood-group determinants.[32] The oligosaccharide chains on mucins have been shown to be important for a variety of their biological properties including their interactions with such animal lectins as the selectins and galectins, as well as their physical properties including their extended linear structures.[33]

In addition to the variety of O-linked carbohydrates found on mucins, there are at least 17 mucin gene products (MUC1–MUC17).[31] These gene products represent two structurally and functionally distinct classes of mucins: secreted gel-forming mucins and transmembrane mucins, although there are a few mucin gene products that do not appear to fit into either category. The transmembrane mucins include MUC1 whose structure includes a cytoplasmic domain, in addition to the extracellular O-glycosylated polypeptide tandem repeat domains.[34] Evidence indicates that the C-terminal cytoplasmic domain of MUC1 is involved in signal transduction mechanisms, including T-cell activation and inhibition, and adhesion signaling responses.[35,36]

Mucins are also useful in the diagnosis of a variety of diseases.[33] In particular, the level of expression of mucin peptide antigens and type of carbohydrate chains of mucins have proved to be useful diagnostic markers for a variety of cancers.[31,37] For example, MUC1 expression as detected immunologically is increased in colon cancers and is associated with a poor diagnosis.[31] Colon cancer-associated mucins also have differences associated with their core carbohydrate structures, often presenting shorter chain versions of normal mucins. Colon cancer mucins often have increased expression of the αGalNAcThr/Ser (Tn-antigen), βGal3GalNAc (T or TF antigen) and αNeuAc6GalNAc (sialyl Tn-antigen).[31] Importantly, recent studies have shown that binding of galectin-3, an endogenous Gal-specific animal lectin, to cancer-associated MUC1 causes increased endothelial cancer cell adhesion.[38] Thus, the molecular recognition properties of cancer related mucins, including their truncated carbohydrates, are important in terms of gaining insight into their structure–activity properties.

Porcine submaxillary mucin (PSM) is a physically well-characterized mucin, and the subject of studies of the regulation of O-glycosylation with glycosyltransferases[cf. 39] and of binding interactions with lectins.[28] The cDNA sequence of PSM has been determined,[40] and the 81 amino acid tandem repeat domain that is present in 100 copies is shown in Fig. 2A. The structures of the carbohydrate chains were determined by chemical[41] and NMR techniques.[42] Gerken and Jentoft[42] isolated the O-glycosylated domain of PSM that possesses a molecular mass of $\sim 10^6$ Da and is fully decorated with naturally occurring carbohydrates (Fd-PSM) (Fig. 2B). The O-glycosylated domain of PSM possessing only αGalNAc residues (Tn-PSM) (Fig. 2C) was also obtained using chemical and enzymatic treatments.[43] The αGalNAc1-*O*-Ser/Thr residue(s) in Tn-PSM is the pancarcinoma carbohydrate antigen Tn that is aberrantly expressed in such mucins as MUC1 in adenocarcinomas.[44] The 81-mer tandem repeat domain of Tn-PSM (81-mer Tn-PSM) (Fig. 2D) and the 38/40-mer digest of this domain (38/40-mer Tn-PSM) (Fig. 2E) also have been obtained using enzymatic digests.[43]

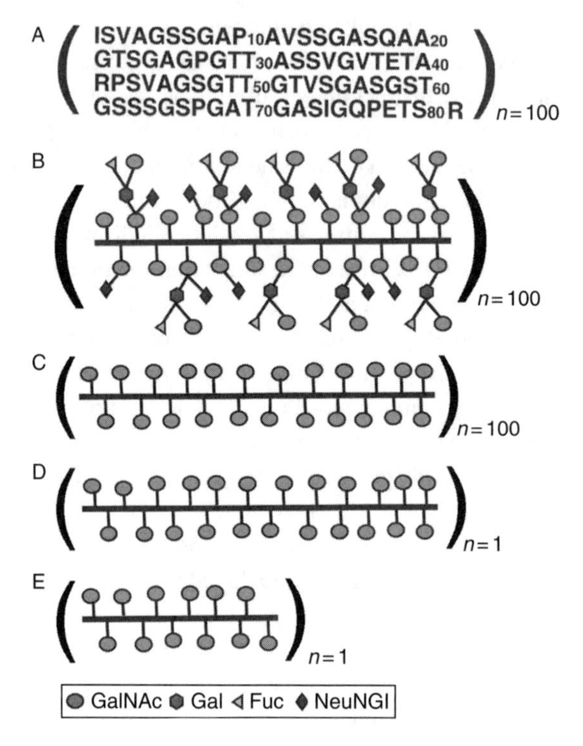

FIG. 2. Structural representations of (A) the amino acid sequence of the 100-repeat 81-residue polypeptide O-glycosylation domain of intact PSM; (B) the fully carbohydrate-decorated form (described in the text) of the 100-repeat 81-residue polypeptide O-glycosylation domain of PSM (Fd-PSM); (C) the 100-repeat 81-residue polypeptide O-glycosylation domain of PSM containing only peptide-linked αGalNAc residues (Tn-PSM); (D) the single 81-residue polypeptide O-glycosylation domain of PSM containing peptide-linked αGalNAc residues (81-mer Tn-PSM); (E) the 38/40-residue polypeptide(s) derived from the 81-residue polypeptide O-glycosylation domain of PSM containing peptide-linked αGalNAc residues (38/40-mer Tn-PSM). The number of glycan chains in Fd-PSM and Tn-PSM is ~2300. The number of αGalNAc residues in 81-mer Tn-PSM is ~23, while the number of αGalNAc residues in 38/40-mer Tn-PSM is ~11–12. (See Color Plate 4.)

The binding and crosslinking of cell-surface mucins and mucin-like glycoproteins by lectins is known to lead to signal transduction effects, including cell growth and cell death.[16,45] For example, galectin-1 crosslinking of CD43, a transmembrane mucin-type glycoprotein receptor that possesses approximately 80 O-linked chains with terminal LacNAc epitopes,[46] along with CD45 induces apoptosis in susceptible T cells.[47] However, details of the energetics and mechanisms of lectin binding and crosslinking of mucins and mucin-type receptors has been lacking.

III. Binding of Lectins to Mucins

1. Affinities of SBA and VML for Mucins

Large increases in the affinities of the GalNAc-specific lectins SBA and *Vatairea macrocarpa* lectin (VML) for modified forms of PSM have been recently reported using isothermal titration microcalorimetry (ITC) and hemagglutination-inhibition measurements.[28] The ITC measurements demonstrate that SBA binds to Tn-PSM possessing ~2300 αGalNAc residues and a molecular mass of ~10^6 Da (Fig. 2C) with a K_d of 0.2 nM (Table I), which constitutes a ~10^6-fold enhanced affinity relative to monovalent αGalNAc1-*O*-Ser. The 81 amino acid tandem repeat domain of Tn-PSM containing ~23 GalNAc residues (81-mer Tn-PSM) (Fig. 2D) binds with ~10^3-fold enhanced affinity, while the 38/40-mer fragment of Tn-PSM containing ~11–12 GalNAc residues (38/40-mer Tn-PSM) (Fig. 2E) shows ~10^2-fold enhanced affinity (Table I). Fd-PSM, the fully decorated form of PSM containing 40% of the core 1 blood group type A tetrasaccharide, and 58% peptide-linked αGalNAc1-*O*-Ser/-Thr residues, with 45% of the peptide-linked GalNAc residues linked α-(2→6) to *N*-glycolylneuraminic acid (Fig. 2B), shows ~10^4 enhanced affinity for SBA (Table I).

Table I

Thermodynamic Binding Parameters for SBA and VML at pH 7.2, 27 °C

Ligand	K_d[a] (μM)	K(rel)[b]	$-\Delta G$[c] (kcal/mol)	$-\Delta H$[d] (kcal/mol)	$-T\Delta S$[e] (kcal/mol)	n[f]
SBA						
αGalNAc1-*O*-Ser	170	1	5.2	7.9	2.7	1.0
38/40-mer Tn-PSM	1.4	120	8.0	32.2	24.2	0.2
81-mer Tn-PSM	0.06	2800	9.8	56.1	46.3	0.12
Tn-PSM	0.0002	850,000	13.1	4310	4297	0.0018
Fd-PSM	0.024	7100	10.4	703	693	0.008
VML						
αGalNAc1-*O*-Ser	130	1	5.3	6.4	1.1	1.0
38/40-mer Tn-PSM	0.20	650	9.1	36.8	27.7	0.2
81-mer Tn-PSM	0.012	11,000	10.8	52.7	41.9	0.1
Tn-PSM	0.0001	1,300,000	13.6	5274	5260	0.0012
Fd-PSM	0.0014	93,000	12.1	1251	1240	0.005

[a] Errors in K_d range from 1% to 7%.
[b] Relative to αGalNAc1-*O*-Ser.
[c] Errors in ΔG are less than 2%.
[d] Errors in ΔH are 1–4%.
[e] Errors in $T\Delta S$ are 1–7%.
[f] Errors in n are less than 4%.

VML displays a similar pattern of affinities for the PSM analogues although there are differences in the absolute affinities[28] (Table I). The higher affinities of SBA and VML for Tn-PSM relative to Fd-PSM indicate the importance of carbohydrate composition and epitope density of the mucins on their affinities for the lectins. The higher affinities of both lectins for Tn-PSM relative to its two shorter-chain analogues demonstrate that the length of a mucin polypeptide and hence total carbohydrate valence determines the affinities of the lectins for the three Tn-PSM analogues.

Kiessling *et al.*[48] reported a similar increase in the inhibitory activity of synthetic polymers of increasing lengths that possess Man residues in a hemagglutination assay with ConA. The authors concluded that the enhanced inhibitory activities of the longer chain polymers were ''largely due to a combination of statistical and chelation effects'' and therefore slower dissociation rates.

2. Thermodynamics of SBA Binding Tn-PSM

The ITC data also include the complete thermodynamics of binding of SBA and VML to the PSM analogues, including their stoichiometries of binding (Table I).[28] The n value for SBA binding to Tn-PSM, which is the binding stoichiometry expressed as number of binding sites per subunit of lectin, is 0.0018 (Table I). The value of $1/n$ has been shown to provide the functional valence of multivalent carbohydrates and glycoproteins binding to lectins.[26,27] The $1/n$ for Tn-PSM binding SBA is 540, which indicates that 540 αGalNAc residues of the ~2300 αGalNAc residues of Tn-PSM bind 540 monomers of SBA since each monomer binds one αGalNAc1-*O*-Ser. The $1/n$ value indicates that the functional valence of Tn-PSM for SBA is less than the structural valence of Tn-PSM. The reason for the fractional binding of SBA to the carbohydrate epitopes in Tn-PSM appears to be the size of the SBA tetramer, which possesses a molecular mass of 120 kDa and one N-linked Man-9 oligomannose chain per monomer.[49,50] Kiessling *et al.* have reported similar fractional occupancies for the binding of ConA, a tetrameric Man-specific lectin similar in size to SBA, to synthetic polymers containing α-mannose residues.[51] Fractional occupancy has also been reported for SBA binding to the nine LacNAc epitopes of ASF.[52]

SBA binding to Tn-PSM yields a ΔH of -4310 kcal/mol as compared to -7.9 kcal/mol for αGalNAc1-*O*-Ser (Table I). If the ΔH for SBA binding to Tn-PSM is divided by the $1/n$ value of 540 αGalNAc residues on Tn-PSM that bind to the lectin, the resulting ΔH per αGalNAc residue of Tn-PSM is -7.98 kcal/mol, which is very similar to the ΔH of -7.9 kcal/mol for αGalNAc1-*O*-Ser. This indicates that each αGalNAc residue of Tn-PSM that binds to SBA possesses the same enthalpy of binding as that of αGalNAc1-*O*-Ser. The observed Δ*H* for SBA binding to Tn-PSM is thus the sum of the individual Δ*H* values of the αGalNAc binding residues of the mucin. Similar observations have been made for the binding of ConA and DGL to synthetic bi-, tri-, and tetra-antennary glycosides.[53]

The calculated TΔ*S* for SBA binding to Tn-PSM is −4297 kcal/mol (Table I). If TΔ*S* is divided by the $1/n$ value of 540, the resulting TΔ*S* value per αGalNAc residue of Tn-PSM is −7.96 kcal/mol, which is more unfavorable than the −2.7 kcal/mol for αGalNAc1-*O*-Ser (Table I). If TΔ*S* were proportional to the number of binding epitopes in Tn-PSM, the observed TΔ*S* value would be −1458 kcal/mol, and Δ*G* would therefore be equal to Δ*H* − TΔ*S* or −2852 kcal/mol, an impossibly large value. The observation that, unlike Δ*H*, TΔ*S* does not scale in proportion to the number of binding epitopes in multivalent carbohydrates binding to lectins, but instead is much more negative has been previously observed in the binding of ConA and DGL to bi-, tri-, and tetra-antennary carbohydrates.[53] The nonproportionality of TΔ*S* is characteristic of different lectin molecules binding to separate epitopes of a multivalent carbohydrate instead of a single lectin binding to multiple epitopes of a multivalent carbohydrate.[54] In the latter case, both Δ*H* and TΔ*S* increase in proportion to the number of binding epitopes in the multivalent ligand, with concomitantly larger increases in affinity.[54] Nevertheless, the observed TΔ*S* value of −4297 kcal/mol when subtracted from the observed Δ*H* value of −4310 kcal/mol for SBA binding to Tn-PSM gives a Δ*G* value of −13.1 kcal/-mol, which is −7.9 kcal/mol greater than the Δ*G* of −5.2 kcal/mol for αGalNAc1-*O*-Ser (Table I).

3. Thermodynamics of SBA Binding 81-mer Tn-PSM

The ITC-derived K_d value for SBA binding to 81-mer Tn-PSM is 0.06 μM. This can be compared with the value of 0.15 μM obtained by hemagglutination inhibition.[28] The K(rel) for 81-mer Tn-PSM is 2800 as compared to αGalNAc1-*O*-Ser. The affinity of SBA for 81-mer Tn-PSM is ~300-fold weaker than that of Tn-PSM.

The *n* value for SBA binding to 81-mer Tn-PSM is 0.12, and $1/n = 8$, which suggests that ~8 αGalNAc residues of 81-mer Tn-PSM bind to ~8 monomers of SBA. Hence, the functional valence of 81-mer PSM is less than its structural valence for SBA binding, as observed for Tn-PSM.

The Δ*H* for SBA binding to 81-mer Tn-PSM is −56.1 kcal/mol. If Δ*H* is divided by the $1/n$ value of 8, the number of bound SBA monomers per ~23 αGalNAc residues, the resulting value is −7.0 kcal/mol per αGalNAc binding residue, which is slightly less than the value of −7.9 kcal/mol for αGalNAc1-*O*-Ser. This indicates that each αGalNAc binding residue of 81-mer Tn-PSM binds with nearly the same ΔH as that of αGalNAc1-*O*-Ser. This finding is similar to that observed for SBA binding to Tn-PSM.

SBA binding to 81-mer Tn-PSM gives a TΔS of -46.3 kcal/mol, which is greater than the calculated value of -21.6 kcal/mol if TΔS were proportional to the number of αGalNAc residues involved in binding to SBA. Thus, TΔS for SBA binding to 81-mer Tn-PSM does not increase in proportion to the number of αGalNAc residues that bind, but rather has a larger negative value. Similar results are observed for SBA binding to Tn-PSM.

4. Thermodynamics of SBA Binding 38/40-mer Tn-PSM

ITC data for the binding of SBA to 38/40-mer Tn-PSM shows a K_d of 1.4 μM (Table I), which can be compared with ~6 μM obtained by hemagglutination inhibition.[28] The K_d for 38/40-mer Tn-PSM is decreased nearly 20-fold relative to that for 81-mer Tn-PSM. Thus, SBA shows lowest affinity for the shortest fragment of Tn-PSM.

The n value for SBA binding to 38/40-mer Tn-PSM is 0.2, and $1/n = 5$. This indicates that ~5 αGalNAc residues of 38/40-mer Tn-PSM binds to five SBA monomers.

The ΔH and TΔS values of -32.2 kcal/mol and -24.2 kcal/mol, respectively, are consistent with the lower affinity of SBA for 38/40-mer Tn-PSM as compared to the longer polypeptide chain analogues. If ΔH is divided by the $1/n$ value of 5, then ΔH per αGalNAc residue of 38/40-mer is -6.44 kcal/mol, which is somewhat lower than the -7.9 kcal/mol for αGalNAc1-*O*-Ser. The lower ΔH per αGalNAc residue may be due to the lower affinity of SBA for 38/40-mer Tn-PSM as compared to Tn-PSM and 81-mer Tn-PSM (discussed in Section 5). If the observed TΔS is divided by the $1/n$ value of 5, the TΔS per αGalNAc residue of 38/40-mer is -4.84 kcal/mol, which is larger than -2.7 kcal/mol for αGalNAc1-*O*-Ser. Similar results are observed for SBA binding to Tn-PSM and 81-mer Tn-PSM, suggesting similar binding mechanisms of SBA with all three PSM analogues.

5. Thermodynamics of SBA Binding Fd-PSM

The ITC derived K_d for SBA binding to Fd-PSM is 0.024 μM (Table I), which can be compared to the inhibition constant of 0.08 μM obtained by hemagglutination inhibition.[28] The loss in affinity of SBA for Fd-PSM relative to Tn-PSM may be due to the lower density of free αGalNAc residues in Fd-PSM, due to Neu5Gl present on ~45% of the total αGalNAc residues.

The n value for SBA binding to Fd-PSM is 0.008 (Table I), and $1/n$ is 125, which is the number of αGalNAc residues of Fd-PSM bound to SBA monomers. This number of αGalNAc residues is consistent with a decreased number of binding sites for SBA on Fd-PSM relative to the 540 αGalNAc residues on Tn-PSM. The factor of ~4 reduction in the number of binding sites on Fd-PSM suggest that in addition to 45% of the total αGalNAc residues capped with Neu5Gl, binding of SBA to the nonreducing αGalNAc residue of the tetrasaccharide chains of PSM may alter the accessibility of SBA binding to the single-peptide-linked αGalNAc residues by a further factor of two. Similar effects are observed for VML binding to Tn-PSM and Fd-PSM.

The ΔH and $T\Delta S$ values for Fd-PSM binding to SBA are −703 kcal/mol and −693 kcal/mol, respectively, and are lower than the corresponding values for Tn-PSM (Table I). If ΔH is divided by $1/n = 125$, the resulting ΔH per binding αGalNAc residue is −5.6 kcal/mol which is somewhat lower than that for αGalNAc1-*O*-Ser (Table I). If $T\Delta S$ is divided by $1/n = 125$, the resulting $T\Delta S$ per binding αGalNAc residue is −5.54 kcal/mol which is greater than that for αGalNAc1-*O*-Ser. Similar results are observed for SBA binding to Tn-PSM, 81-mer Tn-PSM, and 38/40-mer PSM.

6. Thermodynamics of VML Binding Tn-PSM

The ITC data for VML binding to Tn-PSM are similar to those for SBA.[28] The main difference is the stoichiometry of binding of VML, which shows that $1/n = 833$ αGalNAc residues binding to 833 monomers of VML. This can be compared with 540 αGalNAc residues of Tn-PSM binding to 540 monomers of SBA. Importantly, the size of the VML tetramer is similar to that of SBA,[55] and VML also possesses covalently bound carbohydrate similar to SBA.[56]

7. Thermodynamics of VML Binding 81-mer Tn-PSM and 38/40-mer Tn-PSM

The ITC data for VML binding to 81-mer Tn-PSM and 38/40-mer Tn-PSM in Table I are similar to those observed for SBA.[28] Differences between VML and SBA binding to the two fragments of Tn-PSM are principally in their K(rel) values which are somewhat greater for VML. The binding stoichiometries of VML to the two fragments are also similar to those observed for SBA. The results support the conclusion that the affinity of VML, like SBA, decreases with shorter fragments of Tn-PSM.

8. Thermodynamics of VML Binding Fd-PSM

The ITC data for VML binding to Fd-PSM are also similar to those for SBA. Differences between the two lectins are in their *K*(rel) values which are greater for VML. As with SBA, VML binds with greater affinity to Fd-PSM than to 81-mer Tn-PSM and 38/40-mer PSM, but with less affinity to Fd-PSM than to Tn-PSM. The binding stoichiometry of VML to Fd-PSM indicates that there are 200 αGalNAc residues of Fd-PSM that bind to VML monomers. This can be compared to the 125 αGalNAc residues of Fd-PSM that bind to SBA.

IV. Mechanisms of Binding of SBA and VML to PSM: The Bind and Jump Model

The increasing affinities of SBA and VML for 38/40-mer Tn-PSM, 81-mer Tn-PSM and Tn-PSM, respectively, indicate that the lengths of the polypeptide chains and hence total carbohydrate valences of these mucin analogues regulate their affinities for the lectins. The higher affinities of SBA and VML for Tn-PSM relative to Fd-PSM also indicates the importance of carbohydrate composition and epitope density of the mucins on their affinities for lectins. The similarities in the ΔH per αGalNAc binding residue of Tn-PSM and Fd-PSM with that of αGalNAc1-*O*-Ser for the two lectins, respectively, also suggest a common mechanism of binding. Similar correlations with 81-mer Tn-PSM and 38/40-mer Tn-PSM suggest that the binding mechanisms of these analogues are common.

A model to explain these results is similar to that proposed for the binding of lectins to multivalent carbohydrates[53] and globular glycoproteins.[27] However, binding of a lectin to a multivalent carbohydrate or glycoprotein involves internal diffusion ''jumps'' of the lectin from one carbohydrate epitope to another epitope before complete dissociation. Kinetically, this has the effect of decreasing the macroscopic off-rate of the lectin and hence increasing its affinity, since the affinity constant is the ratio of the forward and reverse rate constants. The forward rate constant for lectin binding may also be enhanced as a result of the larger number of binding epitopes on the glycoprotein.

The diffusion-jump model for SBA and VML binding to Tn-PSM and the other PSM analogues can be envisioned as occurring with one subunit of SBA or VML bound to one αGalNAc residue of Tn-PSM at a time (Fig. 3A). (Two subunits of individual SBA or VML molecules simultaneously binding to a single Tn-PSM chain is not supported by the enhanced affinities of both lectins to 38/40-mer Tn-PSM relative to αGalNAc1-*O*-Ser (Table I). If two subunits of an SBA tetramer were bound to 38/40-mer Tn-PSM, the

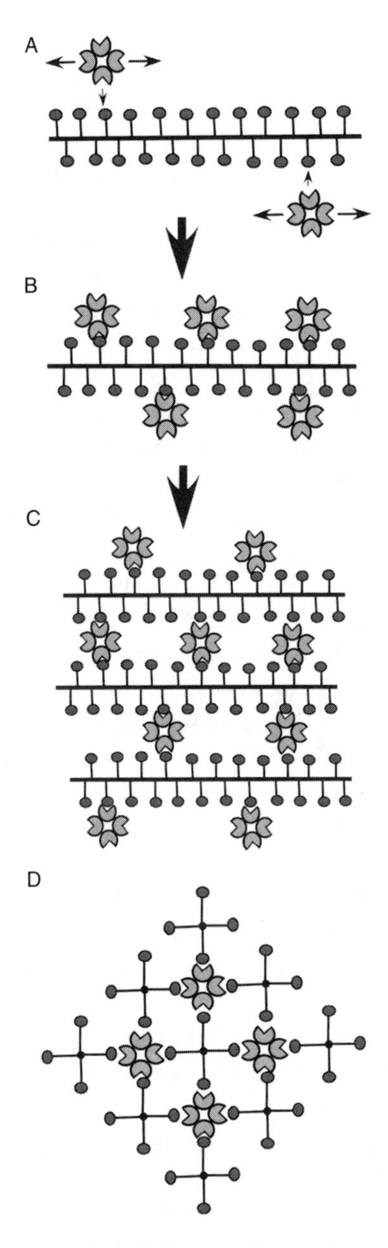

FIG. 3. Schematic representations of (A) SBA or VML binding at low density to Tn-PSM; (B) SBA or VML binding at higher density to Tn-PSM; (C) SBA and VML binding at higher density to Tn-PSM and initiating crosslinking of the complexes; (D) SBA crosslinked complexes with Tn-PSM under saturation-binding conditions. The view is end on of the polypeptide chains of Tn-PSM in Fig. 3C. αGalNAc residues extend out from the polypeptide chains of Tn-PSM in three dimensions. Lectin tetramers are bound to four separate Tn-PSM chains, with staggered binding down the length of the mucin chains. (See Color Plate 5.)

affinity enhancement would be approximately the product of the individual αGalNAc1-*O*-Ser dissociation constant, which would be ~10^{-8} M instead of the observed 10^{-6} M.) This model provides a molecular mechanism to account for the dependence of the affinities of the two lectins on the total carbohydrate valences of Tn-PSM, 81-mer Tn-PSM, and 38/40-mer Tn-PSM. This model further suggests that, as more lectin molecules bind to a mucin chain, the affinity of the lectin will decrease because of steric crowding and shorter diffusion distances on the polypeptide chain of the mucin (Fig. 3B). Indeed, Hill plots for SBA binding to Tn-PSM show evidence for negative binding cooperativity. Similar Hill plots have been interpreted as evidence for increasing negative cooperativity in the binding of galectins to ASF[27] and ConA and DGL to synthetic di-, tri-, and tetra-valent carbohydrates.[53] In these cases, binding of the lectins to the multivalent carbohydrates and glycoprotein were associated with gradients of decreasing microaffinity constants of the lectins for the multiple epitopes of the ligands. This suggest that the observed dissociation constants for SBA and VML binding to the four PSM analogues in Table I represent a composite of the gradient of binding constants present in each case. Such gradients have been estimated to be as large as 3000–6000-fold for galectins binding to ASF.[27] These gradient effects may explain the lower average ΔH values per αGalNAc residue for 81-mer Tn-PSM and 38/40-mer PSM binding to SBA and VML.

The model for SBA and VML binding must also agree with the final saturation density of SBA and VML molecules bound to single PSM analogues, as reflected in ITC n values. For example, the $1/n$ value for SBA binding to Tn-PSM indicates 540 αGalNAc residues bound to 540 SBA monomers. The $1/n$ value for VML binding to Tn-PSM indicates 833 αGalNAc residues bound to 833 VML monomers. Another requirement of the binding model is the observation that at the end of ITC experiments, solutions of SBA and VML with all four PSM analogues begin to precipitate out of solution (data not shown). This indicates that lectin-mediated crosslinking of the mucins occurs following saturation binding.

Both of these observations are accounted for in the schematic representation in Fig. 3C and D of SBA crosslinked with Tn-PSM under saturation conditions. The schematic shows individual SBA molecules crosslinked to four different Tn-PSM molecules. Importantly, to form the crosslinked complex shown in Fig. 3D, lectin molecules bind to αGalNAc residues on all four sides of a Tn-PSM polypeptide chain. This allows staggering of individual SBA molecules along the Tn-PSM polypeptide chain, with concomitant decrease in steric interactions between lectin molecules. This is important, since calculations of the density of SBA molecules bound to Tn-PSM (knowing the diameter of SBA from X-ray crystal studies[50] and the length of the Tn-PSM polypeptide chain) suggests that only ~300 SBA tetramers can bind to the same side of a Tn-PSM polypeptide chain, which is less than the 540 bound monomers of SBA and 833 bound monomers of VML derived from ITC *n* values. The apparent steric crowding is overcome by having lectin molecules bound to all four sides of Tn-PSM.

In summary, the binding models first show a fraction of SBA or VML molecules that bind to Tn-PSM and ''jump'' between different αGalNAc residues of the mucin (Fig. 3A). As the number of bound lectin molecules increases, the affinity of the lectin decreases because of shorter diffusion distances on the mucin chain due to steric crowding and crosslinking by multiple bound lectin molecules (Fig. 3B and C). Finally, upon saturation binding, full lectin-mediated crosslinking of the complexes takes place (Fig. 3D).

V. Thermodynamics of Lectin–Mucin Crosslinking Interactions

1. Hill Plots Show Evidence of Increasing Negative Cooperativity

Hill plots of the raw ITC data for the binding of SBA and VML to Tn-PSM, 81-mer Tn-PSM, 38/40-mer Tn-PSM, and Fd-PSM in Table I show evidence of increasing negative cooperativity, as shown in Fig. 4 for SBA binding to 38/40-mer Tn-PSM. The data in Fig. 4A are shown as a progressive three-point tangent slope plot in Fig. 4B, and the data in Fig. 4B shown as a bar graph in Fig. 4C. The data in Fig. 4B and C show increasing negative cooperativity with increased binding of SBA to 38/40-mer Tn-PSM. Similar results are observed for the binding of SBA to Tn-PSM and 81-mer Tn-PSM. The ITC-derived Hill plot for the binding of SBA to αGalNAc1-*O*-Ser (Fig. 4D) shows a straight line with a slope of 0.92, indicating essentially no binding cooperativity. Thus, the increasing negative cooperativity is due to the polyvalent mucins and not the lectin. Similar results have been observed for the binding of ConA and DGL to synthetic bi-, tri-, and tetraantennary N-linked carbohydrates,[53] and the binding of galectins-1, -2, -3, -4, -5, and -7 to ASF.[27]

2. Analysis of the Stoichiometry of Binding of SBA to the Mucins

Using the $1/n$ value for Tn-PSM indicates that ∼540 GalNAc residues out of the total of ∼2300 GalNAc residues of the mucin bind to SBA under saturation conditions. SBA is a tetramer with four binding sites per molecule.[50] Thus, all four sites of SBA are occupied at the end of the ITC experiment, and hence SBA is completely crosslinked with Tn-PSM. This is also true for the ITC experiments for SBA binding to the shorter mucin fragments. Thus, the increasing negative cooperativity observed in the Hill plots correlates with increasing crosslinking of SBA with the mucins.

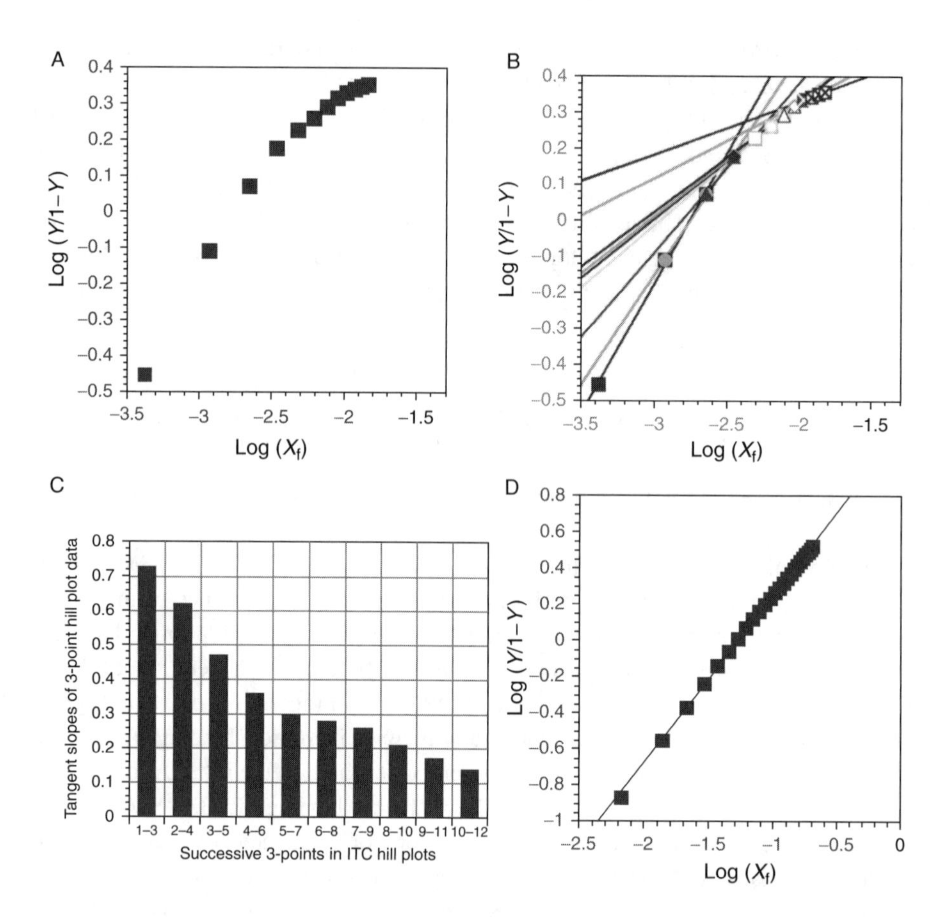

FIG. 4. (A). Hill plot of ITC data for SBA binding to 38/40-mer Tn-PSM. (B). Tangent slopes of progressive three-point intervals of the Hill plot for SBA binding to 38/40-mer Tn-PSM in (A). (C). Bar graphs of the three-point tangent slopes of the ITC data Hill plots of SBA binding to 38/40-mer Tn-PSM in (B). (D). Hill plot of the ITC data of SBA binding to αGalNAc1-*O*-Ser (3.24 mM). The slope value is 0.93. (See Color Plate 6.)

3. Crosslinking of Lectins with the Mucins Correlate with Decreasing Favorable Entropy of Binding

The observed ΔH value for SBA binding to Tn-PSM and its fragments have been shown to be proportional to the number of GalNAc residues in the mucins that bind to the lectin, while the corresponding observed $T\Delta S$ values are nonproportional and are much greater than the predicted proportional values (Section III).[28] These observations have also been reported for the binding of ConA and DGL to bi-, tri-, and tetra-antennary carbohydrates,[53] and galectin-1, -2, -3, -4, -5, and -7 to ASF.[27] The nonproportional behavior of $T\Delta S$ is characteristic of multiple lectin molecules binding to different epitopes of a multivalent carbohydrate instead of a single lectin molecule binding to multiple epitopes of a multivalent carbohydrate.[54] In the latter case, both ΔH and $T\Delta S$ increase in proportion to the number of binding epitopes in the multivalent ligand, with concomitantly larger increases in affinity.[54]

Importantly, the increasing negative cooperativity observed in the Hill plots for SBA and VML binding to the mucins must be due to the nonproportional $T\Delta S$ behavior for these interactions. This indicates that $T\Delta S$ must be very favorable for the initial binding of SBA to the mucins as shown in Fig. 3A. The highly favorable $T\Delta S$ for these interactions can be understood in terms of the suggested bind and jump mechanism for SBA binding to the mucin,[28] which has been used to explain the higher affinity of SBA for Tn-PSM as compared to the shorter mucin fragments.[28] Importantly, the dynamic movement of the lectin on the mucin backbone contributes to the highly favorable $T\Delta S$ of binding.

The observation of increasing negative cooperativity with increasing binding of SBA to the mucins indicates progressive unfavorable $T\Delta S$ contributions. As shown in Fig. 3B, increasing SBA binding to the mucin initially results in reducing the one-dimensional diffusion paths for internally diffusing SBA molecules, as previously suggested.[28] With further increases in the concentration of SBA, lectin/mucin crosslinked complexes begin to occur as shown in Fig. 3C. This leads to even greater unfavorable $T\Delta S$ contributions since the organization of SBA and mucin molecules in the crosslinked complexes is greater than those of the initial interactions in Fig. 3A and B. Hence, SBA/mucin crosslinking interactions are energetically less favorable as compared to the initial binding of the lectin to the mucin.

However, the affinity of bound SBA for GalNAc residues on other mucin molecules during crosslinking is very high compared to that of free SBA to αGalNAc1-*O*-Ser. For example, the observed K_d of SBA for Tn-PSM is 0.2 nM (Table I), at which 50% of the lectin is bound to the mucin. Under these conditions, at least two of the four subunits of SBA are crosslinked with Tn-PSM, and therefore the affinity of the lectin is nearly 10^6-fold greater than that of free SBA for αGalNAc1-*O*-Ser ($K_d = 0.17$ mM) (Table I). The high affinity of SBA in these crosslinked complexes is not due to the ΔH of binding, which is fixed at -8.0 kcal/mol per binding GalNAc residue in Tn-PSM, but instead must be due to favorable $T\Delta S$ contributions per GalNAc residue. Since favorable entropies of binding of SBA to Tn-PSM are ascribed to the bind and jump mechanism, it follows that SBA is dynamically bound to Tn-PSM in the process of crosslinking. As SBA undergoes further crosslinking, increasing negative cooperativity suggests a decrease in the internal diffusion of SBA on Tn-PSM as crosslinking restricts the movement of lectin molecules. Hence, high-affinity crosslinking of Tn-PSM by SBA is driven by the highly favorable entropy of binding of the lectin that is associated with the bind and jump mechanism.

Similar increasing negative cooperativity effects have been observed in the binding of ConA and DGL to bi-, tri-, and tetra-antennary carbohydrates[53] and galectin-1, -2, -3, -4, -5, and -7 to ASF,[27] and much of these effects is likely to be due to crosslinking interactions as well. Indeed, recent experiments with so-called monovalent galectins such as truncated galectin-3 and -5 show evidence of crosslinking activities.

VI. Conclusions and Perspective

1. The Bind and Jump Model for Lectin–Mucin Interactions

The model for lectin binding to mucins suggests that the length of mucin chains and mucin-type domains in glycoproteins modulates the affinities of lectins and related ligands for such glycoprotein receptors. Indeed, the length of mucins and mucin-type domains in glycoproteins is regulated in biological systems[(cf. 30–32)] Furthermore, mucins and mucin-type domains are also expressed with different amino acid sequences, different degrees of clustered carbohydrates, and different decorations of carbohydrates.[(cf. 30)] All of these structural variations are expected to affect the affinities of lectins and other ligands, including glycosyltransferases for mucins and mucin-type domains.

The K_d values of 0.2 nM and 0.1 nM for SBA and VML binding to Tn-PSM, respectively, are among the lowest reported K_d values for glycoprotein–lectin interactions. For comparison, the observed K_d values for galectins-1, -2, -3, -4, -5, and -7 binding to multivalent ASF, which has nine LacNAc epitopes, are in the range of $\sim$1 mM.[27] SBA and VML bind to 38/40-mer Tn-PSM, which has a structural valence of $\sim$11 αGalNAc residues, in the K_d range of the galectins binding to ASF. Thus, the large difference in K_d values for the binding of SBA and VML to Tn-PSM and the galectins binding to ASF appear to reflect the much greater carbohydrate valence of Tn-PSM ($\sim$2300 αGalNAc residues) as compared to ASF (nine LacNAc residues). Indeed, differences in the affinities of the N-terminal domain of galectin-4 for globular glycoproteins and mucins have been reported.[56] The differences in geometry between globular versus linear glycoproteins permits higher carbohydrate valence in mucins that, in turn, is associated with higher affinities toward ligands such as lectins.

Importantly, the internal diffusion model for lectins binding to mucins is distinct from the classical ''lock and key'' model of ligand binding to a receptor. The internal diffusion model is dynamic and allows a small fraction of lectin molecules to bind to the entire length of the receptor (mucin) and facilitate crosslinking interactions with other mucin molecules. For transmembrane mucins, such as CD-43, crosslinking by lectins (galectin-1) can lead to signal transduction effects, including apoptosis.[47] The polyvalence of long, linear chain molecules such as mucins also facilitates the uptake of low concentrations of lectin molecules that are released from the surface of cells.

The observation that SBA and VML bind with higher affinities to Tn-PSM relative to Fd-PSM has implications for the aberrant expression of the Tn epitope in molecules such as MUC1 that is found with overexpression of the Tn epitope in adenocarcinomas.[57] The present data indicate that clustering of the Tn epitope on a mucin such as PSM can result in high-affinity binding of lectins with αGalNAc specificities. Interestingly, a recent study of the binding of galectin-3 to MUC1 from epithelial cancer cells which expresses the TF antigen [βGal-(1→3)-αGalNAc] as well as the Tn antigen causes increased cancer cell endothelial adhesion.[38]

2. Implications of Increasing Negative Cooperativity and Decreasing Favorable Binding Entropy of Lectins–Mucin Crosslinking Interactions

Figure 5 shows a hypothetical plot of the microscopic K_d values versus fractional occupancy of SBA binding to Tn-PSM. The horizontal dashed line is the observed K_d value of 0.2 nM obtained from the ITC experiments (Table I). However, due to

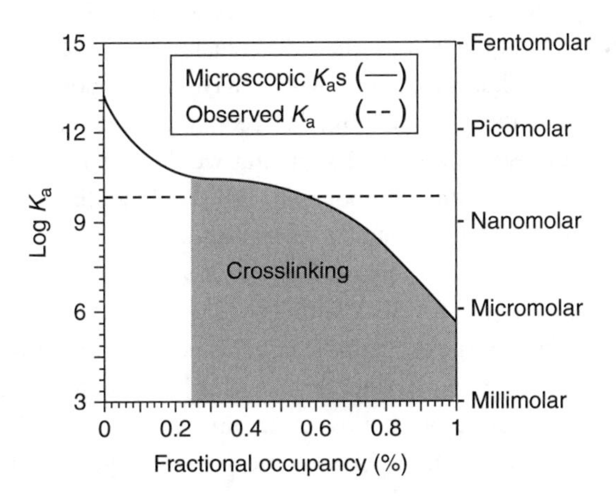

FIG. 5. Hypothetical plot (solid line) of the microscopic K_d values of SBA binding to Tn-PSM from 0% to 100% occupancy (0–540 αGalNAc residues). The dashed line represents the observed K_d value of 0.2 nM for SBA binding to Tn-PSM derived from the ITC data in Table I. Crosslinking begins at about 25% occupancy of Tn-PSM by SBA. (See Color Plate 7.)

increasing negative cooperativity the microscopic K_d values for SBA binding to Tn-PSM increase (decreasing affinity) with increasing fractional occupancy. The solid line is an estimate of the range of microscopic K_d values of SBA with increasing fractional occupany of Tn-PSM from 0% to 100%, which corresponds to 0–540 αGalNAc residues. Estimates of the range of microscopic K_d values for the binding of ConA and DGL to a tetraantennary carbohydrate was ~2500-fold from the first unbound carbohydrate epitope to the fourth epitope.[53] Estimates of the range of microscopic K_d values for the binding of galectins-1, -2, -3, -4, -5, and -7 binding to ASF was ~3000–6000-fold from the first unbound epitope to the ninth unbound epitope.[27] The range of microscopic K_d values for SBA binding to 540 GalNAc residues of Tn-PSM is estimated to be greater than the above examples, and as high as ~10^5–10^6. This suggests that the first few molecules of SBA binding to Tn-PSM possess microscopic K_d values of picomolar or less, as shown in Fig. 5. These SBA molecules have very favorable entropies of binding due their ability to bind and jump along the GalNAc residues of the Tn-PSM polypeptide chain. Thus, the dynamic motion of these bound lectin molecules contributes to their highly favorable entropies of binding. Stiochiometric analysis indicates that approximately one out of four subunits of SBA (tetramer) participates in this initial highly entropically favorable mechanism. With further increases in lectin concentration, crosslinking begins to occur, which is often associated with biological signal transduction effects such as galectin-1-induced apoptosis of T cells.[47]

Importantly, significant increases in the concentration of lectin are required to initiate crosslinking (and signaling) due to increasing negative cooperativity in the interactions (Fig. 5). Thus, signaling would have an important ''threshold'' or ''brake'' mechanism in these types of ligand–receptor interactions. The concentration of lectin required for crosslinking and signaling would depend on the length of the polypeptide chain of the mucin, composition of the polypeptide tandem repeat unit, density of carbohydrate epitopes, and different carbohydrate chains as well as the structure of the lectin. Thus, different mucins will have different binding/crosslinking profiles such as that in Fig. 5 for SBA with Tn-PSM.

In summary, the foregoing results indicate that lectin/mucin crosslinking interactions are entropically unfavorable and are ''driven'' by the highly favorable initial entropy of binding, which involves a dynamic bind and jump mechanism similar to that for proteins binding to DNA. These observations suggest that the binding of ligands to biopolymers, in general, confers enhanced entropic effects that facilitate subsequent complex formation. The magnitude of these effects are large (~10^5–10^6 or larger), and indicate that complex formation (and associated enzymology) proceed much more effectively and at lower concentrations on biopolymers as compared to corresponding complexes in solution. These findings apply to preorganized macromolecular complexes such as cell adhesion molecules, including integrins and cadherens,[cf. 58] macromolecular assembles such as the immunological synapse of the T-cell receptor[59] and antigen–antibody interactions.[60] Multifunctional proteins, including extracellular matrix proteins with repeat domains, are also candidates for enhanced microaffinity gradients and entropic effects.[61] Biopolymers, including glycosaminoglycans with repeating carbohydrate epitopes, may also possess microaffinity gradients and enhanced entropic effects toward ligands, including growth factors and their receptors.[62] These results thus provide evidence for large entropic gains for ligands binding to biopolymers (and macromolecular assemblies in general) that facilitate subsequent complex formation (and enzymology) involving the bound ligand.

References

1. A. Varki, R. Cummings, J. Esko, H. Freeze, G. Hart, and J. Marth, *Essentials of Glycobiology*, Cold Spring Harbor Laboratory Press, Cold Spring Harbor, NY, 1999, p. 653.
2. K. Drickamer and M. E. Taylor, Biology of animal lectins, *Annu. Rev. Cell Biol.*, 9 (1993) 237–264.
3. F.-T. Liu, Galectins: A new family of regulators of inflammation, *Clin. Immunol.*, 97 (2000) 79–88.
4. K. N. Konstantinov, B. A. Robbins, and F.-T. Liu, Galectin-3, a β-galactoside-binding animal lectin, is a marker of anaplastic large-cell lymphoma, *Am. J. Path.*, 148 (1996) 25–30.
5. S. Akahani, P. Nangia-Makker, H. Inohara, H.-R. Kim, and A. Raz, Galectin-3: A novel antiapoptotic molecule with a functional BH1 (NHGR) domain of bcl-2 family, *Cancer Res.*, 57 (1997) 5272–5276.
6. J. M. Rini, Lectin structure, *Annu. Rev. Biophys. Biomol. Struct.*, 24 (1995) 551–577.
7. R. Loris, T. Hamelryck, J. Bouckaert, and L. Wyns, Legume lectin structure, *Biochim. Biophys. Acta*, 1383 (1998) 9–36.
8. N. L. Perillo, K. E. Pace, J. J. Seilhamer, and L. G. Baum, Apoptosis of T cells mediated by galectin-1, *Nature*, 378 (1995) 736–739.
9. N. L. Perillo, C. H. Uittenbogaart, J. T. Nguyen, and L. G. Baum, Galectin-1, an endogenous lectin produced by thymic epithelial cells, induces apoptosis of human thymodcytes, *J. Exp. Med.*, 185 (1997) 1851–1858.

10. G. N. R. Vespa, L. A. Lewis, K. R. Kozak, M. Moran, J. T. Nguyen, L. G. Baum, and M. C. Miceli, Galectin-1 specifically modulates TCR signals to enhance TCR apoptosis but inhibit IL-2 production and proliferation, *J. Immunol.*, 162 (1999) 799–806.
11. M. Demetriou, M. Granovsky, S. Quaggin, and J. W. Dennis, Negative regulation of T-cell activation and autoimmunity by Mgat5 N-glycosylation, *Nature*, 409 (2001) 733–739.
12. J. Kopitz, C. von Reitzenstein, M. Burchert, M. Cantz, and H.-J. Gabius, Galectin-1 is a major receptor for ganglioside GM1, a product of the growth-contolling activity of a cell surface ganglioside sialidase, on human neuroblastoma cells in culture, *J. Biol. Chem.*, 273 (1998) 11205–11211.
13. K. Ohtsubo, S. Takamatsu, M. T. Minowa, A. Yoshida, M. Takeuchi, and J. D. Marth, Dietary and genetic control of glucose transporter 2 glycosylation promotes insulin secretion in suppressing diabetes, *Cell*, 123 (2005) 1307–1321.
14. E. A. Partridge, C. Le Roy, G. M. Di Gulielmo, J. Pawling, P. Cheung, M. Granovsky, I. R. Nabi, J. L. Wrana, and J. W. Dennis, Regulation of cytokine receptors by Golgi N-Glycan processing and endocytosis, *Science*, 306 (2005) 120–124.
15. T. K. Dam and C. F. Brewer, Carbohydrate–lectin crosslinking interactions: Structural, thermodynamic, and biological studies, *Methods Enzymol.*, 362 (2003) 455–486.
16. C. F. Brewer, M. C. Miceli, and L. G. Baum, Clusters, bundles, arrays and lattices: Novel mechanisms for lectin–saccharide-mediated cellular interactions, *Curr. Opin. Struct. Biol.*, 12 (2002) 616–623.
17. J. C. Sacchettini, L. G. Baum, and C. F. Brewer, Multivalent protein–carbohydrate interactions. A new paradigm for supermolecular assembly and signal transduction, *Biochemistry*, 40 (2001) 3009–3015.
18. L. L. Kiessling, J. K. Pontrello, and M. C. Schuster, Synthetic multivalent carbohydrate ligands as effectors or inhibitors of biological processes, in C.-H. Wong, (Ed.), *Carbohydrate-Based Drug Discovery*, Wiley-VCH Weinheim, Germany, 2003, pp. 575–608.
19. T. K. Dam and C. F. Brewer, Multivalent protein–carbohydrate interactions: Isothermal titration microcalorimetry studies, *Methods Enzymol.*, 379 (2004) 107–128.
20. R. T. Lee and Y. C. Lee, Affinity enhancement by multivalent lectin–carbohydrate interaction, *Glycoconjugate J.*, 17 (2000) 543–551.

21. T. K. Dam and C. F. Brewer, Fundamentals of lectin–carbohydrate interactions, in J. P. Kamerling, G.-J. Boons, Y. C. Lee, A. Suzuki, N. Taniguchi, and A. G. Voragen, (Eds.), Elsevier, Oxford, 2007, pp. 397–452.
22. R. T. Lee, Y. Ichikawa, M. Fay, K. Drickamer, M. C. Shao, and Y. C. Lee, Ligand-binding characteristics of rat serum-type mannose-binding protein (MBP-A), *J. Biol. Chem.*, 266 (1991) 4810–4815.
23. P. I. Kitov, J. M. Sadowska, G. Mulvey, G. D. Armstrong, H. Ling, N. Pannu, R. J. Read, and D. R. Bundle, Shiga-like toxins are neutralized by tailored multivalent carbohydrate ligands, *Nature*, 403 (2000) 669–672.
24. D. Page, D. Zanini, and R. Roy, Macromolecular recognition: Effect of multivalency in the inhibition of binding of yeast mannan to concanavalin A and pea lectins by mannosylated dendrimers, *Bioorg. Med. Chem.*, 4 (1996) 1949–1961.
25. R. Roy, D. Page, S. F. Perez, and V. V. Bencomo, Effect of shape, size, and valency of multivalent mannosides on their binding properties to phytohemagglutinins, *Glycoconj. J.*, 15 (1998) 251–263.
26. T. K. Dam, R. Roy, S. K. Das, S. Oscarson, and C. F. Brewer, Binding of multivalent carbohydrates to concanavalin A and *Dioclea grandiflora* lectin. Thermodynamic analysis of the ''multivalency effect'', *J. Biol. Chem.*, 275 (2000) 14223–14230.
27. T. K. Dam, H.-J. Gabius, S. Andre, H. Kaltner, M. Lensch, and C. F. Brewer, Galectins bind to the multivalent glycoprotein asialofetuin with enhanced affinities and a gradient of decreasing binding constants, *Biochemistry*, 44 (2005) 12564–12571.
28. T. K. Dam, T. A. Gerken, B. S. Cavada, K. S. Nascimento, T. R. Moura, and C. F. Brewer, Binding studies of α-GalNAc specific lectins to the α-GalNAc (Tn-antigen) form of porcine submaxillary mucin and its smaller fragments, *J. Biol. Chem.*, 282 (2007) 28256–28263.
29. L. L. Kiessling, J. E. Gestwicki, and L. E. Strong, Synthetic multivalent ligands as probes of signal transduction, *Angew. Chem. Inter. Edit. Eng.*, 45 (2006) 2348–2368.
30. A. Varki, Biological roles of oligosaccharides: All of the theories are correct, *Glycobiology*, 3 (1993) 97–101.
31. J. C. Byrd and R. S. Bresalier, Mucins and mucin binding proteins in colorectal cancer, *Cancer Metastasis Rev.*, 23 (2004) 77–99.
32. M. Fukuda, Role of mucin-type O-glycans in cell adhesion, *Biochim. Biophys. Acta*, 1573 (2002) 394–405.
33. H. C. Hang and C. R. Bertozzi, The chemistry and biology of mucin-type O-linked glycosylation, *Bioorg. Med. Chem.*, 13 (2005) 5021–5034.

34. F.-G. Hanisch and S. Muller, MUC1: The polymorphic appearance of a human mucin, *Glycobiology*, 10 (2000) 439–449.
35. S. Hakomori, The glycosynapse, *Proc. Nat. Acad. Sci. USA*, 99 (2002) 225–232.
36. P. K. Singh and M. A. Hollingsworth, Cell surface-associated mucins in signal transduction, *Trends Cell Biol.*, 16 (2006) 467–476.
37. D. H. Dube and C. R. Bertozzi, Glycans in cancer and inflammation—Potential for therapeutics and diagnostics, *Nat. Rev. Drug Disc.*, 4 (2005) 477–488.
38. L.-G. Yu, N. Andrews, Z. Zhao, D. McKean, J. F. Williams, L. J. Connor, O. V. Gerasimenko, J. Hilkens, J. Hirabayashi, K. Kasai, and J. M. Rhodes, Galectin-3 interaction with thomsen-friedenreich disaccharide on cancer-associated MUC1 causes increased cancer cell endothelial adhesion, *J. Biol. Chem.*, 282 (2007) 773–781.
39. T. A. Gerken, Kinetic modeling confirms the biosynthesis of mucin core 1 (β-Gal (1-3)α-GalNAc-O-Ser/Thr) O-glycan structures are modulated by neighboring glycosylation effects, *Biochemistry*, 43 (2004) 4137–4142.
40. A. E. Eckhardt, C. S. Timpte, A. W. DeLuca, and R. L. Hill, The complete cDNA sequence and structural polymorphism of the polypeptide chain of porcine submaxillary mucin, *J. Biol. Chem.*, 272 (1997) 33204–33210.
41. D. Carlson, Structures and immunological properties of oligosaccharides isolated from pig submaxillary mucins, *J. Biol. Chem.*, 243 (1968) 616–626.
42. T. A. Gerken and N. Jentoft, Structure and dynamics of porcine submaxillary mucin as determined by natural abundance carbon-13 NMR spectroscopy, *Biochemistry*, 26 (1987) 4689–4699.
43. T. A. Gerken, C. L. Owens, and M. Pasumarthy, Determination of the site-specific O-glycosylation pattern of the porcine submaxillary mucin tandem repeat glycopeptide, *J. Biol. Chem.*, 272 (1997) 9709–9719.
44. A. L. Sorensen, C. A. Reis, M. A. Tarp, V. Sankaranarayanan, T. Schwientek, R. Graham, J. Taylor-Papadimitriou, M. A. Hollingsworth, J. Burchell, and H. Clausen, Chemoenzymatically synthesized multimeric Tn/STn MUC1 glycopeptides elicit cancer-specific anti-MUC1 antibody responses and override tolerance, *Glycobiology*, 16 (2006) 96–107.
45. K. S. Lau, E. A. Partridge, A. Grigorian, C. I. Silvescu, V. N. Reinhold, M. Demetriou, and J. W. Dennis, Complex N-glycan number and degree of branching cooperate to regulate cell proliferation and differentiation, *Cell*, 129 (2007) 123–134.
46. M. A. Daniels, K. A. Hogquist, and S. C. Jameson, Sweet 'n' sour: The impact of differential glycosylation on T cell responses, *Nat. Immunol.*, 3 (2002) 903–910.

47. K. E. Pace, C. Lee, P. L. Stewart, and L. G. Baum, Restricted receptor segregation into membrane microdomains occurs on human T cells during apoptosis induced by galectin-1, *J. Immunol.*, 163 (1999) 3801–3811.
48. M. Kanai, K. H. Mortell, and L. L. Kiessling, Varying the size of multivalent ligands: The dependence of concanavalin A binding on neoglycopolymer length, *J. Am. Chem. Soc.*, 119 (1997) 9931–9932.
49. R. Lotan, E. Skutelsky, D. Danon, and N. Sharon, The purification, composition, and specificty of the anti-T lectin from peanut (*Arachis hypogaea*), *J. Biol. Chem.*, 250 (1975) 8518–8523.
50. A. Dessen, D. Gupta, S. Sabesan, C. F. Brewer, and J. C. Sacchettini, X-ray crystal structure of the soybean agglutinin crosslinked with a biantennary analog of the blood group I carbohydrate antigen, *Biochemistry*, 34 (1995) 4933–4942.
51. J. E. Gestwicki, L. E. Strong, C. W. Cairo, F. J. Boehm, and L. L. Kiessling, Cell aggregation by scaffolded receptor clusters, *Chem. Biol.*, 9 (2002) 163–169.
52. D. K. Mandal and C. F. Brewer, Crosslinking activity of the 14-kilodalton β-galactose-specific vertebrate lectin with asialofetuin: Comparison with several galactose-specific plant lectins, *Biochemistry*, 31 (1992) 8465–8472.
53. T. K. Dam, R. Roy, D. Pagé, and C. F. Brewer, Negative cooperativity associated with binding of multivalent carbohydrates to lectins. Thermodynamic analysis of the "multivalency effect"*Biochemistry*, 41 (2002) 1351–1358.
54. T. K. Dam and C. F. Brewer, Thermodynamic studies of lectin–carbohydrate interactions by isothermal titration calorimetry, *Chem. Rev.*, 102 (2002) 387–429.
55. J. J. Calvete, C. F. Santos, K. Mann, T. B. Grangeiro, M. Nimtnz, C. Urbanke, and B. S. Cavada, Amino acid sequence, glycan structure, and proteolytic processing of the lectin of *Vatairea macrocarpa* seeds, *FEBS Lett.*, 452 (1998) 286–292.
56. A. M. Wu, J. H. Wu, J.-H. Liu, T. Singh, S. Andre, H. Kaltner, and H.-J. Gabius, Effects of polyvalency of glycotopes and natural modifications of human blood group ABH/Lewis sugars at the Galβ1-terminated core saccharides on the binding of domain-I of recombinant tandem-repeat-type galectin-4 from rat gastrointestinal tract (G4-N), *Biochimie*, 86 (2004) 317–326.
57. G. F. Springer, T and Tn, general carcinoma autoantigens, *Science*, 224 (1984) 1198–1206.
58. Y. van Kooyk and C. G. Figdor, Avidity regulation of integrins: The driving force in leukocyte adhesion, *Curr. Opin. Cell Biol.*, 12 (2000) 542–547.
59. M. F. Krummel, M. D. Sjaastad, C. Wulfing, and M. M. Davis, Differential clustering of CD4 and CD3d during T cell recognition, *Science*, 289 (2000) 1349–1352.

60. G. Nussbaum, W. Cleare, A. Casadevall, M. D. Scharff, and P. Valadon, Epitope location in the *Cryptococcus neoformans* capsule is a determinant of antibody efficacy, *J. Exp. Med.*, 185 (1997) 685–694.
61. J. Engel, Common structural motifs in proteins of the extracellular matrix, *Curr. Opin. Struct. Biol.*, 3 (1991) 779–785.
62. A. Krufka, S. Guimond, and A. C. Papraeger, Two hierarchies of FGF-2 signaling in heparin: Mitogenic stimulation and high-affiity binding/receptor transphosphorylation, *Biochemistry*, 35 (1996) 11131–11141.

DESIGN AND CREATIVITY IN SYNTHESIS OF MULTIVALENT NEOGLYCOCONJUGATES

BY YOANN M. CHABRE and RENÉ ROY

Department of Chemistry, Université du Québec à Montréal,
Montréal, Québec, H3C 3P8, Canada

ISBN: 978-0-12-380856-1
DOI: 10.1016/S0065-2318(10)63006-5.

Abbreviations

3TC, 2′,3′-dideoxy-3′-thiacytidine (lamivudine); AFM, atomic force microscopy; AGP, human α1-acid glycoprotein; AIBN, 2,2-azobisisobutyronitrile; ASF, asialofetuin glycoprotein; ATRP, atom transfer radical polymerization; BBA, bladder-binding assay; 9-BBN, 9-borabicyclo[3.3.1]nonane; BBV, *N,N*′-4,4′-bis(benzyl-3-boronic acid)bipyridinium dibromide; BIEMA, 2-(2-bromoisobutyryloxy)ethyl methacrylate; bis-MPPA, 2,2-bis(hydroxymethyl)propanoic acid; Boc, *tert*-butoxy carbonyl; BOP, benzotriazole-1-yl-oxy-tris(dimethylamino)phosphonium hexafluorophosphate; Cbz, benzyloxycarbonyl; CCI, carbohydrate–carbohydrate interactions; β-CD, β-cyclodextrin; CD69, cluster of differentiation 69; CFU, colony-forming units; CHO, chinese hamster ovary; CNT, carbon nanotube,; Con A, Concanavalin A; CPS, capsular polysaccharide; CRD, carbohydrate recognition domain; CSLM, confocal laser-scanning microscopic; CT, cholera toxin; CuAAc, Cu(I)-catalyzed azide–alkyne [1,3]-dipolar cycloaddition; DBU, 1,8-diazabicyclo[5.4.0]undec-7-ene; DCC, *N,N*′-dicyclohexylcarbodiimide; DC-SIGN, dendritic cell-specific intercellular adhesion molecule-3-grabbing nonintegrin; DDQ, 2,3-dichloro-5,6-dicyanobenzoquinone; DEG, diethylene glycol; DIBAL-H, diisobutylaluminium hydride; DIC, diisopropylcarbodiimide; DIPEA, diisopropylethylamine; DLS, dynamic light scattering; DMAP, dimethylaminopyridine; DMF, *N,N*-dimethylformamide; DMI, 1,3-dimethyl-2-imidazolidinone; DMPC, dimyristoyphosphatidylcholine; DMSO, dimethyl sulfoxide; DOSY, diffusion ordered spectroscopy; DPPC, dipalmitoyl phosphatidylcholine; EcorL, *Erythrina corallodendron* lectin; EDC or EDAC, 1-ethyl-3-(3-dimethylaminopropyl)carbodiimide; EEDQ, 2-ethoxy-1-ethoxycarbonyl-1,2-dihydroquinoline; EFV, efavirenz; EHEC, enterohemorrhagic *E. coli*; ELISA, enzyme-linked immunosorbent assay; ELLA, enzyme-linked lectin assay; EPR, enhanced permeation retention; ESI-MS, electrospray ionization-mass spectrometry; FITC, fluorescein isothiocyanate; fmoc, 9H-fluoren-9-ylmethoxycarbonyl; FRET, fluorescence resonance energy transfer; Gb_3, glycosphingolipid globotriaosylceramide; GM1, monosialotetrahexosylganglioside; GNA, *Galanthus nivilis* agglutinin; HAI, inhibition of hemagglutination; HATU, (2-(7-aza-1H-benzotriazole-1-yl)-1,1,3,3-tetramethyluronium hexafluorophosphate); HEK293, human embryonic kidney 293; HepG2, hepatocellular carcinoma cells; HIV, human immunodeficiency virus; HOAt, 1-hydroxy-7-azabenzotriazole; HOBt, hydroxybenzotriazole; HP, haematoporphyrin; HPA, *Helix pomatia* agglutinin; HPA-FITC, *Helix pomatia* agglutinin-fluorescein isothiocyanate; HRPO, horseradish peroxidase; HUS, hemolytic uremic syndrome; IC, inhibitory concentration; ICAM, intercellular adhesion molecule; IgG 2G12, immunoglobulin G 2G12 antibody; K_D, dissociation constant;

IT, inhibition titer; ITC, isothermal titration microcalorimetry; KLH, keyhole limpet hemocyanin; LCA, *Lens culinaris* lectin; LFA, *Limax flavus* lectin; LTBh, *E. coli* heat-labile toxin B; NHS, *N*-hydroxysuccinimide; NK, natural killer cell; NMR, nuclear magnetic resonance; MAG, multiple antigenic glycopeptides; MAIG, 3-*O*-methacryloyl-1,2:5,6-di-*O*-isopropylidene-D-glucofuranose; MALDI-TOF, matrix-assisted laser desorption/ionization-time-of-flight; MALS, multiangle light scattering; MBP, mannose-binding protein; Me-αMan, Me α-D-mannopyranoside; MDCK, Madin–Darby canine kidney cells; MHC, major histocompatability complex; MIC, minimum inhibitory concentration; MLCT, metal-to-ligand charge transfer; MM, mucin mimic; MMP, matrix metalloproteinase; MNPG, *m*-nitrophenyl α-D-galactoside; MPPI, mannosylated-polypropyleneimine dendrimers; MPR, mannose-6-phosphate receptor; MRI, magnetic resonance imaging; MSC, mesenchymal stromal cells; MT1-MMP, membrane type-1-matrix metalloproteinase; MUNeuAc, 2′-(4-methylumbelliferyl)-α-*N*-acetylneuraminic acid; MVK, methyl vinyl ketone; MWNT, multiwalled nanotube; OMPC, outer membrane protein complex; P_3CS, tripalmitoyl-*S*-glycerylcysteinyl-serine; PAII-L, *Pseudomonas aeruginosa* lectin II; PAMAM, poly(amidoamine) dendrimers; PAP, poly(*p*-*N*-acryloylamidophenyl); PAP-α-Glc, poly(*p*-*N*-acryloylamidophenyl) α-glucopyranoside; PBS, phosphate buffered saline; PePO, pentaerythrityl phosphodiester oligomer; PET, photoinduced electron transfer; PITC, phenylisothiocyanate; PDT, photodynamic therapy; POSS, polyhedral oligosilsesquioxane; PMBC, peripheral blood mononuclear cells; PPI, polypropyleneimine; PS, photosensitizer; PVK, poly(methyl vinyl ketone); PyBOP, benzotriazol-1-yl-oxytripyrrolidinophosphonium hexafluorophosphate; QSAR, quantitative structure–activity relationship; RCA_{120}, *Ricinus communis* agglutinin; RGP, radial growth polymerization; SCVCP, self-condensing vinyl copolymerization; SEM, scanning electron microscopy; SHC, Sonogashira—Heck—Cassar cross coupling reactions; SLT, Shiga-like toxin; SLT-Iie, Shiga-like toxin II edema variant; SPG-Lac, schizophyllan bearing lactosides; SPR, surface plasmon resonance; SPS, solid-phase synthesis; STEC, Shiga toxigenic group of *Escherichia coli*; Stx, Shiga toxin; SWNT, single-walled nanotube; TBAF, tetra-*n*-butylammonium fluoride; TBAH, tetra-*n*-butylammonium hydroxide; TBTU, 2-(1*H*-benzotriazole-1-yl)-1,1,3,3-tetramethyluronium tetrafluoroborate; TEM, transmission electron microscopy; TEMPO, 2,2,6,6-tetramethylpiperidine-1-oxyl; TF, Thomsen–Friedenreich antigen; TFA, trifluoroacetic acid; TGA, thermogravimetric analysis; THF, tetrahydrofuran; TRIS, tris(hydroxymethyl)aminomethane; VAA, *Viscum album* agglutinin; VT, verotoxin; VV-HRP, horseradish peroxidase-labeled plant lectin *V. villosa*; VVA, *V. villosa* plant lectin; WGA, wheat germ agglutinin; YDS, yolk decasaccharide

I. Introduction

1. Multivalency: Definition and Role

Generally, the valency of a particle (namely a small molecule, oligosaccharide, protein, nucleic acid, lipid or aggregate of these molecules, a virus, bacterium, or cell) can be defined as the number of separate structural units of the same kind that can interact with other particles through ligand–receptor interactions.[1] Thus, one can consider that a molecule having two tethered and identical copies of binding components can be classified as a divalent entity. Similarly, multivalent or polyvalent interactions can be defined as specific simultaneous associations of multiple ligands (or epitopes) present on a molecular construct or biological surface that binds in a cooperative way to multiple receptors expressed on a complementary entity.

The ubiquity of these multivalent interactions at different levels in several biological mechanisms testifies to their essential role. In fact, multivalency in Nature is very often expressed by fractal or ''dendritic'' architectures that represent perhaps the most pervasive topologies observed in vegetal and animal kingdoms.[2] Typical examples of these patterns may be found at different scales of dimensional length (meters to microns), and typical examples can be observed in abiotic systems (such as snow crystals, fractal erosions, manganese dendrites in rock) or in the biological world. The reasons for such extensive mimicry of these dendritic topologies at virtually all scales of dimensional length are not entirely clear. However, one might speculate that these evolutionary architectures have been optimized over the past several billion years to provide structures manifesting maximum interfaces for optimum energy extraction/distribution, nutrient extraction/distribution, information storage/retrieval, and adhesive processes. For example, trees use fractal dendritic patterns above and beneath the ground in order to enhance the exposure of their leaves to sunlight to harvest light and maximize the photosynthesis process, and of their roots to collect water from the soil.

Most notable natural examples of this architecture at the molecular level are probably the glycogen and amylopectin hyperbranched structures that Nature uses for energy storage. Presumably, the many chain ends that decorate these macromolecules facilitate enzymatic access to glucose for high-demand bioenergy events. Another nanoscale example of dendritic architecture in biological systems is found in proteoglycans. These macromolecules appear to provide energy-absorbing, cushioning properties and determine the viscoelastic properties of connective tissues.

In addition, the tremendously complex dendritic respiratory network is composed of bronchioles and alveoli to give the maximum surface for efficient transfer of oxygen into the bloodstream. The arterial and central nervous system networks, together with the kidneys and lung structures also consist of a great number of cells growing into dendritic structures in order to gain the largest exchange of material or information with the surrounding tissues.[3] Recently, the implication of dendritic patterns observed under the feet of the gecko has also been clarified to explain the exceptional ability of these animals to climb rapidly up smooth vertical surfaces.[4] Microscopic examinations have shown that a gecko's foot has nearly five hundred thousand keratinous hairs or setae, each composed of an impressive dendritic network of tiny foot hairs ''spatulae.'' Measurements have revealed that one seta is 10 times more effective at adhesion than predicted from maximal estimates on whole animals. Values of the adhesive force support the hypothesis that individual setae operate by weak attractive quantum chemical forces from molecules in each foot-hair interacting through Van der Waals forces with molecules of the surface.

Finally, a very striking example is also afforded by Nature, particularly in exposing a wide array of complex dendritic glycoconjugates on mammalian and HIV-1 cell surfaces. These carbohydrate structures play critical roles in multiple key cellular events, such as cellular adhesion and recognition, regulation of physiological functions, and pathogenic infections. For HIV-1, several hypothesis have been formulated concerning their role in infection process, and it has been speculated that the N-linked hyperbranched high mannose oligosaccharide ($Man_9GlcNAc_2$), exposed at the exterior envelope glycoprotein gp120, helps the ''hidden'' virus to escape neutralizing antibodies.[5] Indeed, these glycans are produced by the host cell, and are largely unrecognized by the immune system machinery.

2. Multivalency in Protein–Carbohydrate Interactions

Although carbohydrates and their corresponding conjugated glycoforms have long been regarded as only space-filling matrices or post-transcriptional accessory elements in glycoproteins serving to protect them from premature degradation, it has become apparent that such glycoconjugates as glycolipids or glycosaminoglycans exhibit a broad variety of additional biological functions. Indeed, carbohydrates are expressed on the majority of mammalian cell surfaces, and are bound to proteins, glycoproteins, and glycolipids, or conjugated to such cellular constituents as proteoglycans that are entangled in the cell membrane and clustered in multiantennary configurations. In other terms, these oligosaccharides constitute signal transductors between extra- and intracellular media.[6–8] Hence, these glycosylated structures are responsible for the presentation of target structures for microorganisms, toxins and antibodies, control the half-life of proteins, modulation of protein function, or provision of ligands for specific binding events. In addition, they have recently emerged as antigenic determinants in cell–cell recognition, signaling events, and as ligands for bacterial and viral infections.[9,10] As such, they constitute the first line of contact for the adhesion and tissue colonization by several pathogens expressing carbohydrate-binding proteins (lectins). Conversely, several infectious microorganisms use or escape the immune defense mechanisms by masking important receptors or antigenic determinants by exposing self-carbohydrate structures. As mentioned earlier, the consequences of these ''masking'' events are that the bacterial or viral pathogens are transported to target tissues by self cellular systems.

Despite their critical importance, carbohydrate–protein interactions are paradoxically characterized by rather weak association constants (millimicromolar), with limited specificity and selectivity on a per-saccharide basis.[11,12] Nature usually compensates for this situation by exposing numerous copies of the same carbohydrate ligands on the extracellular domains of the cells. Consequently, these interactions are transformed into very potent attractive forces, dramatically and naturally reinforced, when multiple ligand copies are presented to similarly clustered receptors. This phenomenon, resulting from a synergic and cooperative effect, is known as the ''glycocluster or dendritic effect,''[13] and has been initially observed with asialoglycoprotein receptors found on hepatocytes.[14] In its widespread version, it is usually assumed that this effect has its source in the enhanced affinity of a given multivalent glycoside toward a CRD (carbohydrate-recognition domain) by fully occupying one active site at a time.The phenomenon is now widely accepted of having its basis in stabilization by macroscopic ''crosslinking glycocluster effects.'' It has thus come well established that multivalency may offer numerous benefits in terms of affinity and receptor selectivity versus monovalent interaction, and can induce particular clustering organization on the cell surface, notably to provide a strategy for controlling signal transduction pathways within cells.[15]

Surprisingly, however, mammals utilize only nine different monosaccharides, which are organized in a massive amount of structural diversity. The variations of anomeric configurations and linkage positioning between saccharides are thus responsible for unraveling several distinctive ''glycocodes.''[16] As a consequence, the different architectures and topological expressions that result are at the origin of the required high affinity and selectivity for a particular tailored recognition event. Interestingly, even such complex multiantennary glycans as $Man_9GlcNAc_2$ are only partly involved in these binding processes and often, but not unexpectedly, only the peripheral groups are bound in the receptors' active sites. Hence, multivalent interactions are now understood to be a ubiquitous strategy that has evolved in Nature for a wide range of functions, and which provide numerous benefits and unique roles not achievable with monovalent interactions. Consequently, from a simplistic point of view, most multiantennary glycans may be regarded as polyvalent neoglycoconjugates of well-defined architectures and multivalency, designed to mimic the complexity of the multiantennary oligosaccharide structures while emphasizing the recognition of the sole surface.

3. Synthesis and Applications of Multivalent Glycoconjugates

As mentioned before, multivalent carbohydrate–protein interactions mediate many important physiological and pathophysiological processes. However, their thorough understanding suffers from the natural complexity of the carbohydrates resulting from incomplete biosynthesis or subtle attachment of other functionalities at specific positions along the oligosaccharide sequences. Numerous key functions of carbohydrates depend on the observed microheterogeneity, added to well-defined cluster organization. In order to study, characterize, understand, and manipulate these critical interactions, striking advances have been made in isolation, purification, structural analyses, and partial or selective degradation processes. Alternatively, chemical or chemoenzymatic synthesis of multivalent carbohydrate ligands is, however, likely to remain the method of choice to afford tailored multivalent architectures developed as effectors or inhibitors of biological mechanisms. Accordingly, it is highly desirable to tackle the inherent problem of high specificity and increased affinity by simultaneously optimizing factors involved in both multivalency and intrinsic fine tuning of a ligand toward individually targeted carbohydrate-binding protein interactions.

To achieve these goals, glycochemists are actively pursuing an approach that can be summarized in Fig. 1. The steps to be undertaken may be expressed as follows. After assessing the bacterial/viral genomes, proteomic analysis, and search for carbohydrate-binding protein homology through bioinformatics, the presence of a carbohydrate-binding protein is then confirmed. For bacterial or viral lectins, the isolated proteins can be labeled with fluorogenic probes. The ligand-binding specificity of the fluorescent lectins is then determined by glycan microarrays such as the one freely accessible from the Consortium for Functional Glycomics (or the like) to identify the best oligosaccharide "lead."[17] The most recent microarrays are usually constituted of approximately 400 natural and synthetic glycans (version 3.10).[18–22] Once essential carbohydrate residues (epitopes) responsible for the biological activity of interest are

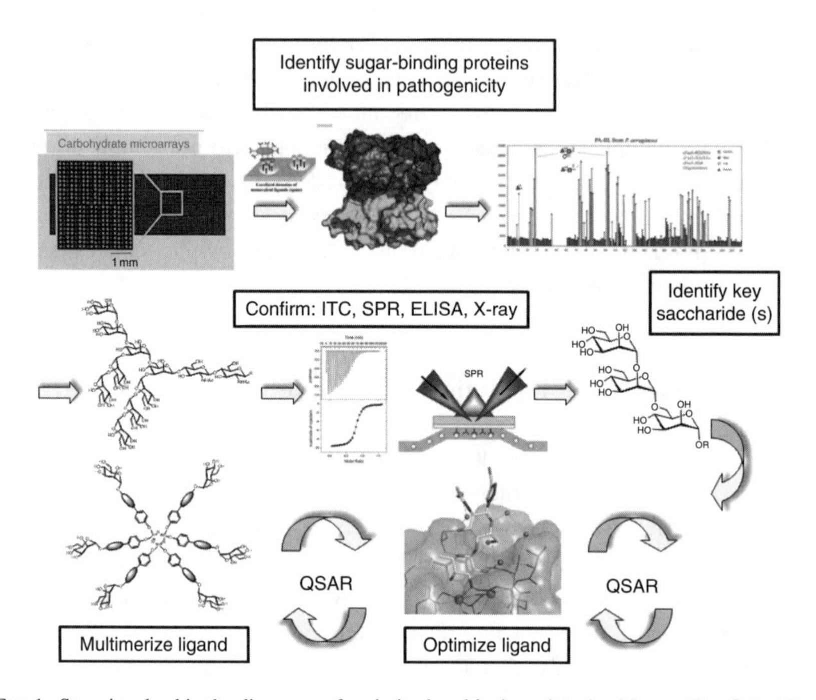

FIG. 1. Steps involved in the discovery of optimized multivalent glycodendrimers. (See Color Plate 8.)

identified, the lead candidates are then validated by using a range of binding assays such as ELISA (enzyme-linked immunosorbent assay), ELLA (enzyme-linked lectin assay), ITC (isothermal titration calorimetry), SPR (surface plasmon resonance), FRET (fluorescence resonance energy transfer), X-ray crystallography, and analogous techniques. The relative binding affinities of confirmed oligosaccharide ligands are next evaluated with a panel of simpler oligosaccharides with, ideally, a monosaccharide as the simplest target to ease potential manufacturing purposes and lead optimization that can be obtained by classical QSAR (quantitative structure–activity relationship).

The resulting ''glycomimetics'' are further transformed into multivalent architectures such as glycodendrimers that should also undergo iterative scaffold optimization. In this context, ''artificial glycoforms'' have played crucial roles in our understanding of multivalent interactions, which encompass chelation,[23] receptor clustering and steric stabilization,[24] subsite binding,[25] and statistical rebinding phenomena.[26] Hence, research on synthetic multivalent macromolecules has intensified, giving rise to a myriad of original glycoconjugate structures which constitute high-affinity multivalent ligands that target surface receptors (namely enzymes, lectins, toxins, and such pathogens as viruses or bacteria). These novel nanometer-size structures have thereupon shown great potential as potent effectors or inhibitors of surface–surface interactions, including cell–cell and cell–pathogen interactions that occur in biological systems.[27]

The following sections illustrate synthetic creativity recently devoted to generating a plethora of multivalent structures used in the biomedical field and with therapeutic goals in particular. Synthetic neoglycoconjugates in which carbohydrate residues are attached to carriers thus present numerous advantages in terms of characterization, structural uniformity, and availability. Hence, presentation of the sugar epitopes as multiple copies on an appropriate scaffold (molecular, dendritic, polymeric) creates a multivalent display that can efficiently mimic the natural mode of affinity enhancement that arises from multiple interactions between the binding proteins and the carbohydrate ligands. In general, the carbohydrate ligands are usually found at the periphery of these macromolecules. Efficient conjugation reactions are required for complete substitution and the structural integrities of the multivalent glycoconjugates are evaluated by such conventional techniques as NMR spectroscopy and mass spectrometry. Progress encountered in terms of their synthetic accessibility and efficiency has allowed optimization of their modulation and activity. These factors remain necessary for investigating how these multivalent structures can best influence the targeted binding activity. These investigations are critical for highlighting the potential of multivalent carbohydrate inhibitors as high-affinity ligands or as effectors capable of clustering cell-surface receptors, and may permit generation of structures having tailored biological activities.

In addition to useful but polydisperse multivalent glycomimetics of prominent components at the surfaces of mammalian cells [such as neoglycoproteins,[28,29] neoglycopeptides,[30–32] neoglycolipids (or glycoliposomes),[33,34] glycopolymers,[35–38] or glyconoparticles,[39–49] new synthetic families of well-defined and monodisperse glycosylated macromolecules have recently emerged, including glycoclusters, such glycosylated nanomaterials as glycofullerenes and glyconanotubes, glycosylated architectures from supramolecular assembly processes, and finally glycodendrimers. The uniformity of these multivalent neoglycoconjugates is ensured by the controlled arrangement of the constitutional building blocks. To this end, a large panel of linkage functions have been used, but the ones most frequently used are amides, thioureas, and recently 1,2,3-triazoles obtained via dipolar cycloadditions of alkyne and azide precursors through a process now termed ''click chemistry.''[50–53] The required functionalities are installed with equal success on either the sugars or on the scaffolds. This particular application has become widely used, since the sugar attachment can be effected with both protected or free sugars. The protecting groups most repeatedly employed on the sugars have been esters because it is usually more difficult to remove large number of ether or acetal-protected sugars (for example, benzyl ethers). Moreover, the lectin-binding efficiency and specificity of multivalent glycoconjugates have been found to be dependent not only on the epitope density but also on the nature of the core and the geometrical characteristics of the multivalent assembly.

The main goal of this chapter consists in the description of the most recently conceived multivalent neoglycoconjugates, emphasizing the synthetic strategy required to afford well-defined glycoclusters, glycofullerenes, glyconanotubes, glycosylated self-assembled systems, and glycodendrimers, together with discussions concerning their respective relevant uses and perspectives in biomedical applications.

II. Glycoclusters

''Glycoclusters'' (or ''cluster glycosides'') can be structurally considered as mimics of oligoantennary oligosaccharides of naturally occurring glycoconjugates. Arbitrarily, this chapter considers these clustered structures as multivalent glycoconjugates, regardless of the number of peripheral saccharides and they are built from ''home-made'' or commercially available scaffolds that do not contain repetitive units. These structural characteristics allow distinction between these generally rather low-valency architectures and the ''true'' multigeneration glycodendrimers that are presented in later sections.

Because of the straightforward synthetic access to such rather small structures, a large variety of glycoclusters, based on multivalent scaffolds and illustrated in Fig. 2, have been described in the past few years. Hence, branched aliphatic or aromatic scaffolds, calixarenes, porphyrins and derivatives, cyclic peptides, carbohydrates and cyclodextrins, or more exotic central cores have been efficiently used as core molecules, to afford small multivalent glycoconjugates having greatly enhanced avidity compared to corresponding monovalent carbohydrates. In particular, the design of such systems, including the valency, the introduction of suitable functions and flexible linkers with optimal lengths, has been tailored. Thus, their optimization has rapidly led to the dramatic enhancement of neoglycoconjugate–protein interactions, and has been useful for determining such critical parameters as optimal association geometries, including optimal distances between saccharide epitopes. It is noteworthy that the majority of these scaffolds have also been used for constructing more complex multigeneration glycodendrimers, in particular orthogonally derivatized AB_x systems, such as TRIS or gallic acid derivatives.

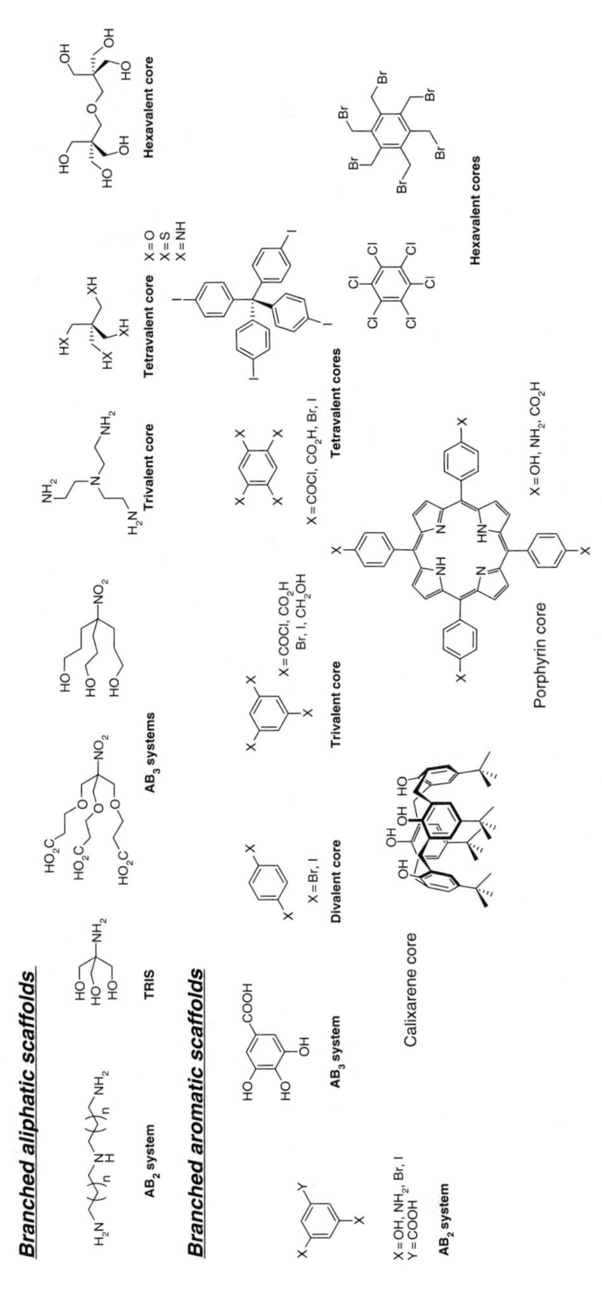

FIG. 2. Common scaffolds used in the design of glycoclusters and glycodendrimers.

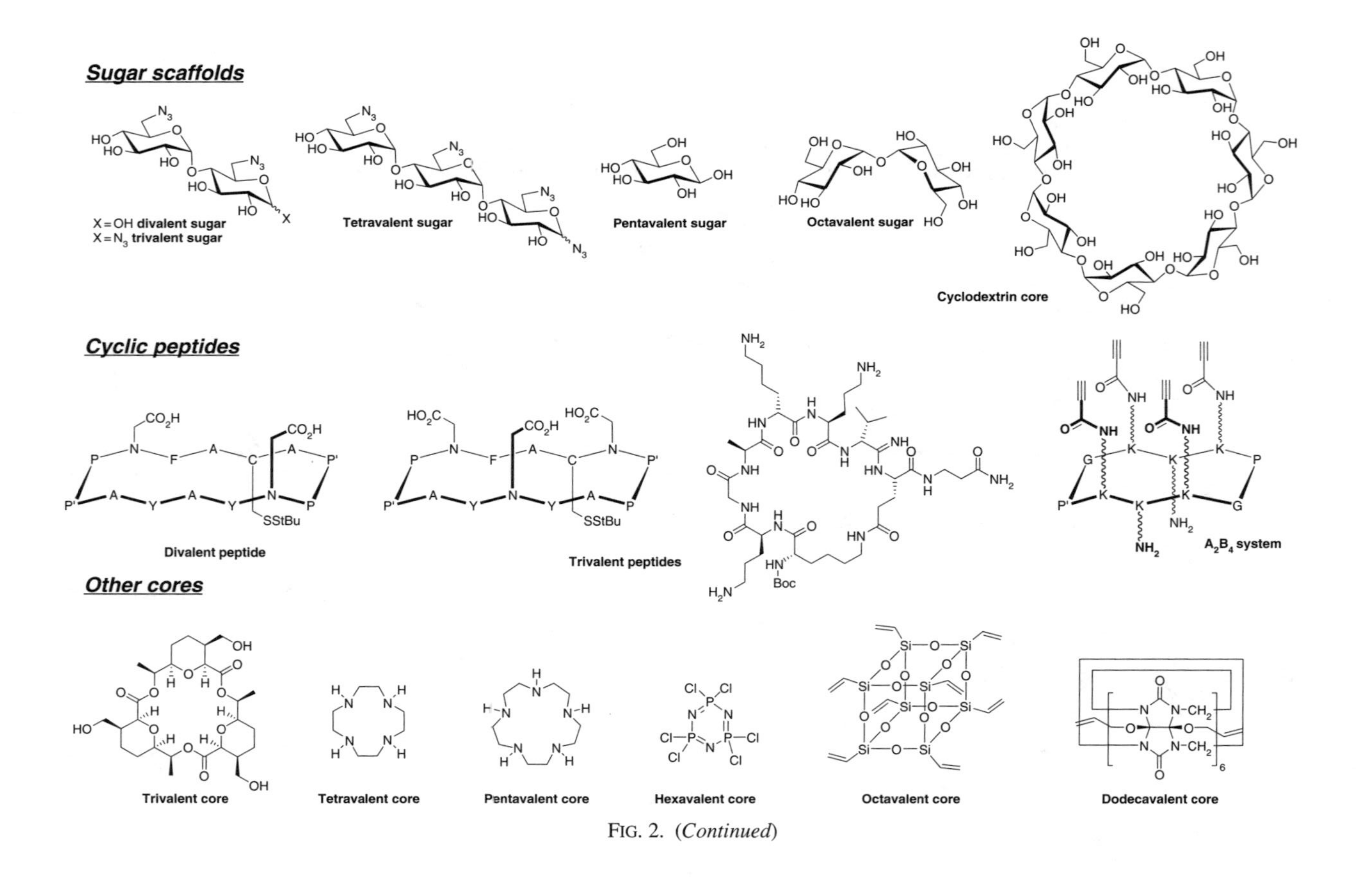

FIG. 2. (*Continued*)

Since a large panel of multivalent scaffolds has recently been reviewed by Roy *et al.*[54–56] and extensively detailed by the group of Santoyo-González for "click chemistry,"[57,58] this section is dedicated to describing the synthesis of optimized glycoclusters, notably highlighting more recent synthetic advances that have led to multivalent candidates that are strikingly biologically relevant.

1. Glycoclusters from Branched Aliphatic Scaffolds

In order to synthesize the target complex glycoclusters, several commercially available or readily derivatized aliphatic scaffolds constituting small multifunctional building blocks have been used to permit rapid access to these multivalent structures (Fig. 2). Historically, one of the first branched glycoside clusters was based on a readily available AB_3 building block, tris(hydroxymethyl)aminomethane (**1**, TRIS), used by Lee in the late 1970s.[59] Owing to the synthetic advantages in terms of symmetry that ensure orthogonal and rapid functionalization, this building block constitutes an ideal candidate for clustering of saccharides. Hence, **1** and its derivatives have been extensively used by several groups for direct glycoside attachment on hydroxyl groups to generate trivalent clusters.[60] However, although the chemistry of glycocluster synthesis was established, the first such glycoconjugates were generally subject to unfavorable steric factors upon binding to proteins. In this context, Kötter *et al.* used extended TRIS derivatives, such as 4-(3-hydroxypropyl)-4-nitroheptane-1,7-diol (**2**) to prepare highly branched glycomimetics (Fig. 3).[61]

Based on these trivalent building block and peptide-coupling methodologies, the rapid synthesis of tri- (**3**, **4**) and nonavalent mannosylated clusters (**5**, **6**) varying in the chemical characteristics of their spacer moieties and lengths was accomplished.[62] The C-6-linked trimannosylated cluster **4** displayed the highest binding potency toward the type-1 fimbrial lectin from *Escherichia coli* (FimH), as tested by inhibition of agglutination and ELISA, for which **4** showed an IC_{50} of 11 μM. Unfortunately, the corresponding nonavalent dendrons **5** and **6** presented only very poor or no inhibitory potencies. To explain these *a priori* unexpected results, it has been postulated that the cluster did not present enough hydroxyl groups for efficient interactions with the lectin surface, or did not present hydroxyl or other functional groups in the appropriate orientation for lectin binding. Alternatively, it is well known that the FimH binds mannosides from the nonreducing end.

By virtue of promising preliminary biological results indicating that tri- and tetravalent glycoclusters constitute potent inhibitors of bacterial binding, which fitted particularly well into the CRD of bacterial lectins (especially for mannose-specific adhesion) straightforward syntheses of low-valency glycoclusters were initiated.

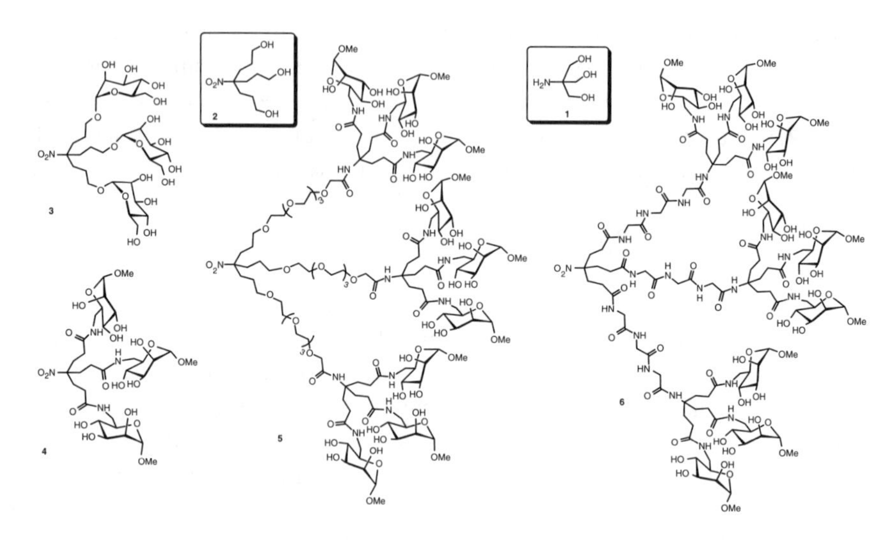

FIG. 3. Early mannoside clusters used as ligands for *E. coli* FimH.[61,62]

In this context, preparation of glycosylamines by direct condensation of amines with reducing sugars could be an appealing strategy. The direct conjugation of reducing sugars with diethylenetriamine (**7**) has been previously reported as allowing efficient and rapid access to divalent carbohydrate derivatives in excellent yields.[63] A few years later, Hayes *et al.* extended this strategy, describing an elegant one-pot methodology that allowed synthesis of higher valent derivatives through reaction of more-highly functionalized amine clusters with unprotected D-mannose (**9**) and dimannosides (Scheme 1).[64] Several linkers of different lengths, flexibility, and valency, including di-, tri-, and tetra-valent amines, were readily incorporated to generate multivalent targets in good yields. Notably, tris(2-aminoethyl)amine (**8**) and pentaerythrityl tetraamine (**13**), [prepared from tetrabromide (**11**) via azide substitution to **12** followed by reduction to **13** with hydrogen on 10% Pd/C], gave the corresponding tri- and tetra-mannoside clusters **10** and **14**, respectively, using the free reducing sugar D-mannose (**9**).

C_3-Symmetrical glycoclusters, based on a trivalent central core, and exhibiting three and six peripheral mannosides have similarly been prepared by Dubber and Lindhorst, who examined the potential of reductive amination for introducing an aldehyde group of a D-mannose derivative as the carbohydrate ligand on tris(2-aminoethyl)amine (**8**) as the branching core.[65] To this end, the glycoclusters were synthesized from (2-mannosyloxy)ethanal (**16**), which was obtained by ozonolysis of

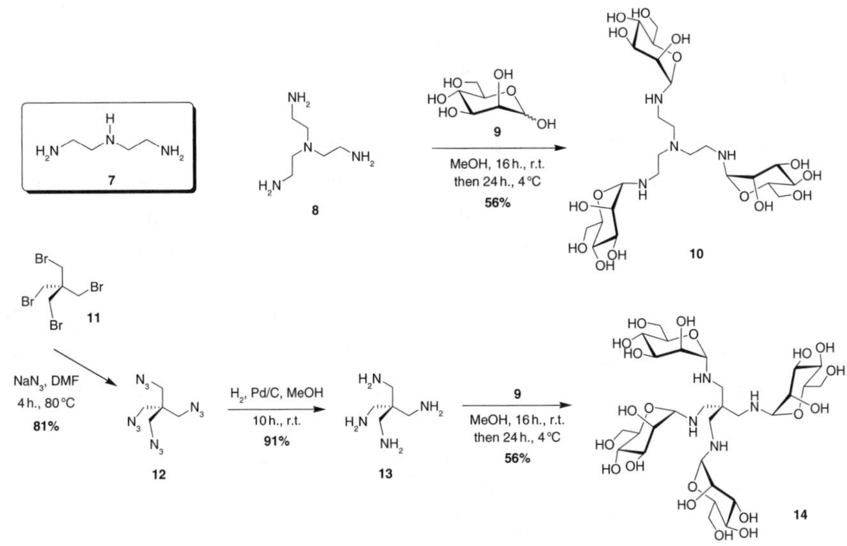

SCHEME 1. Reductive amination leading to tri- and tetravalent mannosylated clusters.[63,64]

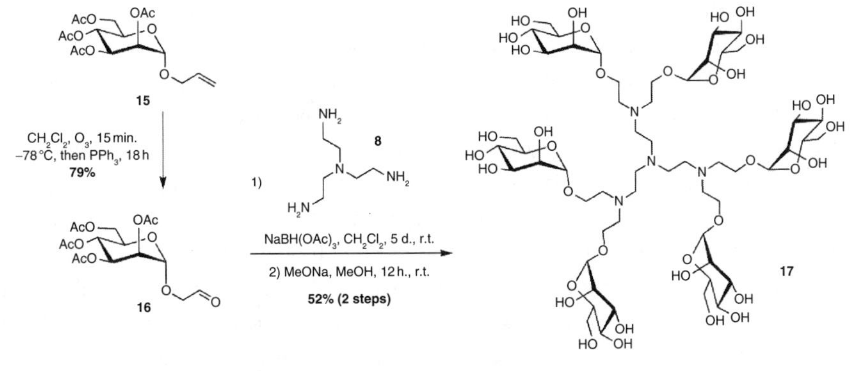

SCHEME 2. Double N-alkylation of a sugar aldehyde by reductive amination.[65]

acetylated allyl α-D-mannopyranoside (**15**), followed by treatment with sodium triacetoxyborohydride $NaBH(OAc)_3$ (Scheme 2). However, the instability of the trivalent conjugate arising from autocatalytic deacetylation led the authors to use an excess of aldehyde to ensure complete double N-alkylation. In more-complex carbohydrate-based multivalent architectures, this situation may afford undesired partial structures.

Another example has been described by Li *et al.*, which proposed an efficient convergent one-pot synthesis of a trivalent mannoside cluster using a Ugi four-component reaction, involving the use of 2-carboxyethyl 2,3,4,6-tetra-*O*-acetyl-α-D-mannopyranoside, benzaldehyde, methyl isocyanoacetate and the tris(2-aminoethyl) amine core **8** in methanol.[66] Biological investigations by ELLA assays indicated efficient inhibition of binding of yeast mannan to the phytohemagglutinin Concanavalin A (Con A) with all the synthesized ligands, with notably an IC_{50} of 30.6 μM for the trivalent derivative, corresponding to about a 10-fold enhancement after valency correction as compared to a methyl α-D-mannopyranoside standard.

As previously mentioned, pentaerythritol and its derivatives constitute another widely used family of aliphatic cores, allowing the construction of multivalent and branched structures via the attachment of four similar or different groups, two pairs of which are tilted at 90°. Accordingly, this tetravalent compound has received considerable interest as an orthogonally protected handle useful for the generation of combinatorial libraries, and as a building block that fits well into the general structure of oligonucleotides and peptides, providing additional functionalities. Hence, through the years, pentaerythritol-based multiantennary glycoclusters, exhibiting such relevant carbohydrates as β-D-galactopyranosides,[67] Galili antigen,[68,69] lactosides,[70] galabiosides, α-D-mannopyranosides or sialic acids, were efficiently prepared. These three last examples that furnished promising antiadhesins toward pathogenic infections will be described in more detail in the following section.

Historically, one of the first example of a glycocluster based on pentaerythritol was furnished in the late 1990s by Hanessian *et al.*, who described the synthesis of di- and tri-haptenic clusters composed of the Tn (GalNAc) and the TF [β-D-Gal-(1→3)-GalNAc] antigens elongated with serine and glycolic acid spacers and attached through amide bonds to pentaerythritol amino derivatives (Fig. 4).[71]

A few years later, Hansen *et al.* proposed the first biological evaluation of galabioside clusters built around thiolated scaffolds with different valencies, including pentaerythrityl derivatives (Scheme 3). Efficient inhibition of hemagglutination (HAI) by the gram-positive bacterium *Streptococcus suis* at nanomolar concentration was achieved with these constructs.[72] Synthesis of the most potent tetravalent soluble inhibitors started with 2-bromoethyl galabioside (**18**), which was treated with sodium azide followed by O-deacetylation under Zemplén conditions and hydrogenolysis to afford 2-aminoethyl galabioside (**19**) in 80% overall yield. Concerning the central core, the commercially available pentaerythritol tetrabromide **11** was treated with methyl 3-mercaptopropanoate and cesium carbonate to give tetraester **20**, which upon hydrolysis led to the corresponding tetraacid **21** in excellent yield. Carboxylic acid activation via formation of a pentafluorophenyl ester, using pentafluorophenol and

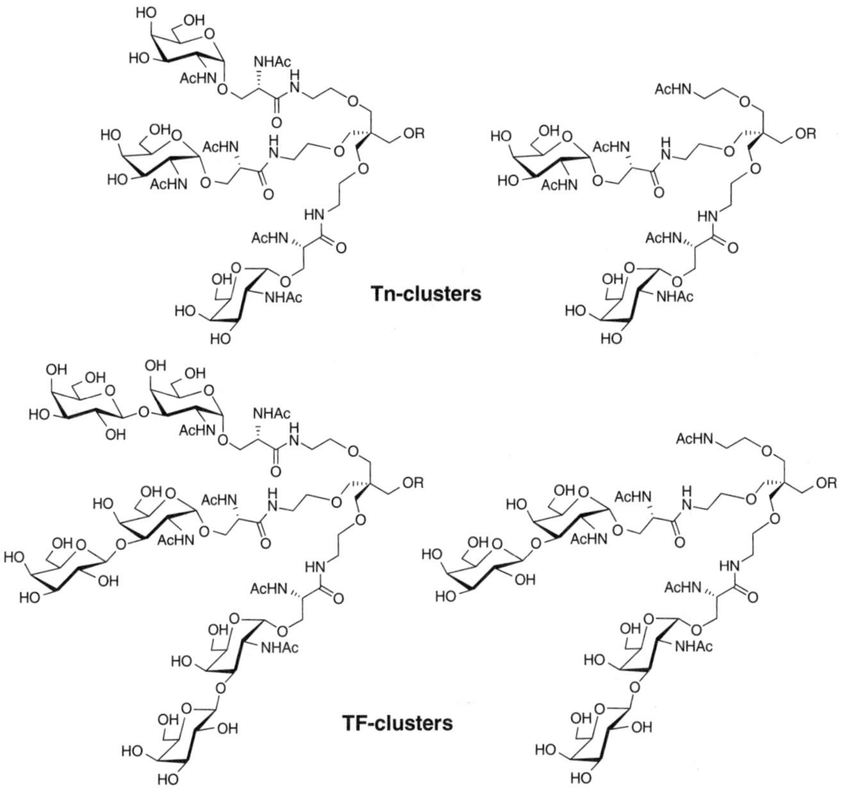

FIG. 4. Structures of trivalent tumor markers based on pentaerythritol derived with the Tn and the TF antigens.[71]

diisopropylcarbodiimide (DIC) and subsequent peptide coupling in the presence of amine **19**, allowed the synthesis in 18% yield of the desired deprotected glycocluster **22** containing four galabioside residues. Biological studies indicated, through the series of multivalent conjugates, a clear connection between inhibitory efficacy, the number of galabiose units present on the potential inhibitor, and the flexibility of the structures. Interestingly, the tetravalent galabioside was several hundred times more efficient than the monomeric galabioside in inhibiting the agglutination of human erythrocytes by the *S. suis* bacterium, resulting in complete inhibition at a concentration as low as 2 nM.

Inhibition of bacterial adhesion of fimbriated *E. coli* to pentaerythritol-based clusters bearing peripheral α-D-mannopyranoside residues has also been often addressed, albeit without systematic structure–activity relationships.[73,74]

SCHEME 3. Tetrameric galabioside having an IC_{50} of 2 nM in the inhibition of hemagglutination of human erythrocytes by *S. suis*.[72]

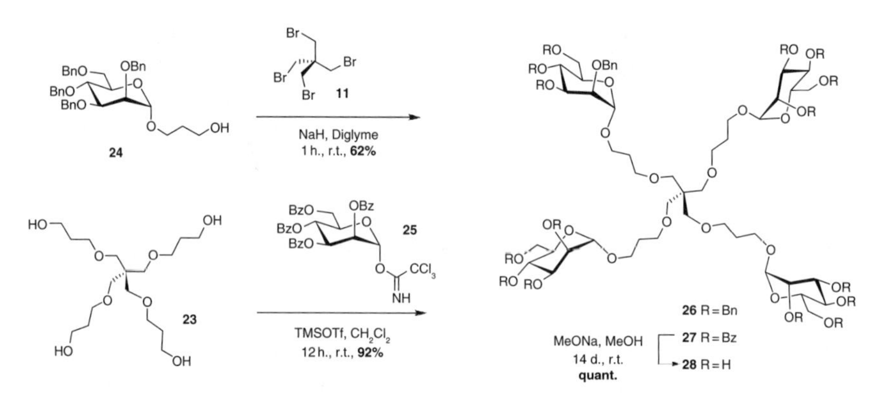

SCHEME 4. Synthesis of tetramannosylated pentaerythritol having an extended linker.[75]

In this context, Lindhorst *et al.* proposed biological investigations of pentaerythritol-based mannoside clusters to test their capacity to block the binding of *E. coli* to yeast mannan *in vitro*.[75] In all of the proposed structures, and according to a rational approach, pentaerythritol itself, as well as the included C_3 spacers, were used as structural components for substituting the inner regions of the monosaccharide moieties (Scheme 4). The tetravalent cluster **27** has been efficiently synthesized according to two different synthetic pathways. In the first route, a C_3 hydroxy linker was introduced within the aglycone moiety of **24**, and a Williamson ether synthesis with pentaerythritol tetrabromide (**11**) led to a mixture of mono-, di-, tri-, and tetra-dentate (**26**) conjugates, even when an eightfold excess of the corresponding alcohol and forcing reaction conditions were used. In the best case, the protected tetravalent cluster **26** was isolated in 62% yield. To circumvent these difficulties, an alternative strategy was investigated, involving the use of a modified pentaerythritol derivative to serve as a longer spacer-equipped tetraol (**23**) for the subsequent glycosylation step, using imidate **25**. To this end, pentaerythritol was initially perallylated and the extended tetraol **23** was obtained by an hydroboration–oxidation sequence on the double bond in the presence of 9-BBN, NaOH, and H_2O_2. Then, the perbenzoylated mannosyl trichloroacetimidate **25** was used as the glycosyl donor for the Lewis acid-assisted mannosylation reaction. Deprotection of the resulting tetramer **27** under Zemplén conditions afforded the desired cluster **28** in excellent yield.

Evaluation of antiadhesive properties of the mannosylated clusters was assessed using ELISA inhibition assays, in comparison to the monovalent reference methyl α-D-mannopyranoside (MeαMan). Results indicated that tetravalent cluster **28** was more than 250 times more potent (thus 62.5 times on a valency-corrected basis) in inhibiting mannose-specific adhesion than MeαMan, with an IC_{50} of 12.6 μM. The observed inhibition of bacterial adhesion was most probably explained by the binding of cluster glycosides to single CRDs, which are distributed along type-1 fimbriae, rather than by multivalent binding, which would reflect interaction of the sugar clusters with more than one CRD.

Subsequently, Touaibia *et al.* described the efficient and systematic synthesis of a family of mannoside clusters built on pentaerythritol and dipentaerythritol scaffolds[76] using regioselective Cu(I)-catalyzed azide–alkyne [1,3]-dipolar cycloaddition (CuAAc) (''click chemistry'').[77–79] The synthetic strategy first involved the use of tetrazide derivative **12** with prop-2-ynyl α-D-mannopyranoside (**29**) under click chemistry conditions, thus providing tetramer **31** in good yield after O-deacetylation (Scheme 5).The conditions under which the Cu(I) catalyst was generated *in situ* from copper(II) sulfate and sodium ascorbate as the reducing agent generally provided slightly better yields than that using the Cu(I) species (CuI) directly. The second cluster was obtained by the treatment of tetrakis(2-propynyloxymethyl)methane (**32**), prepared via nucleophilic substitution of the corresponding pertosylated pentaerythritol and propargyl alkoxide, with 2-azidoethyl 2,3,4,6-tetra-*O*-acetyl-α-D-mannopyranoside (**33**) under the same conditions already described. An excellent yield of the extended cluster **35** was similarly obtained after acetyl-group deprotection. The tetramannoside analogue **38**, bearing a more rigid aromatic spacer, was likewise

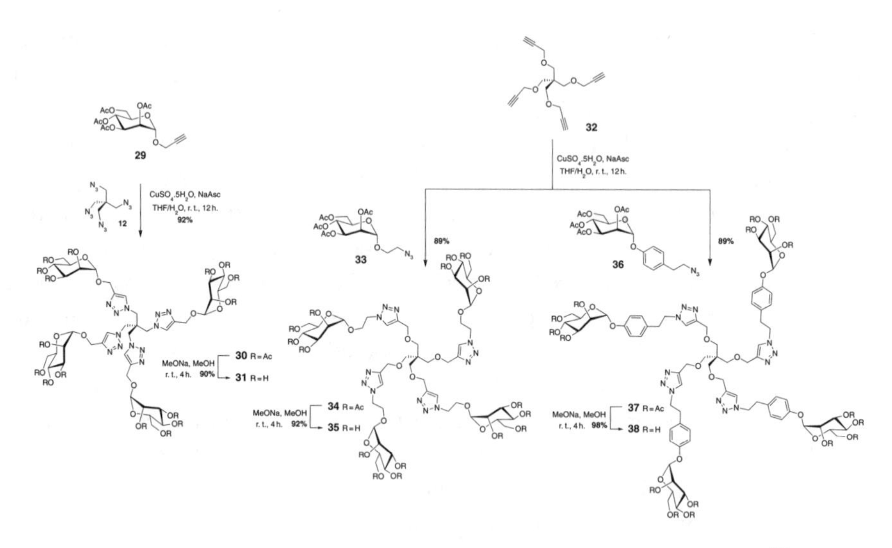

SCHEME 5. Extended tetramannosylated clusters prepared by Touaibia *et al.*[76]

obtained from the tyramine mannoside derivative **36** and the tetrapropargylated core **32**. Interestingly, the presence of the hydrophobic residue in the mannoside aglycone (triazole) was considered to play an important role for an adequate fit into the *E. coli* FimH CRD active site near tyrosine-48 and tyrosine-137.

Further, dendritic growths with new multiarmed clusters having more flexibility and various geometries have also been considered by the authors. Thus, the hexatosylated dipentaerythritol **39** was converted into hexaazide **40**, which upon treatment with propargylated mannoside **29** under standard conditions of click chemistry and subsequent O-deacetylation afforded the hexavalent cluster **42** in good (61%) yield over two steps (Scheme 6).

The elongated analogues **47** and **48** were then synthesized by the reaction of triazide **43** and ditosylates **44** and **45** under basic conditions (KOH, Me_2SO) (Scheme 7). The resulting hexakisazido pentaerythritol scaffolds **46** ($n = 2, 4$) were then independently "clicked" in the presence of prop-2-ynyl α-D-mannopyranoside (**29**), affording hexavalent clusters in 75% yields. Deprotection under Zemplén conditions furnished the corresponding conjugates **47** and **48** having respectively a distance of 11 and 18 Å between each the tripodal mannoside moieties.

Preliminary biological data on this series of mannosylated clusters indicated interesting potency in the inhibition of agglutination of *E. coli* x7122 by baker's yeast, with approximately a hundred times improved efficiency than those obtained with monomeric D-mannose.

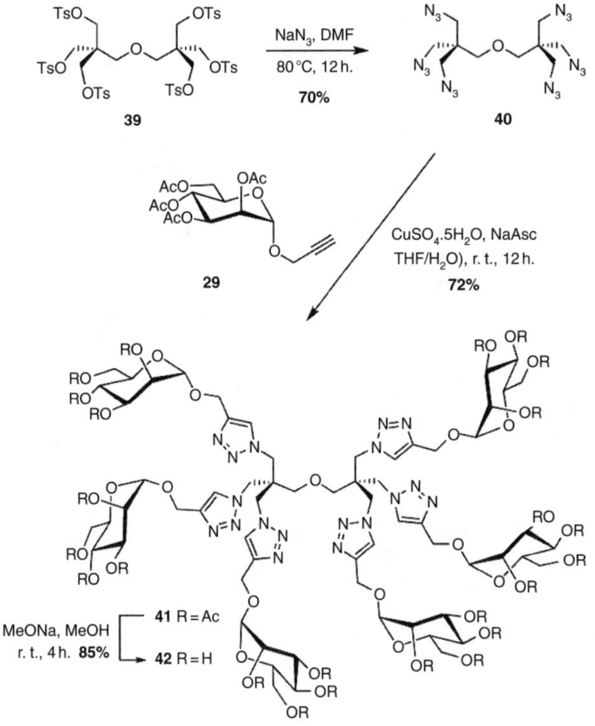

SCHEME 6. Hexakis tetramannosylated clusters for *E. coli* FimH binding.[76]

The pioneering observations of Sharon, who first demonstrated the binding preferences of type-1 fimbriated *E. coli* to mannopyrannosides bearing aromatic aglycones,[80] led to the hypothesis for the existence of ''subsite-assisted aglycone binding.''[81] Based on these premises, the group of Roy *et al.* pursued their investigations to enlarge the panel of mannopyranoside clusters via single-step multiple Sonogashira coupling.[82] The necessary carbohydrate precursors were built with either *p*-iodophenyl, propargyl, or 2-azidoethyl aglycones, whereas the central cores consisted of (di)pentaerythritol-based azide or propargyl derivatives. The first target tetramer (**51**) was synthesized from *p*-iodophenyl α-D-mannopyranoside **49**, previously prepared from peracetylated α,β-D-mannopyranose by glycosidation with triflic acid as a promoter. Then, Sonogashira coupling between tetrakis(2-propynyloxymethyl)methane (**32**) and **49**, followed by subsequent O-acetyl deprotection of **50** provided the key tetravalent cluster **51** in good yield (Scheme 8). Noteworthy is the fact that those transition metal-catalyzed cross-couplings have been optimized during

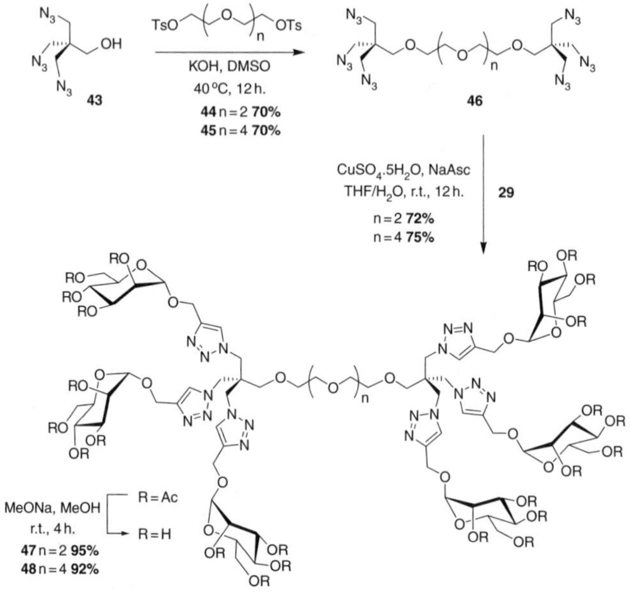

SCHEME 7. Oligoethyleneglycol interspaced, hexakis tetramannosylated clusters for *E. coli* FimH binding.[76]

this study, recommending the use of 5 mol% $Pd(PPh_3)_2Cl_2$ in the presence of 10 mol% Cu(I) catalyst, piperidine as a base in tetrahydrofuran (THF) or *N,N*-dimethylformamide (DMF), with slow addition of the tetrakis alkyne to prevent homocoupling and degradation. Alternatively, tetramer **55**, possessing the reversed linkage functionality, that is, the propargyl group installed on the mannoside residue and the aryl iodide on the pentaerythritol scaffold, was also similarly prepared to investigate the effect of the aryl pharmacophore positioning on binding. In this context, the central tetrakis[(4-iodophenyloxy)methyl]methane **53** was efficiently elaborated by nucleophilic substitution of pentaerythritol tetrabromide **11** using *p*-iodophenol **52** under basic conditions in 78% yield. Treatment of **53** with prop-2-ynyl α-D-mannopyranoside **29** under the optimized Sonogashira coupling conditions just described provided tetramer **54**, which upon further O-deacetylation gave unprotected tetramer **55** in 73% yield over two steps. According to the same strategy, but using Et_3N in DMF at 60 °C instead of the foregoing Sonogashira cross-coupling conditions, the corresponding hexavalent cluster **58** was efficiently obtained, using dipentaerythritol derivative **56** and mannoside **29** after conventional Zemplén deprotection of **57**.

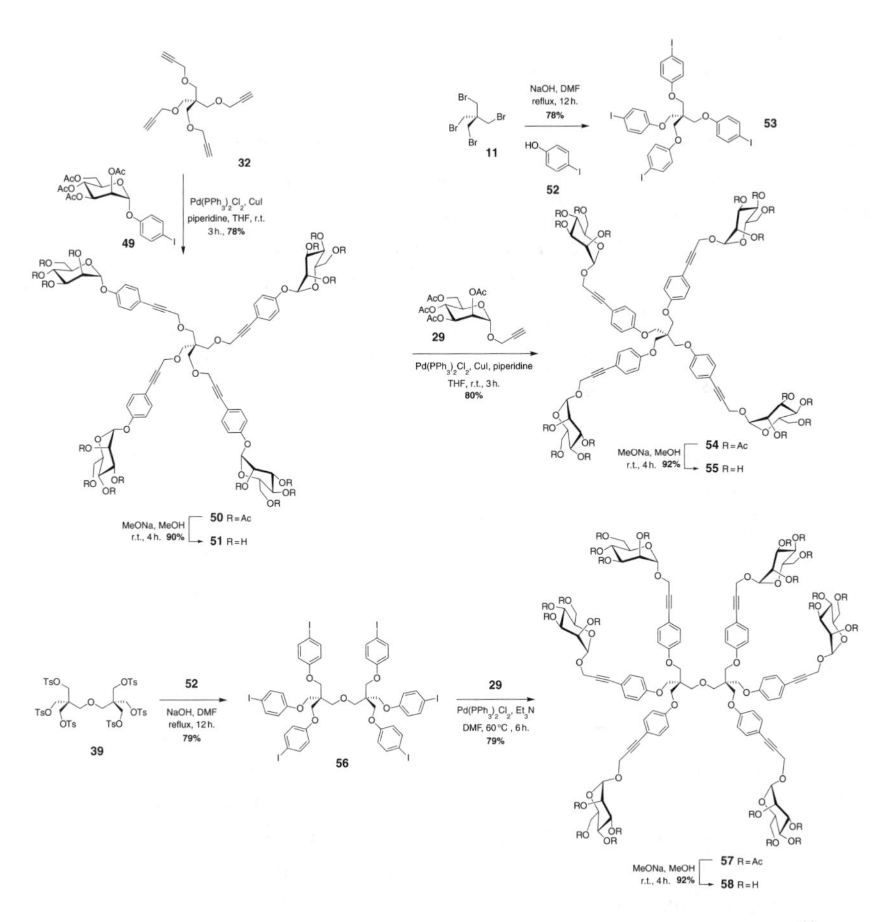

SCHEME 8. Alternative strategy toward the synthesis of hexakis mannopyranosides.[82]

Three different types of biological assays were put in place to evaluate the relative binding properties of these tetra- and hexa-valent mannosylated clusters. Initially, the cross-linking abilities of these molecules were investigated by using a kinetic turbidimetric assay (nephelometry), with the tetrameric phytohemagglutinin Con A from *Canavalia ensiformis* as a model. Significant activity of the glycoclusters was observed when they were used as ligands in interactions with protein receptors, with the rapid formation of cross-linked lattices toward Con A, especially for tetramer **51**. In fact, this last example was the best candidate, and it induced an almost quantitative precipitation of the lectin within 2 min. Obviously, clusters having the alternative triazole heterocycles or the extended series from the Sonogashira coupling were less efficient. These results were rationalized by the authors on the basis of the relative stability of the resulting insoluble complexes, with molecular modeling of tetramer **51** showing that each mannopyranoside residue was at the apex of a tetrahedron in which they were 18.6 and 16.0 Å apart. This distance could easily accommodate the clustering of four different tetrameric Con A lectins.

Each compound was then evaluated for its relative binding affinity by SPR measurements. The affinity of the lectin domain of isolated FimH of *E. coli* K12 toward clusters was obtained by competitive experiments between an immobilized anti-FimH antibody (1C10) and free mannosylated clusters. According to this study, tetramer **51** was designated as the best ligand known, with a K_D of 0.45 nM (1.8 nM/Man) in the subnanomolar range, corresponding to 1222-fold and 3-fold enhancement over the reference monomer methyl α-D-mannopyranoside and the strongest monosaccharide ligand known (HeptαMan), respectively. Once again, the position of the phenyl ring appeared to be rather important with regard to modulating the activity of tetravalent conjugate **55** (K_D = 273 nM), which differed from **51** only by the inverted alkyne–phenyl ring sequence. These results thus further demonstrated that clusters possessing an aryl moiety in the vicinity of the anomeric oxygen atom showed the best overall qualifications. Conjugates obtained by the click chemistry already described have also been studied. Spatial rearrangement of the hexamer having the triazole rings appeared to be a determinant for affinity, as **42** was nearly five times better than the analogous tetramer **31** (K_D of 3 and 14 nM, respectively), thus illustrating the influence of multivalency on this scaffold. The distance between the anomeric oxygen atom and the triazole ring was also a critical factor for affinity. On the other hand, the introduction of four or six mannopyranoside moieties using extended precursors and Sonogashira coupling had only a minor effect on the relative affinity.

Finally, the clusters were tested as inhibitors of hemagglutination of pig and rabbit erythrocytes by type-1 piliated UTI89 clinical isolate *E. coli*. The inhibition titer (IT), that is, the lowest concentration of the inhibitor at which no agglutination occurs, showed tetramer **51** to be the best inhibitor of hemagglutination, with an IT of about 3 μM, or a factor of 6000 as compared to its affinity, and corresponding to 1000-fold better inhibition than that induced by D-mannose. Overall, tetravalent cluster **51** was the best noncovalent cross-linker of Con A and the best ligand known to *E. coli* K12 FimH.

Extended biological investigations concerning structure–function studies were further initiated to evaluate the abilities of these clusters to inhibit Con A-induced membrane type-1-matrix metalloproteinase (MT1-MMP)-mediated pro-MMP-2 activation, cell death, and antiproliferative property in mesenchymal stromal cells (MSC).[83]

Mobilization of MSCs and recruitment by experimental vascularizing tumors involves MT1-MMP functions. Given that the mannose-specific lectin Con A induces MT1-MMP expression and mimics biological lectin–carbohydrate interactions, these clusters were tested to evaluate their potential to block Con A activities on MSC. The results indicated that specific tetra- and hexavalent mannoside clusters, especially **51**, **55**, and **58**, reversed Con A-mediated changes in MSC morphology, and antagonized Con A-induced caspase-3 activity and proMMP-2 activation. They also inhibited Con A, but not the cytoskeleton-disrupting agent Cytochalasin-D-induced MT1-MMP cell-surface proteolytic processing mechanisms, and effects on cell-cycle phase progression. The antiproliferative and proapoptotic impact of Con A on the MT1-MMP–glucose-6-phosphate transporter signaling axis was also reversed by these mannosides. In conclusion, this family of mannosylated clusters very effectively inhibited a spectrum of MT1-MMP-mediated cell responses that could be potentially transposed to target tumor-promoting processes. In addition, their noncytotoxicity allowed their use *in vivo* against experimentally implanted tumors.

Recently, Gouin *et al.* have designed a tetravalent cluster (**60**)[84] based on a potent *E. coli* FimH ligand heptyl α-D-mannoside (**59**)[85] that has been preliminarily recognized as a strong binder to FimH, with a K_D of 5 nM, as determined by SPR measurements (Fig. 5). Furthermore, this derivative inhibited both adhesion of

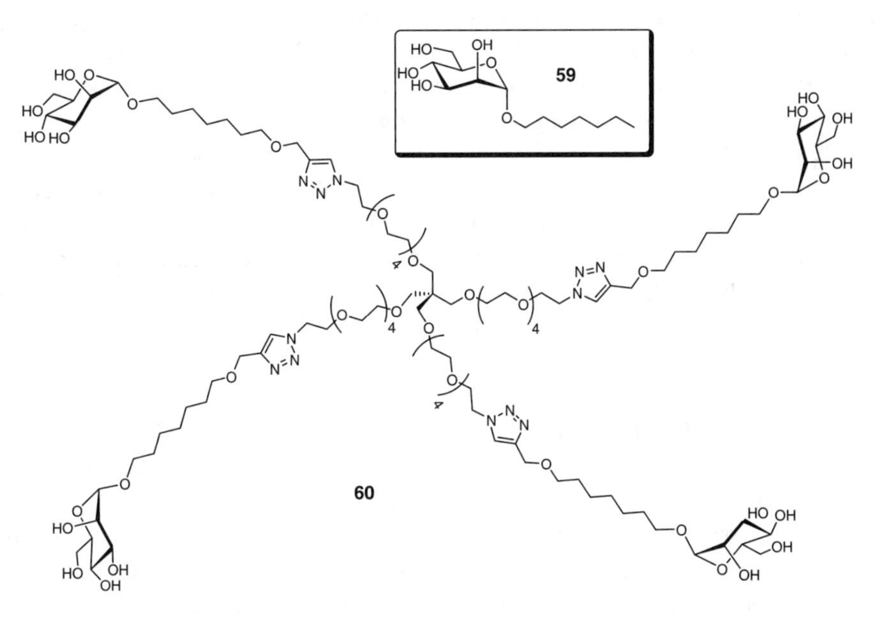

FIG. 5. Best ligands known for the inhibition of *E. coli* binding to murine bladder cells.[84]

type-1-piliated *E. coli* on a bladder cell line and biofilm formation *in vitro*, and also decreases bacterial levels in a murine cystitis model.[86] Suitably functionalized and flexible ethylene glycol linkers were used to tether **59** moieties in order to ensure water solubility and efficient conjugation on a pentaerythritol core via ''click chemistry'' to afford neoglycoconjugate **60**.

Binding affinities of the new flexible synthetic glycocluster toward type-1-piliated *E. coli* were evaluated by HAI and bladder-binding assay (BBA). The results indicated that, with an inhibition of bacterial bladder-cell binding at 12 nM (~6000- and 64-fold lower than mannose and **59**, respectively), the tetravalent compound **60** is currently one of the most promising antiadhesive drugs under development for the treatment of urinary tract infections.

Pentaerythritol-based glycoclusters have also been investigated as potential neuraminidase inhibitors. In this context, Linhardt's group has described a straightforward approach to generate clustered nonnatural *n*-glycosylamines of α-sialic acid, which are known to be resistant to neuraminidase-catalyzed hydrolysis as opposed to the natural *O*-glycosides.[87] Influenza viruses use their hemagglutinin to bind to sialic acid residues located on the surface of the host cell and gain entry into the cell. Once the cell is infected, the new virions use their neuraminidases (or sialidases) to escape from the infected cells. Thus, these neuraminidases have been targeted to stop viral infection by blocking the virus inside the infected cells. To this end, a small library of mono- and divalent 1,2,3-triazole-linked sialic acids has been constructed via click chemistry, together with corresponding tetravalent cluster, generated from tetrapropargylated core and α-sialic acid azide. Preliminary neuraminidase-inhibition assays, involving fluorescence measurements induced by the release of 4-methylumbelliferone produced by the hydrolysis of the substrate [2′-(4-methylumbelliferyl) *N*-acetyl-α-neuraminic acid] by the enzyme [neuraminidase from *Clostridium perfringens* (*Clostridium welchii*)], have been addressed. The results indicated micromolar IC_{50} values, notably for the tetramer with an IC_{50} of 20 μM, comparable to the known sialidase inhibitor Neu5Ac2en (67 μM).

2. Glycoclusters from Branched Aromatic Scaffolds

Inasmuch as only a few of the carbohydrate residues contained in large oligosaccharides can be involved in several binding-recognition process to trigger or inhibit various biological phenomena,[88] it has been assumed that the roles of the remaining sugars within the polymer were limited to a structural matrix, acting simply as spacers maintaining the epitopes at the proper distances to ensure optimal interaction with the receptor-binding sites. Based on these preliminary assumptions, structural analogues of elicitors could result from molecules containing an aromatic core supporting pendant sugar epitopes. By substitution of structural sugar components with benzenoid groups, the rigidity of the polymeric matrix would be preserved with only minimal distortion to the three-dimensional framework. The first and closest analogy to such ''aromatic core'' harboring cluster glycosides was provided by Yariv *et al.* who first described phloroglucinol-based ''artificial antigens'' **61**, with the general formula 1,3,5-tris-(*p*-glycosyloxyphenylazo)-2,4,6-trihydroxybenzene. These early multivalent models had the ability to form brightly colored, cross-linked precipitates with antibodies directed against the appropriate and homologous carbohydrate determinants (Fig. 6).[89]

Through the years, the development of transition metal-catalyzed methodologies, notably involving cyclotrimerization and Sonogashira or Heck cross coupling, has paved the way for rapid and efficient access to aryl glycoclusters with desired and controlled valency.

a. Glycoclusters from Intermolecular Cyclotrimerization.—In 1982, Kaufman and Sidhu described the one-step synthesis of aromatic cluster glycosides via metal-catalyzed cyclotrimerization of appropriate acetylenic sugar precursors.[90] In this context, various 2-propynyl 2,3,4,6-tetra-*O*-acetyl-D-glycopyranosides and their corresponding thioglycopyranosides have been used to explore and validate the

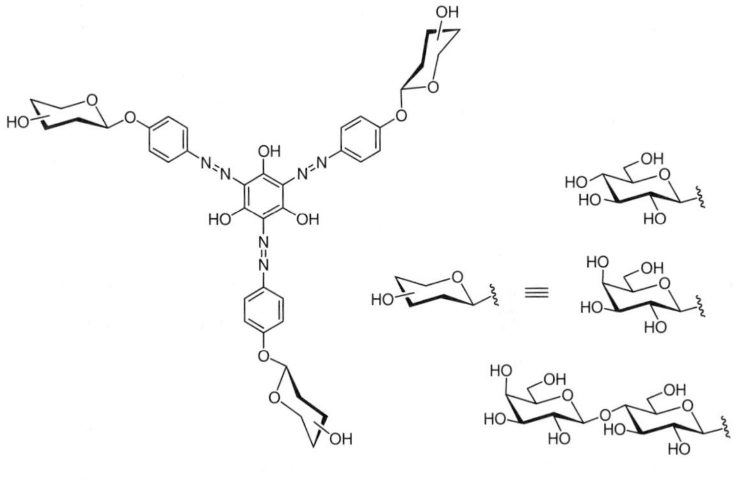

FIG. 6. Early (1962) glycoclusters used in quantititative immunoprecipitation of anticarbohydrate antibodies.[89]

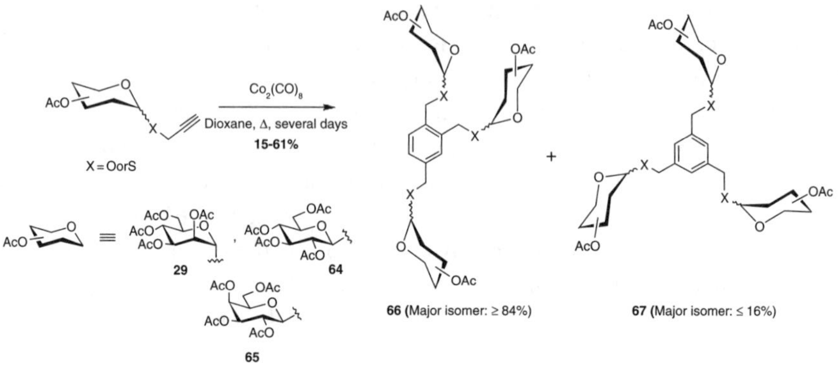

SCHEME 9. Cobalt-catalyzed cyclotrimerization of propargylated glycosides.[90]

feasibility of this type of benzannulation approach in glycochemistry. Treatment of acetylenic precursors (**29**, **64**, **65**) with the conventional cyclotrimerization catalyst, dicobalt octacarbonyl, $Co_2(CO)_8$, gave a mixture of the corresponding regioisomeric trimeric glycosides **66** and **67** (Scheme 9). Although the reaction appears to be quite general, the yields, ranging from 15% to 61%, tend to be markedly influenced by the nature of the starting material, and the thio analogues afforded poor yields, undoubtedly due to poisoning of the cobalt catalyst.

In the same study, the authors also proceeded to the construction of original ''octopus-like'' hexakis glycosides, readily accessible from the corresponding bis (glycoside) monomers (Scheme 10). Thus, cyclotrimerization of 2-butyn-1,4-diyl bis-(β-D-glucopyranoside) octaacetate (**68**) gave benzenehexaylhexamethylene hexakis-(β-D-glucopyranoside) tetraeicoacetate **69** in an excellent yield of 95%. The corresponding thio derivative (**70**) was also prepared according to similar conditions, but in lower yield (29%).

Hexavalent glycocluster **74**, organized around a hexaphenylbenzene core, has also been obtained via a synthetic strategy involving diphenol **71**, which was coupled with tetra-*O*-acetyl-α-D-glucopyranosyl bromide (**72**) using stannate methodology to furnish divalent glycoconjugate **73** in moderate yield. Its cyclotrimerization under the standard cobalt-catalyzed conditions gave the desired glucosylated cluster **74** in 40% yield.

However, these elegant glycosylated nanostructures displayed only weak biological activity relative to their native glucan elicitor counterparts. Nevertheless, such glycoclusters provided unique environments for host–guest chemistry of amphipathic molecules owing to their lipophilic core surrounded by a hydrophilic periphery.

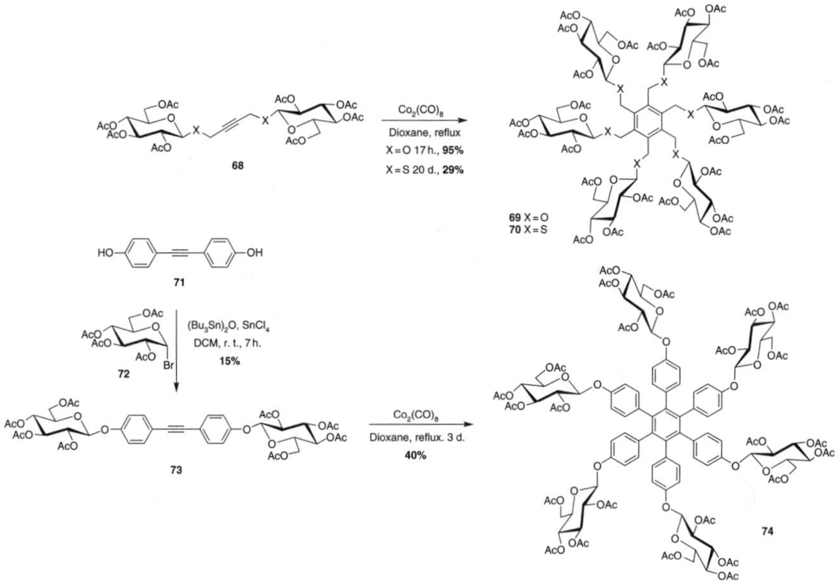

SCHEME 10. Synthesis of ''glycoasterisk'' using cobalt-catalyzed benzannulation.[90]

Subsequently, Roy *et al.* improved the methodology to provide hexavalent mannopyranosides via [2+2+2]-cycloadditions in the presence of the dicobalt octacarbonyl catalyst by refluxing in 1,4-dioxane for only 2 h (as compared to 21 days for the mannoside cluster obtained in the previous example).[91] Regioisomeric mixtures of 1,2,4- and 1,3,5-isomeric clusters were obtained in 63% yield and in 10:1 molar ratio. Interestingly, the reaction was general and could be equally applied to several other saccharides. Prompted by the success of the dicobalt octacarbonyl-catalyzed cyclotrimerization, the same group attempted similar reactions with symmetrical and mannosylated disubstituted alkynes. Under identical conditions as just described, the corresponding hexamer was obtained in 84% yield. The resulting deprotected cluster, obtained quantitatively under Zemplén O-deacetylation conditions, was water soluble and showed excellent cross-linking abilities with tetrameric plant lectins, indicating that the spatial orientation and rigidity provided by the extended inner aryl core offers great potential as neoglycoconjugates.[92]

A few years later, the same group extended this strategy in order to access metabolically stable *C*-glycosyl clusters containing long-arm spacers via a sequence of transition metal-catalyzed transformations (Scheme 11).[93] In this context, cross-metathesis reactions of various *C*-glycosyl compounds with alkenes having available

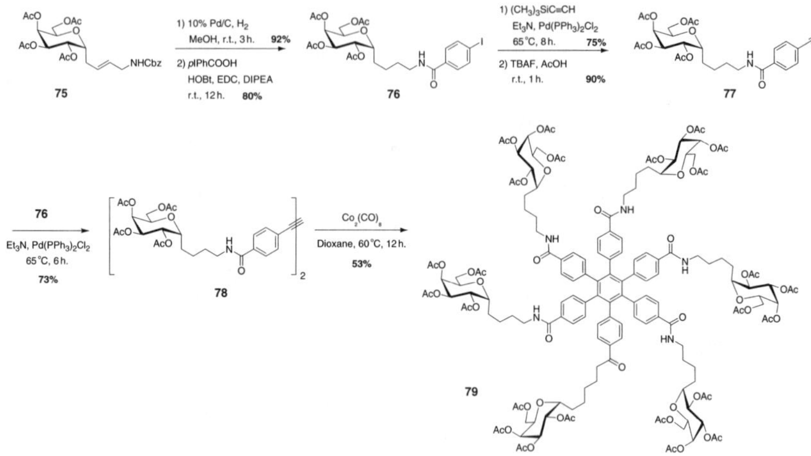

SCHEME 11. ''Molecular-asterisk'' bearing the hydrolytically stable *C*-galactosyl group as synthesized by Roy *et al.*[93]

amine groups were first studied. In particular, a peracetylated *C*-allyl α-D-galactopyranoside analogue, obtained in 81% yield (95:5, α/β) from the reaction of peracetylated galactopyranose in the presence of allyltrimethylsilane and $BF_3.Et_2O$ in acetonitrile,[94] was coupled to *N*-(benzyloxycarbonyl)allylamine using 20 mol% of Grubbs' catalyst (bis(tricyclohexylphosphine)benzylideneruthenium(IV) dichloride [$(Pcy_3)_2Cl_2Ru=CHPh$)] to afford **75** in 45% yield as a single *trans* isomer. The synthesis of the desired hexavalent ''molecular asterisk'' was then initiated by transformation of **75** through hydrogenolysis into its amine derivative in 92% yield. The resulting amine was coupled with *p*-iodobenzoic acid under peptide-coupling conditions to afford aryl iodide **76** in 80% yield. The subsequent palladium-catalyzed Sonogashira reaction between **76** and (trimethylsilyl)acetylene efficiently gave an intermediate which, upon treatment with tetra-*n*-butylammonium fluoride (TBAF), gave the corresponding terminal alkyne **77** in 68% overall yield. With compound **77** in hand, subsequent Sonogashira cross-reaction was carried out with *p*-iodophenyl analogue **76** to afford the key acetylenic dimer intermediate **78** in 73% yield. Finally, the desired cyclotrimerized cluster **79** was isolated in 53% yield after treatment of **78** with dicobalt octacarbonyl in 1,4-dioxane at 60 °C for 12 h. Unfortunately, no biological data involving this deprotected extended cluster was available.

Following these investigations, Das and Roy first reported a Grubbs'-catalyzed version of these intermolecular cyclotrimerizations, using 2-propynyl derivatives and a ruthenium carbenoid catalyst, which thus afforded mixtures of regioisomeric aryl mannopyranosides, in ratios similar to those observed with the dicobalt octacarbonyl-mediated benzannulation but with decreased reaction time and much improved yields.[95] Therefore, 2-propynyl glycosides, including the acetylated α-D-mannopyranoside, β-D-galactopyranoside, and lactoside were treated with Grubbs' catalyst (15 mol%) in dry dichloromethane at room temperature for 12 h. The desired trisubstituted benzene derivatives were isolated as typical mixtures of 1,2,4- and 1,3,5-regioisomers (90:10) in 66–75% yields. By contrast, the access to corresponding hexavalent analogue via cyclotrimerization of disubstituted symmetrical alkynes was not successful, perhaps because of steric hindrance.

b. Glycoclusters from Cross-Coupling Reactions.—Earlier work in the mid 1990s by several research groups highlighted the formation of cross-linked lattices induced when multivalent protein receptors were admixed with synthetic carbohydrate multimers, including dimers.[96,97] All of the reported evidence supported the notion that small-rigidified carbohydrate clusters bearing hydrophobic residues would form stable complexes. The result of the synthetic efforts as presented here was inevitably the formation of glycoside clusters wherein the sugar moieties were linked to the side arms emerging from the central platform through *O*-, *S*-, or *C*-glycosidic linkages.

Based on these observations, Roy *et al.* proposed efficient synthesis of oligomeric carbohydrate clusters, named "sugar-rods," by using Sonogashira–Heck–Cassar (SHC) cross coupling reactions.[91,98] In this context, dimeric conjugates of constrained conformational flexibility, incorporating central hydrophobic aryl residues, were obtained under standard Sonogashira cross-coupling conditions. Hence, heating a mixture of protected propynyl glycosides with a *p*-diiodobenzene core in the presence of catalytic tetrakis(triphenylphosphine)palladium(0) in a 1:1 mixture of DMF–Et_3N quantitatively afforded the corresponding glycosylated bisethynylene derivative. It is worth noting that the reaction was effected in the absence of Cu(I) as cocatalyst, thus preventing the undesired oxidative homodimerization of the propynyl starting material (Glaser reaction). This general strategy, compatible with various glycosides and the usual acetate protecting groups, has also been applied toward the efficient synthesis of divalent "rod-like" thioglycosides, which represent potential enzyme inhibitors because of their resistance to enzymatic hydrolysis.[99]

Sengupta and Sadhukhan adapted this one-step Pd-catalyzed methodology to tri- and tetra-valent aromatic cores in order to generate multiantennary glycoclusters (Scheme 12).[100] A threefold Sonogashira cross-coupling reaction of propargyl β-D-glucoside **64** with 1,3,5-tribromobenzene (**80**) in the presence of $Pd(dba)_2$, PPh_3, in a 1:1 mixture of Et_3N and DMF at 60 °C smoothly gave rise to the centrally planar triantennary glycocluster **81** in 70% isolated yield. The authors also presented an

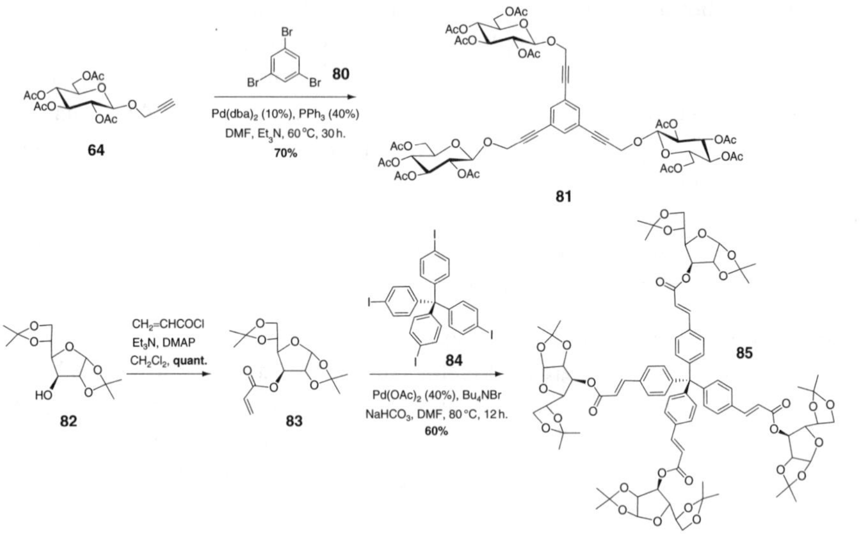

SCHEME 12. Pd(0)-catalyzed Sonogashira and Heck cross-coupling reactions leading to glycoclusters.[100]

example wherein multivalent glycoclusters could be rapidly assembled via a multiple Heck reaction strategy. For this purpose, 2-*O*-acryloyl-1,2:5,6-di-*O*-isopropylidene-α-D-glucofuranose (**83**), derived from alcohol **82**, was used with the centrally tetrahedral core tetra(*p*-iodopenyl)methane (**84**) under phase-transfer-catalyzed conditions ($Pd(OAc)_2$, Bu_4NBr, $NaHCO_3$, DMF) at 80 °C. The resulting tetrahedral glycocluster **85** was obtained in 60% yield.

A few years later, Dondoni *et al.* extended the Pd-catalyzed cross-coupling strategy to *C*-glycosyl compounds, resistant to enzymatic degradation.[101] Dense *C*-glycosylated clusters built around benzene as a rigid platform were thereby synthesized via multiple SHC cross-coupling reactions, using various ethynyl *C*-glycosyl derivatives and polyiodinated benzenes. The limit of this type of conjugation was tested by the use of such crowded systems as those represented by vicinal polyiodobenzenes and short-arm sugar acetylenes. The study elegantly and efficiently afforded di- and tri-valent conjugates from *p*-diiodobenzene and *sym*-triiodobenzene, respectively, involving the use of diversely protected ethynyl glycosides (acetylated or benzylated). In addition, the deprotection via hydrogenolysis of the clusters generated from the conjugation of benzylated saccharides afforded the corresponding ethylene-bridged systems, since the triple bonds were also reduced through this operation. As the use of benzyl-protected ethynyl *C*-glycosyl derivatives turned out to be inefficient as reaction partner in the SHC with the tetraiodobenzene, the desired tetravalent neoglycoconjugate was obtained in only moderate yield (49%). Further cross-coupling attempts under optimized conditions and involving hexaiodobenzene failed, affording only the tetraadduct in 35% yield. Interestingly, O-deacetylation and O-debenzylation, together with the reduction of triple bonds, yielded highly water-soluble structures.

Biological investigations of these rigid clusters were addressed by André *et al.*, who described the preparation of lactoside-bearing glycotope bioisosteres having strong binding affinities to lectins (Fig. 7). According to the optimized Sonogashira conditions already described, and using the required di- and tri-iodinated benzene cores and per-*O*-acetylated 2-propynyl lactoside, they obtained divalent (**86**) and trivalent (**87**) glycosides in yields of 90% and 80%, respectively.[102] Complete O-deacetylation resulted in freely water-soluble clusters in almost quantitative yields.

The relative affinities of these lactoclusters were evaluated in a competitive solid-phase binding assay, using different labeled sugar receptor as probes, notably the *β*-trefoil mistltoe lectin (*Viscum album* agglutinin, VAA) and three mammalian galectins having different modes of presentation of their respective CRDs. Preliminary results indicated that glycoclusters could well surpass the inhibitory capacity of lactose. Of note is the fact that binding of the two homodimeric proto-type galectins-1 and -7 was not effectively influenced by the presence of multivalent compounds.

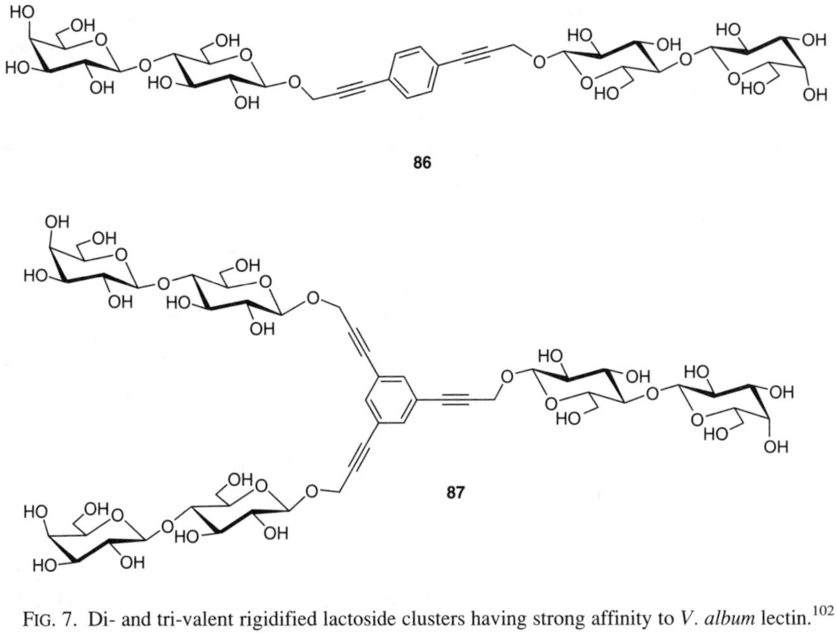

86

87

FIG. 7. Di- and tri-valent rigidified lactoside clusters having strong affinity to *V. album* lectin.[102]

In fact, only a few of them manifested inhibitory capacity in the range of free lactose, demonstrating cases of negative correlation between carrier-dependent presentation of lactose and inhibitory efficiency. However, the trivalent cluster **87** induced the strongest cluster effect with the chimeric-type galectin-3, with an IC_{50} of 30.8 μM (700 μM for free lactose) in contrast to the corresponding divalent lactoside **86**. Its efficiency was independently confirmed by hemagglutination and by *in vitro* cytofluorometric tumor-cell binding analysis. The trivalent compound was a potent inhibitor of galectin-3, blocking its binding to native cell surfaces. Finally, these data underlined the feasibility of galectin-type target selectivity by compound design, despite the use of an identical headgroup for the synthesis.

c. Persulfurated Glycoclusters.—Recently, two research groups have independently described the synthesis of persulfurated glycoclusters, organized around an aromatic core, that showed interesting lectin-binding activities. in an ongoing research program to study mannose-binding proteins (MBPs) by the systematic syntheses and biological evaluation of multivalent glycomimetic inhibitors against bacterial adhesion, Chabre *et al.* have described the synthesis via ''click chemistry'' of dense sulfurated glycoclusters containing up to 18 peripheral α-D-mannopyranoside residues (Scheme 13).[103] The synthetic strategy employed hexavalent sulfurated scaffolds **88**, initially obtained by thioacetylation of a commercial hexabromomethylated precursor with potassium thioacetate, which was further functionalized with TRIS derivatives 2-bromoacetamido-tris[propargyloxy)methyl]aminomethane (**89**). This trivalent moiety was synthesized in an efficient three-step sequence using Boc N-protection of TRIS, followed by propargylation under basic conditions. Quantitative removal of the Boc protecting group under acidic conditions and reaction with bromoacetyl chloride led to precursor **89**, which was then treated with hexathioacetate **88** under basic and reductive conditions to afford the octadecapropargylated dendritic scaffold **90** in excellent yield. Its treatment with 2-azidoethyl tetra-*O*-acetyl-α-D-mannopyranoside (**33**) using ''click chemistry,'' and subsequent deprotection under standard Zemplén conditions provided the densely packed cluster **91** containing 18 peripheral mannoside residues. Initial evaluation against the BclA lectin from *Burkholderia cenocepacia* by ITC revealed promising candidate ligands having strong affinities (Chabre *et al.*, unpublished results).

The second example has been furnished by Sleiman *et al.*, who proposed a new class of persulfurated, semirigid, radial, and low-valent glycosylated ''molecular-asterisks'' possessing dual function as ligands and as probes by virtue of the conjugated electronic system.[104] Moreover, these glyco-asterisks could exhibit conformational preferences for alternating up-and-down patterns around the central core by exposing the phenylthio groups above and below the plane of the benzene ring, which could be

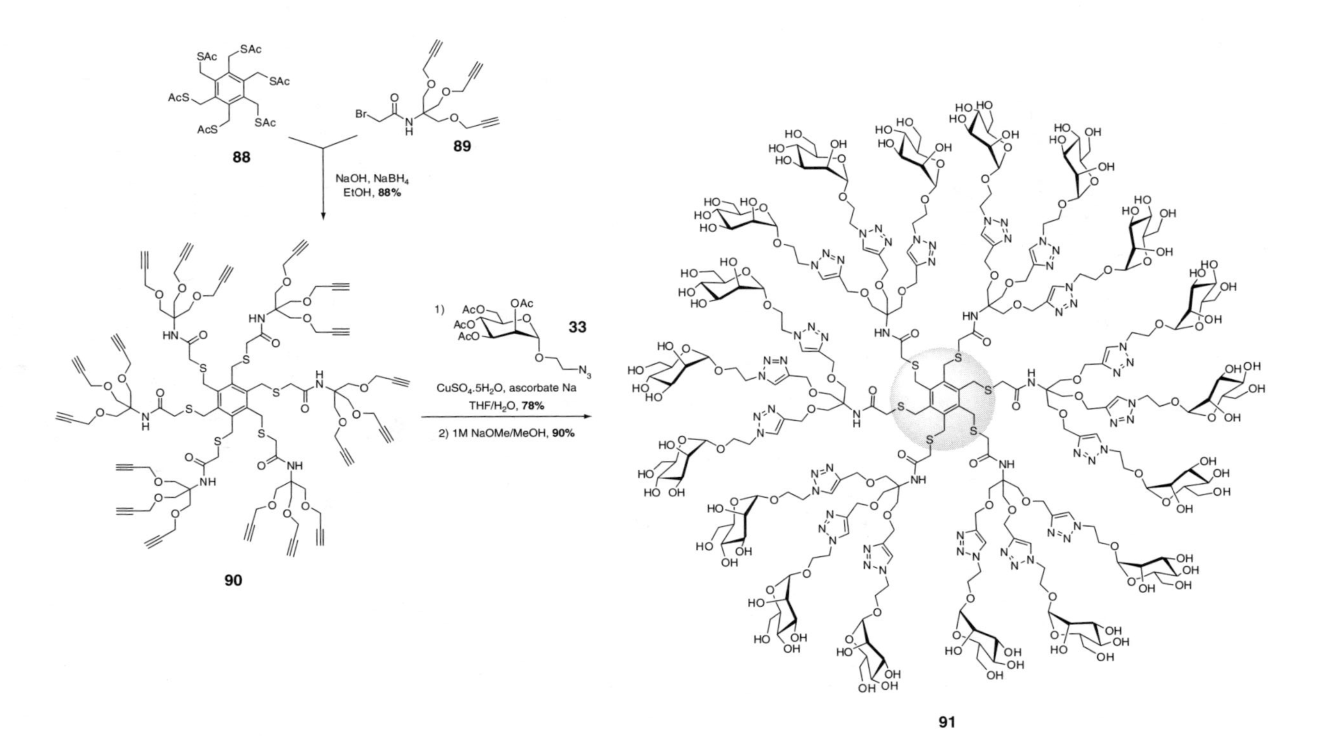

SCHEME 13. Persulfurated glycoclusters bearing 18 α-D-mannopyranoside residues.[103] (See Color Plate 9.)

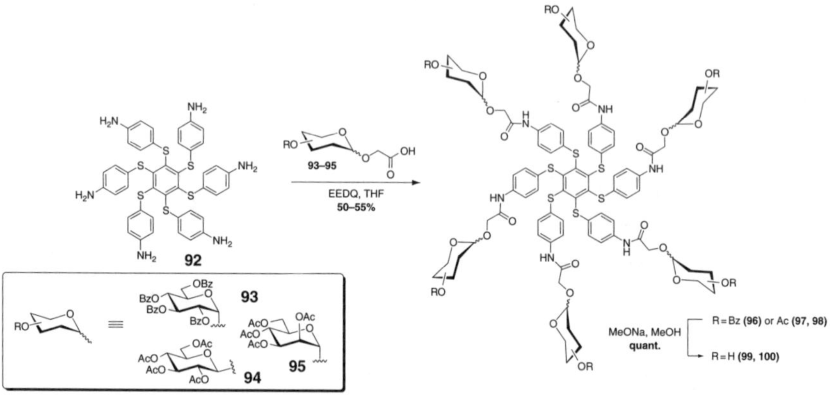

SCHEME 14. Persulfurated "molecular asterisks."[104]

potentially useful for optimized ligand orientation (Scheme 14). These glycoclusters were obtained from the hexaamino persulfurated benzene precursor **92**, readily accessible on a multigram scale by coupling sodium 4-acetamidothiophenolate with commercial hexachlorobenzene in 1,3-dimethyl-2-imidazolidinone (DMI) and subsequent deprotection in concentrated hydrochloric acid. Various protected carbohydrates containing terminal carboxylic acid function (**93–95**) were then introduced onto the multivalent amino-scaffold, using 2-ethoxy-1-ethoxycarbonyl-1,2-dihydroquinoline (EEDQ) as the coupling agent, in yields ranging from 50% to 55%. O-Deacetylation under Zemplén conditions occurred uneventfully to provide the hexavalent clusters **99** and **100**.

In order to assess their biological activity, the authors investigated their ability to inhibit the hemagglutination of rabbit erythrocytes by Con A. Results indicated that the α-glucose asterisk **99** efficiently inhibited hemagglutination at 11 μM, corresponding to a 60-fold enhancement per sugar relative to methyl α-D-glucopyranoside. A more significant impact was recorded for the α-D-mannoside **100**, which showed one of the best inhibitory potencies reported at a minimum concentration of 89 nM, and hence a greater than 3750-fold increase in relative activity per sugar compared to methyl α-D-mannopyranoside. Control experiments proved that nonspecific interactions of the scaffold were not responsible for the observed inhibition effects, which were more probably induced by a powerful cross-linking phenomenon. In addition, the authors observed a 60-fold amplification of selectivity between the mannose and the glucose clusters relative to the monovalent compounds. Dynamic light scattering (DLS) experiments further confirmed the strong aggregating effect of

Con A by the mannoside cluster at concentrations slightly above the nanomolar range, according to a mannose-dependent and partially reversible process. Finally, these results suggested that, rather than an increased avidity for Con A, the unusual potency of these molecular glycoasterisks was due to an efficient and kinetically controlled macromolecular assembly that strongly amplified the effect of these low-valency ligands.

d. Glycoclusters with a Calixarene Core.—Cyclic polyaromatic calix[*n*]arenes, owing to their oligomeric nature and shapes that can be tailored by the size of the macrocycle ring and by the nature of the substituents on the lower rim, are attractive multivalent scaffolds with controllable valency. Hence, they constitute valuable candidates onto which glycosides can be exposed at the periphery. More particularly and analogously to CDs, calix[4]arenes blocked in the cone conformation can, to some extent, mimic a small portion of the multiantennary cell surface, presenting a series of glycosylated residues on the exterior of a lipophilic cavity. Moreover, the cavity and the spacers between the upper rim of the calix[*n*]arenes and the sugar units can be useful for cooperatively binding molecules in a host–guest chemistry process. In this way, the inclusion complexes could be directed toward selected biological targets which specifically recognize the carbohydrate units, with the glycocalix[*n*]arenes acting as novel types of site-directed molecular delivery systems. They thus provide a versatile platform of well-defined shape for the construction of more-sophisticated structures, including neoglycoconjugates that may be named ''calyx sugars.'' Furthermore, because of their low cytotoxicity, calixarene derivatives have found wide applications in the biomedical field as enzyme inhibitors, anticoagulant and antithrombotic agents, antiangiogenic and anticancer, antiviral, antimicrobial, and antifungal products.[105]

Since the first examples of lower and upper rim glycocalixarenes were obtained in 1994 by Marra *et al.*,[106] employing the Mitsunobu reaction or copper(II)-catalyzed glycosylation, the development of efficient synthetic methodologies has allowed the emergence of several examples of *O*-, *N*-, or *C*-glycosyl calix[*n*]arenes, and these have recently been reviewed (**101–106**, Fig. 8).[107,108]

Severe drawbacks resulting from the absence of suitable spacer arms and low water solubility recorded for the first calix[*n*]arene-glycoconjugates have been overcome during subsequent years with the use of adapted chemical ligations and suitably derivatized carbohydrates. As example, 2-thio-α-sialosides have been efficiently conjugated to a calix[4]arene scaffold via nucleophilic substitution, providing extended and water-soluble clusters[109] that may exhibit up to 16 dense clustered epitopes.[110] Wittig olefination,[111,112] Pd-catalyzed Sonogashira cross coupling,[113] click chemistry,[114,115] and azide–nitrile cycloaddition[116] have also given rise to

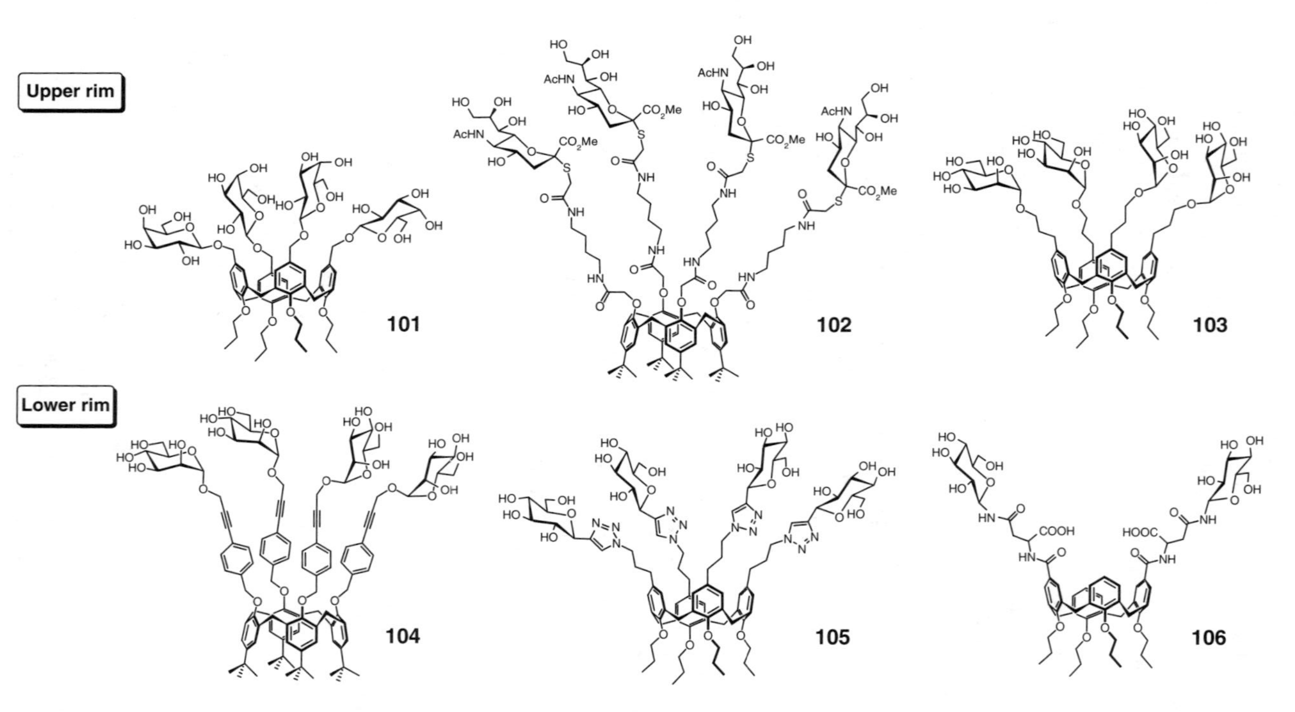

FIG. 8. Typical examples of calyx sugars prepared through various ligation strategies.

diversified calyx sugars. In addition, efficient introduction of biologically relevant carbohydrates on calix[*n*]arenes, through amide[117] or thiourea linkages[118–121] (which can act as hydrogen-bonding groups) have been described and have afforded elongated nanostructures with tailored spacer arms. These have shown promising applications as drug delivery systems via host–guest supramolecular chemistry.

Initial exploratory investigations, addressed by Dondoni *et al.* in the late 1990s,[122] have highlighted the ability of calyx sugar **101** to recognize such neutral and charged molecules as D-glucosamine hydrochloride and tetrabutylammonium dihydrogenphosphate, and showing interesting potential as receptors of phosphate or phosphonate-bearing molecules of biological relevance.

The same group then synthesized thiourea-linked upper rim calix[4]arene glycoconjugates with exposed two or four glucose, galactose, and lactose units, and they further investigated the recognition properties of these glycocalixarenes, together with their interactions with specific lectins.[123] Solution ^{1}H NMR and electrospray ionization-mass spectrometry (ESI-MS) experiments demonstrated their anion-recognition properties, enhanced with the presence of an aromatic ring in the guest, whereas turbidimetric analysis indicated the ability of the glycoclusters bearing four peripheral glucosides and galactosides to interact with Con A and with peanut lectin, respectively. These combined features made these new glycoclusters attractive as possible site-specific molecular delivery systems. Higher valency analogues, such as glucosylthioureidocalix[6]- and calix[8]arenes were further investigated.[124] Supramolecular studies suggested the formation of self-assembled small discoid-like particles in water (3–10 nm in size), in equilibrium with the monomeric macrocycle, which ideally could exhibit a working valency much higher than when compared with that of the monomeric species. Turbidimetric measurements and atomic force microscopy (AFM) assays indicated strong interactions with Con A, especially for the octameric glycoconjugate, causing agglutination with the formation of large supramolecular entities, which progressively evolved toward precipitation because of the extensive lectin cross-linking.

Besides these preliminary studies that have demonstrated the interesting potential of these calyx sugars, recent applications have afforded striking and concrete results. In 2005, elegant work disclosed from combined efforts of the Ungaro and Bernardi groups has described the synthesis of a divalent cholera toxin (CT) glycocalix[4]arene ligand (**109**) having higher affinity than the natural GM1 oligosaccharide **107** (Fig. 9).[125] The bacterial CT is produced by toxicogenic strains of the gram-negative bacillus *Vibrio cholerae*, the causative agent of cholera, the life-threatening acute diarrhea, that mainly affects third-world populations. This heterohexameric AB_5 complex is composed of structurally independent, catalytically active heterodimeric

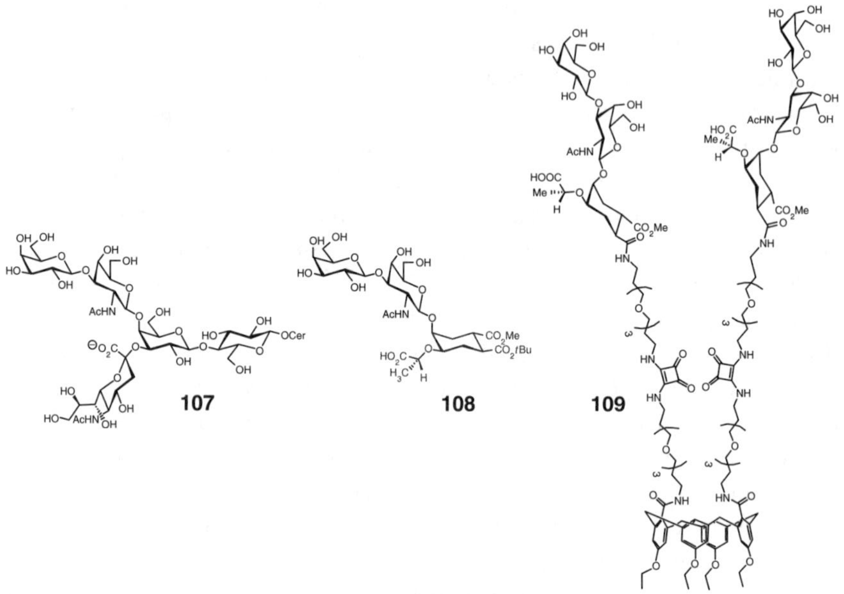

FIG. 9. Divalent GM1 mimic having high affinity against cholera toxin.[125] (See Color Plate 10.)

A (CT-A, enzymatic) and homopentameric B (CT-B, targeting) subunits. This last moiety effects recognition and anchoring to the cell membrane. With 80% sequence homology in common with the heat-labile enterotoxin secreted by some strains of the *E. coli* bacterium (LT), their mode of action remain similar: both toxins exploit the intrinsic complicated trafficking mechanisms of the host cells to gain access to the cytosol, where they exert their detrimental activity. As a prelude to the infection, once in the lumen of the gastrointestinal tract of the human host, the toxins recognize the receptor ganglioside GM1 [βGal1→3βGalNAc1→4(αNeu5Ac2→3)βGal(1→4)βGlc1→1Cer] (**107** with an IC_{50} of 14 nM for CT)) on the surface of epithelial cells through the B subunits arranged in a pentameric pattern which trigger endocytosis.[126] Acute structural studies and biochemical data concerning the fundamental interaction of GM1 with CT demonstrate the critical role of the two sugars at the nonreducing end of GM1, namely galactose and sialic acid. Taking these specific structural properties into consideration, the rational design of artificial receptors for the toxin as monovalent or multivalent ligands has been envisaged.[127] While preparation of the ganglioside GM1 represents a tedious synthetic challenge, one of the strategies of impeding CT-B/GM1 interactions concerns the design and synthesis of functional and structural mimics of the natural CT membrane receptor GM1. In this regard, Bernardi *et al.* described the synthesis of pseudosugar GM1 mimics, designed using molecular modeling techniques and presented as high-affinity binders of CT, including the second-generation conjugate **108**, having a K_D of 190 μM.[128]

In this context, divalent presentation of this promising candidate has been investigated by the authors using the fixed *cone* conformation of calix[4]arene to allow introduction of the molecules onto the *upper rim*, and their projection into the same portion of space, thus mimicking, to some extent, a small portion of the natural cell-surface ganglioside. The synthesis of the divalent neoglycoconjugate was based on initial functionalization of the corresponding diacid with an aminated spacer and further introduction of squaric acid moieties.

The interaction of the deprotected cluster **109** with CT was studied by fluorescence spectroscopy, which indicated that, together with ELISA assays, the ligand displayed a higher affinity for CT (IC_{50} = 48 nM) than the natural GM1 oligosaccharide under the same conditions (IC_{50} = 219 nM). An exceptionally high-affinity enhancement relative to that for the monovalent ligand **108** was thereby obtained, roughly 4000-fold (namely 2000 per sugar mimic). Although complementary detailed thermodynamic analysis is needed to determine the precise role of multivalency in this particular system, the result constituted another striking example highlighting the advantageous utilization of calixarene scaffolds together with the use of a glycomimetic.

Subsequently, Křenek *et al.* proposed a series of calixarenes substituted with 2-acetamido-2-deoxy-β-D-glucopyranose (GlcNAc) directly linked by a thiourea spacer, and tested their binding activity to heterogeneously expressed activation of C-type lectin-like receptors of the rat natural killer (NK) cells NKR-P1, and the receptor CD69 (human NK cells, macrophage).[129] NK cells are important components of the innate immune response against tumors and early protection against viruses and other intracellular pathogens. Interestingly, high-affinity carbohydrate ligands for both major activating receptors, NKR-P1 and CD69, have been found. More particularly, GlcNAc was recognized as a high-affinity receptor in this study. Furthermore, multivalency through multiantennary, branched GlcNAc-containing structures was found to be crucial for a highly specific recognition and binding event to NKR-P1 and CD69 receptors *in vitro*. Hence, this study was initiated to determine differences in the binding capability of calix[4]arene-based glycoconjugates to recombinant receptors, as well as their effects on proliferation and cytotoxic cell-effector function. Results indicated that, in the case of NKR-P1, the binding affinity of β-D-GlcNAc-substituted calixarenes carrying two or four sugar units was comparable to that observed with known linear chitooligomers. The influence of

GlcNAc substitution of the calixarene skeleton on binding affinity for CD69 receptor was more profound, and the corresponding tetravalent cluster containing thiourea functionalities displayed very potent inhibitory activity, with $-\log IC_{50}$ as high as 9.3, making it one of the best ligands for this receptor. The use of glycoconjugates with lower valency led to a dramatic decrease of the binding activity. In addition, the immunostimulating activity of the tetravalent glycoconjugate indicated a proliferation and stimulation of natural cytotoxicity of human peripheral blood mononuclear cells (PBMC) at concentrations of 10^{-4} and 10^{-8} M, as observed for lipopolysaccharide. Moreover, at these concentrations, it induced an increase in the spontaneous death of tumor targets. Generally, calyx sugar derivatives were superior to the corresponding dendritic PAMAM-GlcNAc_8 analogues (see next).

The synthesis of a novel anticancer vaccine candidate built on a nonpeptidic scaffold, in which a cluster of four S-linked Tn antigen glycomimetics (S-Tn) was conjugated to an immunoadjuvant moiety, trihexadecanoyl-*S*-glycerylcysteinyl-serine (P_3CS), through a calix[4]arene scaffold has been described.[130] The glycomimetic *S*-Tn antigen was chosen as antigenic determinant because of its higher metabolic stability and capability to determine the maximal immunostimulating activity of a construct at lower doses with respect to the natural O-Tn analogue.

Preparation of the calix[4]arene platform started from disymmetrical 25,26,27-tripropoxy-28-hydroxy-*p*-*tert*-butylcalix[4]arene (**110**), which reacted with ethyl bromoacetate in the presence of NaH to furnish ester **111** quantitatively (Scheme 15). Direct replacement of the *tert*-butyl groups via aromatic nitration afforded the nitrocalixarene derivative which, upon subsequent reduction with H_2 at atmospheric pressure using Pd-C catalyst, afforded the amino analogue **112**. Afterwards, four Boc-glycine spacers were efficiently introduced into the upper rim of the macrocycle, via standard amide coupling in the presence of PyBOP and DIPEA. Saponification (aq. KOH, THF) of the ester group led to intermediate **113**, containing a single carboxylic acid functionality in the macrocycle lower rim, for amidation to the P_3CS immunoadjuvant unit leading to **114**. Removal of the Boc group with TFA and coupling of the S-Tn acid derivative **115** with HOAt/EDC in dry DMF allowed the formation of the targeted conjugate **116**.

Thus, in the designed construct, the four S-Tn antigen units were linked covalently through the glycine spacers of the calix[4]arene. Three propyloxy groups and one pendant P_3CS unit at the narrow rim block the synthetic core in a rigid cone-shaped conformation, ensuring the preorientation of the antigenic units on the same side with respect to the median plane of the macrocycle. The all-*syn* orientation of Tn antigens was crucial for better mimicking of the cancer cell surface, while the presence of a glycine spacer should avoid steric hindrance and ensure flexibility, both advantageous in the antigen–immune system interaction.

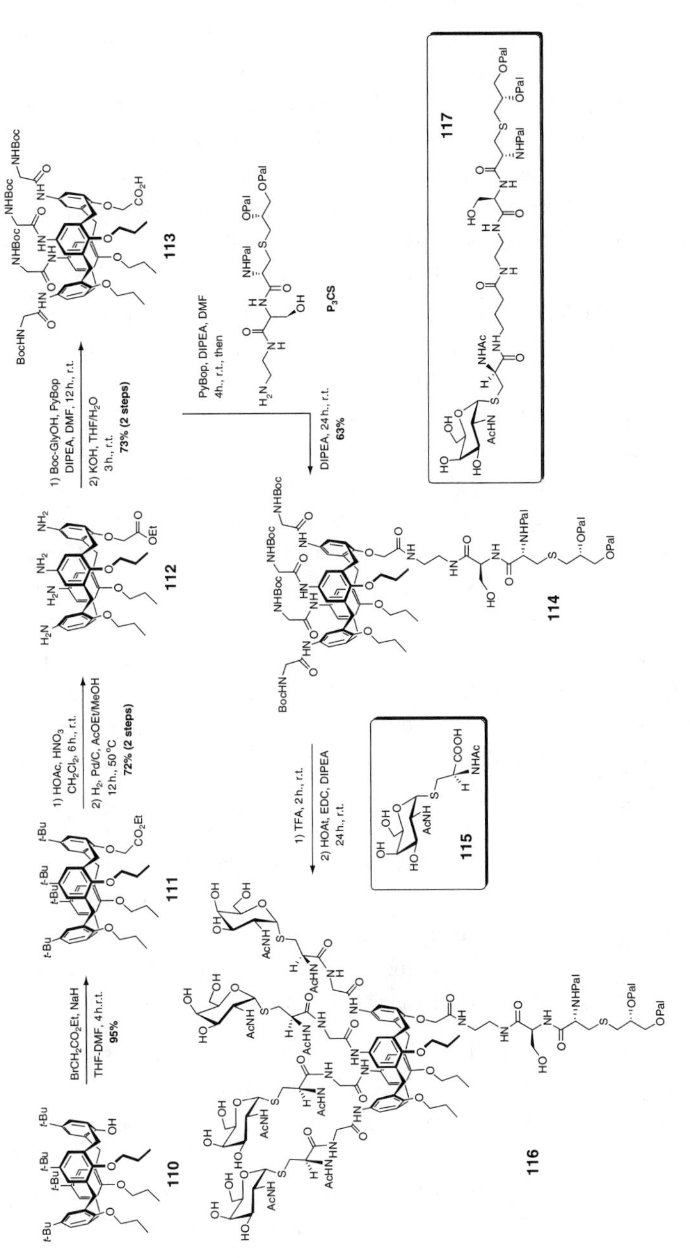

SCHEME 15. Glycosylated calix[4]arene.[130]

The authors evaluated the immunogenicity of the tetra-Tn construct **116** in mice by using a monovalent glycoconjugate **117** as reference compound. The aim was to investigate the role of the cluster effect in amplifying the antibody production with respect to a single presentation of the Tn antigen unit. In the serum-dilution range considered, mice immunized with tetramer **116** showed a substantial and significant production of antibodies. Even when the concentration of monovalent compound was increased fourfold with respect to the parent monovalent derivative **117**, the reactivity elicited by the clustered construct was significantly higher. These data clearly showed that a cluster effect provoked the higher immunostimulating activity of the tetravalent cluster as compared to the monovalent analogue. The authors thus presented still further convincing evidence for a vaccine candidate built on a nonpeptidic platform that induces a more-effective immune response due to the cluster effect and presentation in an optimized arrangement of Tn antigens. The rigid calix[4]arene scaffold was claimed to properly mimic the mucine surface encountered *in vivo*.

In a later investigation, a set of 14 calix[n]arenes (n = 4, 6, or 8) having thiourea-linked galactose or lactose moieties were prepared to analyze their reactivity toward the reference plant lectin from VAA relative to three human galectins (galectins-1, -3, and -4).[131] Despite their high degree of flexibility, the calix[6,8]arenes generally proved especially effective for the plant AB-toxin. In the solid-phase model system involving VAA lectin and using the absorbed glycoprotein asialofetuin (ASF), the inhibitory potency of the hexa- and octavalent galactoside clusters were 5 and 8 μM, respectively, corresponding roughly to a 100-fold enhancement as compared to free galactose. As demonstrated with solid-phase and cell-binding assays, clustering on macrocycles was also particularly effective toward galectin-4. In fact, among the human lectins, different response patterns were registered, the tandem-repeat-type galectin-4 reacted very sensitively to the presence of these test compounds. The IC_{50} value for the lactoside cluster was lowered by a factor of 300-fold relative to the free sugar, when calculating the sugar concentration in the assay. These bioassays underlined clear intergalectin differences and dependence of inhibition on the conformational properties of the calix[n]arenes scaffold, as well as on the shape and valency of the glycoclusters.

e. Glycoclusters with Porphyrin Cores.—The attachment of saccharide components to porphyrin macrocycles gives rise to various derivatives termed glycoporphyrins, which might become of great importance for medicinal and other applications, such as catalysis and molecular recognition.

Since the pioneering synthesis of this type of glycoconjugate first described by Maillard *et al.* in the late 1980s via condensation of 2-formylphenyl tetra-*O*-acetyl-β-D-glucopyranoside with pyrrole,[132] improved synthetic methodologies have generated highly promising water-soluble glycoconjugates showing potential as photosensitizers (PSs) in photodynamic therapy (PDT),[133] or as antibiotics,[134] antiviral agents,[135] and drug sensors.[136]

Interestingly, despite the fact that carbohydrate and porphyrin derivatives constitute two groups of naturally occurring compounds of great significance to all existing organisms, only a few examples of natural glycoporphyrins are known. Tolyporphin A **118**, isolated from the blue-green alga *Tolypothrix nodosa*, whose total synthesis has been reported,[137,138] and the chlorophyll c_2-monogalactosyldiacylglyceride **119**, a derivative isolated from *Emiliana huxleyi*, are two examples of natural glycoporphyrin-like derivatives (Fig. 10).

The biological significance of glycoporphyrins, their limited natural occurrence, and their widespread applications have made the availability of such compounds a scientific challenge for several research groups. Synthetic approaches have consisted in the direct glycosylation of prefunctionalized porphyrins or chemical synthesis from suitable glycoconjugates precursors.

Two general approaches have thus been used to access glycoporphyrins.[139] The first is based on Lindsey's method involving the coupling of pyrroles or dipyrromethanes with glycoaldehydes species through acid-catalyzed condensations, followed by oxidation of the reaction mixture, for instance with 2,3-dichloro-5,6-dicyanobenzoquinone (DDQ) (Fig. 11).[132] By this strategy, O-,[140,141] C-, and S-linked glycosyl compounds[142] can be obtained, and have been examined in detail.[143] Interestingly, higher valency glycoporphyrins can be obtained by following the pyrrole–aldehyde condensation strategy, generating O-glycosylated porphyrin dimers possessing ether linkages.[144]

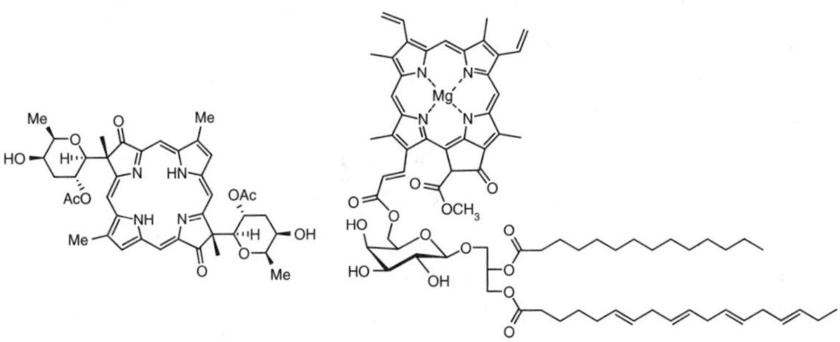

118, Tolyporphin A

119, Chlorophyll c_2-monogalactosyldiacylglyceride

FIG. 10. Natural glycoclusters bearing a porphyrin core.

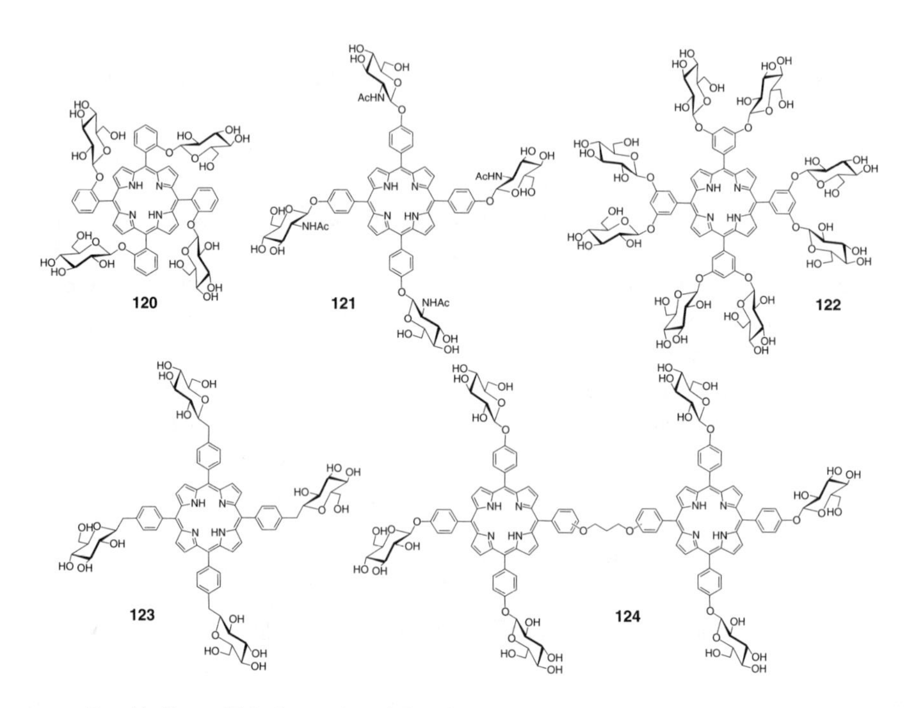

FIG. 11. Sugar aldehydes condensed directly on pyrrole derivatives by Lindsey's chemistry.

The second methodology involves direct introduction of glycosylated moieties onto a suitably functionalized *meso*-arylporphyrin scaffold, accessible from a natural source (protoporphyrin-IX) or by total synthesis. Several O-,[133,145,146] S-,[147] and N-glycoporphyrins[148] (**125–129**) have thus been prepared (Fig. 12). Moreover, in order to explore the influence of the clustered peripheral saccharides around the porphyrin scaffolds, and to evaluate their photophysical properties, the synthesis of dodecavalent porphyrins bearing four trivalent glycodendrons via amide ligation (**129**) has been achieved.[149]

PDT is certainly the most promising application for this kind of neoglycocluster. Its concept is based on the concentration of PS in target cells and upon subsequent irradiation with visible light in the presence of oxygen, leading to specific destruction of the target cells or tissues. There is general agreement that the principal mechanism whereby cell destruction occurs involves the disruption of the cellular, mitochondrial, or nuclear membranes by singlet oxygen ($^{1}O_2$) generated by the action of the PS.[150] Photofrin®, the first PDT formulation introduced, has been approved in several countries including USA, Canada, Netherlands, France, Germany, and Japan for

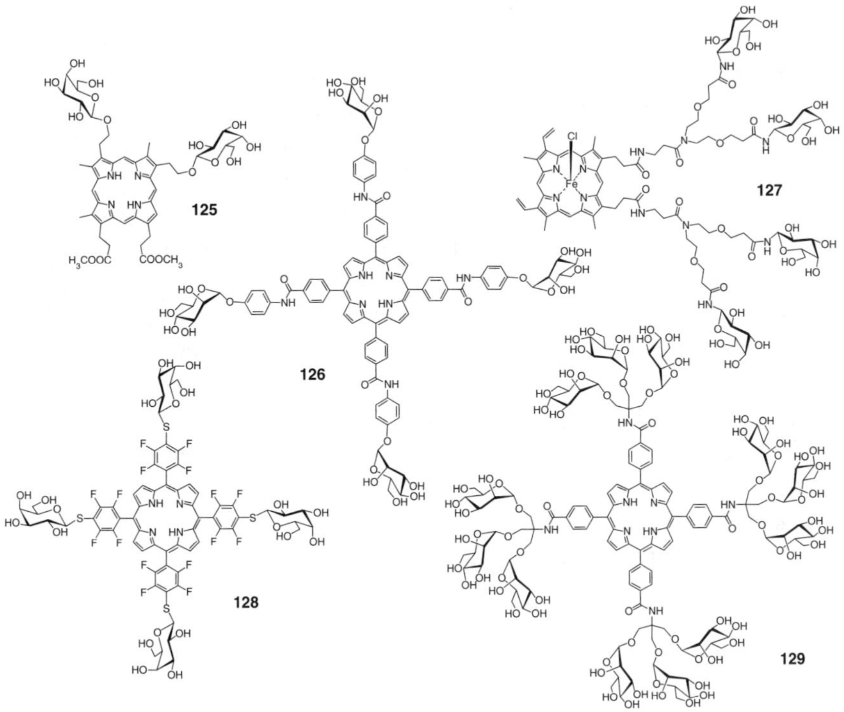

FIG. 12. Glycoconjugates incorporated on porphyrin scaffolds.

treatment of bladder, esophageal, gastric, cervical, and lung cancers, and several other diseases. Although Photofrin® is effective against a number of malignancies, it is not an ideal PS because of the lack of a well-defined structure, its weak absorbance in the red region of the visible spectrum and the long-lasting skin photosensitivity induced. Mainly because of these drawbacks, several other derivatives have been synthesized with the required peripheral substituents, such as carbohydrate residues to control water solubility, biodistribution, pharmacokinetics, and affinity–selectivity for cancer cells.[143,147] Hence, glycoconjugation can advantageously modify the amphiphilic–lipophilic balance of macrocycles and can favor specific interactions of the resulting conjugates with the lectin receptors usually overexpressed in certain malignant cells. Glycoconjugation may thus be a potentially effective strategy for targeting PS toward tumor cells and allow a more efficient and selective PDT treatment. To illustrate this concept, some recent studies leading to efficient glycosylated PS are presented later.

Five novel diethylene glycol (DEG)-linked O- and S-galacto or manno-conjugated *m*-tetraphenyl porphyrins have been prepared and their biological and photobiological properties investigated *in vitro* against a human retinoblastoma cell line (Y79).[133] Preliminary studies established that human retinoblastoma cells express sugar receptors that exhibit a preferential affinity for galactose and mannose residues.[151]

The precursor 5,10,15-*m*-tri(*p*-phenol)-20-phenylporphyrin (**130**) was used for synthesis of all of the glycoconjugates studied (Scheme 16). The trivalent O-glycosylated DEG porphyrins **132a** (β-Gal), **132b** (α-Gal), and **132c** (α-Man) were obtained by treating **130** with the requisite bromo-substituted glycosides (**131**), followed by deprotection under Zemplén conditions, with yields ranging from 22% to 50% over two steps. The corresponding thioglycosylated porphyrin analogues **132d** (β-Gal) and **132e** (α-Man) were obtained according to a slightly modified strategy involving the preliminary synthesis of the bromo-substituted porphyrin glycol **133** via standard Williamson etherification and final introduction of peracetylated 1-thioglycopyranoses **134** under basic conditions, with improved overall yields of 62–75%.

The amount of PS taken up by the Y79 retinoblastoma cells was first determined by fluorescence intensity measurements (reflecting the internalized drug concentration)

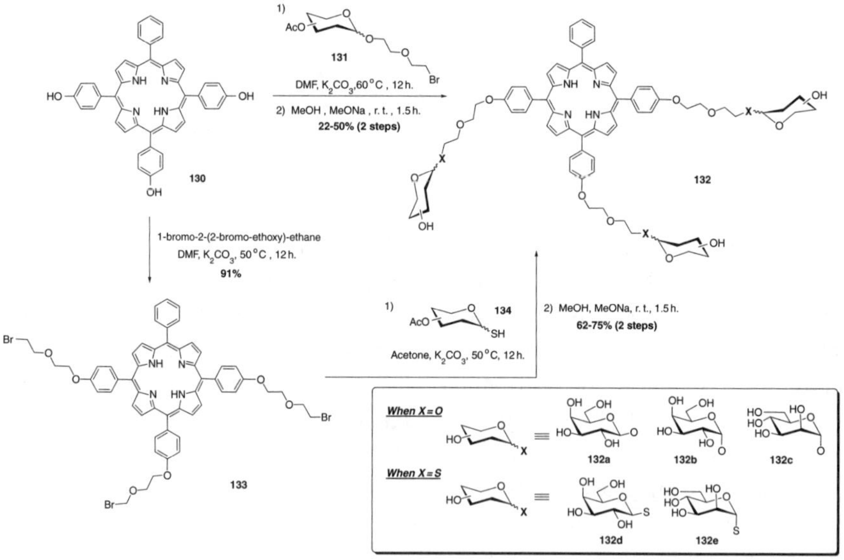

SCHEME 16. Galacto- and manno-substituted tetraphenyl porphyrins used for PDT against human retinoblastomas.[133]

using flow cytometry. The results indicated efficient internalization of the glycoconjugates, preferentially in the membranes of all cytoplasmic organelles, with the extent of uptake dependent upon the nature of the sugar component, its anomeric configuration, and the linker used. An increase in the spacer length linking the tetrapyrrolic ring and the sugar moiety resulted in higher cellular uptake in the case of **132a** (*O*-β-Gal), **132d** (*S*-β-Gal), and **132c** (*O*-α-Man), while a considerable decrease in drug internalization was observed for **132b** (*O*-α-Gal). Hence, these results highlighted the importance of the amphiphilic/lipophilic character of these clusters, indicating that high lipophilicity led to low cellular uptake. The binding affinity toward the sugar-specific receptors on Y79 cells has also been investigated by examining a possible competition effect between the glycoporphyrins and the corresponding glycosylated albumin, indicating inhibition of uptake values of 40–45%, possibly due to cell–sugar–receptor saturation. In addition, their phototoxicity in darkness, evaluated by measurements of the cell survival fraction, was found to be negligible in all cases. High photoactivity was observed for the two α-galacto–manno porphyrins **132b** and **132c** ($LD_{50} = 0.05$ and 0.35 μM, respectively) at 514 nm and low fluence (1 J/cm^2). In addition, glycoporphyrins can also exhibit possible cell–membrane interactions that can affect, for instance, the plasma lifetime of a particular drug.

Promising results have been obtained for both compounds **132b** and **132c** since each one showed high *in vitro* photobiological activity in human retinoblastoma cells. Very low doses of the drug associated with low light intensities were sufficient to observe a marked effect. Moreover, both compounds undergo only very limited cellular metabolic degradation.

The same group further investigated the *in vitro* phototoxicity in order to determine the influence of different structural parameters.[152] Despite lower photodynamic activity than that observed for hydroxylated PSs (in particular the common Foscan® or Temoporphyn®), glycoconjugates displayed phototoxicity against Y79 cells having significant intrinsic cytotoxicity. These results confirmed that the photoactivity could be strongly modulated by the presence of a DEG spacer between the chromophore and the glycoside, and by the anomeric configuration of the sugar. Trivalent α-O-galactosylated porphyrin **132b** was determined to be a better candidate than Foscan® in the clinical application of PDT for conservative therapy of retinoblastomas.

A second example has been addressed in which the synthesis and potential application of a series of neutral O-glycosylated porphyrin dimers and two original O-glycosyl cationic dimers were examined in PDT. In order to understand the influence of the number of glycosyl moieties, the spatial geometry and the ionic character on their photodynamic activity was evaluated.[144] The *in vitro* results concerning their photocytoxicity against K562 human chronic leukemia cells compared favorably with those of the common PS Photoprin II®.

Two strategies for preparation of neutral and monocationic porphyrin models were used. The presence of a spacer arm directly attached to the *meso* phenyl position of the porphyrins gave access to a series of dimers with different substituents on the *meso* position and with different geometries.

Neutral bisporphyrins were synthesized according to Lindsey's methodology, namely the coupling of pyrrole with *p*-formylphenyl 2,3,4,6-tetra-*O*-acetyl-*β*-D-glucopyranoside and *o*- or *p*-hydroxybenzaldehyde with controlled stoichiometry and in the presence of $BF_3.Et_2O$ as catalyst (Scheme 17). Subsequent oxidation of the porphyrinogen intermediates with *p*-chloranil gave porphyrins **135a** and **135b** in 5% and 7% yields, respectively. Treatment of **135a** or **135b** with an excess of 1,3-diiodopropane (K_2CO_3, DMF, reflux) afforded derivatives **136a** or **136b** in 80–85% yields, respectively. Dimers **137a** and **137b** were formed by the reaction of **136a** or **136b** with the monohydroxylated precursors **135a** or **135b** via nucleophilic substitution, followed by O-deacetylation under modified Zemplén conditions. The analogous triglycosylated bisporphyrin **139** was similarly obtained from 5-(4-hydroxyphenyl)-10,15,20 tristolylporphyrin (**138**).

An elegant synthesis of a monomeric porphyrin intermediate bearing one pyridyl and three glycosyl units, precursor of cationic dimers, has also been described (Scheme 18).[153] Thus pyrrole, *p*-formylphenyl 2,3,4,6-tetra-*O*-acetyl-*β*-D-

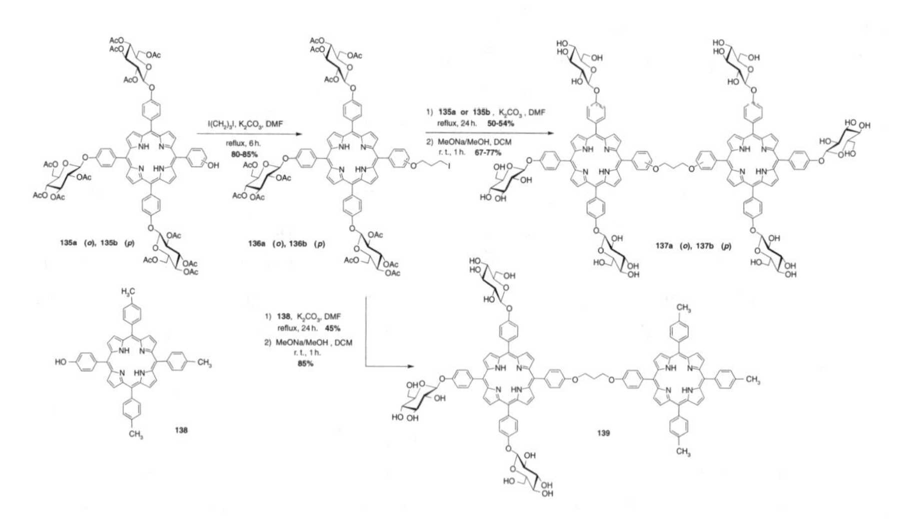

SCHEME 17. Tri- and hexavalent glucosylated porphyrins prepared by Sol *et al.*[144]

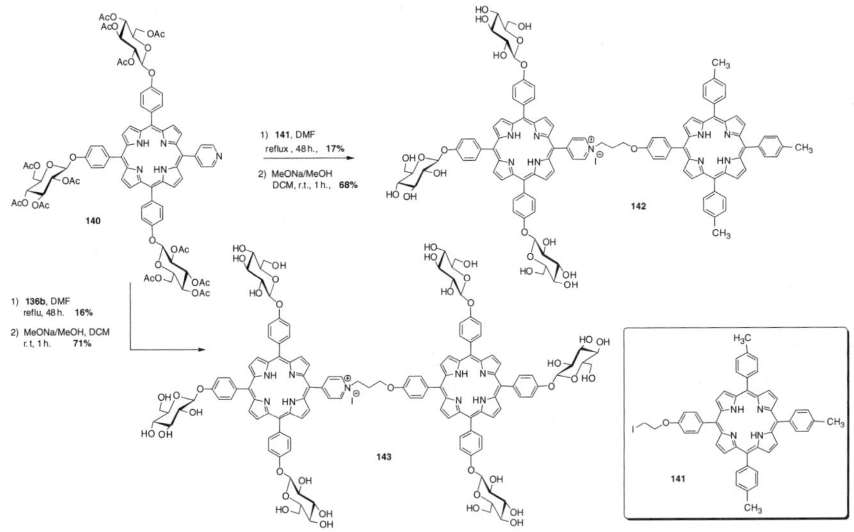

SCHEME 18. Synthesis of glucosylated porphyrins bearing cationic head-groups.[153]

glucopyranoside, and pyridine-4-carboxaldehyde were added to propanoic acid to provide the expected intermediate **140** in 7% yield. Condensation of **140** with the elongated analogue **141** of **136b** in DMF using potassium carbonate afforded adducts in 17% and 16% yields which, following sugar deprotection, provided monocationic dimers, **142** or **143**, respectively.

The authors first determined the partition coefficients of the deprotected bisporphyrin conjugates in order to evaluate their lipophilic character. Both triglycosylated conjugates **139** and **142** were found to be more lipophilic than **137a,b** and **143**. Next, studies concerning their photosensitizing properties, consisting in trapping reactions of $^1O_2$ with ergosterol acetate, indicated production of $^1O_2$ in very high yield, comparable to the reference hematoporphyrin (HP). Their *in vitro* photocytotoxicity was then evaluated against the K562 leukemia cell line. The amphiphilic character was found to be an essential factor for efficient PDT, since the amphiphilic triglycosylated dimers **142** and more particularly neutral **139**, were found to be more active, with a quasi-similar activity to Photofrin® for **142** after 120 min of irradiation, inducing 70–80% cellular death, probably through apoptosis. In addition, it has been shown that the presence of the glucosyl residues on the same side of the dimer conjugates was crucial for optimal activity.

A final illustration of the biological potential of glycoporphyrins has been provided by Ballut *et al.*, who described their incorporation into a dimyristoyphosphatidylcholine

(DMPC) liposome membrane and measured the interaction of the resulting conjugates with Con A lectin.[154] In this context, the authors designed a new family of glycoconjugated PSs bearing only one glycodendron moiety, with variable length for the spacer linking the carbohydrates to the porphyrin scaffold, on the *para* position of one *meso*-phenyl group (Fig. 13). Briefly, the Cbz-glycine spacer was introduced via peptide coupling onto the amino function of a trivalent precursor, followed by deprotection of the three terminal carboxylic acid functions of the scaffold and subsequent introduction of aminoethyl α-D-mannopyrannoside derivative through amide ligation. The suitably functionalized aminodendron (after Cbz-hydrogenolysis) was then coupled to 5-benzoic acid-10,15,20-triphenyl porphyrin through an amide linkage, which after standard O-deacetylation afforded the trivalent glycoporphyrins **144** and **145** in good overall yields.

In order to evaluate the conditions of incorporation of compounds **144** and **145** into a liposome membrane, the two derivatives were mixed with DMPC in a 1:1 ratio and the mixtures were spread at the air–water interface. Interestingly, incorporation of the elongated conjugate **145** in DMPC induced the formation of larger vesicles than the phospholipid ones (diameter of 218 vs. 185 nm) while compound **144** mixed poorly, leading to smaller and less stable vesicles probably because of repulsive interaction between sugar moieties in the vicinity of the phospholipid headgroups. Moreover, 1 h of contact between the mixed liposome containing **145** and Con A, led to a dramatic increase of the vesicle diameter (up to 2510 nm) and polydispersity was observed. These striking results were postulated to originate from the long spacer, which would increase the mobility of the mannoside moieties and thus facilitate their interactions with the lectin. In addition, the existence of Con A dimers and tetramers at the pH studied, allowed lectin interaction with more than one porphyrin molecules possibly borne by different liposomes. Such multiple interactions would lead to the formation of a network of vesicles bridged by Con A molecules, resulting in a dramatic increase

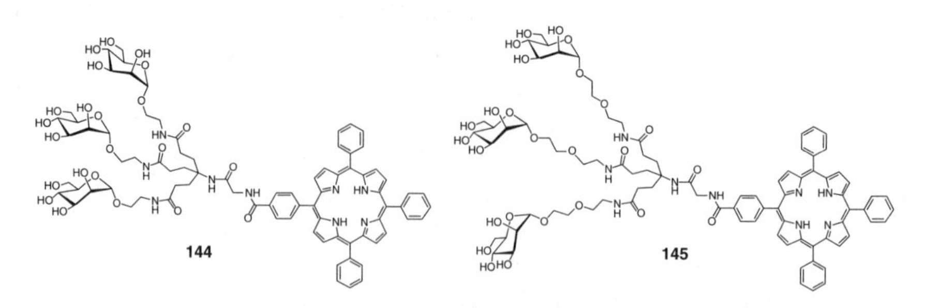

FIG. 13. Mannosylated prophyrin clusters prepared by Ballut et al. for liposome preparation.[154]

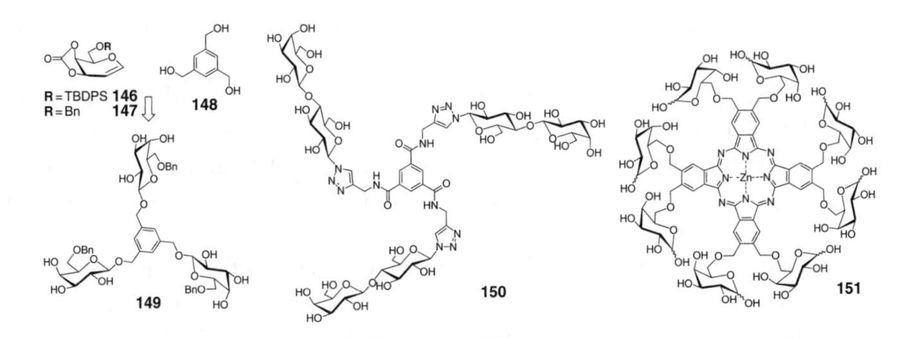

FIG. 14. Varied strategies for the coupling of carbohydrate derivatives to multivalent scaffolds.

in their apparent size. Such liposomes bearing glycodendronized phenylporphyrin could constitute an efficient carrier for drug targeting in PDT.

f. Other Aromatic Glycoclusters.—Clearly, access to other type of clusters has also been addressed, for instance, by using more conventional glycosylation strategy. In this context, the synthesis of trivalent β-D-galactoside **149** using stereoselective Lewis acid-catalyzed tris-glycosylation of 1,3,5-benzenetrimethanol (**148**) with galactal **147** derived from **146** has been described (Fig. 14).[155] A second example concerned the synthesis of trivalent clusters (**150**) containing such disaccharides as lactoside as ligands for galectin-1 and -3,[156] or the more-complex α-L-Fuc*p*-(1→4)-β-D-GlcNAc for the inhibition of binding of PAII-L lectin from *Pseudomonas aeruginosa,*[157] generated by click chemistry between an aromatic tripropargylated core (derived from trimesyl chloride), and the corresponding glycoside azide. Inhibitory properties of 20 μM against galectin-1 were observed for the conjugate **150** in HAI assays. This comprises a 40-fold enhancement as compared to free lactose, and 13.3-fold on a per-saccharide corrected basis. In addition, more-unusual macrocyclic scaffolds, such as phthalocyanine (**151**) have recently emerged to afford dense glycoclusters for which biological relevance remains to be determined.[158,159]

3. Glycoclusters from Carbohydrate Scaffolds

Structurally, carbohydrate derivatives are interesting polyfunctional platforms for the synthesis of oligosaccharide mimetics or glycoclusters. The possibility of selective functionalization of the various hydroxyl groups by well-established methodologies, and control of the relative orientation of the branches by configurational and conformational bias, makes these "full-carbohydrate glycoclusters" ideally suited for mapping the geometrical requirements for efficient lectin binding.

In this context, several synthetic strategies for efficient derivatization of various carbohydrate scaffolds have been described, notably including perallylation of non-reducing mono- or di-saccharides and subsequent ligation of *S*-glycosides by thioether linkages through photochemically promoted radical addition.[160] Controlled introduction of such desired functionalities as hydroxyl or amino from terminal allyl functions can also permit the construction of glycoclusters through ether,[161] or thiourea linkages (**154–156**, Fig. 15).[160,162]

Systematic studies aimed at evaluating their biological properties against uropathogenic *E. coli* have been initially addressed by measuring the relative affinities of Con A towards D-glucose-centered mannosylated clusters **152**α and **152**β and their corresponding analogues, **153**, **154**, and **155** by a competitive ELLA assay.[160] Hence, the known allyl 2,3,4,6-tetra-*O*-allyl-α- and β-D-glucopyranosides were chosen as the core building-blocks for the preparation of pseudosymmetric penta-branched derivatives. The versatile reactivity of the terminal alkene has been exploited in different ways to produce glycoclusters exhibiting a variety of linking functional groups, such as thioether, ether, and thiourea. With this strategy, the higher-valent D-glucose-centered glycocluster **155** containing 15 peripheral epitopes

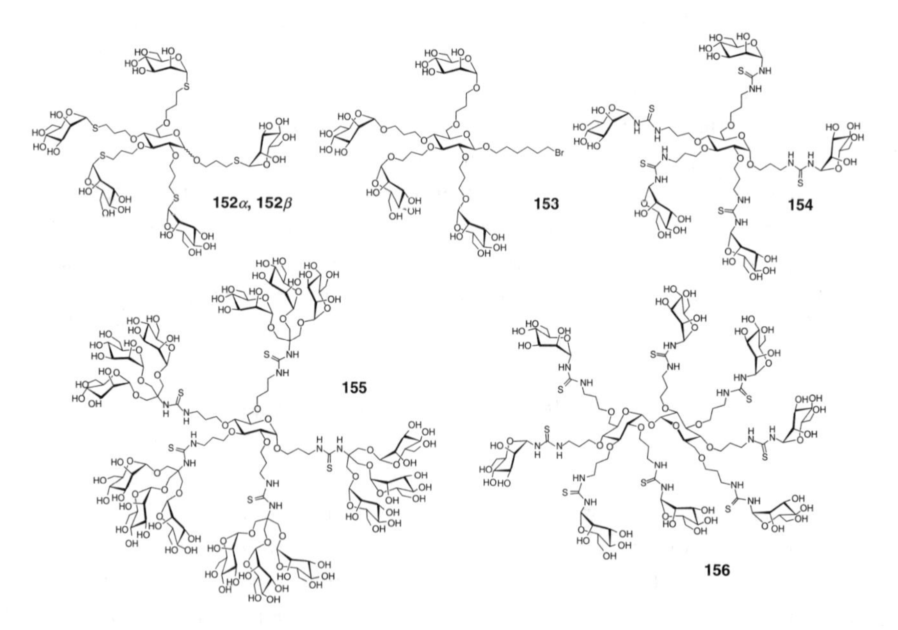

FIG. 15. Various mannosylated glycoclusters built around carbohydrate scaffolds.[160–162]

has also been obtained. The IC_{50} values of the pentavalent clusters **152**α and **152**β (27 and 31 μM, respectively) were indicative of strong Con A affinity relative to the reference methyl α-D-mannopyranoside, corresponding to a 6.4 and 5.5-fold enhancement on a per mannoside basis. A tetravalent analogue (**153**) containing an aglyconic oxygen atom presented identical avidity, indicating that multivalent presentation through four branches is sufficient to elicit a cluster effect. In sharp contrast, replacement of the 1-thiomannose wedges with α-D-mannopyranosylthioureido units virtually abolished any multivalent or statistic effects, along with a dramatic decrease of binding affinity. Alternatively, the 15-valent ligand **155**, possessing classical O-glycosidic linkages, was a very efficient inhibitor of the Con A–yeast mannan association (IC_{50} of 18.5 μM). However, the valency-corrected relative potency was about one-half that of the pentavalent derivatives. Those results illustrated the potential of carbohydrates as multivalent scaffolds for glycocluster synthesis and underlined the importance of careful design of the overall architecture in optimizing glycocluster recognition by specific lectins. As mentioned in the introduction, the TRIS scaffold is NOT optimized for simultaneous binding without the utilization of elongated spacers that allow several lectins to bind the maximum accessible sugar ligands.

In contemporary studies, several others multimannoside glycoconjugates based on carbohydrate scaffolds have been similarly constructed using ''click chemistry.'' The construction of these multivalent glycoclusters involves first the introduction of propargyl or azide functionalities on such mono- or oligo-saccharides as maltoside or maltotrioside as central cores and subsequent reaction with glycosylated ligands containing the complementary functionality (**157–159**, Fig. 16).[163,164]

A remarkable example of a biologically relevant D-glucose-centered glycocluster has been achieved by Kitov *et al.*, who designed an oligovalent, water-soluble carbohydrate ligand named STARFISH having subnanomolar inhibitory activity toward Shiga-like toxin I (SLT-I).[23] Shiga toxin (Stx) and Shiga-like toxins (SLTs or VTs (Verotoxins) are ribosome-inactivating proteins that act as *N*-glycosylases, cleaving several nucleobases from the RNA, and thereby halting eukaryotic protein biosynthesis. Infection by bacteria that produce SLTs results in serious gastrointestinal and urinary tract disorders, and is known to cause a potentially lethal disease, the hemolytic uremic syndrome (HUS), that may result in kidney failure. Stx is produced by *Shigella dysenteria*, whereas SLTs are produced by the Shiga toxigenic group of *E. coli* (STEC), including serotype O157:H7 and enterohemorrhagic *E. coli* (EHEC). The *E. coli* toxins can be further divided into SLT-I and SLT-II having conserved structures in the binding sites, and the SLT-II edema variant (SLT-Iie). Like CT, SLTs belong to a family of bacterial enterotoxins harboring a hexameric AB_5 structure,

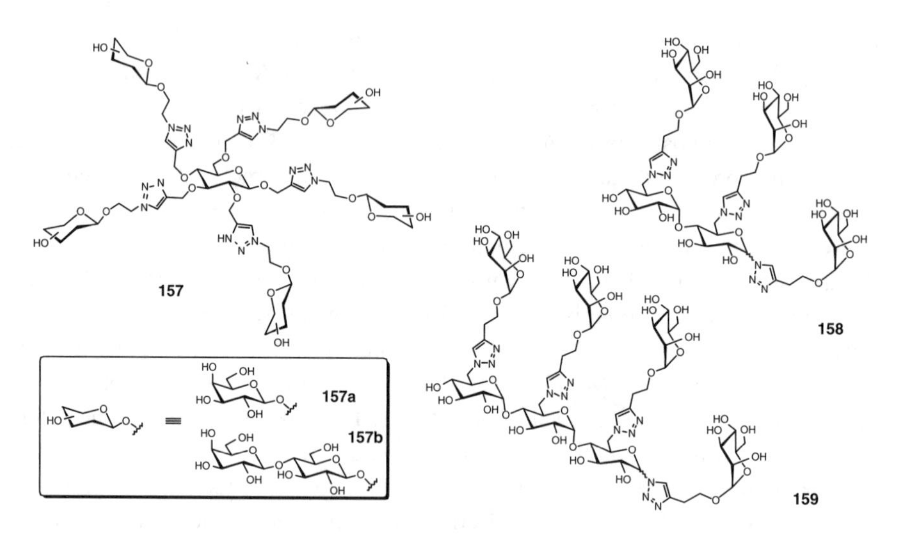

FIG. 16. Multivalent glycoclusters built around saccharide scaffolds using "click chemistry."[163,164]

where A denotes a cytotoxic enzyme located above the center of one face of the B_5-subunit, which represents a symmetrical homopentameric lectin-like carbohydrate-recognizing complex having a doughnut shape that facilitates delivery and entry of the A component into the cell of the host.[165] The SLT-I is virtually identical both in structure and mechanism of action to the toxin expressed by *S. dysenteria*, having similar B-subunits and differing only in one residue (Ser45 vs. Thr45) in their A-subunits. The *in vivo* receptor of Stx and SLTs is the trisaccharide portion (**P^k**) of the neutral glycolipid, globotriosylceramide (GbOse$_3$ or Gb$_3$, αGal-(1-4)βGal(1-4) βGlc(1-1)ceramide), present in greater amounts on the surfaces of kidney glomerular endothelial cells, to which the renal toxicity of Stx may be attributed (Fig. 17).

When the crystallographic structures of the protein or carbohydrate-ligated proteins are known, it is possible to model multivalent oligosaccharide inhibitors. As mentioned, the SLTs possess five noncovalent B-subunits per molecule and each subunit presents three carbohydrate-binding domains. Therefore, oligovalency "tailored" to the structure of the B-subunit pentamer offers the best opportunity for designing higher affinity inhibitors, as 15 binding sites are symmetrically arranged across the toxin surface that engages the cell membrane. Thus, highly selective, potent binding of SLTs to **Gb$_3$** is mainly attributed to the multiple interaction of the B-subunit pentamer with the trisaccharide moiety on **Gb**$_3$. On the basis of these facts, several SLTs antagonists in which this trisaccharide moiety was combined using various dendritic core structures have been reported. Among these, the foregoing study in

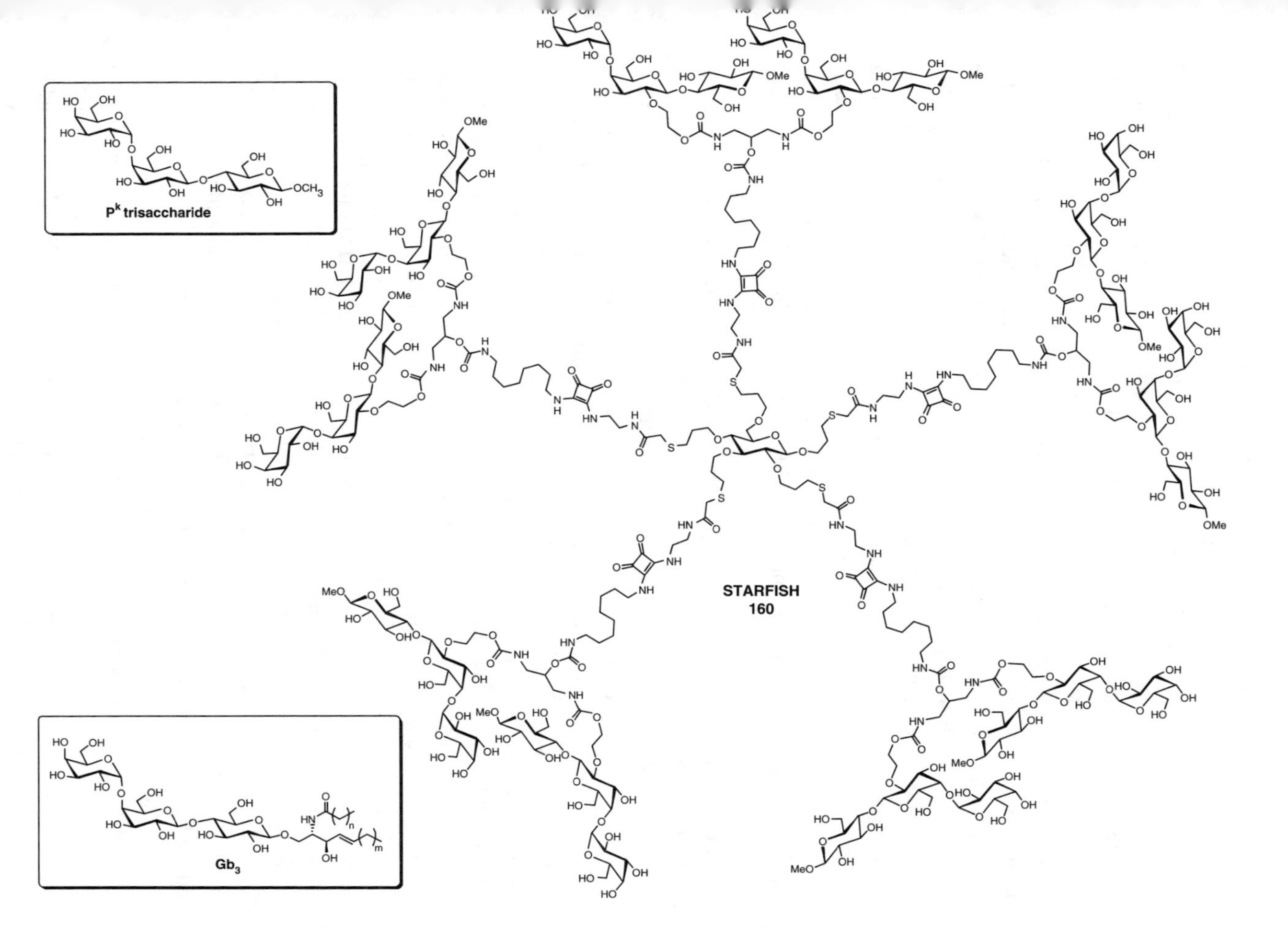

FIG. 17. The STARFISH dodecamer of Kitov *et al.* with the **P**$^{\mathbf{k}}$ *trisaccharide* linked laterally at 2′ and showing subnanomolar inhibitory activity against Shiga-like toxin I.[23]

particular afforded spectacular results and effectiveness in neutralizing SLTs infection. Hence, the use of the crystal structure of the B_5 subunit of *E. coli* 0157:H7 SLT-IB in conjunction with an analogue of its carbohydrate receptor allowed the design of tailored multivalent glycocluster **160** bearing 10 peripheral functionalized $\mathbf{P^k}$ residues. Structurally, this decavalent cluster, named STARFISH, comprised a radially symmetrical star-shaped carbohydrate backbone (D-glucose) with pairs of $\mathbf{Gb_3}$ trisaccharides at the tip of each arm, ideally spaced and oriented to be simultaneously engaged by each the B-subunits and thus to permit pentavalent interaction with SLTs. To achieve this display of tethered ligands, a glucose molecule was chosen as the central core for adequate pentavalent presentation, via its derivatization as pentaallyl ethers, free-radical reaction with thioglycolic acid, and subsequent chain elongation of the radial arms. Each of them was designed to span ~30 Å from the central core of the toxin and the tether was adapted to bridge binding sites 1 and 2 on each B-subunit. ELISA and solid-phase inhibition assays indicated the highest molar activity of any inhibitor reported, with subnanomolar inhibitory activity against STL-I and II (IC_{50} of 0.4 and 6 nM, respectively), thus being 1–10-million-fold higher than that of monovalent ligand. Furthermore, the ability of the inhibitors to protect host cells against a lethal dose of SLT-I in culture could be determined by using Vero cells. Cytotoxicity assays indicated that STARFISH inhibitor **160** provided effective protection of Vero cells cultured in the presence of SLT-I (IC_{50} of 1.19 μM) and SLT-II (IC_{50} of 1.58 μM), even over a 2-day coincubation period. Interestingly, the crystal structure showed that one STARFISH molecule bound to not just one but to two B-subunit pentamers. Instead of binding sites 1 and 2, the tethered $\mathbf{P^k}$-trisaccharides of STARFISH bound to two B-subunit monomers from separate toxin molecules. As, in the experiment, the concentration of STARFISH was sufficient to form a 1:1 complex, it was suggested that the formation of a 2:1 sandwich must be thermodynamically favored, as for the CT. Extended investigations have been made by the same group in order to circumvent the differentiation of SLT-I and SLT-II during biological studies. In fact, STARFISH **160** protected mice when it was injected subcutaneously in admixture with a lethal dose of SLT-I but not SLT-II, and it also reduced the distribution of ^{125}I-STL-I but not ^{125}I-STL-II to the murine kidney and brain. Modification of the nature and length on the linkage between the oligosaccharide component and the backbone (thereby increasing the flexibility of the tether) was then necessary to protect against the more-toxic SLT-II.[166]

Despite a considerable decrease in its inhibitory activity in the solid-phase inhibition assays (with an IC_{50} of 300 nM, corresponding to a ~50-fold increase compared with **160**), an equivalent decrease in performance of the so-called DAISY (**161**) in the more challenging SLT-I and SLT-II verocytotoxicity-neutralization assays was not

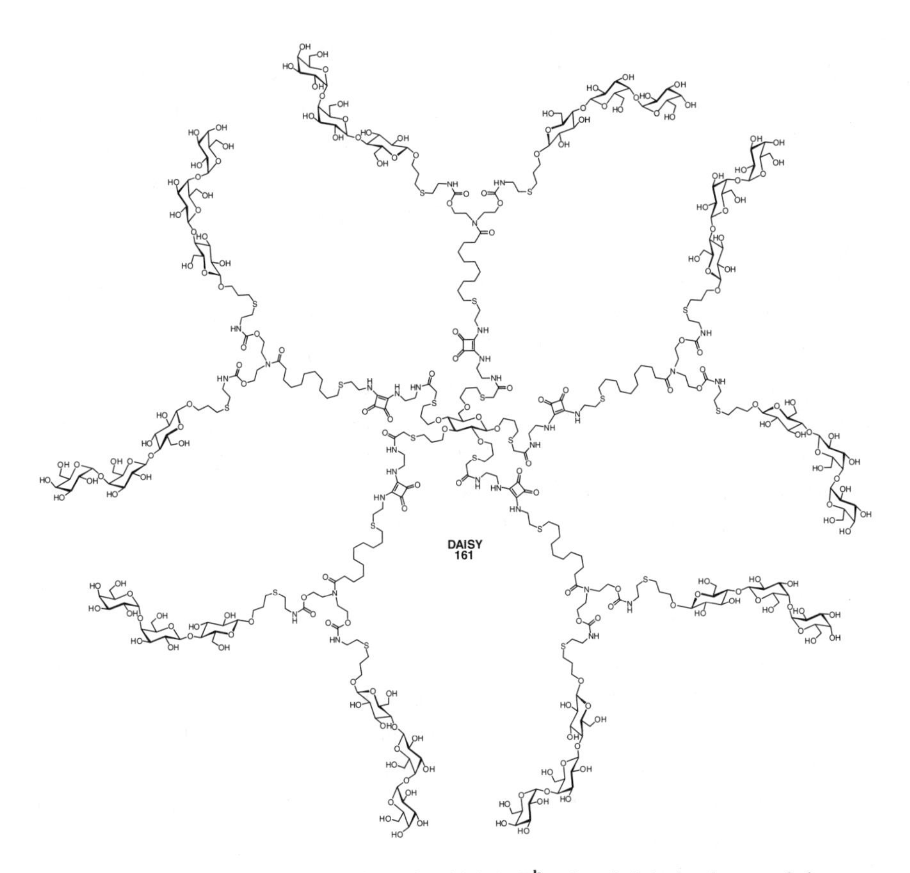

FIG. 18. A variant of the STARFISH 10-mer in which the $\mathbf{P^k}$ antigen is linked to the central glucose core at the anomeric position.[166]

apparent (Fig. 18). Subcutaneous injection of **161** protected mice against oral challenge with SLT-I and also STL-II-producing STEC. Interestingly, **161** did not interfere with the ability of the murine immune system to produce SLTs-specific protective antibodies.

The class of well-defined biocompatible CDs also provides an interesting alternative for construction of carbohydrate-centered clusters. This nonimmunogenic scaffold combines unique structural features, such as high symmetry and a truncated cone-shaped hydrophobic cavity, which can be exploited for efficient and radial epitope presentation, together with complexation of hydrophobic drugs in their cavities for targeted drug delivery.[167] More particularly, perglycosylated CDs showed much improved ability to form stable complexes with an antitumor reagent as compared with the CD itself and they provide one of the most potent candidates among specific carriers toward drug-delivery systems.

In order to introduce peripheral carbohydrates on these macrocyclic scaffolds, a variety of selectively functionalized derivatives has been described, often generated without need for hydroxyl-group protection. Since the first synthesis of perthioglucosylated derivatives of β-CD in 1995,[168] several other groups have described alternative synthetic pathways for the multifunctionalization of CDs, either on their primary faces or on both faces simultaneously. Those methodologies, involving well-documented photoaddition of thioglycosides to polyallylated β-CD ethers in an anti-Markovnikov fashion,[169] followed by selective introduction of such reactive functionalities at the primary positions as iodine,[170,171] amine,[172] and chloroacetamide,[173] gave rise to a panel of neoglycoconjugates named ''glycoCDs.'' These can exhibit simple β-D-glucosides (**162** and **163**) or *N*-acetyl-D-glucosamines (**164**) through ether, thiourea, or amide linkages, or more-sophisticated carbohydrate appendages, such as elongated sialic acids (**165**)[173] or sialyl Lewis^X (**166**)[170] via thioether functions (Fig. 19).

Concerning the general biological relevance of glycoCDs, it was demonstrated that these systems often showed amplified inhibitory effects as compared to their monovalent analogues. For instance, the Roy group synthesized a small library comprising β-D-glucosylated, β-D-galactosylated, α-D-mannosylated, and *N*-acetyl-β-D-glucosaminated CDs. They evaluated their relative binding properties toward different natural carbohydrate-binding plant lectins, using both microtiter plate competitive-inhibition experiments, double-sandwich assays using horseradish peroxidase (HRPO)-labeled lectins, and turbidimetric assays.[174] In general, all persubstituted β-CDs showed good to excellent inhibitory properties, together with abilities to cross-link their analogous plant lectins. Their capacity to anchor both microtiter plate-coated lectins and their corresponding peroxidase-labeled derivatives further confirmed the usefulness of these multivalent neoglycoconjugates in bioanalytical assays.

Another example has been provided by Furuike *et al.* who proposed an efficient practical synthesis of CD-scaffolded glycoclusters via standard nucleophilic substitution of iodide from a heptakis(6-deoxy-6-iodo-β-cyclodextrin) precursor by various unprotected sodium thiolates derived from 3-(3-acetylthiopropanamido)propyl glycosides (Fig. 20).[175] Hence, novel glycoCDs having D-galactose (**167**), *N*-acetyl-D-glucosamine (**168**), lactose (**169**), and *N*-acetyllactosamine (**170**) residues were obtained in high yields ranging from 78% to 88%. In order to evaluate the effect of clustering on the biological potency of these four glycoclusters, the hemagglutination inhibitory activities were examined by means of wheat germ agglutinin (WGA) from *Triticum vulgaris* and *Erythrina corallodendron* lectin (EcorL), which are known as

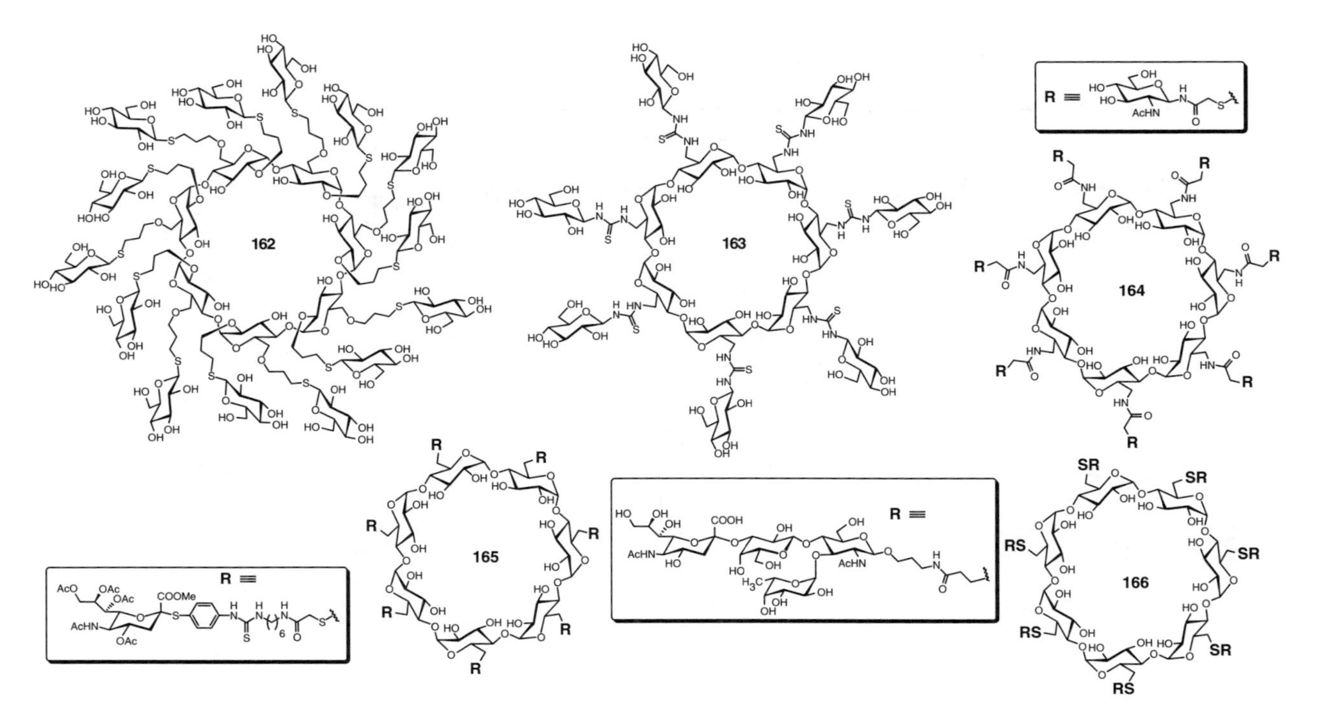

FIG. 19. Cyclodextrins decorated with sugars, including sialic acid and sialyl oligosaccharides.

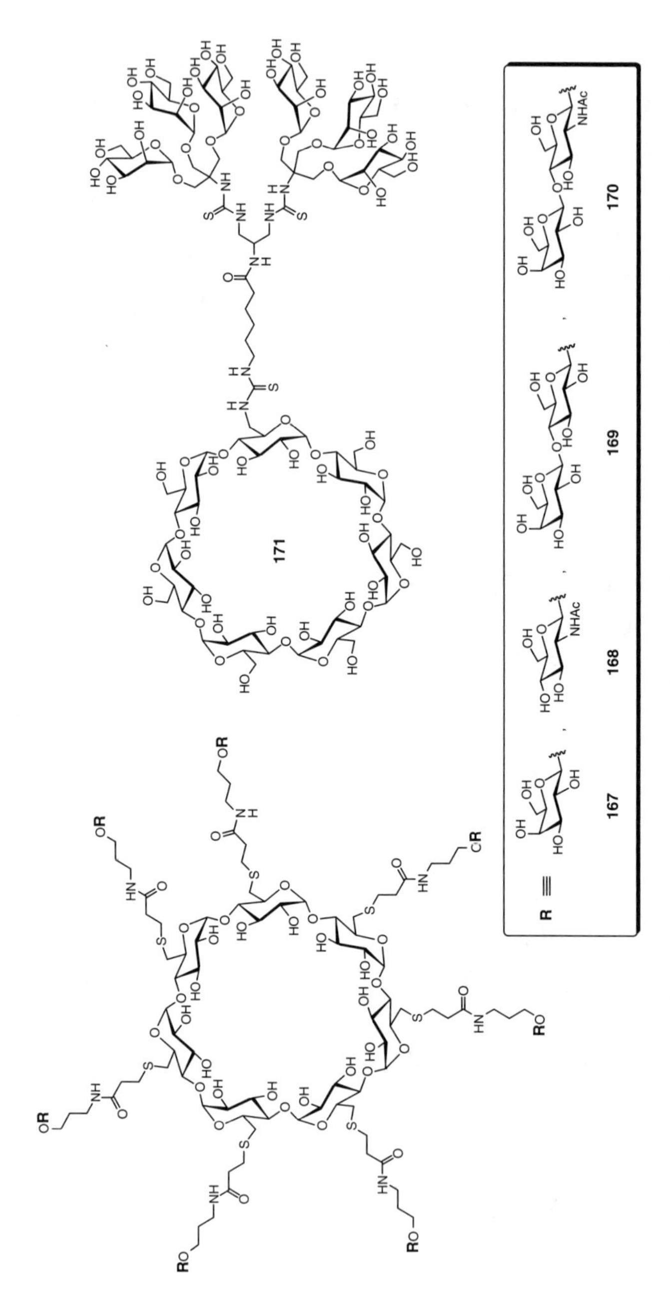

FIG. 20. Glycocyclodextrins and dendronized cyclodextrin-bearing mannose ligands.

N-acetyl-D-glucosamine- and *N*-acetyllactosamine-specific lectins. As anticipated, heptavalent GlcNAc-CD (**168**) showed drastically enhanced affinity against WGA in comparison with the monovalent analogue, corresponding to a 40-fold enhancement of the inhibitory effect. Similarly, the three other glycoconjugates efficiently inhibited hemagglutination induced by EcorL, notably compound **170** bearing *N*-acetyllactosamine residues, which showed the lowest inhibitory concentration (MIC of 89.2 μM).

Finally, another remarkable example has been designed in which a variety of glycoCDs bearing clustered mannosyl ligands were prepared and investigated for binding toward the tetrameric plant lectin Con A.[176,177] In this context, the key template was a 6^I-amino-6^I-deoxy-β-cyclodextrin,[178] the modified 1,2,3-triaminopropane branching component,[179] and an isothiocyanato-functionalized α-D-mannopyranosyl cluster prepared from TRIS. Coupling of the isothiocyanate derivative and the amine-functionalized trimannoside and the branching component gave the thiourea-bridged glycodendrimer–CD conjugates **171**. The monosubstituted hexavalent β-CD mannocluster showed a strong cluster effect in the inhibition of Con A binding to yeast mannan, with an IC_{50} of 10 μM that constituted up to a 22-fold increase on a molar basis as compared to monovalent derivative. In addition, the CD derivatives exhibited extremely high water solubility, more than 20-fold higher as compared to the parent CD (15 mM). Furthermore, the potential utility of such systems in active sugar-directed drug delivery to specific saccharide receptors on biological surfaces has been investigated. Thus, the cavity of β-CD was used to carry the anticancer drug Taxotere, for which the water solubility was greatly improved by the construct. Up to 4.5 and 4.7 g/L of Taxotere was solubilized in 25 mM aqueous solutions of a trivalent mannocyclodextrin derivative at 25 °C, corresponding to more than a 1000-fold solubility enhancement as compared to the water solubility of the isolated drug (0.004 g/L).[180]

4. Glycoclusters from Peptide Scaffolds

The class of peptide-based glycoclusters constitutes a valuable addition to the arsenal of glycoconjugates serving as important tools in glycobiology. Their structural properties allow them to act as mimetics of glycocalyx constituents and permit presentation of the glycans in particular and optimized orientations. Ideally, an unnatural peptide scaffold, cyclic or not, might provide much greater design flexibility than one afforded by the natural sequence. In general, manipulation of the length and nature of the amino acid sequence of the cyclic peptide provides opportunities for designing multivalent ligands that are suitable for different geometric requirements. Moreover, variation in the number of glycan attachments, as well as the distances between glycans, together with subsequent introduction of a suitable chemical handle for conjugation to a carrier protein constitute parameters that could be readily tailored during the scaffold synthesis.

Although standard chemical processes have been used to access linear bioactive glycopeptides,[181–183] solid-phase synthesis (SPS) constitutes one of the most appealing synthetic methodologies and has led to a large variety of well-defined glycoclusters grafted onto synthetic peptide platforms. The numerous successes encountered in the SPS of various peptide-based conjugates, coupled to the ease of preparation and purification of complex derivatives, have motivated chemists to apply this methodology to synthesis of peptide glycoclusters. Hence, linear peptides containing a glycocluster head group (**172**, **173**),[184] multivalent cyclic neoglycopeptides including, for instance, three *N*-acetylglucosamine residues (**174**),[185] and multitopic biotinylated glycoclusters build on a topological cyclodecapeptide template (**175–179**)[186] represent remarkable illustrations of this powerful and straightforward methodology (Fig. 21).

Concerning biological investigations, Zhang *et al.* described the SPS of a pentavalent ligand having a cyclic decapeptide scaffold with built-in linkers in order to create efficient inhibitors for the CT B pentamer.[187] Varying the nature of the flexible amino acids lacking side chains (such as lysine, γ-aminobutanoic acid, and ϵ-aminohexanoic acid) allowed the authors to achieve the desired ring-size variations and to increase the likelihood that the peptides might adopt expanded conformations in solution (Fig. 22). The synthetic peptide-core (**180**) bearing five galactosylamine residues were assayed for their ability to block CT B pentamer binding to ganglioside-coated plates. Interestingly, ligands having longer or shorter linkers than optimal exhibited a loss in inhibitory power, demonstrating that when a ligand's effective dimension is not matching that of its target, there is a decrease in the ligand's affinity. Submicromolar IC_{50} values were obtained for the best derived ligand (**180**), depending on the core size, which ranged from 2.5 to 6 Å. The resulting glycoclusters presented a more than 10^5-fold increase over monovalent galactose, which had an IC_{50} of ~ 100 mM in the same receptor-binding inhibitory assay.

Several glycopeptides, particularly those having mannosides or complex oligomannoside end-groups, have the potential to become entirely synthetic vaccines. More notable are the following examples directed at raising the protective immune response against HIV-1 infection. Noteworthy is the fact that several infectious microorganisms use or escape the immune defense mechanisms by masking important receptors or antigenic determinants through exposing self-carbohydrate structures.

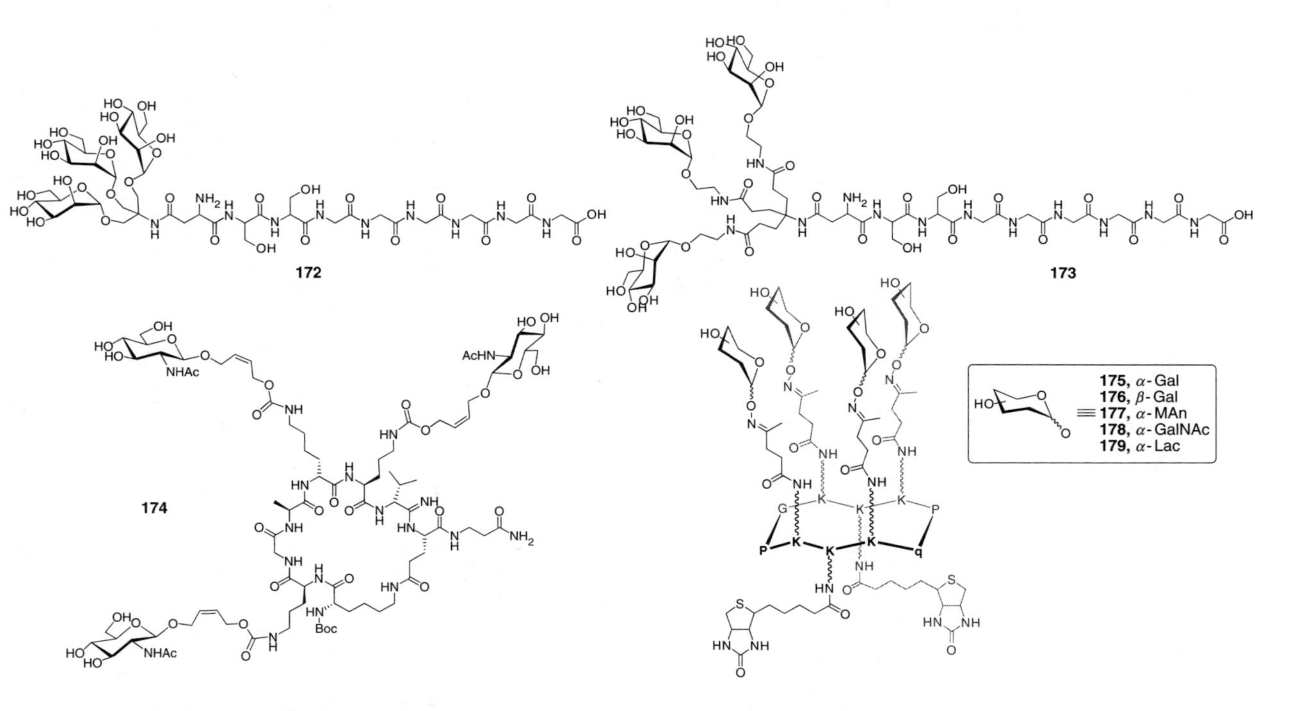

FIG. 21. Peptide-based glycoclusters at the wedge of linear peptide sequences or as cyclic scaffolds.

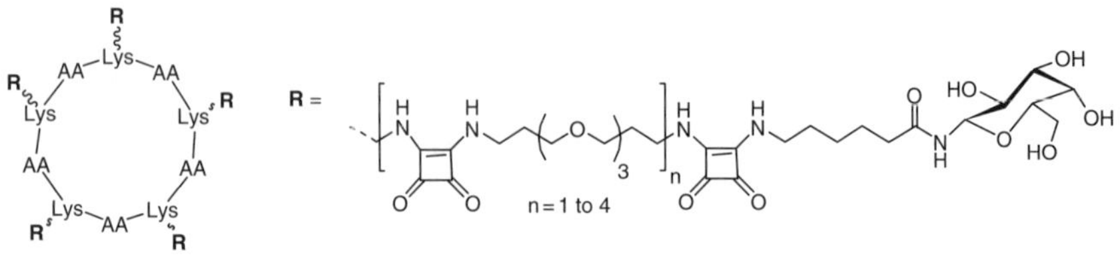

FIG. 22. Potent pentavalent D-galactosylamine inhibitors used by Fan's group for the cholera toxin B pentamer.[187]

The consequences of these ''masking'' events are that the bacterial or viral pathogens are transported to target tissues by self cellular systems. An emerging example of this irregular situation is the recognition, binding, and transport of the hyperbranched high-mannose oligosaccharide ($Man_9GlcNAc_2$) present on HIV gp120 by dendritic cells, with the ensuing transport of the ''hidden'' virus particles to lymph nodes.

Danishefsky's group using 14-residue cyclic peptides bearing β-turns provided by two D-Pro-L-Pro sequences appended at the extremity of the scaffolds (**181**), elegantly constructed the potent HIV-gp120 carbohydrate antigen **182**, which was strongly recognized by the human swapped and protective IgG 2G12 antibody (Fig. 23).[188] The cyclic peptide containing two or three handles for glycan attachment, together with a single cysteine residue for further carrier protein or biological marker conjugation, was built on a prolinated trityl resin using solid-phase peptide chemistry, while the synthetic oligosaccharide $Man_9GlcNAc_2$-NH_2, prepared by amination of the reducing sugar,[189] was attached to the aspartic acid residues properly positioned onto the cyclic peptides **181a** or **181b** using the Lansbury aspartylation procedure, thus providing dimer and trimer (**182**), respectively.[190]

Studies with SPR indicated that the control peptide lacking the saccharide portion, as well as the monovalent glycopeptides, showed no measurable response, thus confirming the importance of multivalent binding event in this critical recognition process. The corresponding divalent and trivalent glycopeptides proved to be strong ligands against the 2G12 bound surface, indicating topographical homology with the natural gp120 structure. With the confirmed antigenicity of both glycopeptides, the authors then used the sulfhydryl anchor of the divalent antigen for its attachment to the highly immunogenic, maleimido-derived outer membrane protein complex (OMPC) carrier from *Neisseria meningitidis*. The conjugate addition, performed at near-neutral pH, afforded a vaccine candidate bearing ~2000 copies of the carbohydrate antigen which was also recognized by 2G12 antibody in an ELISA assay using

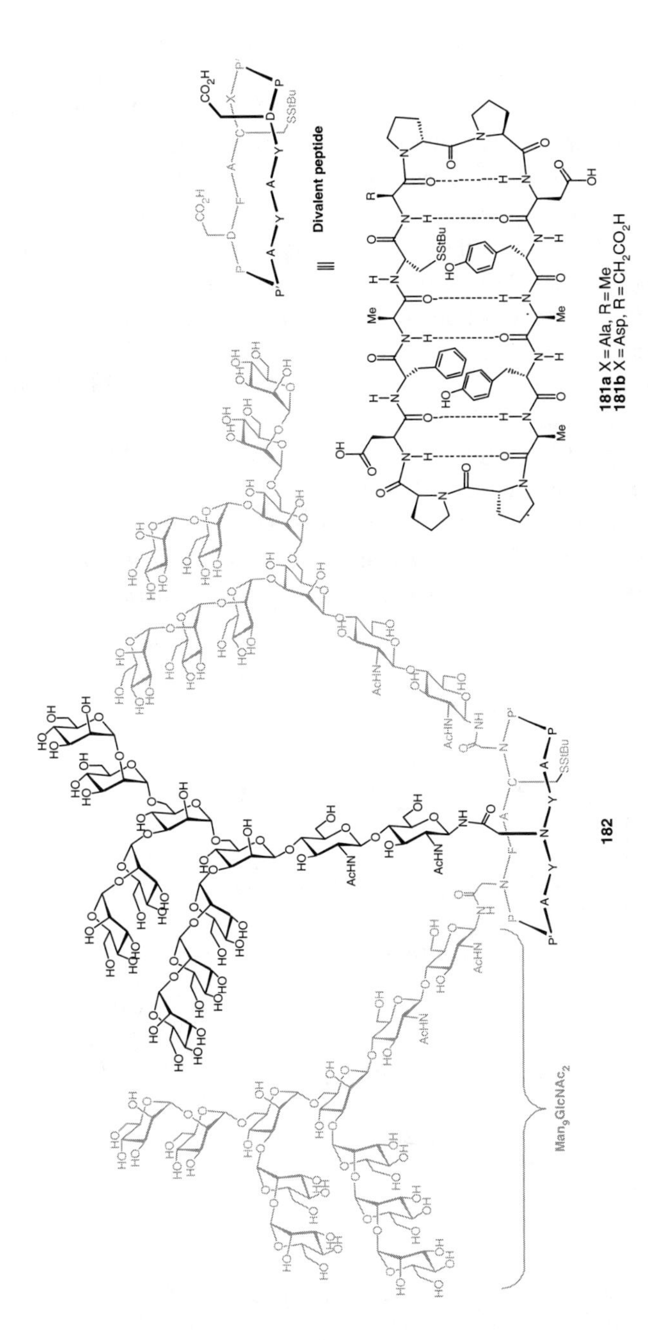

FIG. 23. Representative examples of a potential anti-HIV-1 gp120 conjugate vaccine.[188]

either gp160 or the glycoconjugate as coating antigens. Horse radish peroxidase-conjugated antihuman IgG (HRP-IgG) evidenced detection of the binding event. The results indicated that, while both antigens could bind to the 2G12 antibody, the natural gp160 bound more efficiently. The immunological properties of the semisynthetic vaccine are presently under investigation.

An analogous example has been provided for the construction of a tetrameric gp120 glycan epitope (**186**) by using the determined shortest version derived from the D1 arm of the $Man_9GlcNAc_2$ (**183**), and prepared as an extended azide-bearing aglycone **184** (Fig. 24).[191] The cyclic peptide scaffold **185** was built with 6 lysine, 2 glycine, and 2 proline residues. Four of the lysine ϵ-amino groups were acylated with propynoic acid for glycan attachment by ''click chemistry,'' and the remaining two ε-amines were coupled to the universal T cell epitope derived from tetanus toxoid 15-mer peptide $TT^{830\text{-}845}$ to afford a fully synthetic vaccine candidate **186**.

The binding of vaccine candidate **186** to the human antibody 2G12 was analyzed by SPR technology. While a synthetic monomeric oligosaccharide such as **184** as well as the natural high-mannose type N-glycan $Man_9GlcNAc_2Asn$ (**183**) did not show binding to antibody 2G12, the related synthetic oligosaccharide clusters carrying

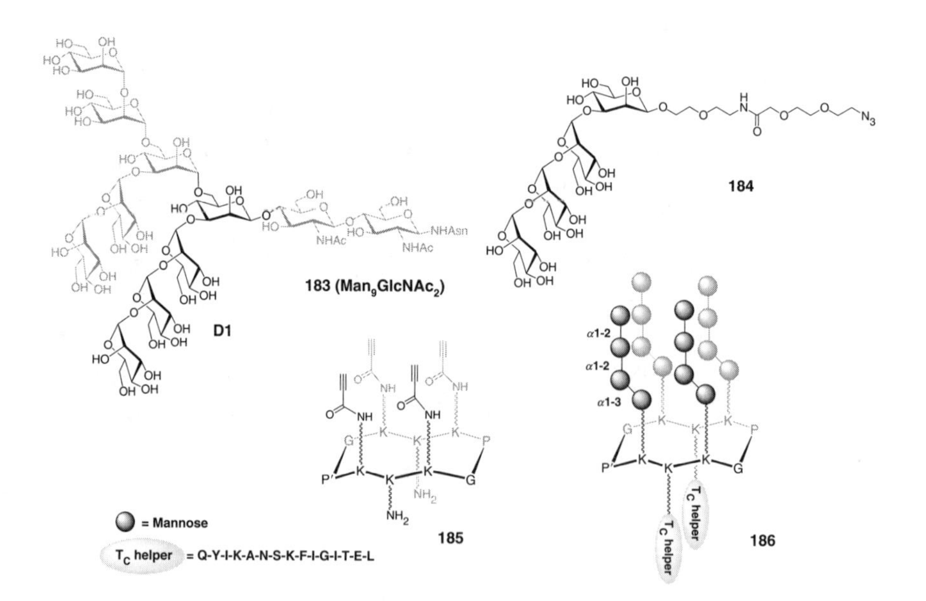

FIG. 24. Cyclic peptide vaccine candidate bearing the minimally epitopic D1 branch of the Man_9 $GlcNAc_2$ antigen of HIV-1 gp120 recognized by the protective human antibody 2G12.[191] (See Color Plate 11.)

four units of the D1 arm tetrasaccharide **186** have demonstrated affinity to this antibody. Furthermore, results suggested that an appropriate spatial orientation of the sugar chains in the cluster was crucial for high-affinity binding to antibody 2G12, and that the introduction of the T-helper epitopes onto the cyclic decapeptide template did not affect the structural integrity of the oligosaccharide cluster formed at the other face of the template. Therefore, compound **186** constitutes a valuable immunogen that might also be able to raise carbohydrate-specific neutralizing antibodies against HIV-1.

5. Other Glycoclusters

To better understand the critical role of oligosaccharide–receptor interactions and their molecular mechanisms through the cluster-effect, and thus access optimized synthetic ligands, several research groups have shown creativity in proposing original multivalent platforms that could allow for tailored valencies, dimensions, and epitope orientations.

For instance, Burke *et al.* investigated a templated ligand array based on a conformationally defined macrocyclic scaffold on which three mannoside residues, appended through a solubilizing linker, were displayed (Fig. 25).[192] The C_3-symmetric hydropyran cyclooligolide core was previously obtained according to an iterative multistep synthetic sequence notably involving key-step macrolactonization under Keck–Steglich high dilution conditions.[193] In particular, the template rigidity conferred a specific orientation of saccharide residues such that they emanate from a single face. In addition, molecular-modeling studies predicted a maximum separation between mannose residues of approximately 35 Å. Given that the binding sites within the Con A tetramer are separated by 65–70 Å, the trivalent ligand **187** could not simultaneously occupy two mannose-binding sites within the tetrameric lectin. Therefore, it could be used to explore mechanisms of multivalent ligand binding in the absence of the chelate effect. SPR competition-binding assay with Con A indicated the absence of nonspecific interactions induced by the template, and highlighted the fact that **187** promoted rather than inhibited binding of Con A to the mannose-substituted lipid surface. The data suggested that the glycocluster was able to bind two or three Con A tetramers simultaneously and to form soluble clusters with high avidity for immobilized ligands. Assay by FRET showed the reversibility of the phenomenon, and corroborated the fact that clustering of Con A by **187** did not depend on subsequent precipitation or cluster binding to a surface. In addition, this synergistic interaction favored the formation of a Con A cluster rather than the formation of a one-to-one complex of Con A and the trivalent ligand.

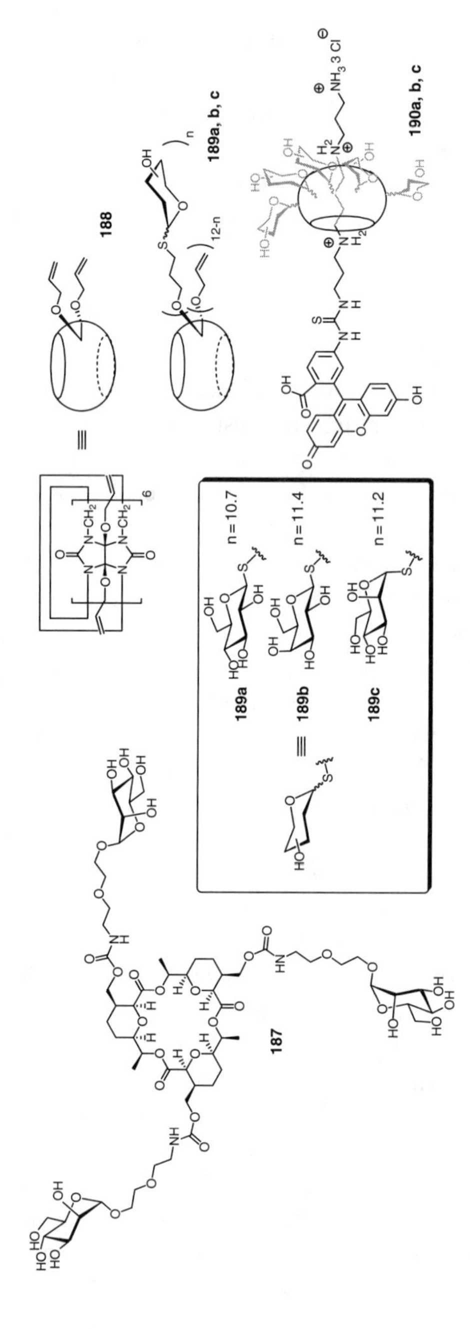

FIG. 25. Cucurbit[*n*]uril cavitand bearing sugar clusters and fluorescein probe.[194]

Another striking example has been provided which exploited cucurbit[n]urils (CB [n], $n = 5$–10) (**188**), a family of macrocyclic cavitands comprising n glycoluril units as multivalent scaffolds for carbohydrates.[194] They possess a hydrophobic cavity accessible through two identical carbonyl-fringed portals, and can form stable host–guest complexes with a wide range of guest molecules. Hence, CB[6]-based clusters **189a-c**, presenting an average of 11 carbohydrate moieties attached to the periphery (**189a**, β-glucose; **189b**, β-galactose; and **189c** α-mannose, respectively) were prepared via photoreaction of (allyloxy)$_{12}$CB[6] (**188**) with corresponding acetylthioglycosides, followed by standard O-deacetylation (Fig. 25). In particular, the mannoside cluster **189c** presented strong avidity for Con A and a 1100-fold excess of Me-αMan was necessary for the disruption of its cross-linking interaction with the lectin. ITC confirmed this affinity and indicated that **189c** behaved predominantly as a trivalent ligand to the lectin with a binding constant $K = (1.9 \pm 0.2) \times 10^{-5}\ M^{-1}$, which was 25 times higher than that for Me-αMan. The authors further exploited the cavity provided by such conjugates to incorporate, through host–guest interactions, the fluorescein isothiocyanate (FITC)–spermine conjugate **190** in order to estimate their potential as drug-delivery vehicles. Targeted delivery experiments *in vitro* were thus carried out with the (FITC)-spermine derivative as a fluorescent probe as well as a drug model, and the HepG2 hepatocellular carcinoma cells that overexpressed galactose receptors as target cells. Intracellular translocation has been estimated by confocal microscopy, and the results confirmed internalization of the CB[6]-galactose-(FITC)–-spermine complex **190b** via galactose-mediated endocytosis. This study demonstrated the potential of glyco-cucurbituryl derivatives as targeted drug-delivery system and paved the way to other interesting therapeutic applications.

Touaibia and Roy have used biodegradable and biocompatible cyclophosphazene derivatives as multivalent platforms to afford densely functionalized α-D-mannoside clusters **191** (Fig. 26).[195] Use of a short and efficient strategy based on single-step Sonogashira and ''click chemistry'' has afforded a variety of glycoclusters around cyclotriphosphazene with different valencies, spacers, and epitope spatial arrangements. Hexavalent derivatives **191a–191d** were obtained from commercial hexachlorocyclotriphosphazene ($P_3N_3Cl_6$) onto which various phenol derivatives had been conjugated via nucleophilic substitution. Direct ligation of α-D-mannoside containing the adequate complementary function via optimized Pd-catalyzed cross-coupling or ''click chemistry'' and subsequent deprotection under Zemplén conditions afforded hexavalent conjugates in excellent yields.

Higher valency glycoclusters were also obtained by reaction of the hexabromomethylated P_3N_3 scaffold derivative with di- or triazide building blocks containing a hydroxy anchoring function (Fig. 27). Corresponding deprotected dodeca- and

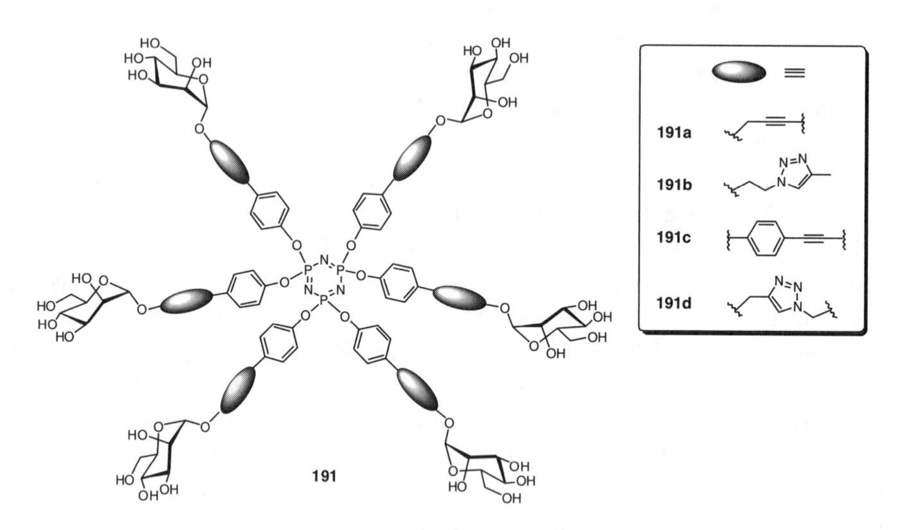

FIG. 26. ‘‘Majoral-type’’ glycodendrimers built around a cyclophosphazene core (P_3N_3).[195]

octadeca-mannoside clusters **192** and **193** were thereby isolated following click chemistry and O-deacetylation. Similarly, decamannoside **194** has been obtained from a dimer bearing an $N_3P_3Cl_5$ fragment prepared from the double monosubstitution of hexachlorocyclotriphosphazene with bisphenol A, using *n*-BuLi as a base and appropriate stoichiometry. Subsequent replacement of the remaining chlorides by iodoaryl groups allowed the preparation of the Sonogashira adduct **194**.

To evaluate the influence of structural parameters governed by the cyclophosphazene core concerning the valency and the spatial orientation of epitopes, as well as the nature of linkers directly related to the ligation technique used for the mannoside incorporation, the authors performed preliminary kinetic turbidimetric assays with Con A. Insoluble cross-linked complexes formed rapidly for all compounds, without marked difference for the hexavalent analogues. On the other hand, the incorporation of additional mannosyl units led merely to statistical binding-affinity enhancements, notably for the less-dense decamer **194**, which presents favorable extended intersugar distances.

The class of azamacrocycles also constitutes an interesting platform for controlled multivalent presentation of epitopes (Fig. 28). Although examples are quite scarce, their use has generated biologically relevant glycoclusters. For instance, the tetravalent α-mannosyl clusters **195** scaffolded on 1,4,8,11-tetraazacyclotetradecane (cyclam) have been described and showed interesting inhibitory properties in the adhesion of type-1 fimbriated *E. coli* to guinea pig erythrocytes[196] or binding properties toward Con A (**196, 197**).[197]

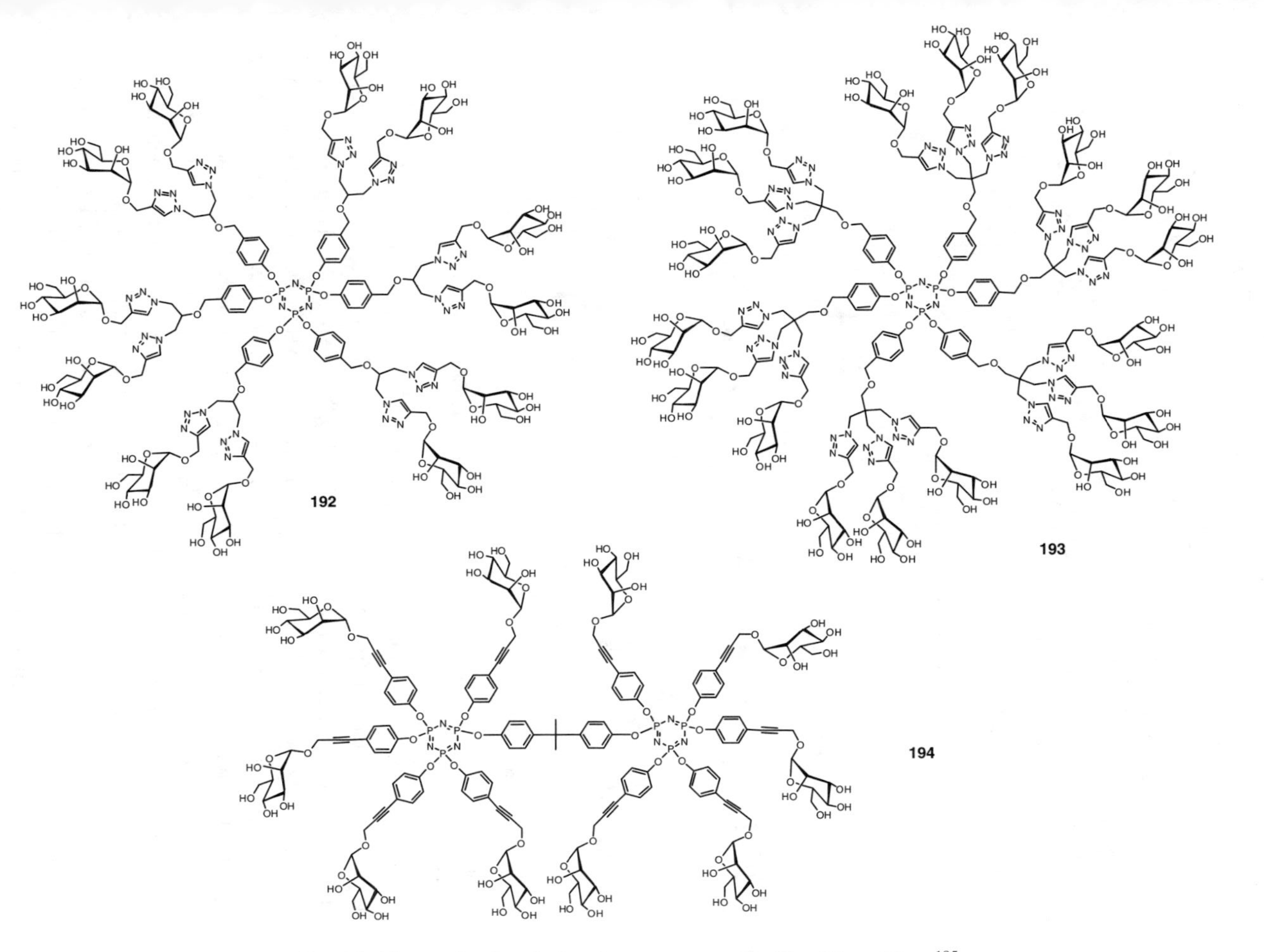

FIG. 27. Mannosylated cyclophosphazenes prepared by Touaibia and Roy.[195]

FIG. 28. Mannosylated clusters built around the azamacrocycle cyclam.[196,197]

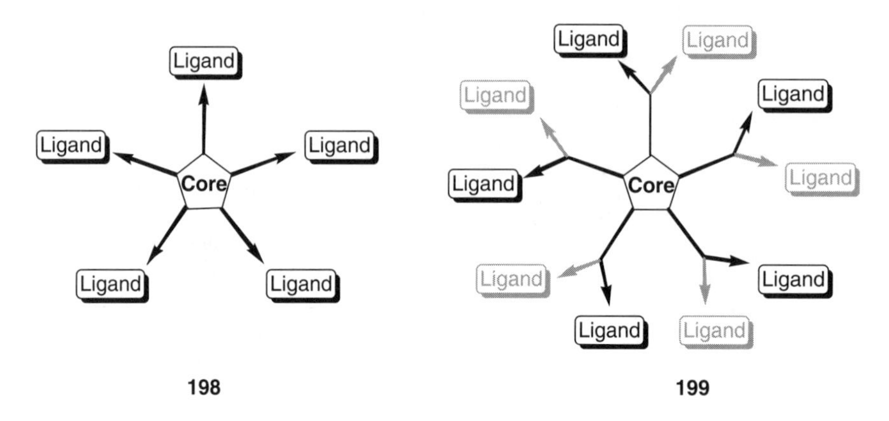

FIG. 29. Potent antiadhesins against the AB5 subunit of bacterial toxin.[198]

Significant enhancements have been obtained by Fan *et al.*, who similarly included the pentavalent analogue (pentacyclen) into their arsenal of templates toward the high-affinity multivalent ligands **198** and **199** (Fig. 29).[198] The modular design of multivalent antagonists (antiadhesins) targeting members of the AB_5 bacterial-toxin family has been achieved, including CT and *E. coli* heat-labile enterotoxin (LT). The strategy exploited the fivefold symmetry of the binding sites on the toxins B pentamer.

This synthetic plan was based on difficulties in distinguishing ligand-mediated aggregation of the protein from an actual gain in effective affinity when highly multimeric ligands were studied. In this context, the authors expected substantial enhancement in affinity by using multivalent systems that allowed geometrically restrained presentation of the exposed ligands to match the specific arrangement of the target protein's binding sites. With this in mind, synthesis of various pentavalent derivatives **198** and **199** containing three distinct modules has been achieved.

A semirigid core, having "fingers" projecting outward in the direction of the receptor's binding sites and fitting perfectly into the active site have been assembled.

A study describing the syntheses of β-D-galactosylated pentavalent ligands built around a pentacyclen core and adopting a fivefold symmetry, and the systematic effects of flexible linkers has been published (Fig. 30).[198] Preliminary results detailed the affinity for LT, and the best pentavalent ligand **200** ($R = R_1$ with $n = 4$) showed an IC_{50} that was 10^5-fold higher than galactose. A few years later, the same group optimized the structure, incorporating five copies of the more potent *m*-nitrophenyl α-D-galactoside (MNPG) derivatives rather than five copies of β-D-galactose.[199] The ~100-fold higher single-site affinity enhancement of MNPG for the binding site of the toxin relative to galactose was found to yield a proportionate increase in the affinity and IC_{50} measured for the respective pentavalent constructs. Indeed, ELISA experiments involving CTB pentamer revealed an IC_{50} of 0.9 μM for **201** ($R = R^2$), corresponding to about an 18-fold affinity enhancement over **200** ($R = R^1$). These results demonstrated that improved affinity for a single-site ligand confers a parallel improvement in the affinity of a pentavalent construct presenting five copies of the same ligand. Furthermore, from DLS studies and crystallographic data, it resulted that a 1:1 CTB/**201** complex was the major mode of association in solution between the ligand and the toxin. The lead structure **201** was then affined with the incorporation of guanidine-bridged poly(ethylene glycol)s of various lengths, in order to enhance its water solubility. Thus, compound **202** ($R = R^3$) having an optimized effective

FIG. 30. The best ligands against *E. coli* heat-labile enterotoxin.[198–201]

dimension for strong affinity, displayed an IC_{50} of 6 nM, being lower than that of the natural receptor GM1, and which represented the first multivalent ligand based on a simple galactoside having nanomolar affinity against CTB.[200] During parallel investigations, spectacular improvement has been observed with the construction of a family of complex ligands having five flexible arms, each ending with a bivalent ligand **203** ($R = R^4$).[201] ELISA tests revealed that the improved decavalent derivatives were significantly more potent, showing affinities for CTB an order of magnitude better than the corresponding nonbranched ligands. More precisely, a more than 10^6-fold enhancement in inhibitory power over the monovalent ligand was achieved with the best decavalent candidate **203** (with $n = 4$) with an IC_{50} of 40 nM, a value that lies in the same range as the IC_{50} of the natural receptor (50 nM). Dynamic light-scattering studies demonstrated the formation of concentration-dependent unique 1:1 and 1:2 ligand–toxin discrete complexes in solution, with no sign of formation of large aggregates. Crystallographic studies confirmed that the decavalent inhibitor resulted in a ''sandwich arrangement'' of two B pentamers facing each other and bridged by the ligand. The improvement in IC_{50} displayed by the decavalent ligands might be attributed to a substantial difference in affinity between the galactose fragment within the pentavalent ligand and the rather short, nonspanning bivalent galactose moiety present in the decavalent ligand.

Finally, two other exotic scaffolds has been described that can be added to the large panel of glycoclusters described so far. The first consists in linear pentaerythrityl phosphodiester oligomers (PePOs) onto which galactosyl clusters were attached via combinatorial and automated synthesis on a solid support (Fig. 31).[202] The propargylated scaffold was synthesized by standard DNA solid-phase supported phosphoramidite chemistry, and azido-galactoside residues were conjugated by ''click chemistry.'' Microwave-assisted acceleration of the reaction time afforded the corresponding multivalent clusters **204a,b** with controlled length and valency. The elongated fucosylated analogues **205** have also been described by the same authors, and showed increased affinity in ELLA competition assay for *P. aeruginosa* lectin (PA-IIL), with IC_{50} values 10–20 times higher than the monovalent L-fucose but corresponding to only a modest twofold increase on a per saccharide basis.[203]

The second one concerns polyhedral oligosilsesquioxanes (POSS) that have been functionalized with carbohydrate moieties **206–208** via standard amide bonds[204] or more efficiently via photolytic thiol addition from a perallylated precursor.[205] Biological evaluation of the octavalent POSS-glycocluster **208** exhibiting elongated lactoside residues was investigated by measuring the inhibitory effect on the binding of asialo-oligosaccharides from human α1-acid glycoprotein (AGP) by RCA120 (a β-galactose specific lectin), using capillary-affinity electrophoresis. The first

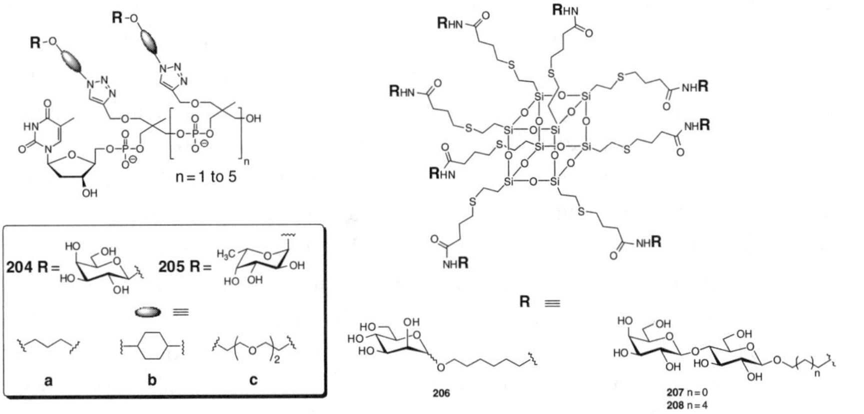

FIG. 31. Various multimeric glycoclusters.

results indicated strong inhibition, ~200 times higher than that induced by free lactose and being thus attributable to the cluster effect.

III. Glycosylated Carbon-Based Nanostructures

The past few years have witnessed the discovery, development and, in some cases, large-scale production and manufacturing of novel materials that lie within the nanometer size scale. More particularly, the incorporation of nanotechnologies in early-stage development of new drugs, diagnostics, or therapeutics constitutes a powerful addition to the arsenal of classical macromolecular structures. Consequently, this strategy opens new avenues for the original and efficient designs of adapted nanomedicine. In this context and owing to their three-dimensional architectures, the fullerenes and nanotubes (both carbon-based nanomaterials), constitute promising candidates for novel nanometric constructs having biological properties and offering challenging scaffolds toward multivalent presentation of saccharide units.

1. Glycofullerenes

Fullerenes, the third allotropic form of carbon along with graphite and diamond, are a novel class of spheroidally shaped molecules made exclusively of carbon atoms. They have generated much enthusiasm and numerous research efforts during the past few years.[206] Hence, the chemical and physical features of C_{60}, also named

Buckminsterfullerene, the most representative example among the fullerenes, have been extensively explored. Their intrinsic properties such as their size, hydrophobicity, three-dimensionality, and electronic properties have made them extremely promising nanostructures, offering interesting features at the interface of various scientific disciplines, ranging from material sciences[207] to biological and medicinal chemistry.[208–210]

In order to demonstrate and illustrate their attractive potential in the foregoing applications, an array of studies has been recorded, including cytotoxicity investigations on both unmodified and functionalized fullerenes. Initial results have indicated that these novel and fascinating architectures were not carcinogenic when applied to the skin, nor did they affect the proliferation and the viability of cells when they are internalized. Hence, despite an observed dose-dependent toxicity phenomenon reported for certain related derivatives, the early observations indicate great promise for applications in DNA cleavage, PDT, enzymatic inhibition, antiviral, antibacterial, and antiapoptotic activity.[209–211] However, the total lack of solubility in aqueous or physiological media is a severe drawback for their quick and efficient development as suitable carriers. In order to circumvent the natural repulsion of fullerenes for water, several methodologies have been adopted, including their entrapment into tailored microcapsules, their suspension with the help of cosolvents, and their chemical derivatization, notably their introduction onto peripheral solubilizing appendages. Furthermore, it has been shown that the multivalent presentation of polar groups around the fullerene spheres can prevent clustering phenomena in reasonably dilute solutions and consequently increase the hydrosolubility of the resulting conjugates. In this context, a variety of chemical functionalities have been utilized to increase both the hydrophilicity (with groups such as OH, CO_2H, NH_2, quaternary ammonium, and CD) and to prepare novel compounds possessing biological and pharmacological activity.

Among the panel of known fullerene derivatives, the fulleroglycoconjugates (also termed glycofullerenes) exhibit a combination of interesting properties related to water solubility and biological relevance. Spherical topology of the fullerenes has furnished suitable scaffolds for multivalent presentation of peripheral carbohydrate residues, and chemistry has been adapted for their efficient and controlled conjugation. The first examples of glycofullerenes were monovalent conjugates which were subsequently tailored and optimized to generate multivalent glyconanostructures based on C_{60} where biological properties were enhanced by taking advantage of the "glycoside cluster effect."

a. Monovalent Structures.—In the early 1990s, Vasella *et al.* reported the first synthesis of a monoglycosylated fullerene, introducing one carbohydrate residue on C_{60} via the nucleophilic glycosylidene carbene precursors **209** and generating the

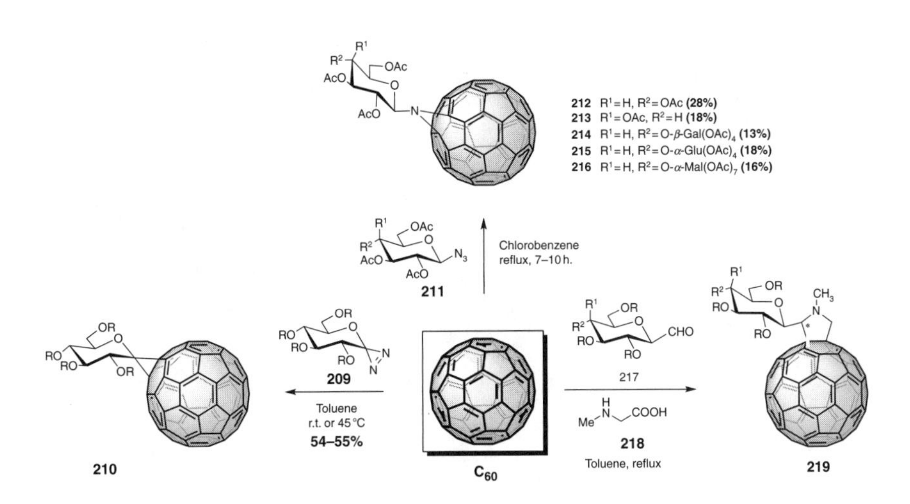

SCHEME 19. Synthesis of monovalent fulleroglycoconjugates directly introduced onto C_{60}.

enantiomerically pure spiro C-linked glycosyl-C_{60} derivatives **210** (Scheme 19).[212] Another approach permitted synthesis of fullerene glycoconjugates via the thermal cycloaddition of the per-*O*-acetyl glycosyl azides **211** in boiling chlorobenzene.[213] Although this methodology generated a mixture of two inseparable stereoisomers of *N*-β-glycopyranosyl [5,6]azafulleroids in relatively poor yields (13–28%), the generality of the reaction has been demonstrated with a series of mono- (**212**, D-glucopyranose and **213**, D-galactopyranose), di- (**214**, lactose and **215**, maltose), and trisaccharide (**216**) conjugates to C_{60}.

Soon afterwards, Dondoni *et al.* developed a three-component approach involving C_{60}, a carbohydrate aldehyde (**217**), and *N*-methylglycine **218** (sarcosine) leading to the formation of glycofulleropyrrolidine monocycloadducts (**219**) in poor yields (10–14%), via 1,3-dipolar cycloaddition reaction of the intermediate azomethine ylide to C_{60}.[214] The feasibility has nevertheless been demonstrated with the use of a series of 1-deoxy-1-*C*-formyl derivatives of galactopyranose, glucopyranose, and mannofuranose. Moreover, this methodology has been advantageously adapted for synthesis of C_{60} derivatives containing a 6-(β-D-glycopyranosylamino)pyrimidin-4-one unit.[215]

Alternative strategies utilizing the initial introduction of suitable chemical functionality onto the fullerene have been adopted by several research groups, allowing conjugation of the saccharides with complementary functions in the final synthetic steps. This approach has led to improved flexibility with the chemical functions on the carbohydrate moieties, especially the anchoring ones, and a better control on the number of attached saccharide residues.

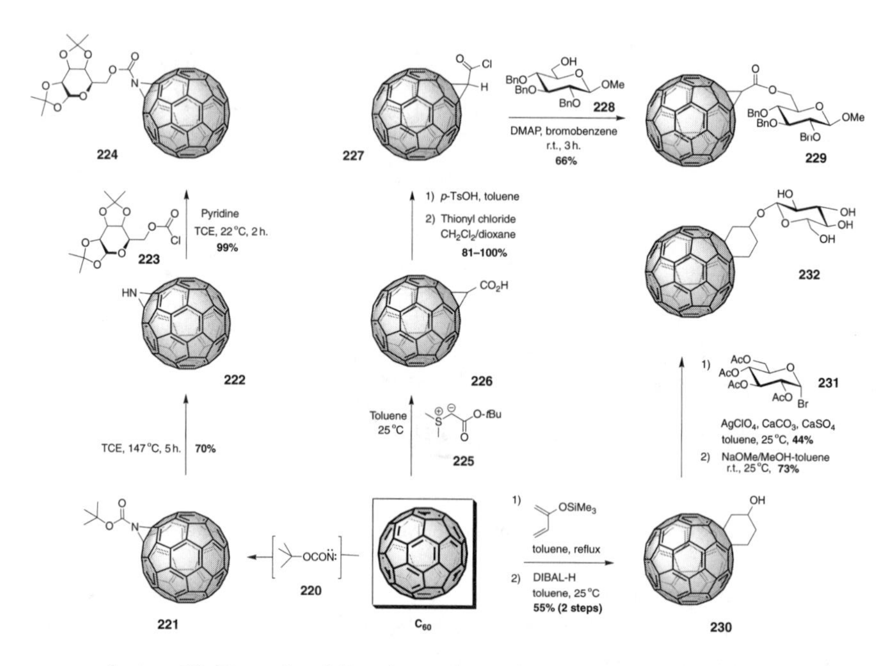

SCHEME 20. Monovalent fulleroglycoconjugates introduced by preactivation of C_{60}.

In this context, Banks *et al.* proposed the synthesis of glycofullerenes based on the use of aziridino[2′,3′:1,2][60]fullerene ($C_{60}NH$) (**222**) (Scheme 20).[216] Its preparation in good yield (55–60%) involved the initial formation of *N-tert*-butoxycarbonylaziridinol[2′,3′:1,2][60]fullerene **221**, generated by *in situ* trapping of the intermediate nitrene ($tBuO_2CN$:) **220** by C_{60}. Use of the D-galactose chloroformate derivative **223** through direct acylation under mild conditions led to the desired galactofullerene **224** in quantitative yield.

Another example has been provided by Ito *et al.*, who described the use of methanofullerene derivatives as powerful and stable precursors for glycofullerenes.[217] Their study was based on the use of [60]fullerenoacetyl chloride (**227**), obtained from the *tert*-butyl [60]fullerenoacetate derivative **226**, which had been prepared in 56% yield by treatment of corresponding stabilized sulfonium ylides **225** with C_{60}.[218] Subsequent transformation with *p*-TsOH in toluene gave [60]fullerenoacetic acid, which was directly converted into the corresponding acyl chloride **227** by using thionyl chloride. Standard ester formation with methyl 2,3,4-tetra-*O*-benzyl-*β*-D-glucopyranoside (**228**) and 4-(dimethylamino)pyridine (DMAP) afforded the desired hybrid derivative **229** in 66% yield.

An additional study investigated the DNA- and protein-degradation properties of the novel glycofullerene derivative **232**.[219] Its synthesis was based on a Diels–Alder reaction of 2-trimethylsilyloxy-1,3-butadiene with $\mathbf{C_{60}}$, followed by the reduction of the cyclohexanone derivative with DIBAL-H in toluene to afford the corresponding racemic cyclohexanol **230** in excellent yield.[220] Glycosylation with glycosyl bromide **231** in the presence of $AgClO_4$, $CaCO_3$, and $CaSO_4$ in toluene and subsequent deprotection under Zemplén conditions gave the desired glycofullerene **232**, which was shown to effect selective degradation of the HIV-protease by photoirradiation.

Synthetic aspects for access to monovalent fullerene–carbohydrate hybrids were highlighted, but only a few biological applications were mentioned. In contrast, multivalent presentation of saccharides by multiple anchorages to the same structure, or their presentation as antennary glycodendrons, has generated promising results.

b. Multivalent Structures.—Divalent hybrid glyconanostructures have been prepared via the [3+2]-cycloaddition reaction between 2-azidoethyl glycopyranoside derivatives and C_{60}. The first example described a strategy for synthesis of [60] fullerenols carrying mono- and bis-α-D-mannopyranosides (Scheme 21).[221] The methodology was based on initial thermal coupling of 2-azidoethyl mannopyranoside (**33**) and an equimolar amount of $\mathbf{C_{60}}$ in chlorobenzene, giving a mixture of two isomeric

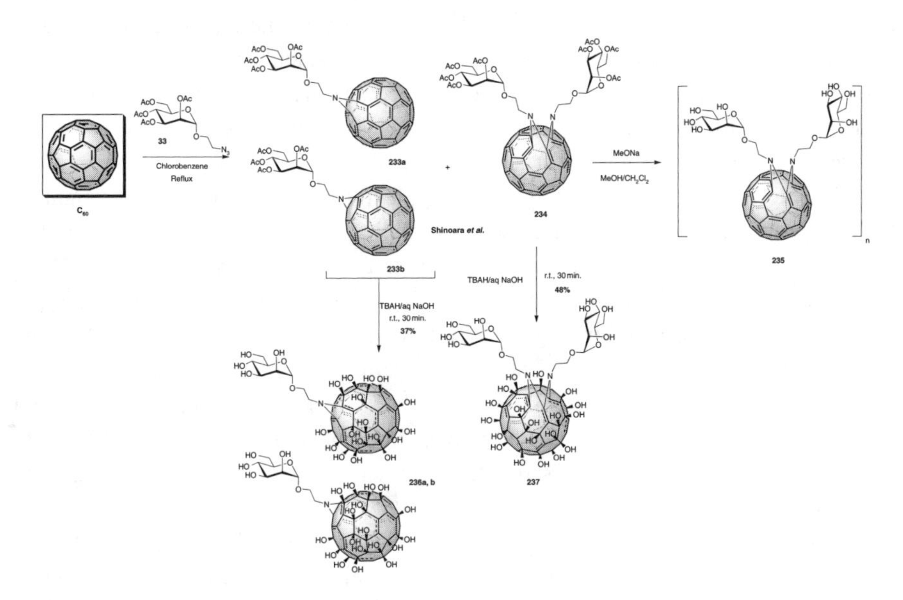

SCHEME 21. Synthesis of mannosylated fulleroglyco hybrids by [3+2]-cycloaddition of glycosyl azide onto C_{60}.[221–224]

monoadducts (α-D-mannosyl [5,6]-azafulleroid, **233a**, and α-D-mannosyl [6,6]-aziridinofullerene, **233b**)[222] and a bismannosylated adduct **234**. Subsequent simultaneous deacetylation and polyhydroxylation were then conducted in the presence of tetra-*n*-butylammonium hydroxide (TBAH) in aqueous NaOH, to afford deprotected mono- (**236**) and bisfullerenols (**237**) with an average of heterogeneously distributed 29 OH groups per C_{60}. Biological evaluations, notably involving lectin-induced HAI assay with an α-D-mannoside-specific lectin (Con A), revealed that mono- and bis-mannosyl fullerenols exhibited diminished activity for both binding to Con A and aggregating erythrocytes as compared to the activity of mannosylated neoglycopolymers. A few years later, a similar grafting methodology was used to generate a series of mono- and bis-sugar-pendant [60]fullerene derivatives prepared from a variety of carbohydrate-linked azides.[223] The phototoxicity of the corresponding deprotected glycoconjugates was evaluated for potential applications in PDT. Substantial production of singlet oxygen ($^1O_2$) under laser irradiation (355 nm) was observed for both adducts, notably for the monosugar derivatives, which shown a better photosensitizing ability. In addition, *in vitro* studies involving HeLa cells confirmed their photocytotoxicity and indicated a carbohydrate-dependent efficiency. In line with previous work, another application has been proposed which described the formation of stable self-assembling structures in aqueous solvents from the deprotected bis(α-D-mannopyranosyl)-[60]fullerene conjugate **235**.[224] The diameters of the resulting large aggregates (100–300 nm), resembling bilayer vesicles or unadulterated liposomes, were determined by DLS and AFM. These supramolecular structures were able to encapsulate Ba^{2+} ions and such organic molecules as Acridine Red, constituting for instance, promising candidates for slow drug delivery. The ability of these mannofullerenes to bind to a mannoside-recognizing lectin (Con A) was also investigated. Colloidal suspensions of the stable self-assembling structure **235** presented higher blocking activity than the two monoadducts, with an interesting submicromolar MIC (minimum inhibitory concentration) value in a lectin-induced hemagglutination assay. This strong protein-binding activity has been rationalized by the authors by the fact that the fullerene scaffold allowed a spatial arrangement of the bis(mannopyranoside) moiety that could mimic 3,6-branched α-D-trimannoside, a natural ligand of Con A. Unfortunately, this methodology often generates mixture of diversely substituted adducts, difficult to separate, in rather low yields and presenting only very limited solubility in water.

An alternative synthetic approach, first developed by Bingel[225] allowed the efficient nucleophilic cyclopropanation of fullerenes via their reaction with bromomalonate derivatives in the presence of base. This approach, the most reliable method for the synthesis of functionalized methanofullerenes, combined the advantages of mild reaction conditions, good yields, and exclusive formation of [6,6]-bridged adducts. Furthermore, the degree of functionalization could be controlled by the stoichiometry of the malonate derivatives for access to higher adducts in one step (bis up to hexakis). This methodology suffers, however, from the complexity of preparing the bromomalonates, often generating mixtures of mono- and di-bromo derivatives with similar chromatographic properties. An optimized methodology has then been proposed by Camp and Hirsch that avoids the use of bromomalonates by direct treatment of the fullerene with malonates in the presence of CBr_4 (or I_2) and DBU (1,8-diazabicyclo-[5.4.0]undec-7-ene).[226] It has been exploited in collaborative work describing the synthesis of amphiphilic glycofullerodendrimers and their incorporation into Langmuir and Langmuir–Blodgett films (Scheme 22).[227] The synthesis of the first C_{60}-dendrimer conjugate containing one glycodendron headgroup started from the treatment of C_{60} with diethyl malonate (**238**) in the presence of CBr_4 and DBU to furnish the monoadduct **239a** in 57% yield (Scheme 22). A clean formation of **239b** from its diester **239a** was accomplished in toluene with a 20-fold molar excess of NaH at 60 °C.[228] DCC (*N,N*'-dicyclohexylcarbodiimide)-mediated double amide bond formation between **239b** and the O-acetylated trisglucoside wedge **240**,[229] accompanied by monodecarboxylation of the starting material allowed the formation of glycofullerodendron **241** in 32% yield. The bisadduct **247** was synthesized via the same synthetic pathway through preliminary coupling with DCC–DMAP-mediated esterification of *m*-benzenedimethanol (**242**) with malonic acid monoester **243** in 83% yield.

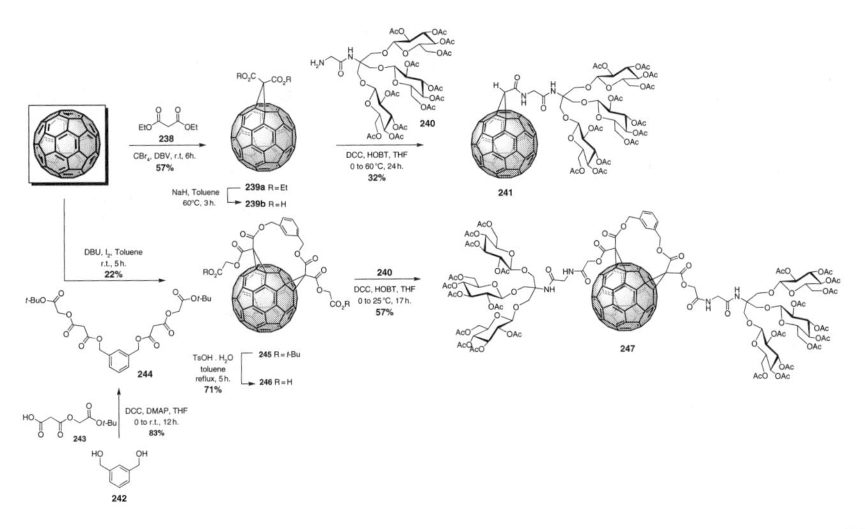

SCHEME 22. Synthesis of C_{60} malonates by Bingel's procedure, followed by glycodendronization.[227]

The resulting bismalonate derivative **244** was then engaged in double-Bingel macrocyclization with C_{60} to afford the C_s-symmetrical *cis-2* bisadduct **245** in 22% yield. Subsequent selective cleavage of the *tert*-butyl ester under acidic conditions provided the desired diacid core **246**,[230] which was finally treated with the glycodendron **240** under the foregoing conditions to yield glycofullerodendrimer **247** exhibiting six peripherally protected glucopyranoside residues. With this work, the authors highlighted the preponderant role of the bulky glycodendron headgroups on C_{60} that formed a compact insulating layer around the carbon sphere, avoiding the general propensity of amphipilic fullerene derivatives to aggregate irreversibly. Hence, both amphiphilic fullerenes **241** and **247** were able to form stable, ordered monomolecular Langmuir layers at the air–water interface and showed reversible behavior in successive compression–expansion cycles. In addition, the monolayers were transferred successfully as X-type Langmuir–Blodgett films onto quartz slides for anticipated applications as biosensors.

The group of Hirsch has reported the synthesis of ''sugar balls'' according to the two distinct methodologies already mentioned, but involving the direct grafting of glycosylated dendrons onto C_{60} (Scheme 23). They also described their supramolecular assembly in aqueous solution.[231] In this context, bis(α-D-mannopyranosyl)fullerene (**250**) has been synthesized by a sequence involving the bis-(α-D-mannopyranosylated malonate) **248**. This resulted from a peptide-coupling reaction between the corresponding 2-aminoethyl 2,3,4,6-tetra-*O*-acetyl-α-D-mannopyranoside, quantitatively obtained from correrspondoing azide derivative **33** after Staudinger reduction, and malonic acid, in 56% yield. The subsequent nucleophilic cyclopropanation of C_{60} with **248** under typical Bingel–Hirsch conditions, followed by O-deacetylation of **249** under Zemplén conditions afforded fully deprotected [6,6]-monoadduct **250** in an overall isolated yield of ~6%. It is noteworthy that the protected architecture, presenting robust amide linkages around the malonate anchor points, tolerated basic cleavage of the acetyl protecting groups without side reactions or decomposition of the fullereno sugar.

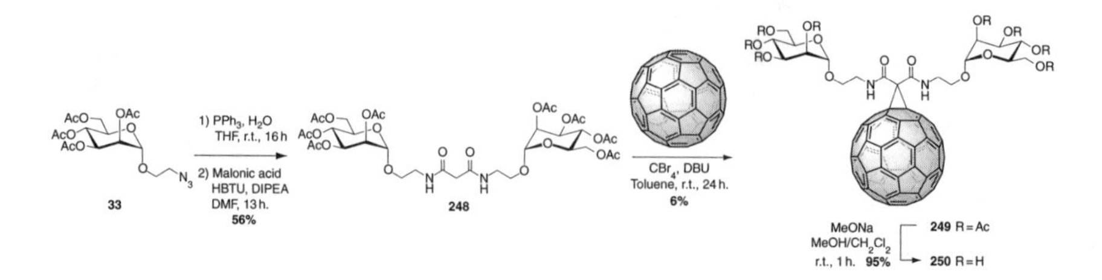

SCHEME 23. Synthesis of a bismannoside-C_{60} adduct using the Bingel–Hirsch procedure.[231]

Unfortunately, its relatively low solubility in aqueous media compelled the authors to enhance the number of peripheral saccharidic units. Toward this goal, they developed fullerene glycoconjugate **258**, where two dendritic branches terminated by six deprotected α-D-mannopyranosyl building blocks were connected through two adjacent imino bridges to the all-carbon framework (Scheme 24). Moreover, in this type of C_{60} adduct, which constitutes a 1,9-dihydro-1a-aza-1(2)a-homo(C60-I_h)[5,6]fullerene derivative, the entire 60-π-electron system of the fullerene core was retained. For its preparation, AB_3 glycosylated dendritic building blocks were first synthesized by preliminary protection of Newkome's dendron **251**[232] with benzyl chloroformate (CbzCl), then treatment with formic acid for orthogonal and quantitative cleavage of the *tert*-butyl ester to give triacid **252** upon which, peptide coupling with 2-aminoethyl 2,3,4,6-tetra-*O*-acetyl-α-D-mannopyranoside **253** provided trimannoside **254** in 83% yield. Selective removal of the Cbz protecting group by hydrogenolysis gave the corresponding amine in quantitative yield, which was engaged without further purification with a suitable linker bearing a terminal azide group, namely 2-[2-(2-azidoethoxy) ethoxy]acetyl fluoride (**255**), in the presence of DMAP in CH_2Cl_2 to provide the azido oligo(ethylene glycol)-terminated glycodendron **256** in 94% yield. Its subsequent thermal coupling with C_{60} in toluene afforded a mixture of mono- and bisglycodendron adducts. Separation and purification by column chromatography gave regioselectively the twofold cluster-opened diazabishomofullerene adduct **257**,

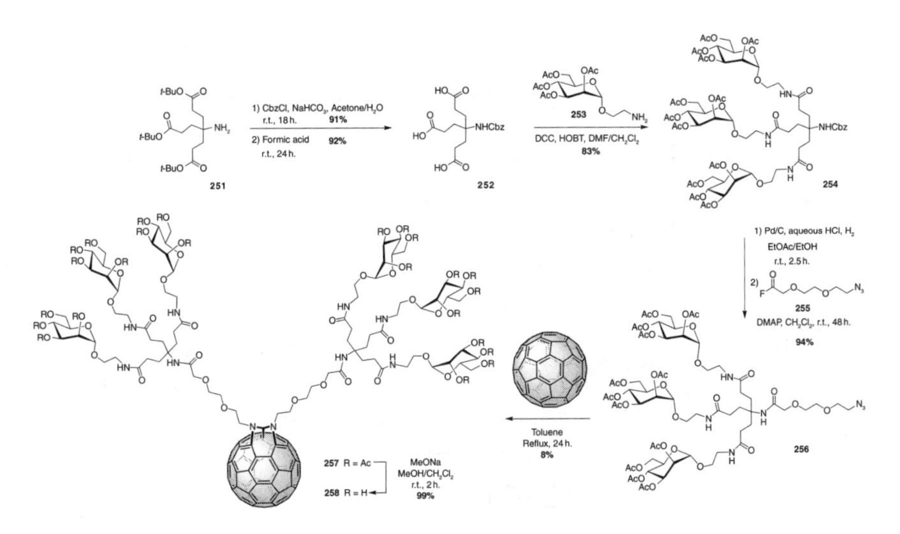

SCHEME 24. Direct anchoring of azide-bearing mannosylated dendrons onto C_{60}.[231]

but in only 8% yield. Finally, quantitative deprotection under standard Zemplén conditions provided **258**. This C_{60} glycoconjugates having six peripheral mannoside residues is the most water-soluble fullereno sugar (> 40 mg/mL) known. Its amphiphilic nature, with a cone-shaped structure, triggered the formation of small supramolecular aggregates in aqueous solutions (observed in DOSY NMR and corroborated by TEM investigations), as uniform spherical micelles with an extremely narrow size distribution. These micellar sugar balls, with a diameter below 5 nm, furnish an original arrangement of saccharides that opens the gate to new biomedical applications.

Another fruitful methodology, initially developed by Nakamura's group[233] allowed the introduction of five carbohydrate moieties onto a C_{60} scaffold using the efficient synthesis of C_5 symmetric fullerene derivatives (Scheme 25).[234] A thiolate–alkyl halide coupling reaction in aqueous basic media provided their one-step synthesis in good yields, taking advantage of the high nucleophilicity of the thiolate anions generated. Synthesis of the key fulleropentathiol intermediate **261** involved the fivefold addition reaction of the copper derivative **259**, prepared *in situ* from the corresponding Grignard reagent and CuBr·SMe$_2$. The desired fullerene cyclopentadiene bearing tetrahydropyranyl-protected thiophenol moieties (**260**) was obtained in 93% isolated yield, which was then deprotected in *o*-dichlorobenzene in the presence 2-mercaptoethanol and trifluoroacetic acid to afford the pentathiol derivative **261** in

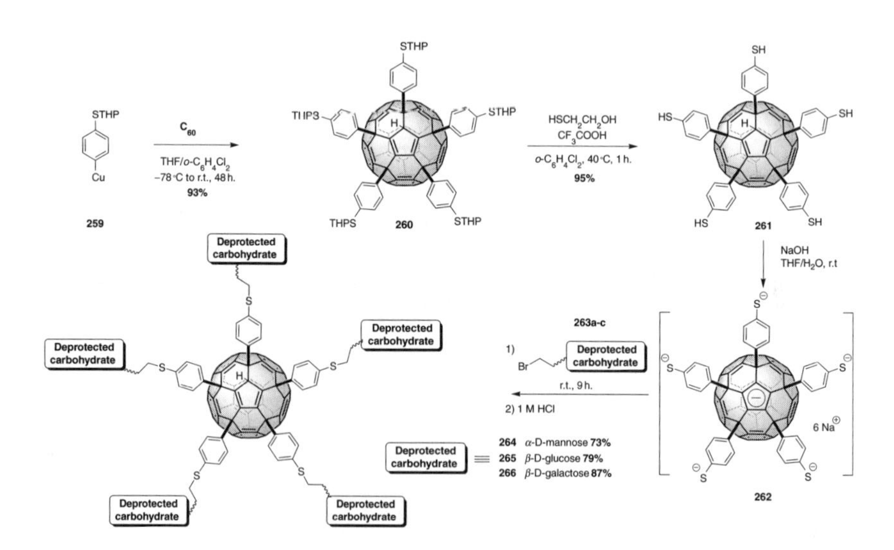

SCHEME 25. Pentavalent fulleroglycoconjugates.[234]

95% yield. Subsequent installation of five saccharide moieties via a nucleophilic substitution reaction in aqueous THF with 6 equivalents of sodium hydroxide and 2-bromoethyl glycopyranosides **263a–c** afforded the fullerene glycoconjugates α-D-mannoside **264**, β-D-glucoside **265**, and β-D-galactoside **266**, in 73%, 79%, and 87% yields, respectively.

Despite the generality of this methodology for simple carbohydrate derivatives, its application for introduction of larger oligosaccharides via direct nucleophilic substitution produced side products and led to lower yields. To circumvent this situation, the same authors described the optimization and adaptation of this methodology for the quantitative synthesis of fullerenes bearing five saccharides, such as the glucose **270a**, maltotriose **270b**, and globotriaosylceramide Gb3-$\mathbf{P^k}$ trisaccharide **270c** derivative (Scheme 26).[235] Hence, more-sophisticated nanometer-scale pentavalent molecular architectures could be readily prepared by use of the key pentaalkynylfullerene intermediate **268** obtained directly by the derivatization of fulleropentathiol **261** with 7-bromohept-1-yne **267** in 73% yield under the foregoing conditions. Click chemistry allowed efficient conjugation of the functionalized saccharide moieties with a terminal complementary azide function (**269a–c**) on the pentaalkynylated core **268** under mild conditions. Hence, in the typical presence of $CuBr.SMe_2$ and DIPEA in Me_2SO, deprotected fullero-derivatives **270a–c** were obtained in nearly quantitative yields through optimized conditions involving controlled heating or microwave irradiation.

Similarly to the STARFISH design already described (Fig. 17), the peripheral and radial presentation of five $\mathbf{P^k}$-trisaccharide residues presenting tailored spacers could

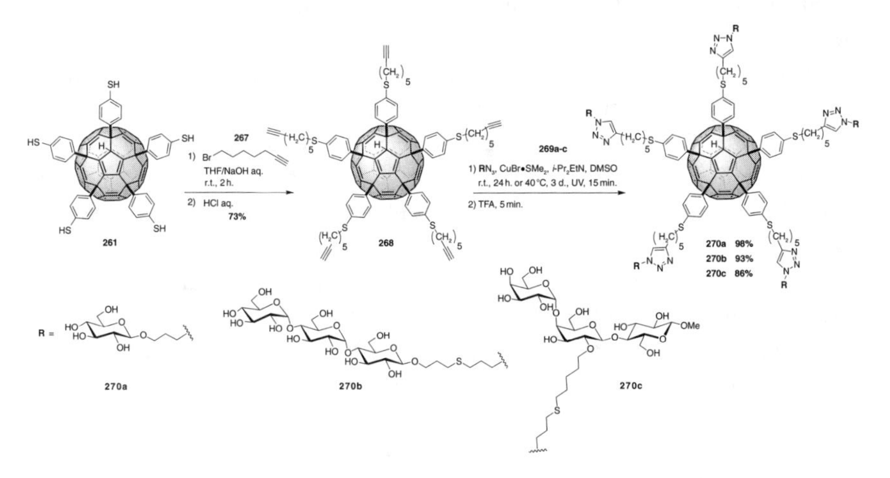

SCHEME 26. Alkylation of fulleropentathiol by an alkyne spacer followed by ''click chemistry'' with various azido sugars, including the $\mathbf{P^k}$ trisaccharide.[235]

furnish adapted structures for efficient and specific interactions with SLTs. In line with promising results obtained with the functionalization of fullerenes, several research groups have recently described the derivatization of carbon nanotubes (CNTs) with biologically relevant carbohydrates.

2. Glyconanotubes

CNTs are members of the fullerene structural family and consist exclusively of carbon atoms arranged in a series of condensed benzene rings, organized in graphitic sheets that are rolled up into a tubular structure.[236] CNTs made up of a single graphene layer wrapped into a cylindrical structure constitute single-walled CNTs (''SWNTs''), whereas multiwalled CNTs (''MWNTs'') are generated from a central tubule of nanometric diameter and surrounded by several graphitic layers spaced by a distance of about 0.34 nm. Most commonly, the diameter range of CNTs varies between 0.4 and 2 nm for SWNTs and between 1.4 and 100 nm for MWNTs, while their length is in the range of micrometers (up to 100 nm). Moreover, CNTs typically form bundles that are entangled together in the solid state, giving rise to a highly complex ''spaghetti'' network.[237] Intrinsically, CNTs possess very interesting and unique physicochemical properties such as: high surface area, ordered structure with a high aspect ratio, ultralight weight, excellent chemical stability, high electrical conductivity, high thermal conductivity, and metallic or semiconducting behavior according to the arrangement of the hexagon rings along the tubular surface. For these reasons, these nano-objects have raised great enthusiasm and expectations in many different applications, including material, biological, and medical sciences.[238] More particularly, their intriguing structures and properties have led to the emergence of functionalized CNTs as new biologically relevant alternatives for therapeutic and diagnostic applications,[239] including their use as vectors for delivery of therapeutic molecules[240] (plasmid gene delivery,[241] peptides/proteins,[242] and antibiotics[243]). Indeed, prior to become viable and effective nanomedicines, the toxicological and pharmacological profiles of CNTs had been determined by several research groups and reviewed by Lacerda *et al.*[244] As with biological studies on fullerenes, functionalized CNTs offer potential as nanomedicines, whereas nonfunctionalized CNTs usually contain amorphous carbon, carbon nanoparticles, and residues from metal catalysts, regardless of the procedures for their production. Indeed, their inherent hydrophobic properties, associated with their innate propensity to form bundles via aggregation through van der Waals forces, constitute major technical barriers for their utilization in medicinal applications. To overcome these drawbacks, modification of their surface is typically achieved by adsorption, electrostatic interaction, or covalent bonding of different molecules and chemical procedures that render them more soluble in biological media.[245] Such modifications generally improve the water solubility of CNTs, decrease the aggregation phenomenon, and transform their biocompatibility and biodistribution profiles.

It has been shown experimentally that such bioactive proteins, nucleic acids, genes, and more particularly carbohydrates can be successfully conjugated to CNTs. In particular, because of versatile modes of chemical modification and solubilization, the one-dimensional nanostructure of SWNTs, with high surface area-to-weight ratio and some structural flexibility, can be exploited as platforms for multivalent carbohydrate arrays under physiological conditions. The resulting densely attached sugars, as pendant groups on the CNT hybrids, serve as multivalent carbohydrate ligands and show potential biological activities as well as behaving as glycoconjugate polymers. Multivalent presentation of saccharidic residues has been realized via noncovalent stabilization processes through favorable hydrophobic, π–π stacking or electrostatic interactions, or via sidewall- or defect-targeted covalent functionalization. Several groups have reported multivalent carbohydrate–CNTs conjugates, such schizophyllan polysaccharides, lipid-terminated glycopolymers, and carbohydrate ligands toward bacterial toxins and viral proteins.

a. Glyconanotubes from Noncovalent Interactions.—The initial challenge of obtaining SWNTs soluble in organic or aqueous media by a supramolecular approach was motivated by the necessity for retaining their intrinsic physical properties without affecting the all-carbon scaffold. The first successful example was realized through noncovalent functionalization by wrapping synthetic and biocompatible water-soluble polymers around the SWNTs. The formation of supramolecular complexes in solution between SWNTs and the preformed helical structure of starch (a polymer composed of the linear amylose and branched amylopectin) has been investigated.[246] The enhanced water solubility of the supramolecular complexes was significant, and further investigations suggested that the presence of the preformed helical structures of amylose may not be required.[247] Other studies examplifying complexation of the SWNTs with CDs have also yielded some interesting results.[248]

As a natural extension of these preliminary studies, promising applications in the area of glyconanotechnology have been directed toward artificial glycoconjugates carrying bioactive carbohydrates, densely distributed onto CNT platforms, to mimic cell-surface oligosaccharides. A striking example describes a biomimetic surface modification of CNTs, using glycosylated polymers designed to mimic cell-surface mucins (Scheme 27).[249] These high molecular weight glycoproteins are involved in

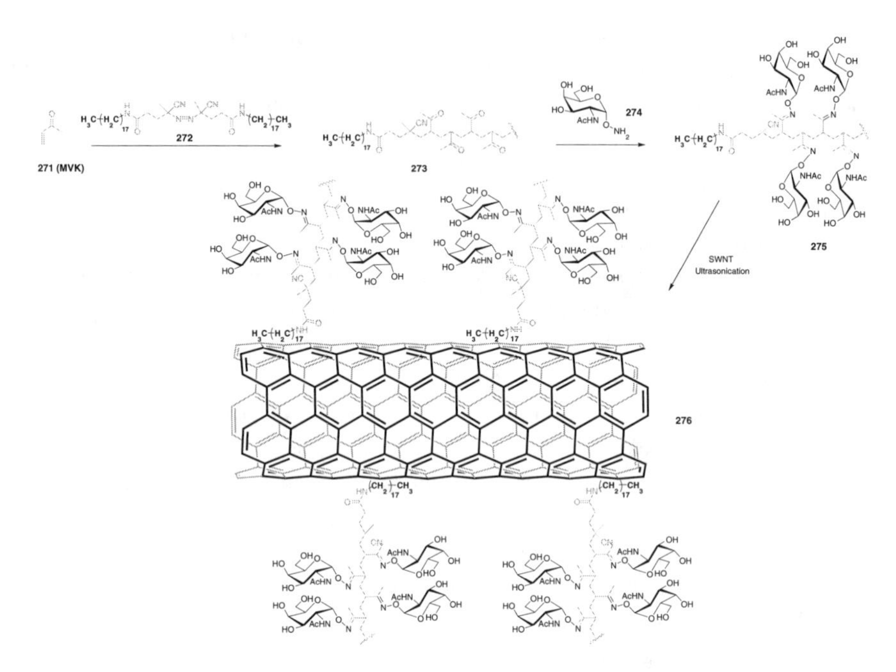

SCHEME 27. Noncovalent functionalization of SWNTs by using an amphiphilic glycopolymer, ending with a lipid tail for mucin mimicry.[249]

specific molecular cell–cell recognition processes. Additionally, the dense clusters of O-linked glycans confer rigidity, provide strong hydration and passivation against biofouling (cell adhesions), thus ensuring protection of structural functions against nonspecific biomolecular interactions. It has been demonstrated that, in native mucins, the clustered *N*-acetyl-α-D-galactosamine (α-GalNAc) residues proximal to the peptide were the major contributors to the rigidification of peptide backbones.

Based on these observations, the authors prepared synthetically tractable mucin mimic (MM) in which α-GalNAc residues were linked through an oxime bond to a poly(methyl vinyl ketone) (PVK) lipid-teminated backbone. Its preparation started with the initial introduction of a C_{18} lipid chain at the extremities of 4,4′-azobis(4-cyanopentanoic acid) via peptide coupling to give **272**. Polymerization of the functionalized initiator via radical polymerization of methyl vinyl ketone (1-buten-3-one, **271**, MVK) afforded C_{18}-poly(MVK) **273** onto which aminooxy-functionalized α-GalNAc **274** residues were grafted through oxime bonds, affording C_{18}-α-MM polymer **275**. An aqueous solution of the resulting amphiphilic glycopolymer was then subjected to ultrasonication in the presence of SWNTs which were thereupon fully solubilized, suggesting the formation of hydrophilic surface coating through the self-assembly of the terminal lipid chains on the nanotubes via hydrophobic interactions, affording glycoconjugates C_{18}-α-MM-SWNTs (**276**). Evidence for mucin mimetic coating was provided by three different imaging techniques, namely AFM, scanning electron microscopy (SEM), and transmission electron microscopy (TEM).

As anticipated, substantial enhancement of water solubility was observed and the previously entangled SWNT bundles dissociated to form much finer bundles, stable in aqueous solution for several months. Despite the inherent difficulties from the imaging technique conditions, the formation of individual nanotubes with fairly uniform diameters of 65–70 nm was observed. The recognition and resistance properties of these nanotube–mucin mimetics were investigated with the use of *Helix pomatia* agglutinin (HPA). Incubation of **276** with a solution of HPA conjugated with fluorescein isothiocyanate (HPA-FITC), and appropriate subsequent treatment led to significant fluorescence that was attributed to a dependent receptor–ligand interaction. Moreover, studies under similar conditions but involving glycoconjugates bearing the β anomer of GalNAc manifested no significant fluorescence labeling. Hence, the lectin did not interact with the coated nanotubes in the absence of its favored ligand. These results clearly demonstrated that SWNTs coated with a MM could engage in specific molecular recognition with protein receptors and resist nonspecific protein binding.

A few years later, the same group extended studies with functionalized CNTs–cell interactions.[250] In order to exploit their ability to interact with cells, the advantage of the hexavalency of HPA and its capability to cross-link cells and glycoproteins was explored. Two parallel experiments were conducted, on one hand involving the preformed complex of HPA-FITC with C_{18}-α-MM-coated CNTs incubated with chinese hamster ovary (CHO) cells, and on the other the preincubation of CHO cells with unmodified HPA and subsequent treatment with modified C_{18}-α-MM-coated CNTs containing the fluorescent dye Texas Red. In both instances, analysis by fluorescence microscopy and flow cytometry suggested the formation of the tricomponent α-GalNAc-HPA–CNT–cell surfaces complex, and with a dose- and precomplexation-dependent labeling for the second study. Control CNTs modified with C_{18}-β-MM-CNTs indicated no significant fluorescent labeling at low concentration. In conclusion, demonstration that lipid-terminated poly(MVK)-based glycopolymers could coat CNT surfaces and promote their binding to cells through receptor–ligand interactions has been demonstrated. Modified CNTs were nontoxic to cultured cells (CHO and Jurkat cells), but irregular surface and nonuniform thickness of the CNTs coating might be explained by the high polydispersities (>1.7) of the polymers employed. The surface heterogeneity generated could constitute a drawback to the systematic use in a reproducible way of glycopolymer-coated CNTs as sensors of protein binding.

In 2004, Hasegawa *et al.* demonstrated that SWNTs could be literally wrapped within a helical polymeric superstructure composed of schizophyllan, a polysaccharide bearing lactoside appendages (SPG-Lac, **278**), in aqueous solution (Scheme 28).[251] Water-soluble SPG-Lac–SWNTs nanocomposites (**279**) were obtained by mixing the glycopolymer in Me_2SO with SWNTs dispersed in water. Their formation was unequivocally observed and confirmed by a battery of such analytical techniques as UV–vis spectroscopy, colorimetry, thermal gravimetry, and AFM. Molecular recognition of the resulting composite was assessed by SPR, using

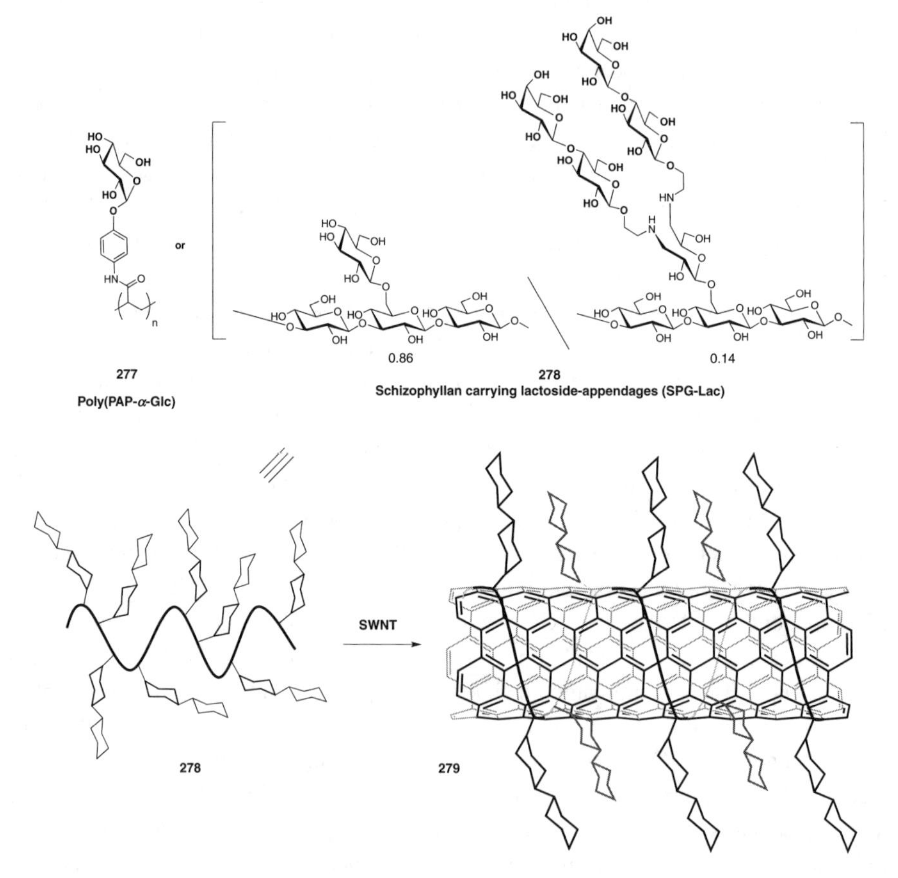

SCHEME 28. Solubilization of SWNTs (**279**) with artificial (**277**) or natural (**278**) schizophyllans.[251,252]

immobilized RCA_{120} lectin (*Ricinus communis* agglutinin, *β*-Lac-specific) on gold surfaces, and the results indicated highly specific affinity toward the lectin. This specificity was also confirmed by confocal laser-scanning microscopic (CSLM) observations using FITC-labeled RCA_{120}. Promising applications as SWNTs-based sensory systems have been demonstrated by the authors, who succeeded in constructing a layer-by-layer structure composed of the SPG-Lac–SWNTs composite and RCA_{120}.

By the same methodology, an individual dispersion of water-soluble SWNTs randomly wrapped by a helical glycoconjugate polymer (poly(*p*-*N*-acryloylamidophenyl)-α-D-glucopyranoside, PAP-α-Glc, **277**) has been obtained.[252] The characteristic sharp photoluminescence signal observed by near-infrared fluorescence spectroscopy, associated with the multivalent presentation of peripheral carbohydrates, made this system promising as new sensing approach for detecting carbohydrate-recognition proteins.

The preparation of biocompatible SWNTs, noncovalently functionalized with bioactive glycodendrimers, has been reported (Scheme 29).[253] A bifunctional dendritic scaffold **284** was built using the 2,2-bis(hydroxymethyl)propanoic acid as a building block **281**, which was linked to an azidopyrene tail (**280**) capable of binding the surface

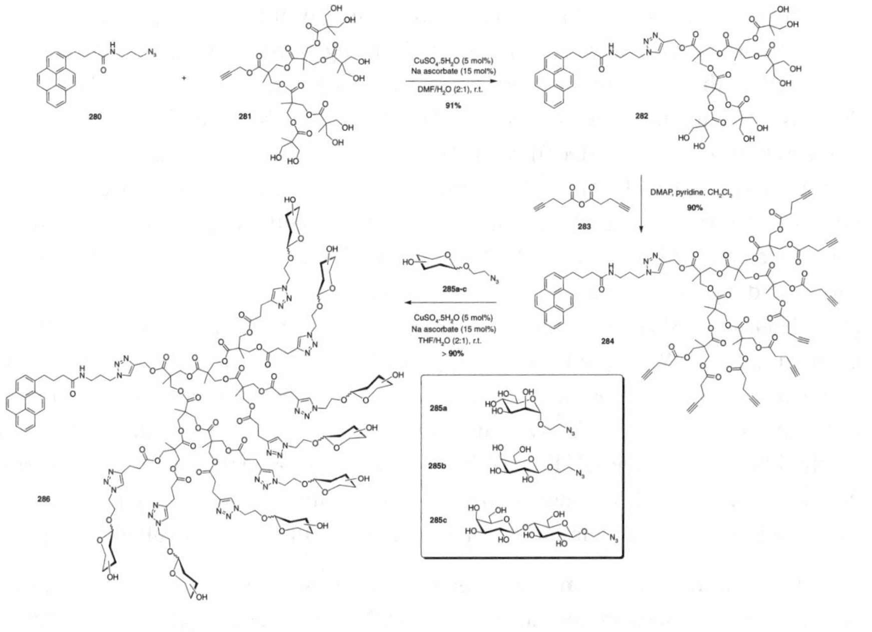

SCHEME 29. Syntheses of glycodendrons with hydrophobic and fluorescent pyrene head-groups for π–π stacking to SWNTs.[253]

of SWNTs through π–π interactions. The orthogonal synthetic approach was based on sequential condensation of the dendrititic polyol **282** with 2,2-bis(hydroxymethyl)-propanoic anhydride (**283**).[254] At the desired generation, azide-functionalized pyrene **280** was chemoselectively ligated to the propargylated and hydroxylated dendritic scaffold **281** by CuAAC methodology. A panel of deprotected 2-azidoethyl mono- or disaccharides (**285a**, α-mannoside; **285b**, β-galactoside; and **285c**, β-lactoside) were then efficiently grafted onto the multivalent dendrimer **284** by click chemistry. The uniformity and purity of the resulting amphiphilic glycodendrimers **286a–c** were confirmed by NMR spectroscopy and MALDI-TOF mass spectrometry. The glycodendrimers were then adsorbed onto SWNTs by ultrasonication in aqueous solution.

Analysis by SEM and TEM revealed small bundles and individual SWNTs coated entirely with a thin uniform layer of glycodendrimers, contrasting markedly with the thick and heterogeneous coatings obtained by use of the glycopolymers mentioned earlier. Specific binding of SWNT-bound glycodendrimers to receptors was observed with the FITC-lectins: *C. ensiformis* agglutinin (Con A), *Arachis hypogaea* agglutinin (PNA), and *Psophocarpus tetragonolobus* agglutinin (PTA), which recognized [G(3)] Man-SWNTs **287a**, [G(3)] Gal-SWNTs **287b**, and Lac **287c**, respectively (Fig. 32). Similar results were obtained with the corresponding [G(2)] glycodendrimers. In addition, experiments in which SWNTs were initially functionalized with a mixture of [G(2)] Man **287a** and [G(2)] Lac **287c** using various molar ratios, and then incubated with a 1:1 mixture of Texas Red-conjugated PNA and FITC-Con A, demonstrated that multiple epitopes displayed on SWNTs could bind simultaneously to discrete proteins. The [G(2)] Man-SWNTs **287a** were then successfully engaged with FITC-Con A and CHO cells in the experimental protocol earlier employed to promote specific binding of modified SWNTs to the cell membranes. Similar results were obtained with the other [G(2)] glycodendrimer-functionalized SWNTs with corresponding labeled lectins. Finally, incubation of HEK293 cells with [G(3)] glycodendrimer–SWNTs conjugates did not affect their proliferation, in contrast to the unmodified SWNTs, indicating that glycodendrimers mitigated the cytotoxicity.

Besides the preparation of glyconanotube conjugates by supramolecular interactions, another synthetic pathway commonly used involved covalent attachment of the saccharidic units on the CNT scaffolds. The progress recently recorded concerning the derivatization of unfunctionalized CNTs, allowing introduction of suitable anchoring functions at their surface, are the basis of this synthetic alternative.[255]

b. Glyconanotubes from Covalent Interactions.—Multivalent presentation of peripheral saccharides around functionalized CNTs has been mostly assessed through covalent processes. In most applications, the use of CNTs containing carboxylic acid

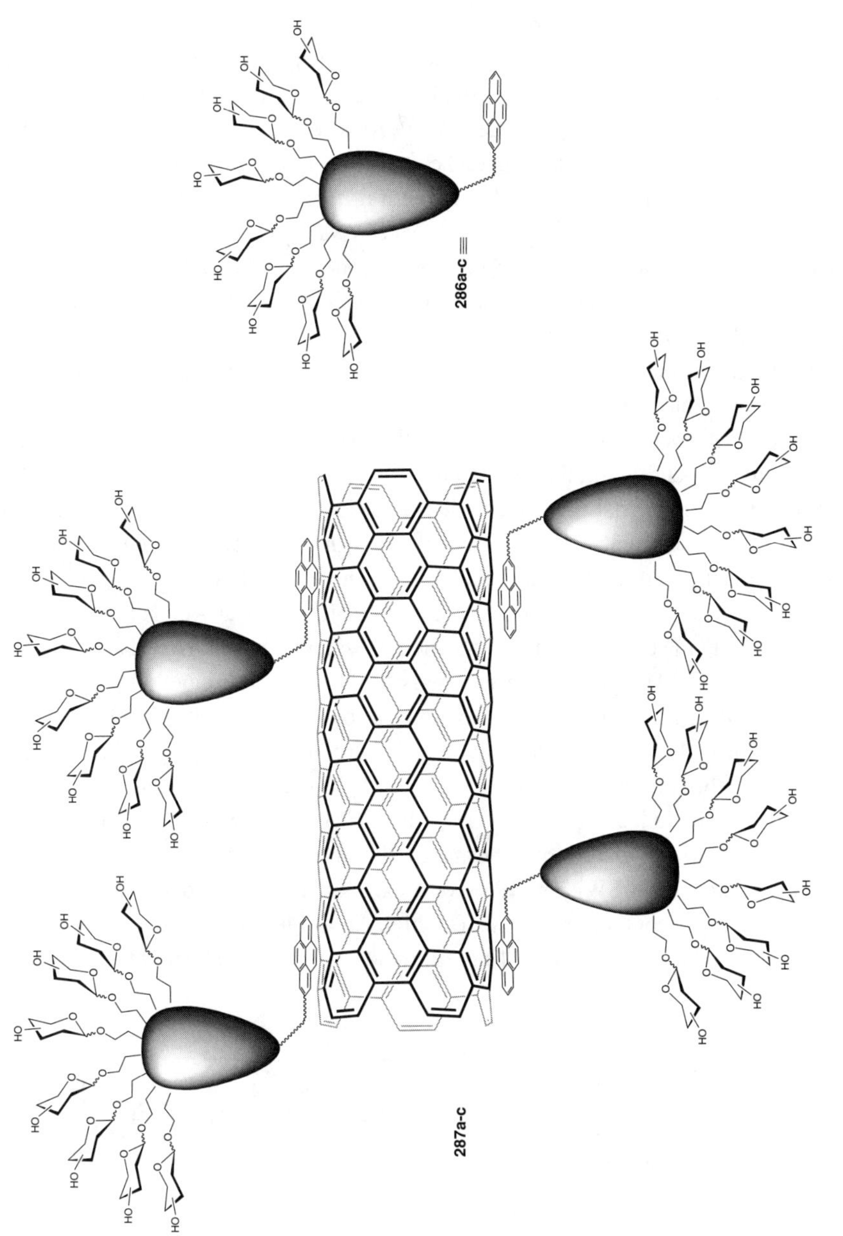

FIG. 32. Pyrene-based glycodendrons adsorbed onto SWNTs.[253]

groups as suitable anchoring sites constitutes the favored covalent methodology for installing suitable functional group that render the CNT soluble in appropriate solvents.[256] Typically, carboxylic acid functions are generated on CNTs through the oxidation of surface defects, by suspending the CNTs in 3 M HNO_3, and then treatment with aqueous HCl (Scheme 30).[257] As already presented in this chapter, in addition to their biological relevance, the conjugation of unprotected carbohydrates onto all-carbon scaffolds can solve the problem of their intrinsic low solubility.

One of the first examples of a water-soluble SWNTs glycoconjugate was obtained by covalent grafting of glucosamine via amide bonds.[258] The carboxylic acid functions of **288** were first converted into the corresponding acyl chlorides by suspending the functionalized SWNTs in a solution of thionyl chloride ($SOCl_2$) in DMF. The resulting solid was then mixed with an anhydrous solution of glucosamine in THF at reflux to afford glucosamine grafted-SWNTs. The new glycoconjugates were characterized by UV–vis and AFM analysis, and their solubility in water ranged from 0.1 to 0.3 mg/mL, depending on the temperature.

The synthesis of the first β-D-galactoside-modified SWNT (Gal-SWNT, **296**) with high water solubility has been described, along with its interactions with galactose-specific lectins that formed supramolecular junctions at the surface of the glycocoated SWNTs (Fig. 33).[259] Using a similar strategy, analogous Man–SWNTs conjugates (**297**) effective in the capturing of pathogenic *E. coli* under physiological solution was also synthesized.[260] The synthesis was based on the use of 2-aminoethyl β-D-galactopyranoside (**290a**) grafted onto SWNT-bound carboxylic acids (**288**) under sonication conditions, through carbodiimide-activated amidation with 1-ethyl-3-(3-dimethylaminopropyl)carbodiimide (EDAC) in aqueous KH_2PO_4 buffer. Complete instrumental characterization included solution-phase NMR, SEM and TEM, Raman, and near-IR spectroscopy, optical absorption, thermogravimetry, and spectrophotometry. The resulting galactosylated-SWNTs (**296**) were exfoliated and well dispersed, and presented a total sugar content of ~65%. The high sugar ratio and large surface area of the covalently galactosylated SWNTs permitted the display of abundant sugar arrays

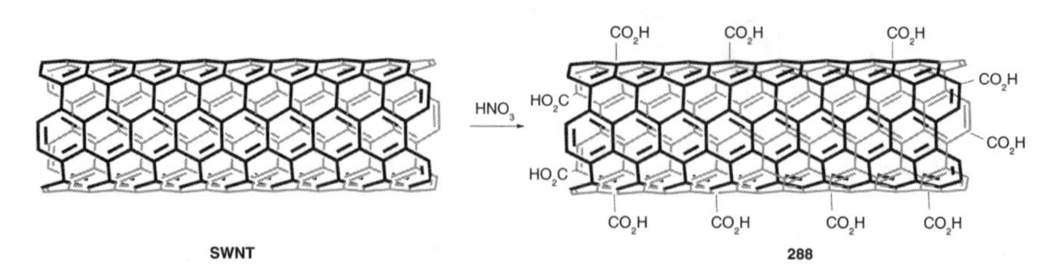

SCHEME 30. Oxidation of defect sites of SWNTs to provide carboxylated SWNTs.[257]

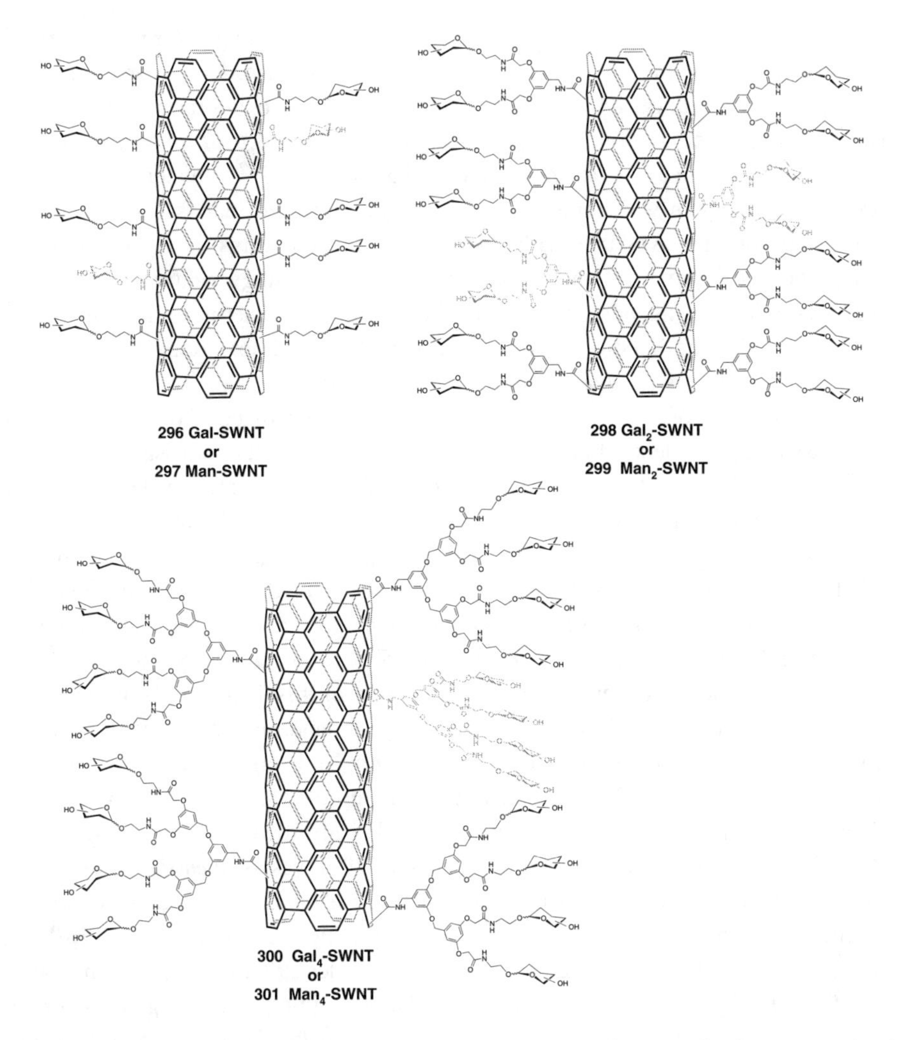

FIG. 33. Carboxylic acid-functionalized SWNTs bearing series of galactosylated or mannosylated mono-, di-, or trivalent dendrons.[262]

implicated in their strong binding interaction with receptors on pathogenic *E. coli* 0157:H7 cells. Significant cell agglutination was observed by SEM, with multiple nanotubes binding to one cell and some nanotubes ‘‘bridging’’ adjacent cells.

The same research group went further into detail concerning the antimicrobial properties of monosaccharide-coated SWNTs, advantageously exploiting their application for efficient and specific recognition of a nonvirulent strain of *Bacillus anthracis*.[261] In this context, similar nanostructures containing a large number of monosaccharides (mannose or galactose) have been synthesized by conjugation of the corresponding 2-aminoethyl glycopyranosides (**290a,b**) onto **288** through amidation. The resulting water-soluble conjugates both strongly bound to *B. anthracis* spores, leading to significant aggregation with the mediation of the Ca^{2+} divalent cation. This binding phenomenon was unique to the nanotube-displayed monosaccharide molecules, and was not available with other displaying platforms such as polymeric nanoparticles (polystyrene nanobeads for instance), indicating specific arrangements of the carbohydrate ligands on the nanotube scaffolds. Furthermore, the associated substantial reduction in colony-forming units (CFU) could potentially find valuable applications in efforts for detection and decontamination.

Another series of hydrophilic glycodendron-functionalized SWNTs was also synthesized and the lectin-binding interactions studied.[262] It was assumed that the limited defect sites on the original nanotube surface could generate only a limited number of carboxylic acid functions. The desired dense population of multivalent ligands could therefore be reached by covalently grafting sugar dendrons containing a suitable anchoring functionality. Toward this goal, β-D-galactopyranoside dendrons Gal_2-NH_2 (**292a**), Gal_4-NH_2 (**295a**), and their analogous α-D-mannopyranoside dendrons Man_2-NH_2 (**292b**), and Man_4-NH_2 (**295b**) were efficiently prepared by a systematic synthetic sequence based on the use of trisubstituted benzene cores (Scheme 31). Divalent **292a,b** and tetravalent glycodendrons **295a,b** were obtained from brominated derivatives **291a,b** and **294a,b**, respectively. Synthesis of the required dendritic building blocks was initiated with the trisubstituted benzene core **289a**, from which *t*-butyl ester removal with TFA afforded brominated precursors **289b**. The tetravalent bromine-ending analogue (**293**) was synthesized by selective etherification of 3,5-dihydroxybenzyl alcohol with **289a** in order to double the surface functionalities, followed by the introduction of bromine function with the PPh_3/NBS tandem. The acetylated amine-tethered monosaccharides **290a,b** were then conjugated to **289b** or **293** by carbodiimide-activated amidation yielding protected glycodendrons **291a,b** and **294a,b** in 99% and 67% yields, respectively. Finally the amine-ending glycodendrons were prepared by a common synthetic protocol involving treatment with NaN_3, O-deacetylation under Zemplén conditions, and subsequent hydrogenolysis with H_2 on Pd/C.

The amine-derivatized glyco wedges were finally used for the functionalization of SWNT acids **288** by peptide coupling as before, to afford **298–301** (Fig. 33). The resulting larger number of displayed ligands per SWNT was responsible for the significant increase of water solubility, increasing with the functionalization, and the enhancement of biocompatibility. In addition, they enabled more quantitative

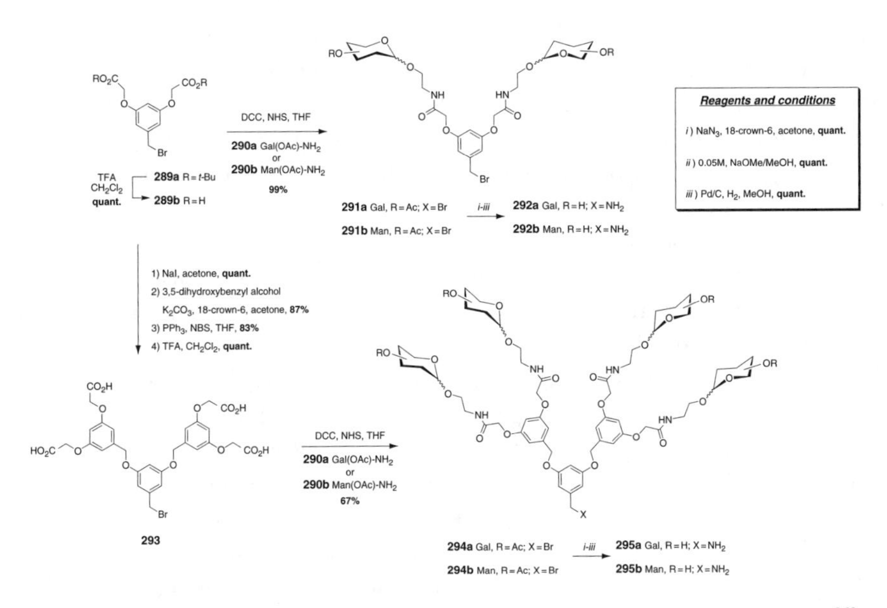

SCHEME 31. Amine-ending glycodendrons bearing either galactoside or mannoside residues.[262]

characterization, permitting a better understanding of the structural details of SWNTs exhibiting multivalent carbohydrate ligands. Solution and gel-phase NMR studies, SEM, Raman, optical absorption studies, spectrophotometry, and thermogravimetric analysis (TGA) have all been employed in order to determine the morphology, the arrangement of glycodendrons around the SWNT scaffolds, and the sugar content. No meaningful differences in sample composition between the Gal- (**296**) and Man- (**297**), Gal_2- (**298**) and Man_2- (**299**), Gal_4- (**300**), and Man_4- (**301**) glyconanotubes as functional moieties were observed. More sugar residues were displayed on the same amount of SWNTs for the divalent conjugates than for the monovalent ones. Furthermore, and interestingly, the electronic properties of the CNTs were largely preserved during the functionalization.

These carbohydrate-functionalized SWNTs were evaluated in binding assays with pathogenic *E. coli* and with *Bacillus subtilis* (a nonvirulent form of *B. anthracis* or anthrax). As compared with the binding properties of Gal-SWNT (**296**) to *E. coli* O157:H7 and the induced decrease of CFU evoked earlier, considerable improvements were recorded with divalent Gal_2-SWNT (**298**). In the presence of the same *E. coli* strain, the amount of recovered cellular aggregates was clearly larger, and with a more significant reduction of CFU. This more favorable binding process by

Gal_2-SWNTs (**298**) seemed to suggest that the paired β-D-galactosides are more effective in the specific interactions. Moreover, Man_2-SWNTs (**299**) and Man_4-SWNTs (**301**) were both capable of binding and aggregating *B. subtilis* spores in the presence of calcium cations. Despite preliminary encouraging observations, quantitative evaluations of the foregoing binding assays were complex: a lack of major tendencies and inconsistent results over a same series rendered the interpretation difficult and inaccurate.

Finally, a subsequent study furnished a clear demonstration that CNTs could provide suitable support for rapid and efficient polymeric growth for the synthesis of novel and biocompatible linear and hyperbranched glycopolymers by the ''grafting from'' strategy, with good and high reproducibility (Fig. 34).[263] The successful and controlled linear glucopyranoside-polymer grafting onto MWNTs was achieved by atom-transfer radical polymerization (ATRP), while hyperbranched glycopolymers were introduced by self-condensing vinyl copolymerization (SCVCP) of the corresponding functionalized 1,2:5,6-di-*O*-isopropylidene-3-*O*-methacryloyl-D-glucofuranose (**305**) and inimer 2-(2-bromoisobutyryloxy)ethyl methacrylate (**306**) via ATRP. The MWNT-CO_2H precursor was generated by oxidation of the crude MWNT with 60% HNO_3, and then modified with ethylene glycol, generating MWNT-OH. The CNT-based macroinitiator MWNT-Br (**302**) for ATRP was formed by treating MWNT-OH with 2-bromo-2-methylpropanoyl bromide. The morphology and nanostructures of the resulting multihydroxylated conjugates **303** and **304** were observed by SEM, TEM, and SFM, indicating the formation of nanotube–supported polymer brushes for linear polymers and a polymer layer in a core-shell structure for hyperbranched structures, implying their uniform growth on the surfaces of the MWNTs. These original water-soluble glycopolymer-grafted CNTs offer promise in the fields of tissue engineering and bionanomaterials.

In conclusion, such three-dimensional carbon-based nanostructures as fullerenes and nanotubes constitute scaffolds whose efficient functionalization and derivatization

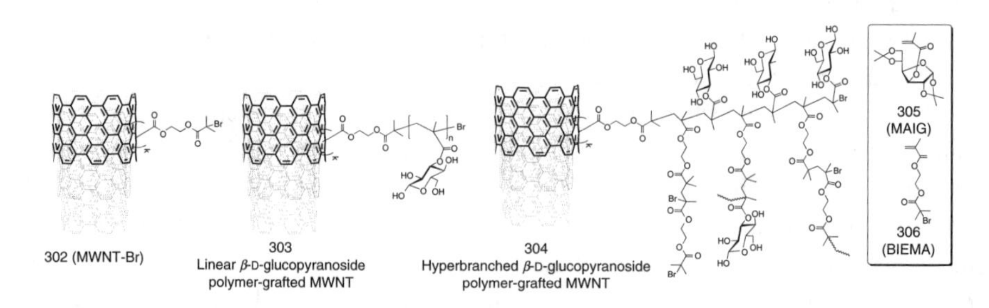

FIG. 34. Glycopolymer grafted onto SWNTs.[263]

has been developed in recent years. Accurate control of the introduction of biologically relevant moieties has been achieved, giving rise to original nanostructures with intriguing and promising properties. In particular, monovalent carbohydrates, more complex oligosaccharides, or glycodendrons have been successfully anchored at the surfaces, through covalent or noncovalent methodologies. Although practical applications remain as yet limited, preliminary results suggest that these nanomaterials offer unique and favorable arrangements of multivalent carbohydrate arrays that are not readily available with other displaying platforms.

IV. Multivalent Glycoconjugates by Self-Assembly

To enhance the number of saccharidic units exhibited, construction of multivalent scaffolds through supramolecular chemistry provides an interesting alternative for the rapid synthesis of glycodendrimers. Dynamic and reversible self-assembly process consisting in coordinating metal-assisted association of ligands is one of the most striking examples of supramolecular organization.[264] Over the years, optical and electrochemical properties of various complexes have been widely studied and exploited in different domains for material and biological applications.[265,266]

To target specific biological events, cells or tissues, the introduction of relevant recognition molecules such as carbohydrates have been investigated during recent years. In this context, metalloglycoclusters-initiated self-assembly of ligands containing simple saccharides or more dense glycodendrons around a coordinating metal has contributed valuable additions of original and promising structures to the range of available multivalent glycoconjugates. It is rather surprising that only a few examples have been described, because the metal-assisted association of carbohydrate components offers a straightforward access to carbohydrate clusters in which the number and the relative orientation of the carbohydrate residues can be modulated almost at will by changing the structure of the ligand and the nature of the metal. In addition, the intrinsic photophysical properties of the metal complexes can be advantageously exploited, for instance to create efficient biosensors. Furthermore, a second example of multivalent presentation of carbohydrates through self-organization process resulting from lipophilic cores has been furnished through the aggregation of glycodendrimers in aqueous solution, allowing the homogenous and controlled formation of hypervalent structures with distinct recognition properties as compared to the individual precursors.

1. Self-Assembly Using Coordinating Metals

Historically, the first multivalent glycoconjugate obtained by metal-assisted self-assembly was described in 1994 by Sakai and Sasaki (Fig. 35).[267] The methodology was based on the trimerization of unsymmetrical Bipy-GalNAc (**307**) induced by the presence of Fe(II) salts, to give the tridentate Fe^{II}(bipy-GalNAc)$_3$ cluster **308** as a mixture of four diastereomeric isomers (Λ-*fac*, Δ-*fac*, Λ-*mer*, and Δ-*mer*).

A comparative study involving bipy-GalNAc (**307**), Fe^{II}(bipy-GalNAc)$_3$ dendrimer (**308**), and asialoglycophorin A using the GalNAc-specific plant lectin, *Vicia villosa* B_4, has been assessed by the authors to determine and compare their relative binding potential. Better binding affinities were observed for the trivalent complex **308** over the monovalent precursor **307**, indicating the role played by the cluster formation of GalNAc residues around the metal template. More interestingly, the similar strong binding affinities observed for asialoglycophorin A, a natural glycoprotein containing several repeating units of GalNAc linked to Ser or Thr residues, and Fe^{II}(bipy-GalNAc)$_3$ were attributed to the adapted organization of GalNAc residues in the complex with adequate intercarbohydrate distances.

A few years later, the same group proposed further investigations to explain the strong binding affinities earlier observed with the self-assembled complex, describing the dynamic molecular recognition of their synthetic ligands by lectins.[268] Studies involving *V. villosa* B_4 lectin and the complex highlighted the fact that the ratio of the four diastereoisomers gradually changed during the binding experiment. After 32 h at room temperature, the authors recorded enrichment in the Λ-*mer* isomer, which in the end comprised 85% of the total isomers (vs. 15% in the native mixture). These experimental observations suggested that the dynamic equilibrium seen at room

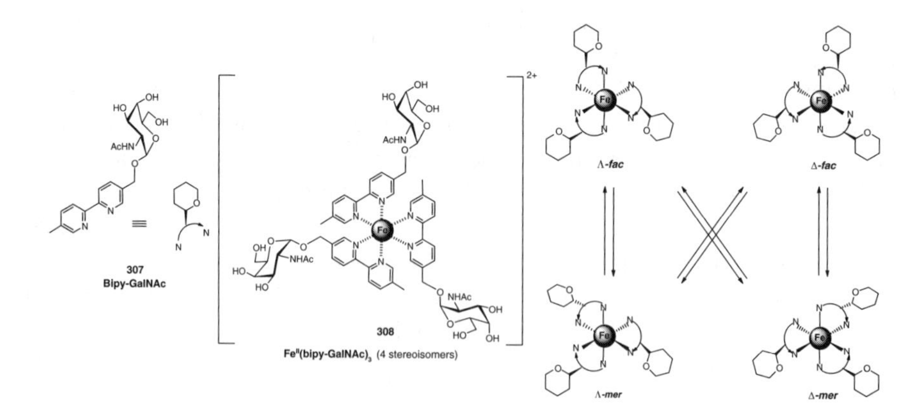

FIG. 35. Self-assembled glycodendrimers around a Fe(II)-bipyridyl complex.[267]

temperature allowed the spatial arrangement of the three GalNAc residues to self-adjust in order to provide a better complementarity to the multivalent carbohydrate-binding sites of the lectin.[269] This tendency was confirmed with studies involving other GalNAc-specific lectins, such as *Glycine max*, specific for a terminal α-D-GalNAc residue attached to the 3′-OH group of galactose. Selective enrichment of other diastereoisomers has been observed, indicating the adaptability of the process for lectins presenting different shape and functionnalities in their carbohydrate-binding pockets. The intramolecular lectin's binding sites, being too distant from one another to be reached by the close GalNAc residues, strongly suggest a dynamic cross-linking process.

Kobayashi *et al.* employed hexavalent metal complexes based on the self-assembly of three symmetric 2,2′-bipyridine ligands tethering two functionalized glycopyranosides around ruthenium(III)[270] or iron(II)[271] as the core component (Fig. 36). They undertook complete comparative studies involving glycosylated Tris-bypyridine iron and ruthenium complexes in order to develop robust biosensors for monitoring saccharide-binding phenomena. Synthesis of the ligands was based on the use of a 2,2′-bipyridine-4,4′-dicarboxylic acid core onto which amino glycopyranoside derivatives were conjugated via amide functions. Hence, self-assembled glycoclusters

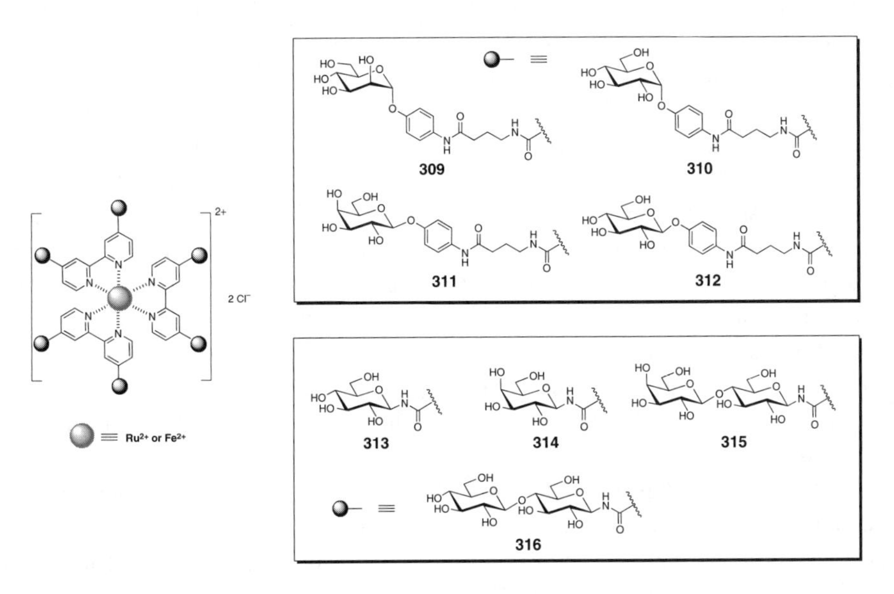

FIG. 36. Self-assembled glycodendrimers using chelating metal cations and divalent bipyridine scaffolds.[270,271] (See Color Plate 12.)

consisted of direct *N*-linked glycosylamine appendages to the complex, or through a O-linked *p*-butanamidophenyl spacer, were produced. Three equivalents of the resulting conjugates were then treated with $RuCl_3$ or $FeCl_2$ in a minimum amount of boiling water–alcohol mixture to afford the corresponding tris-bipyridineruthenium and iron complexes as mixtures of separable diastereoisomers.

To determine the importance of the structural parameters, the first study involved ferrous O- (**309–312**) and N-(**313–316**) glycoclusters. In this context, the influence of the spacers on the lectin affinity was investigated. Their relative binding properties to specific lectins were evaluated by a HAI assay and by SPR, using α-glucoside- and α-mannoside-specific ConA and β-galactoside-specific RCA_{120} lectins. Lectin affinities observed for the N-linked-glycoclusters **313–316** were low, but high affinities (IC_{50} in the micromolar range) were recorded for the elongated conjugates, thus highlighting the importance of the flexibility and the density of the carbohydrate scaffolds. Flexibility seemed essential for reaching high and specific lectin-recognition, and this was probably induced through optimized conformational organization of the epitopes to fit the binding sites of the lectins. On the other hand, strong binding avidities to lectins were not observed, because of the lability of the iron complexes, which readily dissociated in dilute aqueous solution ($<10^{-6}$ M).

For this reason, the authors further investigated more stable elongated ruthenium O-linked glycoclusters, and showed their avidity to the corresponding lectin to be enhanced, reaching IC_{50} in the nanomolar range, comparable to known potent glycopolymers. The results indicated that dense saccharide clusters played essantial roles in their specific and strong binding, as compared to monovalent precursors. Furthermore, the complexes exhibited intense emission spectra after excitation of the metal-to-ligand charge transfer (MLCT) band in water, as compared to nonglycosylated ruthenium complex. The strong luminescence was attributed to the peripheral carbohydrate shell caused by closely packed saccharide clusters, which could possibly isolate the luminescent core from that lost to the outer aqueous environment. The structures were considered by the authors to resemble natural proteins having redox and luminescent cores embedded in the polypeptide shells. Interestingly, binding of lectin to these glycoconjugates decreases their luminescence by disrupting their outer carbohydrate shell structure.

In 2003, combined efforts of the two research groups just mentioned developed a one-pot transglycosylation strategy for the construction of complex-type disialooligosaccharides starting from an α-D-glucopyranoside residue self-assembled onto a tris-bipyridineruthenium complex $\Delta[Ru(\alpha\text{-Glc-3-bpy})_3]Cl_2$ (**317**) (Scheme 32).[272] They succeeded in this commendable chemoenzymatic one-pot transformation by

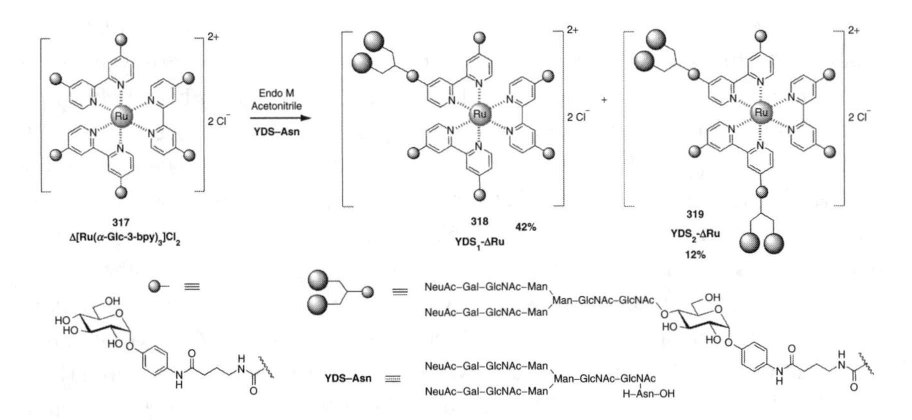

SCHEME 32. Enzymatic transglycosylation of a preformed bisglucobipyridine core self-assembled as a hexavalent cluster around a ruthenium cation.[272] (See Color Plate 13.)

using endo-β-D-GlcNAc-ase from *Mucor hiemalis* (Endo M), which effectively catalyzed transfer of the egg-yolk decasaccharide (YDS) from the asparagine-bonded complex-type disialooligosaccharide (YDS-Asn) onto the 4-OH position of a α-D-glucoside unit of **317**. Therefore, a mixture of two separable products, consisting of mono- (YDS$_1$-ΔRu) (**318**) and bisadducts (**319**), was obtained in 42% and 12% yields, respectively.

The binding properties of **318** and **319** in the presence of influenza type-A viruses (A/Memphis/1/71) were evaluated, using an inhibition test with virus-infected MDCK cells. The results revealed that, despite their small molecular weight, the YDS-adducts and more especially **319**, exhibited strong inhibitory potency, with an IC$_{50}$ of 8.4 μM, being values of two orders of magnitude higher than that of the parent YDS-Asn. The origin of this excellent virus affinity was unknown, but speculations concerning favorable electrostatic and hydrophobic interactions between the cationic and aromatic-rich complex center and certain amino acid residues of the hemagglutinin have been formulated. In addition, **318** and **319** exhibited strong luminescence at 605 nm, the intensity of which increased with the number of YDS residues. As evoked in the earlier work, this enhancement was attributable to the bulky oligosaccharides spanning around the luminescent core. Interestingly, the addition of influenza type-A viruses to the **318** and **319** in PBS resulted in luminescence quenching, probably caused by disruption of the saccharide shell and the resulting exposure of the luminescent core to the aqueous environment. Although the exact mechanism was not clear, this strategy could allow for the construction of robust and efficient sensitive biosensors for various toxins, viruses, and bacteria.

This strategy constitutes an extension of the work of Constable *et al.*[273] who initially described the self-assembly of peripheral saccharides covalently linked to terpyridine moieties. Divalent iron(II) and ruthenium(II) complexes bearing functionalized β-D-glucopyranosides or β-D-galactopyranosides have been efficiently prepared by this approach (Scheme 33). The peracetylated glucosylated terpyridine ligand **322** was obtained under phase-transfer conditions involving the reaction between the nucleophilic 4′-hydroxy-2′2′:6′,2″-terpyridine **321** and tetra-*O*-acetyl-α-D-glucopyranosyl bromide (**320**). The elongated analogue **326** was synthesized by simple nucleophilic substitution using K_2CO_3, KI, **321** and 2-bromoethyl tetra-*O*-acetyl-β-D-glucopyranoside (**325**). 1,2:3,4-Di-*O*-isopropylidene-6-*O*-tosyl-α-D-galactopyranose (**329**) was similarly treated with **321,** using sodium hydride in THF to afford **330**. Divalent complexes (**323**, **327**, and **331**) were typically formed by treating their terpyridine intermediates with $[NH_4]Fe(SO_4)_2.6H_2O$ in acetonitrile or ethanol and precipitation with ammonium hexafluorophosphate. Subsequent deprotection of the saccharide portions with aqueous methanolic K_2CO_3 or trifluoroacetic acid in water, respectively, afforded the glycoadducts **324**, **328**, and **332**.

The authors conducted preliminary biological investigation on these complexes, studying their stability toward enzymatic hydrolysis. Interestingly, the glucoside residue was not released from the complexes under the action almond β-glucosidase,

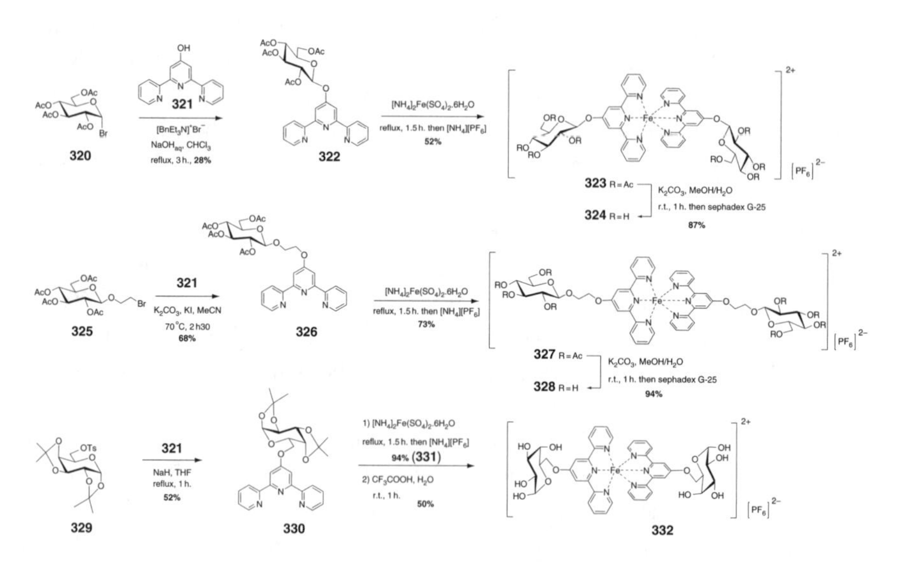

SCHEME 33. Self-assembled glycoclusters using a monosubstituted terpyridine ligand.[273]

while free ligands were hydrolyzed under similar conditions. The precise molecular origin of these effects was, however, not determined.

Later, the same group attempted to improve the efficiency of coupling of the carbohydrates onto the terpyridine scaffold by introducing a reactive pentaflurophenyl group onto the 4′-position (**333**) with the use of simple alcohols or saccharides (Scheme 34).[274] Somewhat surprisingly, it appeared that the nucleophilic $S_N$aryl substitution could not deliver the expected conjugates. They then relied on an alternative strategy using a 4′-tetrafluorophenoxy spacer (**334**), which allowed synthesis of the intermediate **335**, via attack of the electrophilic sugar **325**. Complex formation between **335** with ferrous chloride furnished the divalent iron complex **336** in good yield. The lengthy synthetic sequence, together with the modest yields obtained for the starting materials, rendered the strategy unsuitable for further development.

In order to prove the generality of the concept and enhance the stability of the resulting complexes, several research groups investigated the use of various ligands or coordinating metals. With this concept in mind, a versatile synthetic methodology leading to the self-assembly of galactose–oligopyridine conjugates around various metal cores has been disclosed.[275] The simple and successful strategy allowed structural variation in the components and linkers, generating multivalent adducts of different sizes, shapes, valencies, and conformational mobilities. The synthesis of the necessary galactosylated precursors started from tetra-*O*-acetyl-*β*-D-

1) MeOH or EtOH, $NaOH_{aq}$, reflux, 3 h. 65–73%; 2) Pyridinium chloride, 210 °C, 3 h. 31%

K_2CO_3, KI, acetonitrile, 70 °C, 2 h30, 31%

$FeCl_2.4H_2O$, EtOH, r.t., 1 h. then $[NH_4][PF_6]$, 71%

333 334 325 335 336

SCHEME 34. Alternative strategy used by Constable et al. for the construction of a dimeric iron(II) complex built around a terpyridine ligand.[274]

galactopyranosyl trichloroacetimidate (**337**), further functionalized under standard trimethylsilyl triflate-promoted glycosidation reaction conditions and subsequent transformation with various alcohols (Scheme 35). Three peracetylated galactoside conjugates (**338**, **340** and **342**), containing terminal amino, phenol, and aryl iodide functions, respectively, were thereby generated.

Functionalized oligopyridines consisting in bi- (**343**, **345**, **347**, **349**) and terpyridine (**351**) scaffolds were synthesized according to classical chemical strategies to allow efficient anchorage of the complementary functions of the β-D-galactosylated moieties **339**, **341**, and **342** (Scheme 36). The bisimine conjugate **344** was first prepared by condensation of 2-aminoethyl 2,3,4,6-tetra-*O*-acetyl-β-D-galactopyranoside (**339**) with the bisaldehydo bipyridine **343** in excellent (90%) yield. In order to obtain the elongated analogues **346** and **348**, 3-(4-hydroxyphenyl)-1-propyl 2,3,4,6-tetra-*O*-acetyl-β-D-galactopyranoside (**341**) was anchored to the dibromomethyl bipyridines **345** and **347** via nucleophilic substitution in 55% and 30% yields, respectively. Finally, Pd-(0)-catalyzed Sonogashira coupling was used efficiently to introduce 4-iodophenyl 2,3,4,6-tetra-*O*-acetyl-β-D-galactopyranoside (**342**) on the ethynylated bi-**349** and ter-pyridine **351** to afford divalent glycoconjugates **350** and **352**, in 33% and 60% yields, respectively.

Complexations of the bipyridine ligands with CuOTf and terpyridine ligand with $Zn(OTf)_2$, afforded the corresponding 2:1 (**353**) and 1:1 (**354**) complexes, respectively, as single species (Fig. 37). Depending on the metal ligands, spectroscopic evidence indicated the formation of tetrahedral structures in the Cu complexes, regardless of the length or the nature of the spacers, while a trigonal-bipyramidal geometry was observed for the self-assembled glycoconjugates resulting from the

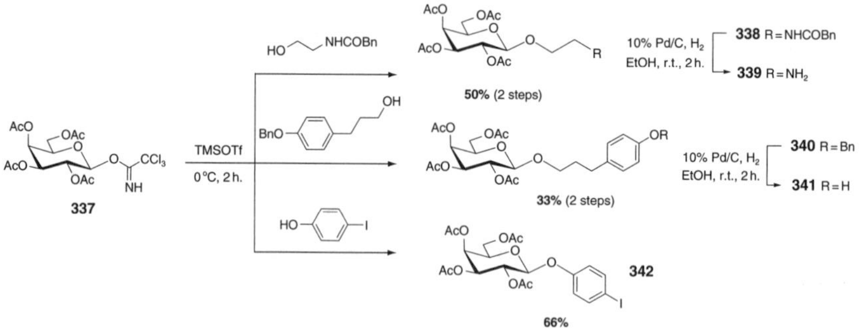

SCHEME 35. Synthesis of β-D-galactopyranoside precursors, bearing amine, alcohol, and iodide as anchoring functionalities, for metal-based self-assembly.[275]

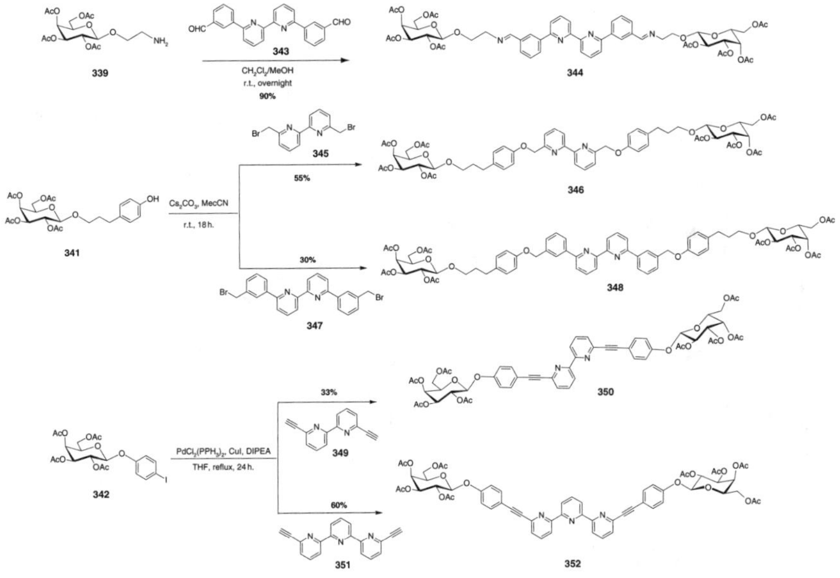

SCHEME 36. Divalent β-D-galactopyranosides built on oligopyridine metal ligand scaffolds.[275]

Zn metal. In addition, unprotected bipyridine–Cu(I) complexes were obtained via O-deacetylation under Zemplén conditions, in contrast to the terpyridine–Zn(II) adduct, where the complex decomposed.

A later investigation describes the attachment of different 1-thiosaccharides to 2,2′-substituted bipyridines to obtain biologically stable ligands and radiolabelled metal complexes of Re and Tc (Scheme 37).[276] The novel ligands were obtained by coupling 4,4′-dibromomethyl-2,2′-bipyridine **355** with 2,3,4,6-tetra-*O*-acetyl-1-thio-β-D-glucopyranose (**356**), 2,3,4,6-tetra-*O*-acetyl-1-thio-β-D-galactopyranose (**357**), and 2,3,4,6-tetra-*O*-acetyl-1-*S*-acetyl-1-thio-α-D-mannopyranose (**358**), under basic conditions. The ensuing thioglycosidically linked bipyridine derivatives were O-deacetylated under standard Zemplén conditions to afford the water-soluble ligands **359–361**. Luminescent rhenium complexes (**362–364**) were formed by refluxing a methanolic suspension of the corresponding ligand with rhenium pentacarbonyl chloride [$Re(CO)_5Cl$]. New glycoconjugated complexes of the general formula [$Re(L)(CO)_3Cl$], considered as luminescent probes, were characterized by mass spectrometry, elemental analysis, ^1H and ^{13}C NMR, IR, UV–vis, and fluorescence spectroscopy. One of the structures was unequivocally characterized by X-ray crystallography, indicating a surprising nonsymmetric metallocomplex architecture. In addition, the rather stable radiolabelled ^{99m}Tc

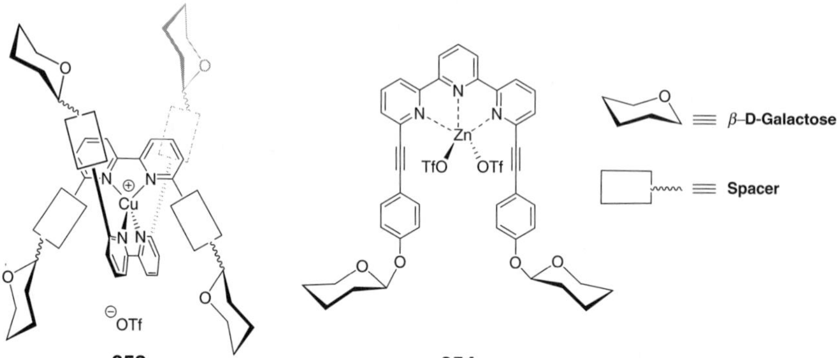

FIG. 37. Metal complexes resulting from the glycosylated bi- or ter-pyridine ligands.[275]

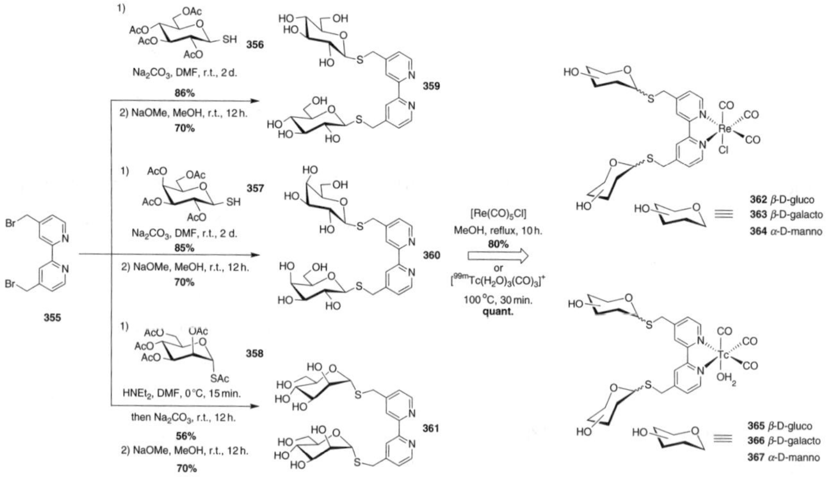

SCHEME 37. Preparation of luminescent Re and ^{99m}Tc glycoclusters obtained by self-assembly.[276]

complexes **365–367** were synthesized similarly in quantitative yields, using $[^{99m}Tc(CO)_3(H_2O)_3]^+$ as the metal ion source.

In order to increase the complexity of the peripheral epitopes resulting from the self-assembly process, glycodendrons have been successfully grafted onto various ligands, allowing the formation of dense glycocomplexes. Indeed, in most of the examples, branching of the dendrimers produces a microenvironment that encapsulates the coordinating metal core and can drastically modify its photochemical and electrochemical properties, and these can be modulated at will for suitable applications.

The first such example was proposed by Roy and Kim[277] who described the synthesis and the relative lectin-binding properties of square planar complexes organized around a Cu(II) coordinating core, and glycodendrons containing four or eight peripheral Tn-antigen units (α-GalNAc known as a cancer marker), covalently linked to a bipyridine core (Scheme 38). 2-Aminoethyl 2-acetamido-3,4,6-tri-*O*-acetyl-2-deoxy-α-D-glucopyranoside (**368**) and its analogue **371** containing an elongated linker (obtained by peptide coupling with *N*-Boc-protected 6-aminohexanoic acid followed by acid deprotection with TFA, CH_2Cl_2, 76%) were used for anchoring. Compounds **368** and **371** were then coupled with 2,2′-bipyridine-4,4′-dicarboxylic acid chloride (**369**) to provide divalent Tn antigens **370** and **372**, respectively, after Zemplén deprotection.

In order to provide access to self-assembling structures of increasing carbohydrate valencies, the authors described a convergent strategy for dimeric GalNAc building blocks having both short- and long-space arms. Thus, aminated derivatives **368** and **371** were treated with bromoacetyl chloride in the presence of DIPEA to afford bromoacetylated GalNAc derivatives **373** and **378** in excellent yields (Scheme 39). N,N-Dialkylation involving mono-*N*-Boc-1,4-diaminobutane (**374**) and bromoacetyl derivatives **373** and **378**, followed by trifluoroacetolysis afforded dimers **375** and **379** in 72–73% yields. By a synthetic pathway similar to that described earlier, their

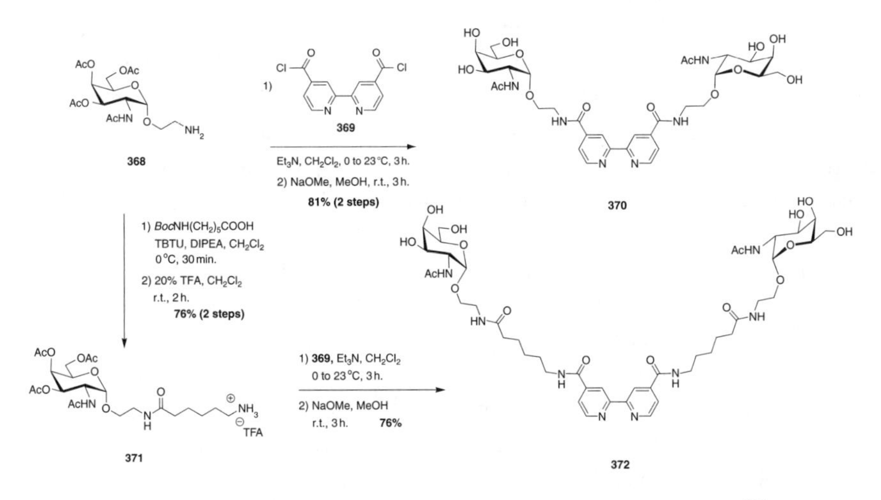

SCHEME 38. Tn-Dimers built on 2,2′-bipyridine core for self-assembly.[277]

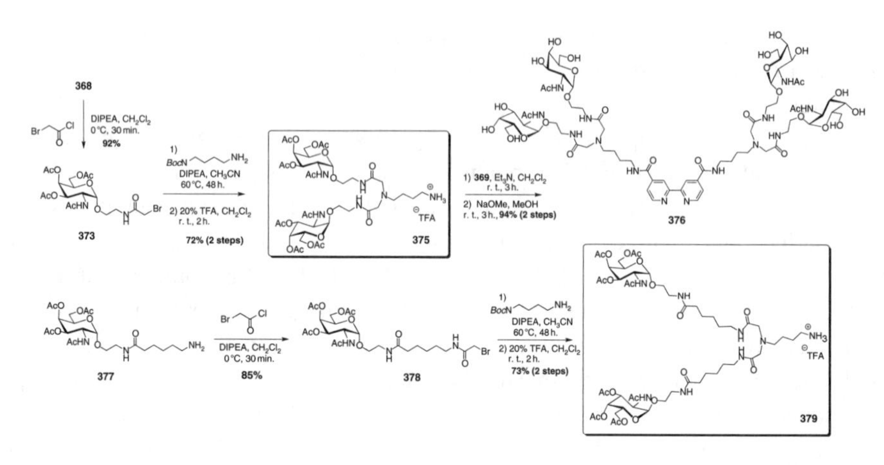

SCHEME 39. Other elongated Tn-antigen (α-D-GalNAc) clusters for self-assembly.[277]

amidation with 2,2′-bipyridine-4,4′-dicarboxylic acid chloride (**369**) in the presence of Et_3N, followed by O-deacetylation yielded the elongated bipyridyl tetramer **376**.

The preformed dendrons **370**, **372**, and **376** were clustered around copper(II) ($CuSO_4$ $5H_2O$) in aqueous media to provide efficiently the symmetric and hydroxylated glycoclusters and glycodendrimers **380–382** (Fig. 38), respectively, exhibiting four and eight GalNAc residues. These metallated α-D-GalNAc-bearing glycodendrimers were fully characterized by standard spectrometric analyses, ^1H and ^{13}C NMR, MALDI-TOF mass spectrometry, and UV–vis spectroscopy.

The potential of these self-assembled complexes to cross-link and to precipitate horseradish peroxidase-labeled plant lectin *V. villosa* (VVA-HRP) was confirmed by a solid-phase competitive inhibition assay (ELLA) with the use of asialoglycophorin as coating material. All of the neoglycoconjugates, including simple divalent and tetravalent ligands, inhibited the binding of asiaglycophorin to VVA-HRP with greater efficacy than the monomeric standard inhibitor allyl α-D-GalNAc (IC_{50} of 158.3 μM). As anticipated, the elongated dimeric analogue **381** was the most potent (IC_{50} of 1.82 μM) and the tetravalent ligand **380** showed less potency (IC_{50} of 4.09 μM), probably because of the shorter intrasugar distances between each of the branches. As compared to their nonmetal complexes, the metallic clusters clearly exhibited enhanced inhibitory potencies following the expected ''cluster effect.'' Indeed, the Cu(II)-(bipy-GalNAc) tetramers, **380** and **381** were 61- and 251-fold, respectively, more potent than the monovalent control, while octamer **382** exhibited a 259-fold enhanced avidity. Interestingly, the best binding affinity was recorded for the tetramer Cu(II) nucleated derivative **381** (63-fold on a per saccharide basis) that has

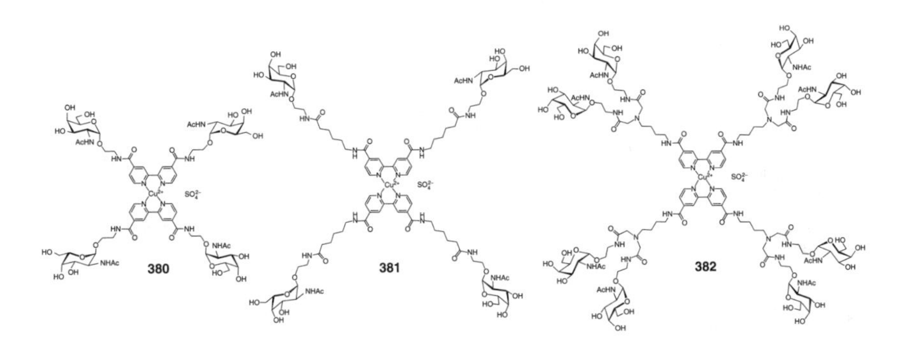

FIG. 38. α-D-GalNAc 2,2′-bipyridine oligomers self-assembled around a copper(II) salt.[277]

the longest intersugar distances. These results clearly demonstrate that the structural arrangement and flexibility of the molecules play crucial roles in their relative binding affinities. This work confirmed that the aglycone spacer and valency enhancement of sugar residues in neoglycoconjugates were responsible for an increase in binding affinity in carbohydrate–protein interactions.

This supramolecular assembly process has also been directed toward the synthesis of metallic and fluorescent glycodendritic architectures.[278] Thus, homogenous fluorescent and sugar-functionalized metallic dendrimers, containing various numbers and types of monosaccharides have been prepared. The chelation of transition or lanthanide metals was ensured by the presence of 8-hydroxyquinoline (Scheme 40). Metallic glycodendrimers containing up to 18 peripheral mannoside residues have been prepared starting from *N*-(tris[(2-cyanoethoxy)methyl])methylamine (**383**), which was treated with concentrated HCl in ethanol to yield triester **384**. Successive peptide couplings with Boc-*β*-alanine and then with 8-*O*-benzylquinoline-2-carboxylic acid (**385**) yielded the functionalized tripod **386**, which after hydrolysis and further coupling with pentafluorophenol afforded the active triester **387** in 71% yield. Tetraacetylated mannose containing an anomeric 2-aminoethoxy linker (**290b**) was then coupled to **387** to generate the trivalent neoglycoconjugate **388**. O-Deacetylation and further hydrogenolysis afforded the corresponding complex precursor **389** in 24% yield over two steps.

Second-generation mannosylated dendrons were synthesized by the process employed for **389**, starting with the introduction of Boc-*β*-alanine on triester **384** to provide **390** in 63% yield. Saponification and coupling with pentafluorophenol under standard conditions afforded triester **391**. The 2-aminoethyl mannoside **290b** was then coupled to **391** to give **392**. *In situ* removal of the *N*-Boc-protecting group (TFA) and treatment with **387** afforded, after deprotection, the nonavalent glycodendron **393** in

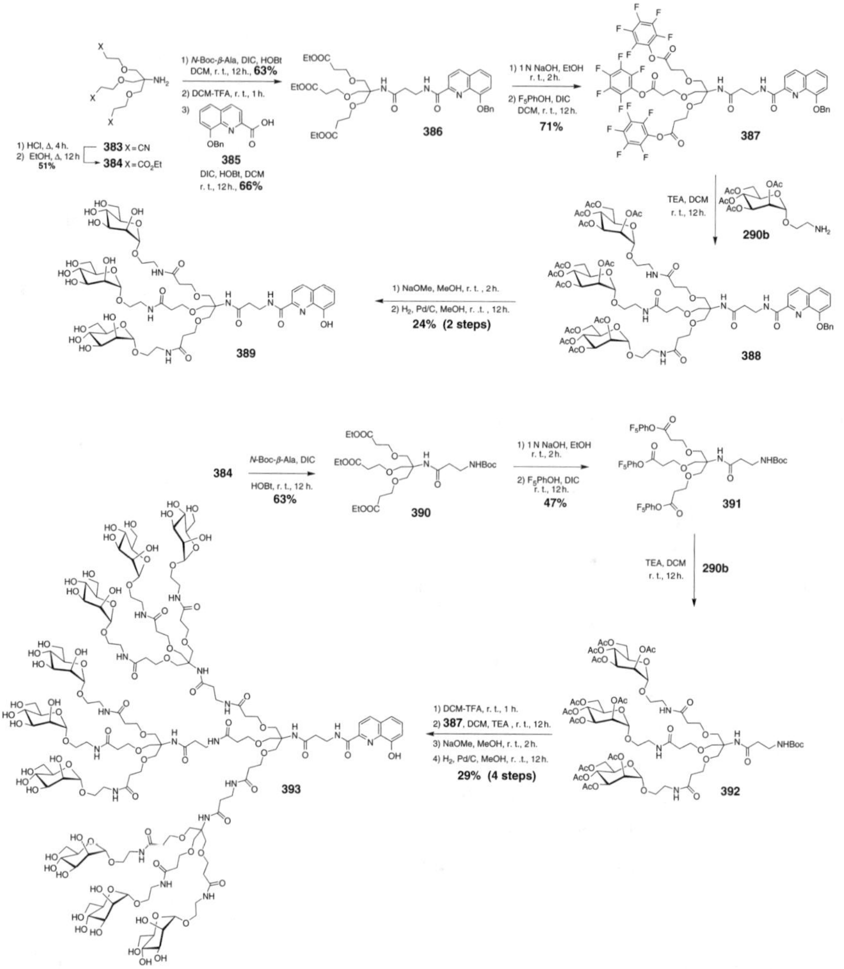

SCHEME 40. Mannosylated dendrons having an 8-hydroxyquinoline ligand at the focal point for self-assembly.[278]

29% yield over four steps. Finally, treatment of the foregoing precursors under appropriate stoichiometry with $Zn(OAc)_2$, [or $Al(OAc)_3$ or $GdCl_3$ for lower generation] in hot methanol afforded the self-assembled complexes **394** and **395** in 75–82% yields, containing either 6 or 18 peripheral mannopyranosylated residues (Fig. 39).

After assessing the interesting optical properties of the complexes, specific protein interactions with ConA as a model lectin were investigated in preliminary

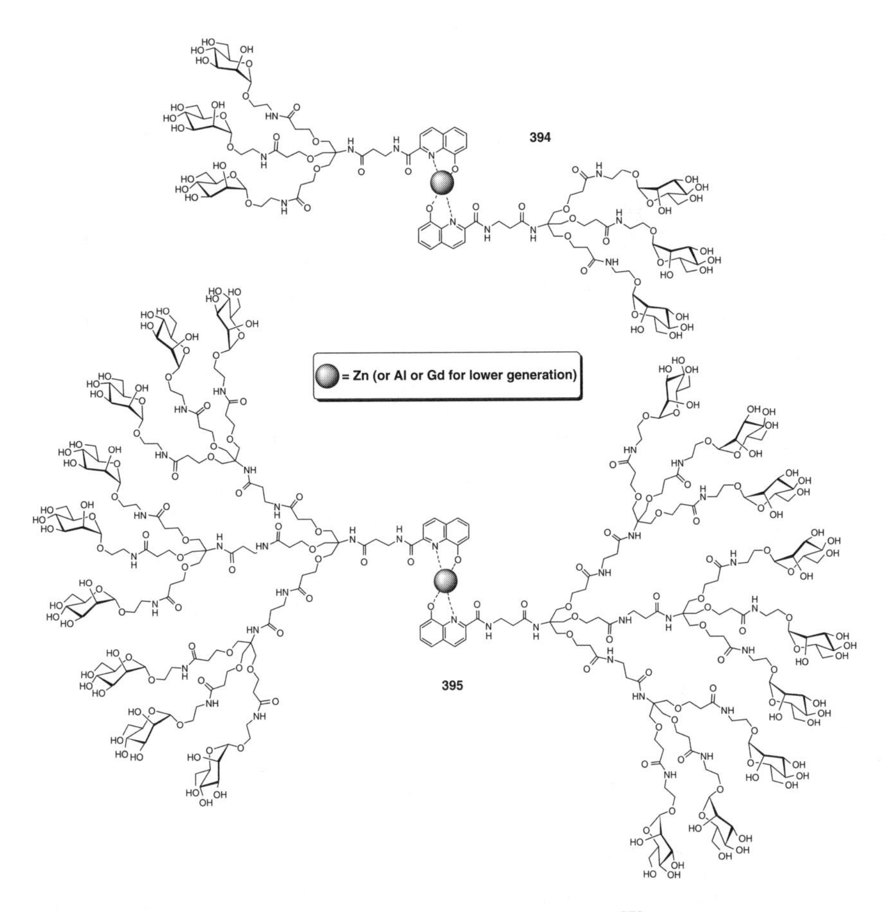

FIG. 39. Self-assembled mannodendrons.[278]

turbidimetry assays. Initial results indicated that metallic second-generation glycodendrimer **395** induced strong binding because of the large and dense glycocluster at the surface, and was a highly efficient lectin sensor. This approach to lanthanide-glycodendrimers [for instance with Gd(III)] might thus provide tunable fluorescent or MRI reagents for imaging.

The same research group extended their studies, proposing a systematic investigation of the core activities for different carbohydrate densities during biosensing processes.[279] Three new mannosylated dendrimers with a $Ru(bipy)_3$ core unit were synthesized, and the influence of the number and size of dendritic branches on the rate of electron and energy transfer, as well as the lectin-biosensing abilities (Scheme 41),

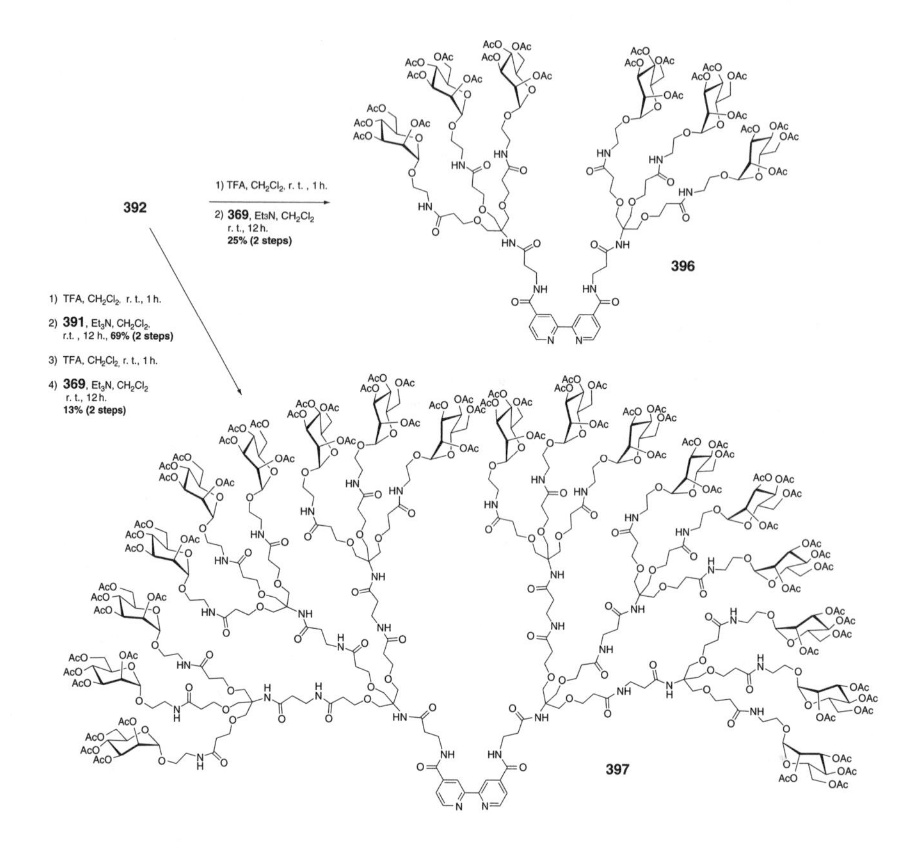

SCHEME 41. Mannosylated glycodendrimers on a bipyridine core for ruthenium complexation.[279]

were investigated. Mannose-capped dendrimers **398**, **399**, and **400** were prepared using glycosylated bipyridine precursors by treatment with $RuCl_3$ or *cis*-Ru-$(bipy)_2Cl_2$. A first-generation dendron (**396**) was prepared in modest yield (25%) via the previous strategy, by removal of the *N*-Boc protecting group of **392** and subsequent peptide coupling with 2,2′-bipyridine-4,4′-dicarboxylic acyl chloride (**369**). A second-generation dendron, containing nine saccharidic units (not shown), was synthesized in 69% yield by the introduction of an amine derived from **392** and coupling to **391**. Following *N*-Boc deprotection (TFA) and subsequent peptide coupling with bipyridine derivative **369** gave **397** in 13% yield over two steps.

Boiling the first- and second-generation glycodendrons **396** and **397** in ethanol in the presence of *cis*-$Ru(bipy)_2Cl_2$ or $RuCl_3$ followed by O-deacetylation under Zemplén conditions yielded dendritic complexes **398–400** (Scheme 42) containing 6 or 18 peripheral mannopyranosides, respectively.

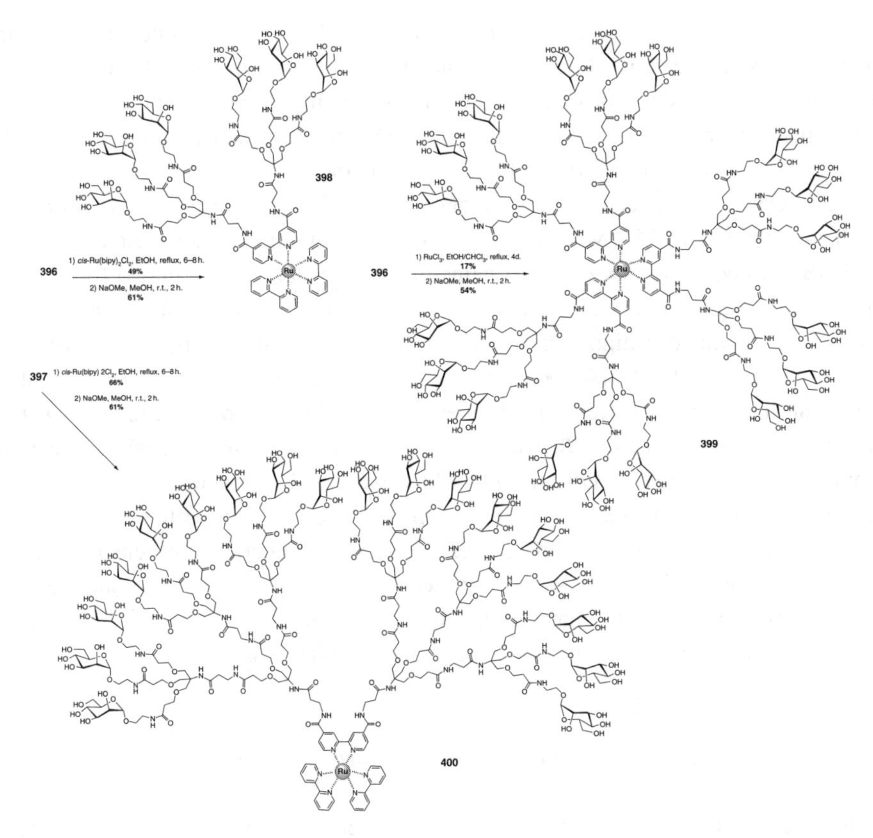

SCHEME 42. Preparation of photoinducible electron-transfer Ru(II)-complexes.[279]

The rate of electron transfer within the complexes was investigated by photoinduced electron transfer (PET) between photoexcited Ru(II)-templates and a suitable quencher (*N*, *N*′-4,4′-bis(benzyl-3-boronic acid)bipyridinium dibromide, BBV) having high affinity for sugar. As expected, all compounds exhibited different extents of quenching, depending on the degree of peripheral carbohydrate density that efficiently insulated the core properties. The rate of energy transfer induced by photoexcitation of the complexes was also evaluated by quantification of molecular oxygen in the singlet state trapped by a quencher (TEMPO). The rate of appearance of the generated stable species (TEMPO) decreased from **398** to **400**, thus supporting the notion that dense carbohydrate insulation of the Ru(II) template presents a shield that stopped effective energy transfer to dissolved oxygen. In addition, the selectivity and sensitivity of these processes have been studied by use of complexes as lectin biosensors. ConA and *Galanthus nivilis agglutinin* (GNA), both tetramers that contain one and three mannose-binding sites per subunit, respectively, were selected for these investigations. The concept was based on the reconstitution of the fluorescent signal when preliminary complexes between the BBV quencher and mannosylated-complexes were disrupted in favor of the tight interactions that take place between glycodendrimers and lectins. A spontaneous gain in fluoresecence was observed with the addition of 75 nM of ConA to the initial mixture of hexavalent complex and BBV. In contrast, more modest gains in fluorescence were obtained for dendrimers containing 18 mannosides under the same conditions, but a steady and linear increment continued at higher concentrations of ConA. Similar experiments with the higher valency lectin GNA induced the same tendency. Based on these results, detection limits for each Ru-complex were evaluated and compared, indicating that hexavalent glycoconjugate **398** was the most sensitive; whereas a higher detection limit was determined for **399** and **400** which exhibited a broader limited range of linear response. In conclusion, although complex **398** represented the best compromise between encapsulation and efficient quenching properties for sensitive lectin sensing, glycodendrimers **399** and **400** were also suitable sensitive biomarkers to study lectin–carbohydrate interactions, owing to their high quantum yield and excellent lectin-affinity through their high carbohydrate density.

2. Self-Assembly of Glycodendrons in Solution

In the early 2000s, Thoma *et al.* presented an alternative concept for the polyvalent presentation of ligands, based on the supramolecular chemistry of rather small molecules that fulfill single-molecule entity criteria.[280] They described the noncovalent and dynamic self-assembly of original and functionalized dendrons capped with carbohydrate ligands in aqueous solution, generating polyvalent glyconanoparticles. Dendritic scaffolds were prepared by a convergent ''outside-in'' approach.[281] This iterative methodology was based on a single building block (**403**) obtained by the coupling of methyl 3,5-diaminobenzoate and 4-(*tert*-Boc-aminomethyl)benzoic acid. Selective cleavage of the methyl ester group to give **404**, and subsequent removal of the *N*-Boc group liberated **405**, comprising the G(1) dendrimer generation with two amine end groups. Its coupling with **404**, followed by orthogonal methyl ester cleavage furnished **408**, a G(2) dendrimer generation containing four terminal anchorage functions. The third-generation dendrimer, with eight end groups, was obtained by treatment of **408** with **404**. The same procedure was repetitively applied to yield dendrimers with up to 64 end groups, corresponding to the sixth generation.

The peripheral amines of the glycodendrons of each generation were isolated as TFA salts and further transformed by introduction of chloroacetamide groups, using a large excess of chloroacetic anhydride in DMF, leading to complete conversion. Subsequent introduction of thiolated oligosaccharides such as **401** (Galili antigen) and **402** (Lac-SH) onto the dendritic cores was realized in the final step of the synthetic sequence in the presence of DBU, to afford water-soluble glycodendrimers **406**, **409**, **410a**, and **410b** (not shown; sixth generation) presenting a controlled number of peripheral oligosaccharidic residues based on the generation (Scheme 43).

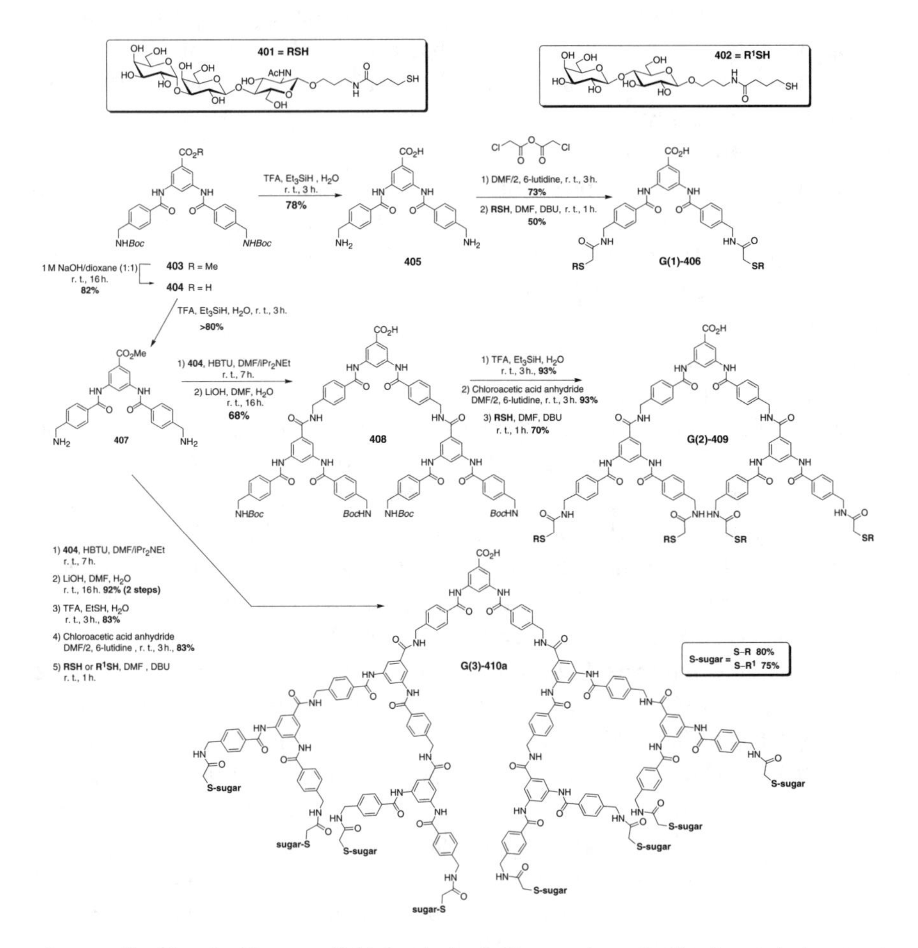

SCHEME 43. Glycodendrimers scaffolded on hydrophobic repeating-units, forming stacked aggregates in aqueous solutions.[280,281]

Aggregation of all glycodendrimers in water was observed by ^{1}H NMR and quantified by using multiangle light scattering (MALS). Measurements indicated the formation of small aggregates of 50 kDa for the first generation (**406**) whereas the tetravalent analogue **409** formed large particles of 7000 kDa, corresponding to the aggregation of more than 1500 individual molecules per particle. Surprisingly, the particle weight obtained with higher generation dendrimers decreased with increasing mass of the individual molecule. Furthermore, the third-generation glycodendrimers **410**, containing eight (RS- or R^1S-) residues, exhibited similar size and weight, indicating that particle structure depended primarily on the dendritic core architecture, and less on the size and the nature of the carbohydrate group. The synthesis of several dendritic cores containing more branched aliphatic components in their structures, such as cyclohexyl rings, linear alkyl chains, or methylated amides demonstrated that only highly aromatic architectures induced aggregation as a result of π-stacking, hydrophobic interactions, and rigidity.[281] The authors assumed that intramolecular π-stacking led to preorganization, allowing efficient core–core contacts. An optimal core–core interaction was recorded for the second generation, whereas the lowest generation dendrimer was too small and the largest ones were too shielded by the dense carbohydrates. In addition, both the shape and the size of these homogenous nanoparticles (within a factor of 2) was investigated by AFM, indicating a disk-like morphology, having average diameters decreasing with increasing mass of the individual molecule, thus resembling their tendency in solution.

The biological activity of these noncovalent multivalent ligands was investigated via the inhibition of decavalent IgMs directed against the αGal (Galili) xenoantigen. To assess the compounds as polyvalent IgM ligands, the authors employed two *in vitro* assays in which the inhibition of both the anti-αGal IgM binding to the xenoantigen and the αGal-mediated lysis of pig erythrocytes were measured. In agreement with previous results, monomeric αGal was inactive at 100 μM in both assays. Whereas no activity was observed for the divalent conjugate, the second- and third-generation dendrimers **409** and **410** were highly potent in both assays (0.025, 0.035 μM and 0.010, 0.010 μM, respectively). The fourth-generation analogue, containing 16 peripheral αGal residues, also showed high potency in the binding assay (0.019 μM) but was significantly less potent in the hemolysis assay (0.18 μM). Generally, potency decreased drastically for the larger dendrimers which did not induce aggregation. The avidity thus correlated with the size of the aggregates, but not with the size of the individual particles. Preliminary *in vitro* results highlighted the involvment of polyvalent aggregates in antibody binding, without nonspecific interactions of the dendrimer backbone. The most potent compound (**410**) bearing eight trisaccharidic epitopes, was selected for *in vivo* profiling in cynomolgus monkeys.

Within 5 min after injection, the anti-αGal IgMs detected by ELISA were diminished to 20% of the initial value determined prior to the administration, and remained at low levels for more than 4 h. Most importantly, anti-αGal antibody-mediated hemolytic activity was completely eliminated.

Even though examples remain rather scarce, the preparation of multivalent structures via supramolecular chemistry involving the self-organization of ligands around a coordinating core has afforded glycoconjugates in which the valency, the size, and the structures can be efficiently tailored with the nature of the coordinating metal and the chemical composition of the ligand. In most instances, the stable complexes generated constitute promising organizations with enhanced biological properties as compared to the corresponding monomeric ligand. Moreover, this noncovalent, dynamic, and reversible self-assembly process can, ideally, allow the utilization of the polyvalent receptor (lectin) as a template to optimize its own polyvalent inhibitor with the organization of carbohydrate moieties to fit perfectly into the recognition sites. This adaptability has also been emphasized by the significant improvement of biological activities as compared to those of individual species that has been observed for homogenous noncovalent glycoparticles that self-assembled in aqueous media via a similar dynamic equilibrium process.

In the light of promising applications offered by these supramolecular glycoconjugate structures, notably for the intrinsic optical and electrochemical properties of metal complexes, this underexploited research area is undoubtedly in its infancy and holds promise for such systems as biomarkers or sensitive biosensors.

V. Glycodendrons and Glycodendrimers

1. Introduction

a. Definition and History.—A dendrimer is a synthetic highly branched monodisperse and polyfunctional macromolecule, constituted by repetitive units (so-called ''generations'') that are chemically bound to each other by an arborescent process around a multifunctional central core.[282,283] Thus, as opposed to traditional polymers, which often have poorly defined molecular structures that are clearly an important drawback for medical application in terms of product characteristics, dendrimers are structurally well defined and can be synthesized from a fully controlled iterative approach. Although differences exist in terms of rigidity and compaction, dendrimers are often compared to ''artificial proteins'' with their semiglobular or globular structures, mostly with a high density of peripheral functionalities and a small molecular ''volume.''[284,285] It is now accepted that dendritic polymers are the fourth major class of polymeric architecture, consisting of three subsets that are based on the degree of structural control, namely: (a) random hyperbranched polymers, (b) dendrigraft polymers, and (c) dendrimers. The concept of repetitive and controlled synthetic growth with branching was first reported by Vögtle, who achieved the construction of a low molecular weight ''cascade'' polyamine.[286] However, it was not until 1985 that the groups of Newkome[287] and Tomalia[288] independently described a divergent macromolecular synthesis, giving birth to the first well-characterized true dendrimers, named ''arborols'' and ''PAMAM'' [poly(amidoamine)] dendrimers, respectively. Their strategy thus efficiently avoided problems of low yields, purity, or purification encountered by Vögtle in his cascade synthesis. In their chapter, Tomalia *et al.* paved the way to dendritic structures: their definition and construction, and introducing for the first time the term ''dendrimer,'' which arises from the Greek *dendron* meaning ''tree'' or ''branch,'' and *meros* meaning ''part.'' Their original and efficient methodology still constitutes the preferred commercial route to the trademarked Starburst® dendrimer family, with molecular weights ranging from several hundred to over 1 million Daltons (namely, generations 1–13). Until the mid-1990s, the synthetic challenge of such aesthetic structures stimulated numerous research groups to investigate intensively a range of synthetic strategies. These efforts gave rise to original dendritic structures emerging from two main synthetic strategies used to construct perfectly branched dendrimers: the divergent and the convergent approach. Over time, an accelerated version of the convergent strategy has been developed in order to increase its throughput and efficiency by using clever adaptations of cores or dendrons.

Initial efforts gave rise to well-characterized dendritic macromolecules, but applications remained limited because of the lack of specific functionalities. An exponential increase of publication volume observed for about 15 years testified the growing interest for dendrimers and has led to versatile and powerful iterative methodologies for systematically and expeditiously accessing complex dendritic structures. The perfect control of tridimensional parameters (size, shape, geometry) and the covalent introduction of functionalities in the core, the branches, or the high number extremities, or by physical encapsulation in the microenvironment created by cavities confer such desired properties as solubility, and hydrophilic/hydrophobic balance. Thus, creativity has allowed these structures to become integrated with nearly all contemporary scientific disciplines.

Undoubtedly, biology and nanomedecine, more particularly biomedical and therapeutic applications, are the domains that have generated the highest infatuation for these architectures. However, the complexity of mechanisms involved in biological processes presents an important challenge for efficient structure design. In fact, historically, problems concerning toxicity, hydrosolubility, degradability, targeting specificity, pharmacokinetic, and biodistribution profiles for this kind of applications have been recurrent when using monomeric systems. Application of polymeric systems has been exploited advantageously to enhance hydrosolubility, biocompatibility, lack of toxicity, immunogenicity in order to improve drug stability, and selectivity toward malignant tissues. This enhancement of therapeutic properties observed with the use of "prodrugs" has been explained by the physical properties of these polymeric structures, presenting high hydrodynamic volumes that facilitate blood persistence and accumulation in tumoral tissues in particular.This phenomenon, called the "EPR effect" (i.e., enhanced permeation retention), may be roughly explained by physiological and biochemical differences observed between tumoral and healthy tissues. In fact, tumor tissues present specific vascularization with a defective lymphatic drainage system that allows macromolecules (with molecular weight $\geq$ 20 kDa) permeability, retention, and accumulation.[289] However, a high polydispersity index and low drugs loading are often responsible for critical lack of reproducibility and efficiency.

Dendrimers, combining several of the advantages described for polymers, along with monodispersity and a high density of functionalities with a small molecular volume, have been exploited for about 15 years to surpass the problems usually encountered. Thus, their particularly unique structures and properties have motivated their use in numerous applications as drug or gene delivery devices in anticancer therapy, and as antibacterial, antiviral, or antitumoral agents. The use of dendrimers in biological systems, and also systematic studies of the most common dendritic scaffolds to determinate their biocompatibility, such as *in vitro* and *in vivo* cytotoxicity, their biostability or immunogenicity have been extensively reviewed.[290,291] One typical example concerns the use of dendrimers as "glycocarriers" for the control of multimeric presentation of biologically relevant carbohydrate moieties that are useful for targeting modified tissue in malignant diseases for diagnostic and therapeutic purposes. In such molecules, termed "glycodendrimers," the saccharide portions are conjugated according to the principles of dendritic growth or are ligated to preexisting highly functionalized and repetitive dendritic scaffolds having varied molecular weights and structures. Since they first appeared in the literature in 1993,[292] glycodendrimers and related glycodendrons, with their spheroidal or dendritic wedge structures, have been initially designed as bioisosteres of cell-surface multiantennary glycans that stimulated wide interests within the scientific community.[293–298] As with conventional dendritic structures, glycodendrimers can be obtained as dendrons, spherical or as globular architectures, or "hybrid dendronized polymers'' according to divergent, convergent, or accelerated approaches. All of these original synthetic clusters were constructed in such a way that their valencies, shapes, and carbohydrate contents could be varied at will with a controlled integration of dendritic building blocks. Hence, recent progresses in synthesis of dendritic structure can allow easier optimization, to afford tailored glycoconjugates with biologically adapted and optimized properties.

b. Glycodendrimer Syntheses.—Historically, the divergent strategy was used to prepare the first dendritic structures. Dendrimers are built iteratively out from a central core, layer by layer, requiring activation/addition steps to afford the desired dendritic structures: the focal and multifunctional molecules are systematically expanded outward using various chemical linkages. The first-generation dendrimer is simply formed by attaching branching units to the core molecules. To form the second-generation dendrimer, the peripheral functional groups then react with a complementary chemical function presented on the branched building blocks. To avoid hyperbranched polymerization, this step has to be carefully controlled by using protected (or inert) groups on the building blocks. Activation (deprotection or chemical transformation) of the newly attached surface groups leads to the second-generation dendrimer. The generation growth quickly allows exponential multiplication of active terminal functions, and the process is repeated until the required degree of branching is obtained. For glycodendrimers, the sugars are then appended at the periphery of the molecules. Although this approach is conceptually straightforward, synthetic problems are sometimes encountered, involving the use of very large excess of reagent (or monomer) in each synthetic step to ensure complete functionalization and the necessity of efficient coupling in regard to the exponentially increasing number of functions. To accentuate the difficulty, and although monomers are generally easy to remove, separation of the required dendrimer from the structurally flawed by-products usually remains challenging, because of their mass, size, and general properties that are very close to the perfect dendrimer. For the sake of simplicity and diminished cost, the inner scaffold portion of the glycodendrimer can be either synthesized by a one-pot procedure using hyperbranched polymer methodologies, or purchased directly. Indeed, several dendrimers having various surface functionalities and building blocks are commercially available: polyamidoamine dendrimers (PAMAM, Starburst®, Dendritic Nanotechnologies), polypropyleneimine (PPI, Astramol®, DSM Fine Chemicals), polyglycerol dendrimers, and hyperbranched dendritic polymers (Boltorn®), are most commonly used as multibranched dendritic core or glycodendron precursors, and most of them are known to be nontoxic and nonimmunogenic.

An alternative and more efficacious convergent strategy was reported in 1990 by Hawker and Fréchet,[299] using the symmetrical nature of these structures, in order to overcome some of the synthetic and purification problems associated with the divergent methodology. It involves the preliminary synthesis of peripheral branched dendritic arms, named ''dendrons'' or ''glycodendrons,'' from the ''outside-in.'' This concept can be described by envisaging the attachment of X terminal units containing one reactive group to one polyfunctional monomer possessing orthogonally protected functionalities, resulting in the first-generation dendron. Transformation of the unique focal site, followed by treatment with $^{1}/_{X}$ equivalent of the protected monomer or a polyfunctional central core affords the next higher generation dendron or a dendrimer. The advantages of the convergent strategy lie with the diminished number of reactions carried out in each step. Moreover, purification of the desired dendrimer becomes less problematic than in the divergent case, fewer by-products are generated and they are structurally very different from the perfect target dendrimer. However, the fact that the focal functionalities of the wedge may be sterically inaccessible from within the infrastructure (depending on the generation) causes difficulties toward subsequent linkage to the core, thus resulting in slower and less efficient reactions as the generation grows. Nevertheless, it is now well established that a large number of surface glycan moieties impede accessibility of the carbohydrate by carbohydrate-binding receptors. Considering these synthetic advantages, this methodology has been used successfully for access to dissymmetric dendritic structures.[300]

Cumbersome purifications and synthetic disadvantages observed with both iterative approaches have motivated investigations toward accelerated synthetic methodologies. To improve synthesis efficacy and limit synthetic steps, while preserving monodispersity and functionalization versatility, new strategies have been successfully addressed. These include the development of larger building blocks involved in ''multigeneration'' coupling, as largely described by Fréchet *et al.* In this way, high molecular weight dendrimers and dendrons have been synthesized using highly polyfunctional dendritic cores (''hypercores''),[301] and high-generation dendrons for subsequent coupling reaction (''hypermonomers''),[302] which can also be obtained by an orthogonal protected functions approach (''double exponential growth strategies'').[303] An elegant application has been described[304] based on this design to allow for more rapid dendritic construction, starting with small glycoclusters, which are attached to branching units to form glycosylated hypermonomers and then finally to a suitable central part. More recently, another clever strategy based on ''orthogonal monomer systems'' has been designed.[305,306] The judicious use of functionalized building blocks that are coupled together without need of protection/deprotection steps has allowed rapid and easy access to homogeneous and heterogeneous dendrimers.

2. Glycodendrons

The convergent alternative for the synthesis of glycodendrimers is attractive, since early synthesized glyco-coated molecular wedges present readily available and interesting multivalent candidates for biological investigations. Obviously, by analogy to classical synthesis of globular structures, their growth can emanate from the controlled succession of suitable functionalized building blocks provided by standard AB_2 or AB_3 systems.

a. AB_2 Systems.—Aromatic AB_2 dendritic building blocks have been widely used to construct glycodendrons possessing interesting biological activities. One of their first applications was to prevent tissue infection by the pentameric bacterial toxin from *V. cholerae*. The approach for preventing binding of CT-GM_1 consisted in considering the terminal galactose residue (Fig. 9) as the anchoring fragment, to which various pharmacophores could be attached to provide optimized small-molecule antagonists against CT. This strategy was based on the fact that this galactose residue bound very specifically at a buried pocket of the receptor-binding site. However, the rest of the binding site is very shallow and lacks well-defined hydrophobic pockets that can be exploited using traditional structure-based drug design to arrive at potent inhibitors. Initial studies were made by screening commercially available galactose derivatives. Using ELISA, it resulted that *m*-nitrophenyl α-D-galactopyranoside (MNPG) was the best inhibitor identified, with an IC_{50} of 720 μM, corresponding to a 100-fold better affinity for CT than that of the "lead" galactoside.[307] In addition, a small library of antagonists showing up to 14-fold improvement as compared to the best MNPG candidate was designed. The library consisted of 3,5-disubstituted phenyl galactosides (as α/β anomeric mixtures), in which a hydrophobic ring-system was linked via a short, flexible aliphatic linker through an amide linkage at the remaining *meta* position of the MNPG core.[308]

Further improvements were obtained by using a simplified strategy based on the use of relatively simple lactose derivatives (Fig. 40).[309] The potent monovalent

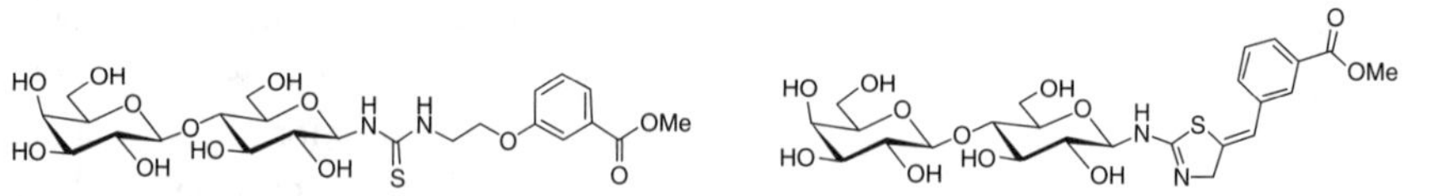

411

412

FIG. 40. Simple lactoside analogues as surrogates for the more complex GM_1 oligosaccharide ligand against the pentameric bacterial toxin from *Vibrio cholerae*.[309]

inhibitor (**411**) for CT was developed and had a K_D of 248 μM from a direct fluorescence-binding assay. This rather low value was attributed to the presence of the thiourea moiety, and the aryl group that seemed to contribute to the 70-fold binding enhancement as compared to unsubstituted lactose (K_D = 18 mM). An improvement in binding affinity was further observed by the same group with a new lactose 2-aminothiazoline conjugate (**412**), formed by a cyclization of the thiourea sulfur atom onto a triple bond and containing a more rigid spacer between the sugar and the aryl group.[310] Fluorescence studies revealed one order of magnitude enhancement in its affinity ($K_D = 23$ μM) for the CTB subunit, as compared with that of thiourea derivative.

In order to improve their respective inhibitory potencies against CT, the optimized monovalent inhibitors just described were linked to various AB_2 building blocks, thus adding the concept of multivalency. Therefore, an improvement in receptor-binding activity was expected with the use of the glycodendron approach, taking advantage of the symmetrical arrangement of five identical binding sites on the toxin B pentamer, as mentioned in earlier sections. This context allowed the synthesis of generation 1, 2, and 3 of lactosylated dendrons based on a 3,5-bis(2-aminoethoxy)benzoic acid repeating unit and containing 2 (**413**), 4 (**415**), and 8 peripheral lactoside residues (**417**), respectively, bound to the dendritic scaffold via thiourea linkages (Fig. 41).[309] Binding affinities with the CT subunit were determined by fluorescence assay (FRET) and ranged from 18 mM for lactose to 33 μM for the octavalent glycodendron **417**.

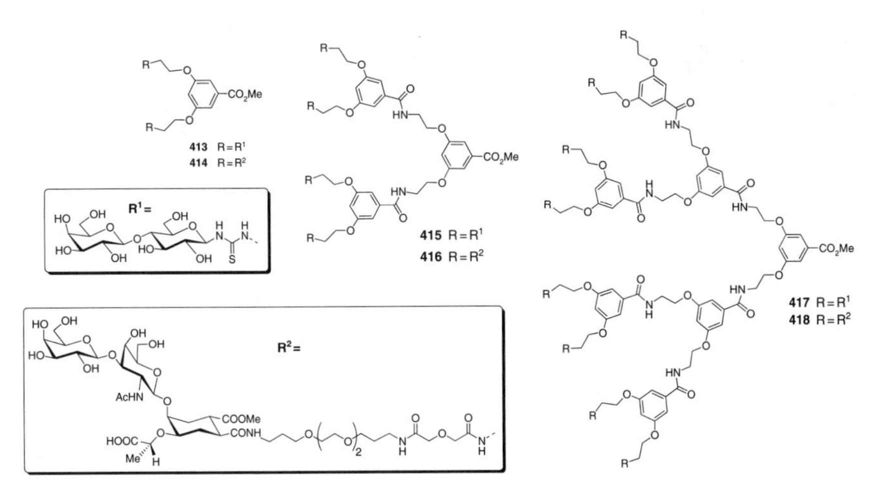

FIG. 41. Glycodendrimers bearing lactoside and Bernardi's GM_1os surrogates.[311]

This value corresponded to an eightfold enhancement as compared to corresponding monovalent lactoside derivative under the same conditions. However, an increase in the branching of the dendron (namely valency) provided only a modest increase in the potency of the ligand, corresponding to a rather constant relative potency per lactose, regardless of the number of peripheral epitopes.

To further advance the effectiveness of such glycodendrimers, the same group incorporated two modifications to the system.[311] The first alteration implicated glycodendrons having a slightly modified scaffold, outfitted with elongated arms, and peripheral attachment of the GM_1 mimics proposed earlier by Bernardi *et al.*[125] using the modified GM_1 analogue **108** (Fig. 9). Unlike GM_1, the modified **108** derivative was obtained on a gram scale and the synthetic sequence was adapted to achieve differentiation of the carboxyl group in the cyclohexanediol moiety to allow further functionalization. Using the SPR inhibition assay, the products showed good inhibition, with IC_{50} of 13 μM for the divalent **414** and 0.5 μM for the tetravalent **416**, corresponding to a 440-fold improvement over its monovalent counterpart. The octavalent analog **418** was the most potent compound, as determined using an ELISA assay.

In ongoing efforts, highly effective CT inhibitors were obtained by using similar architectures with a combination of several critical factors in their design: the use of authentic or modified GM_1 oligosaccharide sequences as the optimal monovalent ligand (**419–421**), bound to multivalent dendritic scaffolds presenting elongated spacer arms of optimal length (Fig. 42).[312]

Compound **419** and its galacto-modified analogue **421**, each bearing an azido group, can be conjugated to polyalkynic dendritic scaffold by "click chemistry." To evaluate the potency of the inhibitors **419** and **421** and their corresponding di-, tetra-, and octavalent derivatives, an ELISA assay was used which indicated unprecedented affinities and potencies, notably for octavalent **422**, which exhibited a very

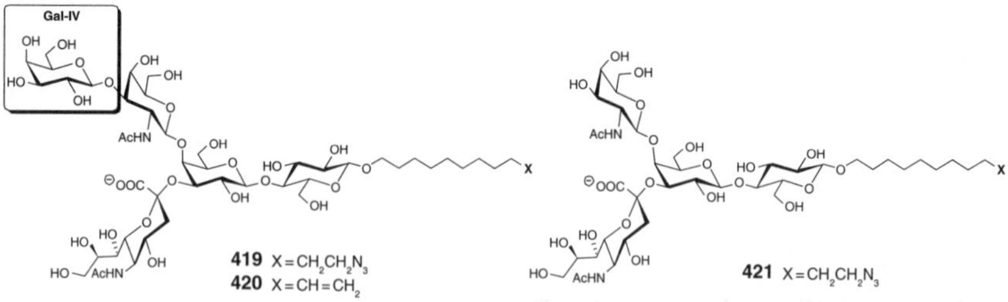

FIG. 42. GM_1 and agalacto-GM_1 ligands.[312]

low IC_{50} (around 50 pM), meaning that each GM_1os moiety bound 47,500-fold more strongly than the corresponding monovalent **419**. Di- and tetravalent GM_1os systems had IC_{50} values roughly 4–5 orders of magnitude lower than the octamer. In the agalacto dendrimer series related to **421**, the octavalent **423** was a weaker inhibitor than its tetravalent counterpart, having an IC_{50} of 0.1 μM.

Further studies concerning the activity of glycodendrons containing two or four peripheral GM_1os against the *E. coli* heat-labile toxin (LTBh) B-pentamer have been similarly described (Fig. 43).[313]

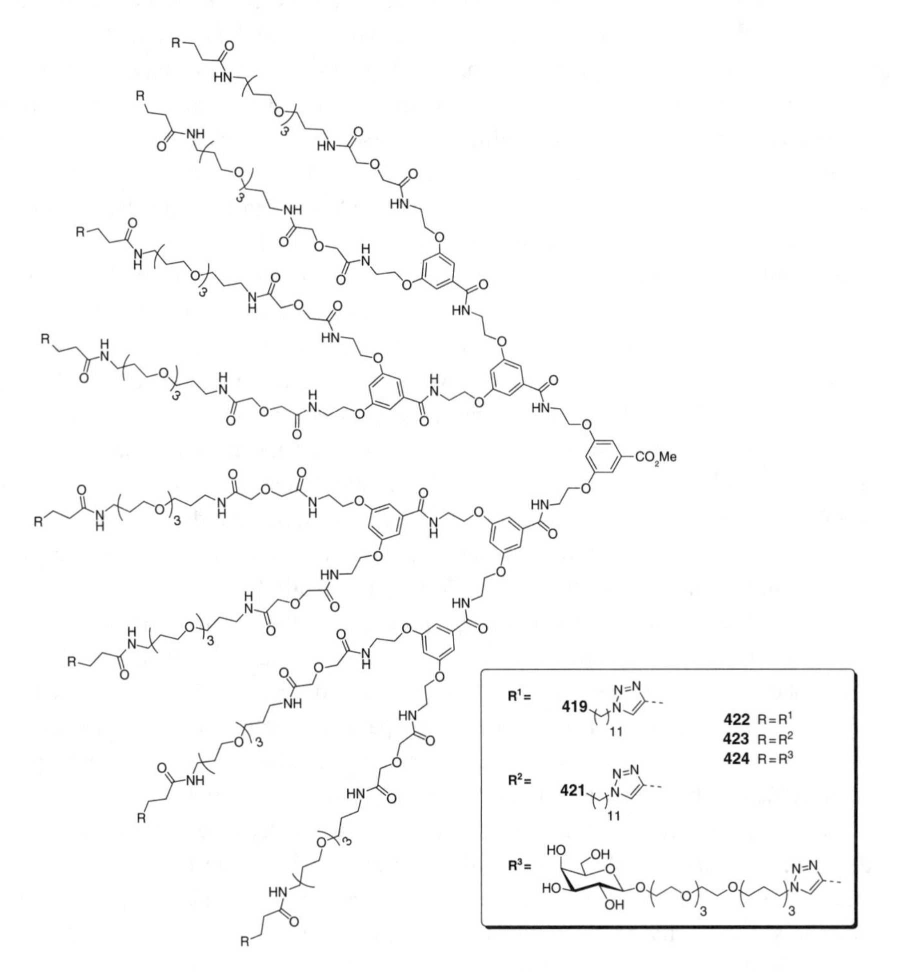

FIG. 43. Octameric glycodendrons bearing GM_1 analogues.[312–314]

Analytical ultracentrifugation and DLS have been used to demonstrate that the multivalent inhibitors induced protein aggregation and the formation of space-filling networks. This aggregation process appeared to take place when using ligands that did not match the valency of the protein receptor. Interestingly, the valency of the inhibitor had a dramatic effect on the mechanism of aggregation, influencing both the kinetics of aggregation and the stability of intermediate protein complexes. In addition, structural studies employing AFM have revealed that a divalent inhibitor induced head-to-head dimerization of the protein toxin, which either prevented the LTBh pentamer from sitting flat on the surface, or gave rise to a protein bilayer. Considering that tetravalent structures were shown to be more potent inhibitors than pentavalent analogues of similar size, the development of a strategy based on the use of mismatched valencies may provide more relevant multivalent therapeutics against pentameric bacterial toxins, thus adding to the arsenal of multivalent strategies.

To avoid the tedious preparation of GM_1 derivatives, the direct and efficient preparation of dendritic inhibitors based on a simple galactoside was also proposed by the same group.[314] Hence, a simple β-D-galactopyranoside bearing a poly(ethylene glycol) unit, crudely mimicking the other sugar rings of GM_1os and ending with a lipophilic part having a terminal azido function, was attached to keep this factor the same as in the systems just mentioned. The compounds of highest valency (**424**) and the corresponding tetravalent system showed IC_{50}s in the same range as the GM_1os derivative **422**. The multivalency effect, as expressed by the relative potency per sugar, still increased from di- to tetra-valent (923 vs. 2400), while remaining basically the same for the octavalent analogue (2500). Although results obtained with dendrimers coated with a sole galactoside residue were less spectacular than those observed with the agalacto-GM_1os derivatives, they again constituted an important step toward potent ligands against CT of low cost and ready availability.

Parallel investigations by the same group described the construction of new and rigidified multivalent structures bearing up to four lactose-2-aminothiazoline units at the periphery.[315] The synthesis of the dendritic scaffold started with aromatic diiodide **425**, which was treated with Boc-protected propargylamine under modified Sonogashira coupling conditions, to yield the branching unit **426**, fully orthogonally protected as the key building block (Scheme 44). Part of this material was exposed to a CH_2Cl_2–TFA mixture to afford quantitatively the N-deprotected compound **427**. The other part was smoothly hydrolyzed with Tesser's base to lead to carboxylic acid derivative **428**. The two fragments were then coupled under standard peptide conditions and subsequently treated with TFA to yield the tetraamino compound **428**.

Coupling di- (**427**) and tetraamines (**429**) to the lactosyl β-isothiocyanate (**430**) in the presence of iPr$_2$NEt and subsequent acid treatments to facilitate formation of the

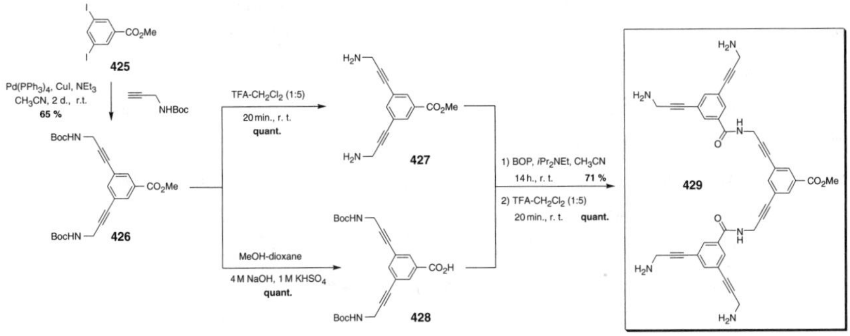

SCHEME 44. 3,5-Diaminobenzoic acid scaffolds prepared using Sonogashira chemistry.[315]

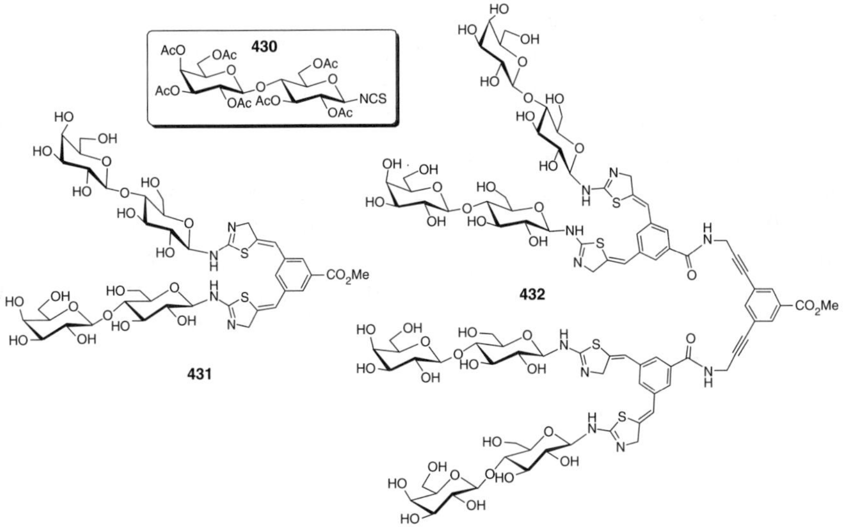

FIG. 44. Dendrons with 2- and 4-aminothiazole lactosides.[315]

aminothiazole afforded the glycoclusters in moderate yields of 65% and 33%, respectively (Fig. 44). Standard deacetylation under Zemplén conditions yielded di- (**431**) and tetra-valent dendrons (**432**).

The deprotected lactosides were evaluated as inhibitors against lectin binding in a solid-phase inhibition assay with immobilized ASF on the surface of microtiter plate wells, mimicking cell-surface presentation, while mammalian galectins-1, -3, and -5 were in solution. Strong multivalency effects and selectivity were observed for the tetravalent lactoside in the inhibition of galectin-3 binding, even better than for ASF, with an IC_{50} of 70 nM, corresponding to a 4300-fold enhancement compared to lactose (thus a factor of 1071 in the relative potency of each lactose unit). On the other hand, although rigidified glycodendrons generally generated more interesting results than the corresponding analogues **415** and **417**, no marked multivalency effects in the inherent binding affinities to galectin-3 were observed by fluorescence spectroscopy with all components in solution.

A biotinylated glycopeptide dendron, based on a dendritic L-lysine scaffold and bearing four T-antigen tumor markers [β-Gal-(1→3)-αGalNAc, T-Ag **433**] (Fig. 45), was first proposed by Baek and Roy in 2001.[316] The doubly associative binding interactions between the heterobifunctional biosensor **433** and the coating streptavidin, together with mouse T-Ag monoclonal antibody, were demonstrating using conventional solid-phase double sandwich ELISA. Hence, the virtue of the T-Ag glycodendrimer used as coating agent was a very effective anchoring motif of high avidity (nanomolar coating), and constituted an efficient cell-surface model.

Using analogous aromatic AB_2 building blocks, a modular approach leading to glycoconjugates with multiple copies of **Gb3** analogues that can induce differentiation between structurally homologous Shiga 1 (Stx1) and Shiga 2 (Stx2) toxins from complex samples has been developed.[317] To this end, divalent systems bearing **Gb3** (Fig. 17) analogues, or those of a neutralizing polysaccharide corresponding to serogroup O117 (O117 LPS, **434**), have been constructed (Fig. 46). Interestingly, O117 LPS resembles **Gb3**, but there are significant structural differences since **Gb3**

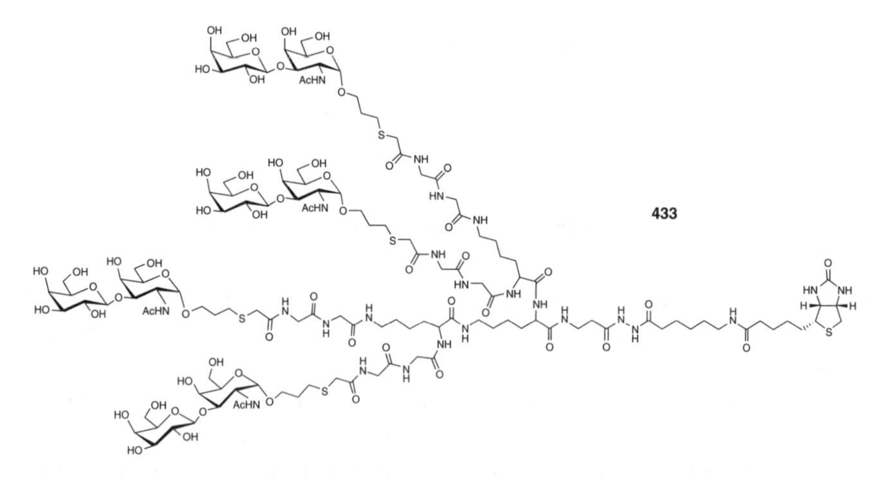

FIG. 45. Biotinylated dendritic T-Antigen synthesized by Baek and Roy.[316]

FIG. 46. Dimers of Shiga toxin analogues.[317]

has a terminal α-(1→4)-digalactoside moiety, whereas the neutralizing polysaccharide has a modified terminal digalactose moiety possessing a bulky *N*-acetyl group at each 2-position. In contrast to **Gb3**, which binds to both Stx1 and Stx2, O117 LPS was not able to bind Stx1 or to neutralize its effect on Vero cells, suggesting that the *N*-acetyl group sufficiently modifies the binding specificity toward Stx1. Tailored biantennary glycoconjugates (**435–437**) consisting of three structural components including peripheral carbohydrate-recognition components, flexible aliphatic spacers, and a biotinylated dimer have been synthesized with varied N-acetylation patterns. Biotin was used because it afforded ready access to multivalency as one streptavidin tetramer binds to four biotin molecules, and it can be conjugated to commercial streptavidin matrices for toxin capture. Molecules were designed such that the biotin and, hence, streptavidin were attached to the opposite end of the rigid scaffold to minimize interference by biotin in the binding event.

Binding of Stx1, Stx2, and Stx2c, a variant found in human clinical samples, was assessed by ELISA analysis. Results indicated that Stx2 bound to the di- and mono-*N*-acetyl-substituted galactosamine **435** and **437**, respectively, whereas Stx1 failed to bind to either compound. More precisely, **437** appeared to be a better substrate for Stx2 than **435**. In contrast and surprisingly, **436**, constructed with the **Gb3** analogue, bound exclusively to Stx1, probably due to the biantennary architecture with a short spacer that constrained binding to Stx2. Finally, the authors proved the ability of **436** to capture Stx1 in clinical applications, without any interference from a complex sample, indicating the feasibility of highly selective and sensitive synthetic glycoconjugate-based detection reagents for Stx by introducing simple manipulations in the structure of known saccharide receptors.

Using the same versatile modular synthetic strategy, the same group developed biotinylated bi- (**438**) and tetra-antennary (**439**) mannosylated glycoconjugates to capture and detect *E. coli* cells, and compared the relative capturing ability of these molecules to commercial polyclonal antibodies (Fig. 47).[318] Instead of aliphatic spacers, tetraethylene glycol linkers were used to diminish nonspecific binding and to impart flexibility for a better fit in the active sites.

Biotinylated glycoconjugates or antibodies were grafted on commercial streptavidin-coated magnetic beads and the resulting material was used to capture, isolate, and quantify bacterial recovery by using a luminescence assay. In this context, ''glyco-magnetic'' beads completely covered with glycans were incubated with two isogenic strains of *E. coli*, ORN178 and ORN208. The ORN178 *E. coli* bears numerous fimbrial adhesins (FimH) possessing binding preferences to α-mannosides. The second strain is a mutant lacking pilus expression. In initial experiments, strain ORN178 mediated the aggregation of beads coated with mannose-bearing divalent compound **438** within minutes of addition to the beads, whereas strain ORN208 did not. Bacterial aggregation has been shown to be dependent on multivalency, and these results suggested that a single bacterium could bind to multiple beads. Moreover, at higher concentrations of *E. coli*, the tetravalent conjugate **439** was responsible for an increase in capture effectiveness. The authors further compared their competence relative to standard antibody-coated beads for the capture of bacteria. The results indicated that the glycoconjugate-coated magnetic beads outperformed traditional antibody-coated magnetic beads in sensivity and selectivity when compared under identical experimental conditions. In addition, these systems could capture *E. coli* from environmental samples of stagnant water, with the possibility for targeting specific pathogenic

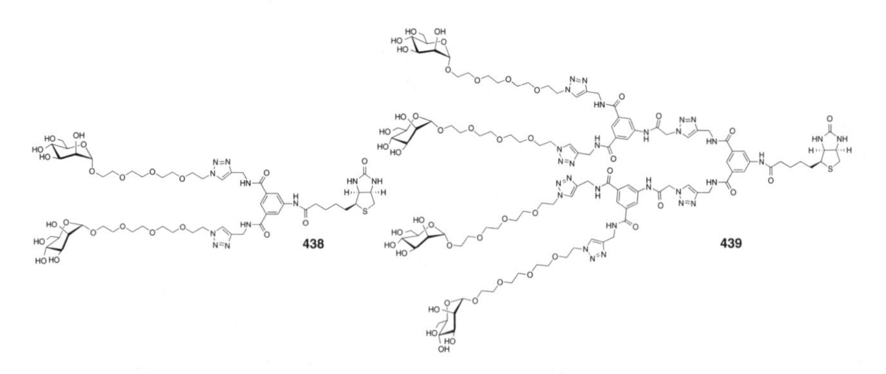

FIG. 47. Mannosylated dendrons with biotin end group for *E. coli* capture.[318]

bacteria modulating the nature of the carbohydrate recognition component. These experiments thus clearly revealed the power of glycans in biosensing, and demonstrated that these stable and inexpensive glycomagnetic beads could be used for capture and isolation of pathogens from other complex matrices.

In 2000, Saladapure and Lindhorst reported the synthesis of glycopeptide dendrons in which the peptide coupling of orthogonally protected AB_2-type carbohydrate units formed the basis for an iterative sequence leading to the various generations of glycodendrons in either a divergent or convergent manner.[319] The synthetic strategy was based on the use of efficient peptide chemistry in the linking step, avoiding sophisticated glycosylation techniques for each coupling step, and allowing possible adaptation to solid-phase chemistry (Scheme 45). The key AB_2-type glucoside building block **440** was used for the preparation of multigeneration glycopeptide dendrons from its orthogonal deprotection, to yield complementary building blocks **441** and **442**. Hence, mildly alkaline hydrolysis of the methyl ester groups led to dicarboxylic acid **441**, whereas removal of the *N*-Boc protecting group with a TFA–dimethyl sulfide mixture afforded amino-functionalized glucoside **442**. HATU-mediated

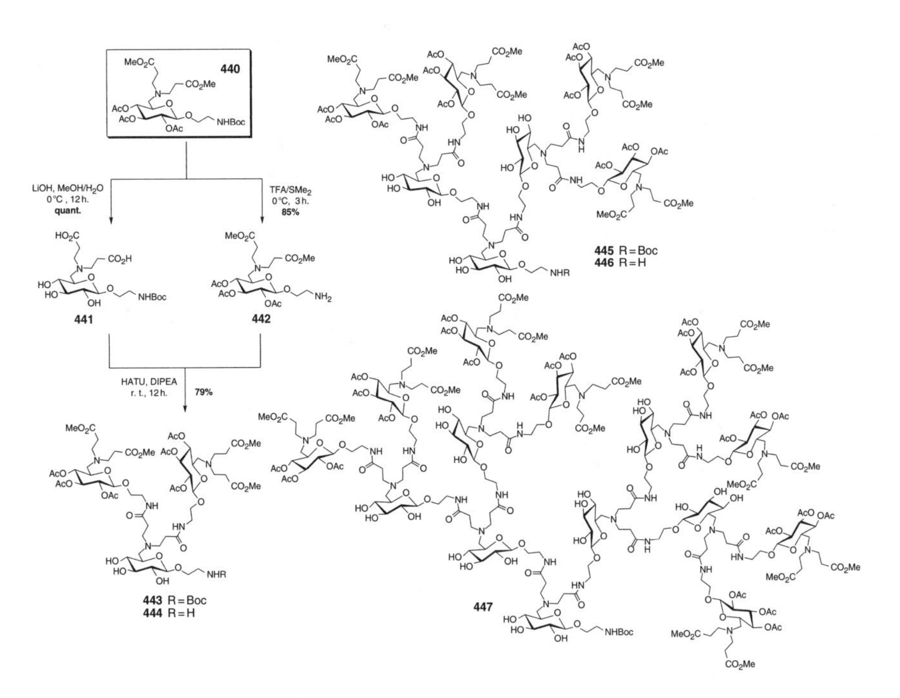

SCHEME 45. AB_2-sugar scaffold used in glycodendrimer synthesis.[319]

peptide coupling, together with DIPEA gave protected first-generation glycodendron **443** in 79% yield. Using the *N*-Boc-deprotection and peptide-coupling reaction sequence, the synthesis of higher generations of glycopeptide dendrons was carried out convergently. Hence, **443** was converted into its amine derivative **444**, whose coupling with **441** led to protected G(2) glycopeptide dendron **445** in 60% yield. The same convergent strategy was applied for the construction of the third-generation glycodendron **447** containing eight peripheral glucoside units, via peptide coupling between **446** dendron and **441**. Although no biological investigation was proposed by the authors, the stability of O-deacetylated G(1) glycodendron against β-glycosidases from almonds was evaluated and no degradation was observed over several hours.

Four years later, Nelson and Stoddart designed di- and tetra-valent lactosylated dendrons according to a convergent pathway and under mild and chemoselective conditions (Scheme 46).[320] The strategy was based on the photochemical addition of hepta-*O*-acetyl-1-thio-β-lactose (**448**) onto bisallyl trisaccharide **449** to form divalent adduct **450**. Two nearly quantitative deprotection steps consisting in standard

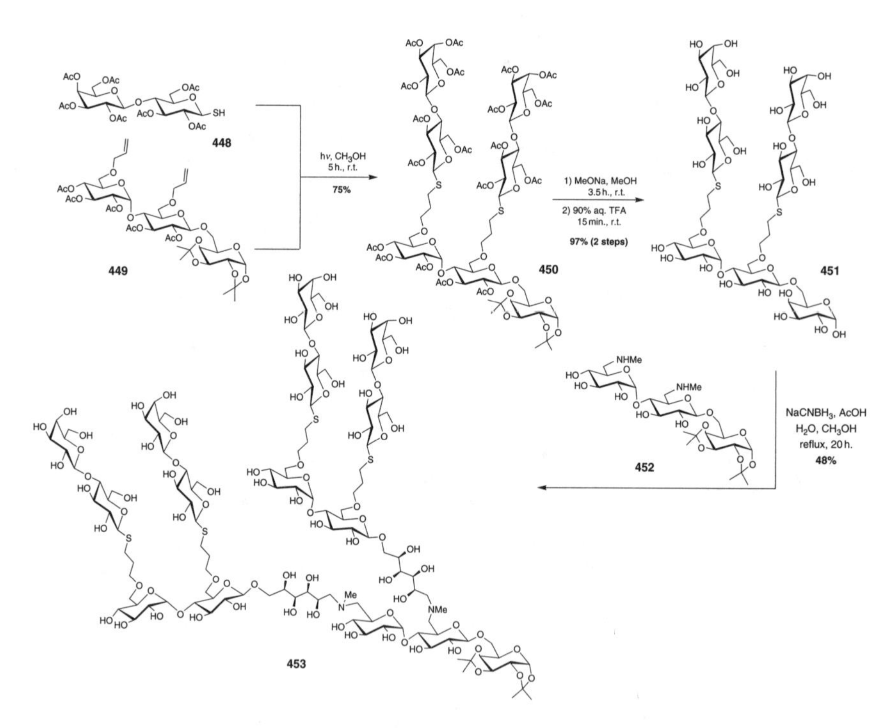

SCHEME 46. Photochemical construction of lactosylated dendrons.[320]

Zemplén O-deacetylation followed by cleavage of the acetal protecting groups with aqueous TFA afforded G(0) dendron **451**, whose reducing end was reductively aminated with bismethylamino trisaccharide **452** using cyanoborohydride in 1:1 MeOH–H_2O to furnish the G(1) dendron **435** in 48% yield.

In 2007, Heidecke and Lindhorst detailed an original approach toward glycodendron synthesis using a 3,6-diallylated precursor serving as an AB_2 system, and in which a hydroboration–oxidation sequence or radical addition of mercaptoethanol as an activating step, and subsequent glycosylation with branched or unbranched sugar trichloroacetimidates constituted key steps toward dendritic growth (Fig. 48).[321] A collection of six new hyperbranched mannosylated glycodendrons (**454**–**459**) was thereby prepared, and tested using ELISA for their potential as inhibitors of type-1 fimbriae-mediated bacterial adhesion of *E. coli* to yeast mannan polysaccharide from *Saccharomyces cerevisiae*.

The branched oligomannosides differed with regard to both their carbohydrate content and to their spacer characteristics. Binding data to *E. coli* indicated that all these glycodendrons performed better than the monovalent MeαMan, exceeding its inhibitory potency by one or two orders of magnitude. The small conjugate **454**, consisting of three α-D-mannosyl moieties, presented the weakest inhibition (IC_{50} of

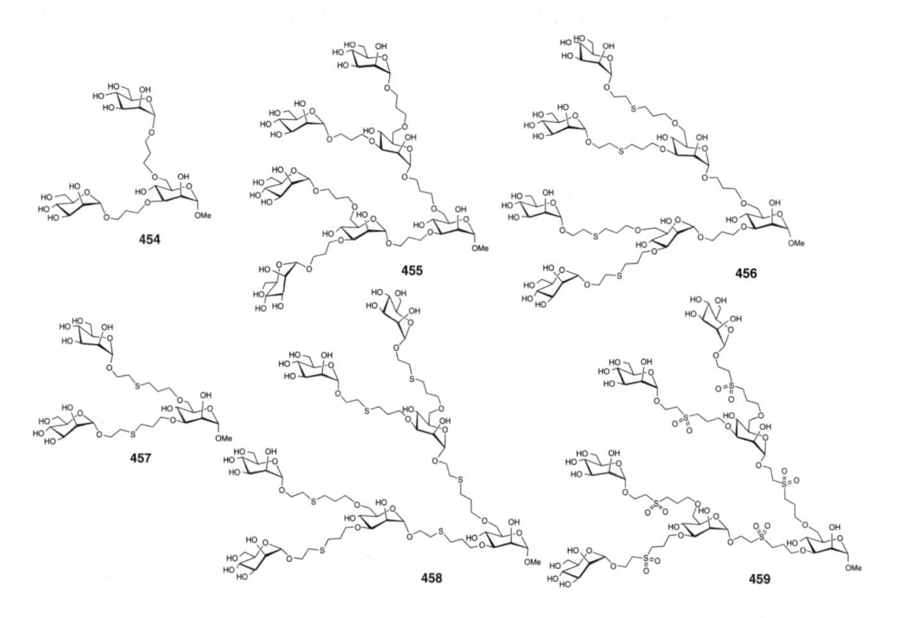

FIG. 48. Mannosylated dendrons built on an AB_2-mannoside scaffold.[321]

5.9 mmol). Its extended analogue **457**, in which two α-D-mannoside units were tethered on thiahexyl spacers, performed unexpectedly well. For instance, it showed average inhibitory potency (89 μM) exceeding that of the larger glycodendron **455** (IC_{50} of 0.14 mmol) in which the carbohydrate moieties were spacered by propyl units. Thus, with regard to the influence of the spacer characteristics, it was concluded that conformationally flexible glycodendrons containing four peripheral mannosyl units and the longer thiahexyl spacers (such as **456** and **458**, IC_{50} of 31 and 55 μmol, respectively) showed increased inhibitory potencies relative to their counterparts bearing propyl spacers (such as **455**, the shorter analogue of **456**). Furthermore, oxidation of the sulfide groups of **458**, providing sulfone **459**, had a pronounced negative effect on its inhibitory potency, indicating that the lipophilic properties of the spacers might also promote the affinity of a given glycoconjugate, as previously concluded.

Finally, a polyether AB_2 system, initially developed by Jayaraman *et al.*[322] have been used for the synthesis of di- and tetravalent clusters decorated with β-D-galactoside moieties (Scheme 47).[323] In addition, the convergent construction allowed the preparation of more complex systems consisting of ''mixed'' glycodendrons carrying both galactoside and mannoside moieties as biologically important ligands. This strategy involved a sequence of repetitive simple or double Williamson etherifications with methallyl dichloride (**461**), followed by generation of alcohols **463** and **464** via

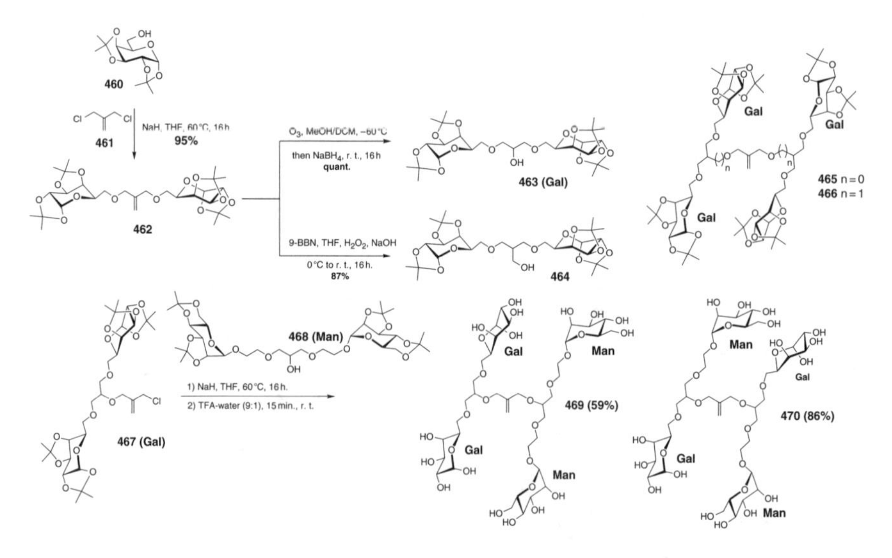

SCHEME 47. Polyether glycodendrons built from methallyl dichloride (**461**).[323]

subsequent ozonolysis–reduction or by hydroboration–oxidation of the double bond with 9-BBN transformations. To this end, double etherification of **461** with 1,2:3,4-di-*O*-isopropylidene-D-galactopyranose (**460**) under basic conditions afforded the symmetrical alkene **462** bearing two galactos-6-yl residues in 95% yield. Ozonolysis followed by reductive work-up with sodium borohydride proceeded quantitatively to yield the corresponding alcohol **463**. Next-generation glycodendron **465**, containing four peripheral galactose units was then obtained in excellent yield by Williamson etherification with **461**. Application of the hydroboration–oxidation sequence with 9-BBN, leading to **464** together with the same subsequent etherification afforded the more flexible analogue **466**. Deprotection of both glycodendrons in TFA–water proceeded in good yields within 15 min.

Toward better mimetics of highly complex natural oligosaccharides, the authors also embarked on the synthesis of "mixed" glycodendrons bearing carbohydrate moieties of different kinds, employing two different routes. The first one involved the bisgalactose-substituted alcohol **463** with an equimolar amount of **461** to allow the synthesis of chloride **467** in 67% yield. Subsequent etherification with bismannose-modified alcohol **468** furnished **469** which after acidic deprotection provided an example of a mixed-type polyether glycodendron. The second way was based on the preliminary desymmetrization of **461** with one of the saccharides, followed by attachment of the other one on the remaining reactive function. Generation of the alcohol group from the double bond and subsequent Williamson etherification on **461** afforded, after TFA–water deprotection, the second "mixed" glycodendron **470**.

c. AB_3 Systems.—Among the most widely used AB_3 dendritic building blocks, derivatives of the aliphatic TRIS and of the aromatic 3,4,5-trihydroxybenzoic acid (gallic acid) provide systems of choice for constructing dense glycodendrons according to an iterative and orthogonal synthetic strategy. Since the pioneering work of Newkome and coworkers in the early 1980s describing the synthesis of "arborols," the widespread use of TRIS and its derivatives has afforded highly functionalized structures.[287,324] TRIS thus offers synthetic advantages in terms of symmetry for ensuring an accelerated dendritic growth: the amine can serve as an anchoring function and the hydroxyl groups can allow efficient Tris-functionalization.

For instance, TRIS-based glycodendrons have recently been designed by the group of Wong, based on an efficient synthesis of structures displaying multivalent oligomannosides in high density, notably to mimic the glycans on HIV-1 gp120 (Fig. 49).[325] Their interaction with the antibody 2G12 and DC-SIGN lectin has been characterized by a glycan microarray binding assay. An AB_3 type dendritic skeleton **473** (functionalized TRIS) was chosen as a precursor for constructing

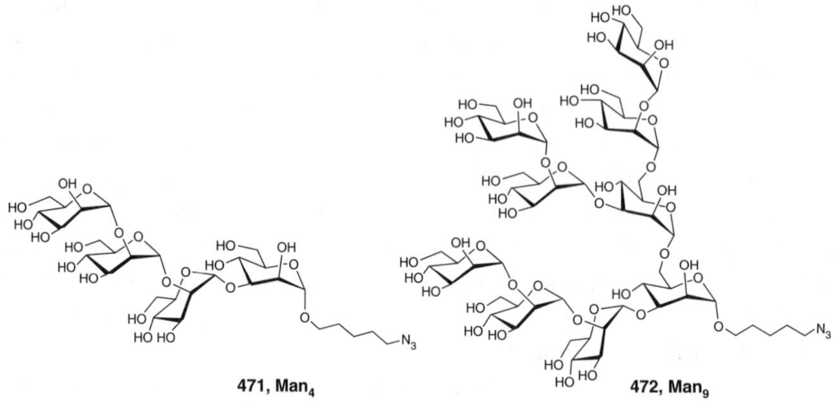

FIG. 49. Oligomannoside-ending azides used in "click-chemistry" toward HIV-1 gp120 mimetics recognized by human antibody 2G12 and DC-SIGN.[325]

densely packed glycodendrons that were achieved in a few generations. The versatile ligation was ensured by the use of catalyzed alkyne–azide 1,3-dipolar cycloaddition reaction (CuAAC) to conjugate the sterically demanding azido oligomannosides **471** and **472**, designated as Man$_4$ and Man$_9$, respectively, to the polypropargylated dendrons (Scheme 48). A convergent approach was also designed to facilitate the homogeneity of the dendritic scaffolds, using an iterative sequence based on *N*-Boc removal and subsequent EDC–HOBt-mediated peptide coupling on tricarboxylic acid derivative **473**, to afford polypropargylated compounds **475–477** for the first, second, and third generations, respectively.

Interesting results from biological studies indicated that G(2)-Man$_9$ **479** appeared to be an effective mimic of the HIV-1 gp120 surface glycan, suitable for conjugation to a carrier protein as a vaccine candidate. Furthermore, evaluation of inhibition of DC-SIGN with **479** was studied via gp120/Fc-DC-SIGN ELISA tests. Excellent inhibition activity in the nanomolar range was demonstrated, in contrast to the millimolar range from the reference mannoside. In these experiments, no inhibition was observed for the corresponding nonglycosylated alkynyl dendron **477** (up to 0.1 mM), indicating that the multivalent oligomannose residues were responsible for DC-SIGN binding. The inhibition of glycodendrons interacting with antibody 2G12 and DC-SIGN indicated that these dendritic architectures, especially for G(2)-Man$_9$ glycodendron **479**, had the potential for use in the development of both carbohydrate vaccine candidates and as antiviral agents.

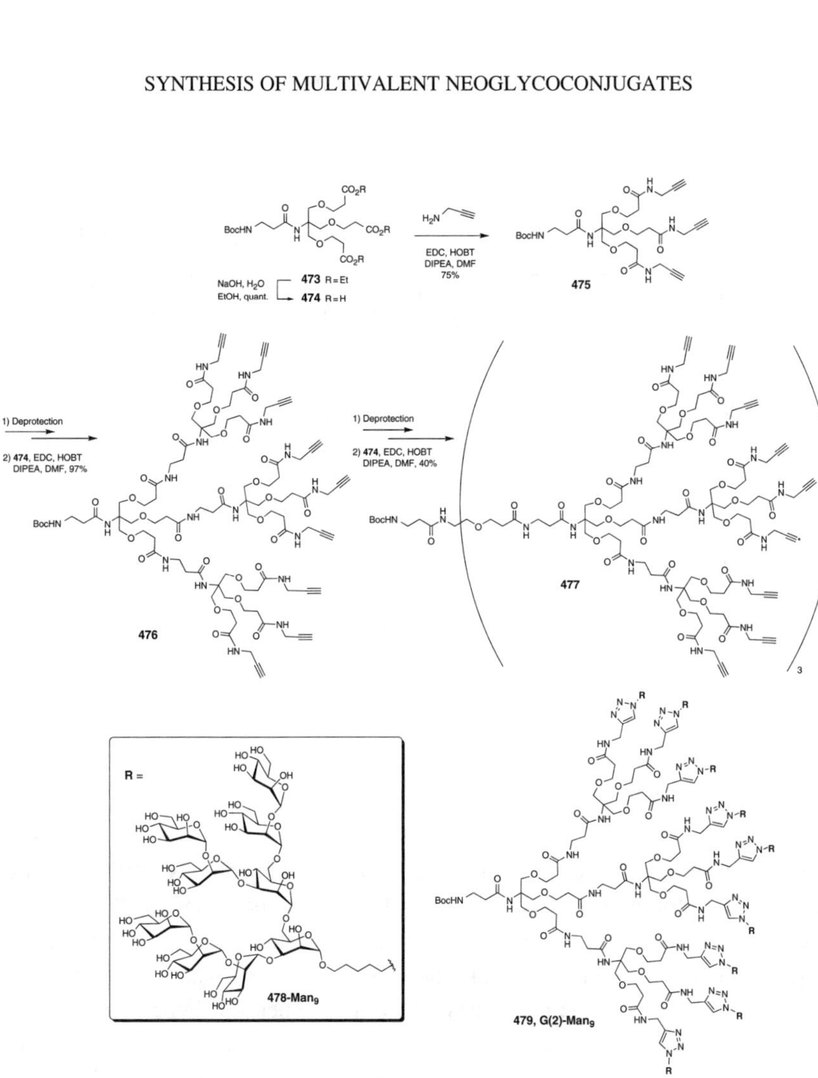

SCHEME 48. Propargylated dendrons using a modified TRIS scaffold.[325]

As with TRIS, the commercially available gallic acid **480** constitutes an ideal candidate as branching unit for rapid dendritic growth. It allows dendron and dendrimer scaffolding to reach 3^n surface groups at the nth generation. Thus, gallic acid and its derivatives can afford highly functionalized glycosylated structures, taking advantage of its geometry for orthogonal transformation. It was initially used by Roy *et al.* for the construction of hyperbranched dendritic lactosides presenting up to nine peripheral saccharide residues (Scheme 49).[326] The convergent strategy described relied on the synthesis of a thiolated lactoside derivative **482**, to be added to a

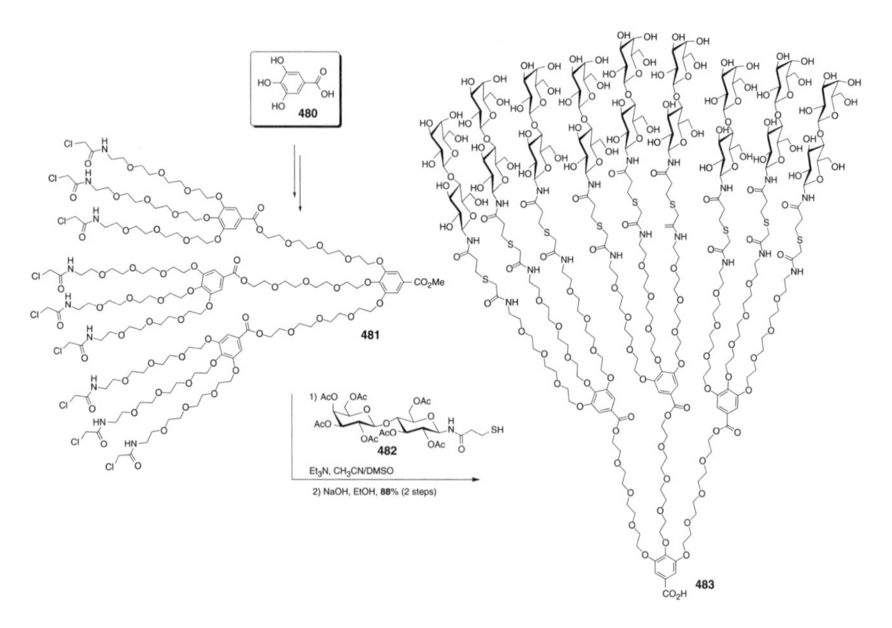

SCHEME 49. First gallic acid-based glycodendrons synthesized by Roy *et al.*[326]

preformed gallic acid derivative containing G(0) or G(1) dendrons (**481**) capped with functionalized tetra(ethylene)glycol, to afford such glycodendrons as **483**. The hydrophilic spacer was chosen to ensure advantageous water solubility of the resulting dendrimer and to counteract the hydrophobic effect of the aromatic gallic acid.

Two years later, the same group described the synthesis of hyperbranched glycodendrimers containing sialic acid residues, according to a similar strategy involving gallic acid derivatives and oligo(ethylene)glycol as dendritic backbones.[327] The foregoing conditions were used to conjugate α-thiosialosides onto an N-chloroacetylated dendritic precursor (**481** for instance) by nucleophilic substitution, affording the anticipated sialodendrimers in high yields. Interestingly, turbidimetric analysis confirmed the strong potential of G(1) sialodendrimers having nine readily accessible sialic acid residues to efficiently bind, cross-link, and precipitate two different lectins: the wheat-germ agglutinin WGA and the lectin from the slug *Limax flavus* (LFA).

Similar structures were later employed to create original dendronized polymers **485** and **486**, based on a chitosan backbone and using such sialodendrons as **484** (Fig. 50).[328] Chitosan itself is nontoxic, biodegradable, and has widespread biological activities, but major intrinsic drawbacks such as low solubility in both organic solvents and water have hampered its development as a bioactive polymer. Thus, the synthesis of water-soluble

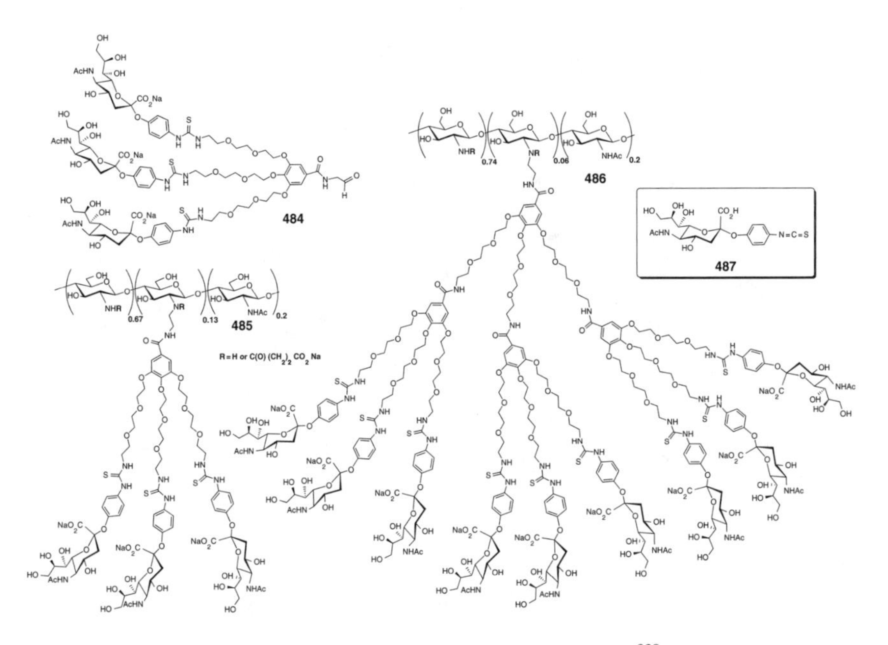

FIG. 50. Polysialic acid dendronized chitosan.[328]

dendronized chitosan–sialic acid hybrids was successfully achieved, using gallic acid as focal point and tri(ethylene)glycol as hydrophilic spacer arm, in order to investigate their potential to inhibit viral pathogens, including the flu virus.

Sialodendrons bearing a focal aldehyde end-group (**484**, for a trivalent dendron) were synthesized by a reiterative amide-bond strategy, based on the use of polyamine-ending trivalent or nonavalent dendritic scaffolds having gallic acid as the branching unit and capped with an acetal as a precursor for the aldehyde function. Sialic acid *p*-phenylisothiocyanate (PITC) derivative **487** was conjugated via thiourea linkages, followed by hydrolysis of the resulting aldehyde acetal with TFA to provide aldehyde **484**. The same procedures were followed efficiently to access the next generation of dendrimers. Finally, the focal aldehyde sialodendrons were grafted convergently onto the chitosan polysaccharide backbone by reductive amination in good yields, with the degrees of substitution indicated in Fig. 50. The water solubility of these original hybrids was further enhanced when unmodified amino groups of the chitosan backbone were succinylated with an excess of succinic anhydride.

An improved strategy using microwave-assisted synthesis, involving a gallic acid core and copper-catalyzed [3+2] cycloaddition (CuAAc), afforded a series of glycodendrons.[329] The straightforward synthesis of this series of glycodendrons was

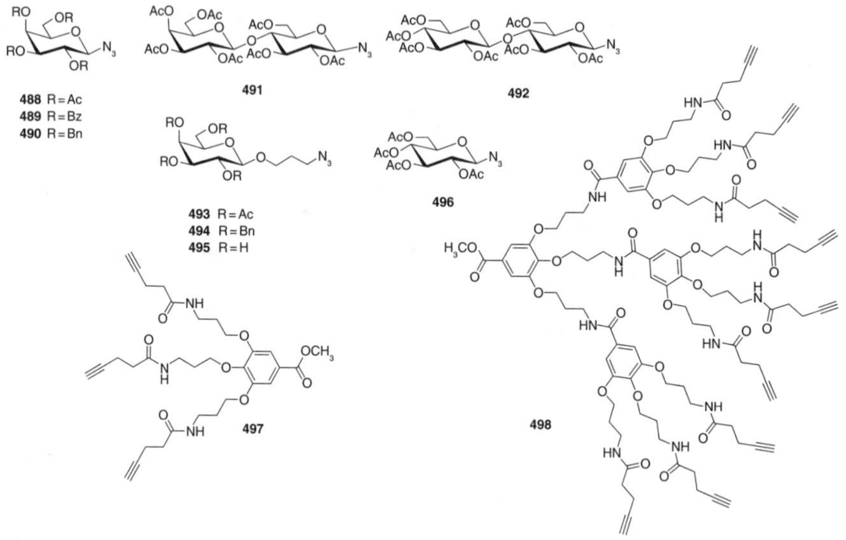

FIG. 51. Azides and propargylated dendrons for the rapid assembly of glycodendrimers.[329]

achieved in high yields, starting from azido sugar derivatives (**488–496**) and their subsequent Cu(I)-catalyzed cycloaddition with acetylene-bearing dendrimers **497** and **498** (Fig. 51). This strategy allowed the rapid preparation of triazole glycodendrimers up to the nonavalent level and the successfull use of unprotected carbohydrates. The direct introduction of unprotected carbohydrates provides an interesting approach, avoiding tedious final deprotection steps and allowing, for steric reasons, more efficient couplings.

An additional example provided by Fernandez-Megia *et al.* describes a quick, efficient, and reliable multivalent conjugation of unprotected alkyne-derived carbohydrates to three generations of azido-terminated gallic acid-triethylene glycol dendrons (Fig. 52).[330] In this work, azide-terminated dendrons were favored over those incorporating terminal alkynes because of the potential bias of the latter to Cu(II)-catalyzed intradendritic oxidative coupling. Under aqueous conditions and employing typical ''click chemistry,'' glycodendrimers containing up to 27 [G(3)] unprotected fucose (**499a**), mannose (**499b**), and lactose (**499c**) residues were efficiently isolated in high yields, after practical purification of the reaction mixture by ultrafiltration.

Further investigations by the same group led to the synthesis of three generations of a new family of block copolymers PEG-(**R**)-saccharide **500a**, **500b**, and **500c** with very good to excellent yields.[331] Interestingly, an NMR relaxation study of the

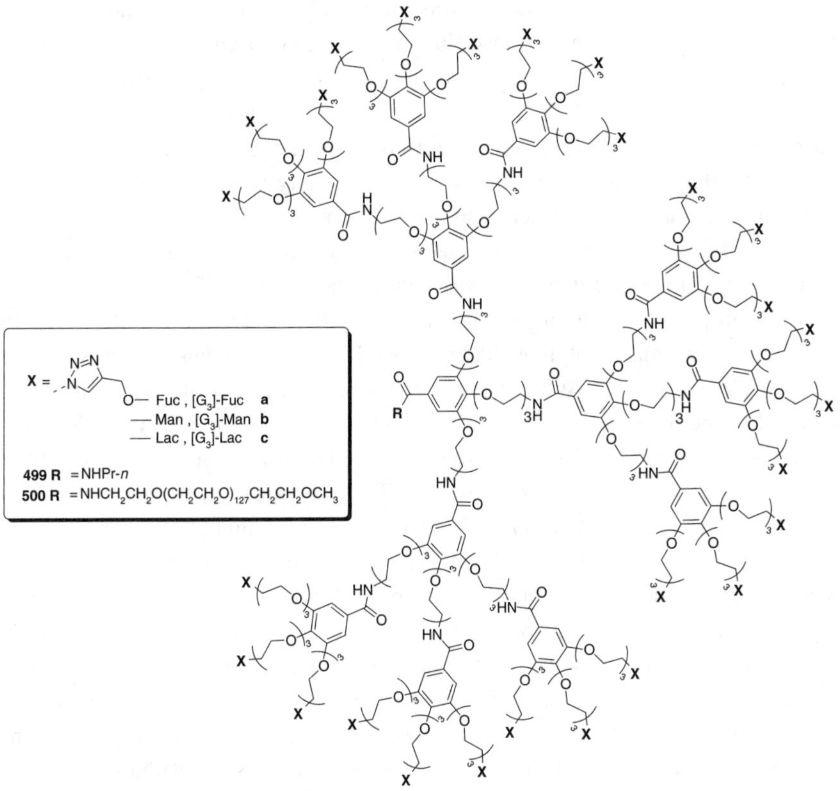

X = triazolyl-CH2O— Fuc , [G3]-Fuc a
— Man , [G3]-Man b
— Lac , [G3]-Lac c

499 R = NHPr-n
500 R = NHCH2CH2O(CH2CH2O)127CH2CH2OCH3

FIG. 52. Gallic acid-based glycodendrimers prepared using "click chemistry."[330,331]

azido-terminated PEG-dendritic block copolymer precursors revealed a radial decrease of density from the core to the periphery, becoming more intense on increasing the generation of the dendritic block. Furthermore, the resulting PEGylated and mannosylated glycodendrimers demonstrated an increased capacity to aggregate lectins with increasing generation.

3. Glycodendrimers

a. Glycopeptide Dendrimers.—As stated earlier, the first glycopeptide dendrimers were described in the literature in 1993.[292] They were built using divergent solid-phase peptide Fmoc-chemistry and L-lysine as a repeating amino acid on a

Wang resin. The initial sugar attached was sialic acid, which was introduced to confer strong inhibitory properties against flu virus hemagglutinin, a lectin-like protein recognizing α-sialosides on respiratory mucins. This early hyperbranched L-lysine scaffold was elongated with *N*-chloroacetylglycylglycine and efficiently coupled to a peracetylated α-thiosialoside. Using dendrons bearing eight surface α-thiosialosides, Roy *et al.*[332] demonstrated that each saccharide moiety was a 1000-fold better, on a per saccharide basis, than the corresponding monomer, while being stable to viral neuraminidase. These interesting architectures were, however, not as potent as sialopolymers since they could not properly cover the entire surfaces of the spherical virions, a property well exploited by random-coiled polymers.[333,334] It was later argued that dendronized polymers had improved potencies in this respect (see following PAMAM section).[335] When the same poly-L-lysine scaffold was utilized with mannoside residues bearing an optimized arylated aglycon, the resulting 8- and 16-mer glycodendrons showed 100,000-fold increased inhibitory potency against fimbriated *E. coli* K12 on a per mannoside basis.[336] The structure of the 8-mer mannosylated dendrimer (**501**) is illustrated in Fig. 53. The octameric aminophenyl α-D-mannopyranoside had an IC_{50} of 2.8 nM (22.4 nM/Man). It was found that a competitive binding assay measuring the binding of ^{125}I-labeled, highly mannosylated neoglycoprotein (BSA) to type-1 fimbriated *E. coli* (K12) strain in suspension gave much lower IC_{50} values than the equivalent values obtained by hemagglutination or in assays that involved microplate immobilization. Two key features strongly influencing the affinity to *E. coli* adhesin were: (1) the presence of an α-oriented aglycone bearing a hydrophobic group, and (2) the presence of multiple mannoside residues that can span a distance of 20 nm or longer. The two best inhibitors were a highly mannosylated neoglycoprotein with the longest linking arm between a mannose and protein amino group, and the 16-mer mannosylated dendrimer (fourth generation), with an IC_{50} of 0.9 nM (14.4 nM/Man).

First described and patented by Denkewalter *et al.* in the late 1970s,[337] dendronized lysine is one of the most widely used core structures in dendrimer[338] and glycodendrimer synthesis. An excellent series of reviews by the group of Ježek on the synthesis and biological applications of lysine-based glycodendrimers has been published.[339–342] Such glycosylated lysine dendrimers have been prepared both on solid supports[343–345] and in solution.[346–348] Reaction of the peripheral amino groups with a variety of electrophiles carrying pendant sugar residues (such as isothiocyanates, carboxylic acid, and reducing sugars) have lead to glycodendrimer thioureas, amides, and amines, respectively. Sugars bearing activated ester groups have also been coupled to dendritic lysines to provide galactoside- and *N*-acetylglucosamine-capped glycoclusters,[316,349] including the important Thomsen Friedenreich (TF) disaccharide

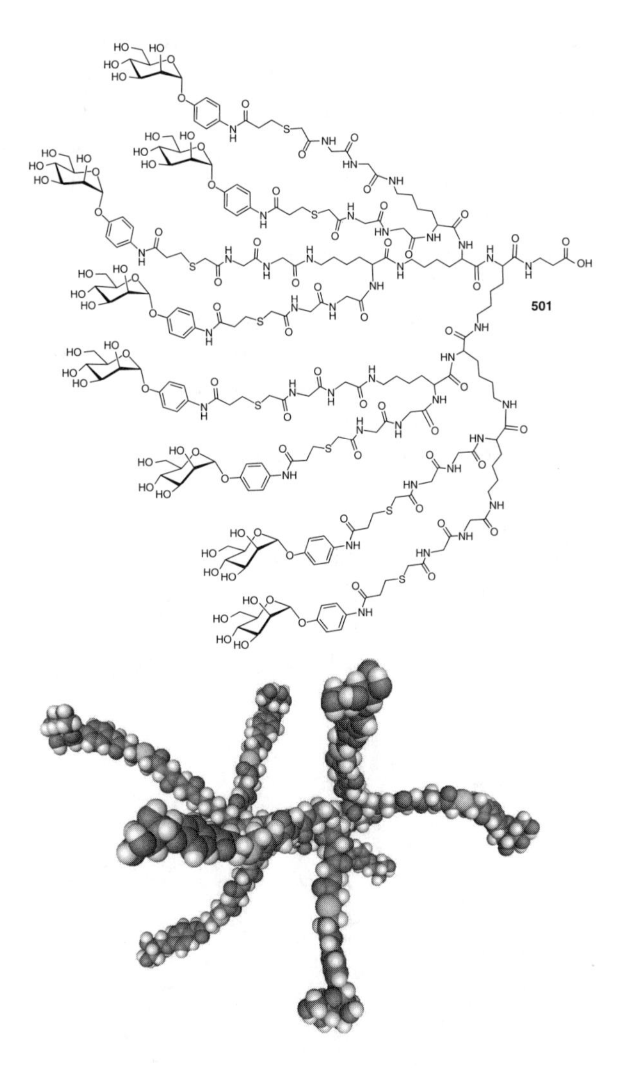

FIG. 53. Mannosylated dendron based on a poly-L-lysine scaffold. This construct leads to subnanomolar inhibitory potency against uropathogenic *E. coli*.[336] (See Color Plate 14.)

antigen [β-D-Gal(1$\rightarrow$3)-α-D-GalNAc] known as a breast cancer marker and against which monoclonal antibodies were raised (see, for instance, Fig. 45).[350,351] The corresponding TF-bearing glycodendrimers bound to the antibodies, and were shown to be adsorbed strongly onto the surface of microtiter plates. The analogous

β-D-GlcNAc glycodendrimer was also further elaborated as the Lewis^{X} tetrasaccharide antigens **502**, using multiple chemoenzymatic processes (Fig. 54).[352]

Analogous mannosylated architectures have also been proposed. MBP acting as receptors can mediate uptake and internalization of both soluble and particulate glycoconjugates, and as such they take part in innate immunity.[353,354] The broad pattern-recognition displayed by mannose receptors, together with their implication in adaptive immunity, has stimulated considerable efforts toward the selective delivery of enzymes,[355,356] drugs,[357–359] oligonucleotides or genes,[360–363] and antigens[364–366] to cells expressing them, for therapeutic and vaccine strategies. Another important mannose receptor is a membrane-associated protein restricted on sinusoidal liver cells, peripheral and bone marrow macrophages, and dendritic cells. It also recognizes and internalizes mannosylated glycoconjugates from pathological microorganisms, tumor and yeast cells, such glycoproteins as type-I procollagen, tissue-type plasminogen activator, or various lysosomal enzymes. As such, this mannose receptor contributes to the nonimmune host-defense system. In addition, the macrophage receptor is implicated in major histocompatability complex-mediated antigen (MHC) presentation by dendritic cells (DC-SIGN) (Dendritic Cell-Specific ICAM-3 grabbing nonintegrin) (ICAM-3 = intercellular adhesion molecule). DC-SIGN also belongs to the family of C-type lectins able to bind high-mannose glycoproteins of HIV-gp120,

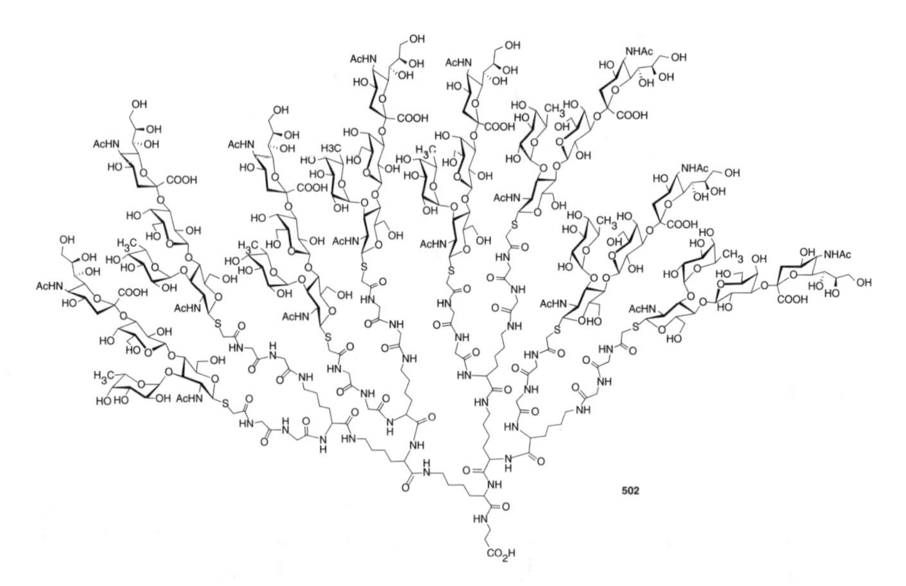

FIG. 54. Chemoenzymatically prepared sialyl Lewis^{X} glycodendrimer.[352]

Ebola-gp 1, or Dengue-gp E. DC-SIGN is also known to oligomerize, and it is therefore particularly important to understand the intrinsic binding and multivalent binding requirements of this lectin.

In this context, Biessen *et al.* conceived oligomeric linear L-lysine mannosides bearing the same arylated aglycone just described for **501**.[367] Its associated *p*-isothiocyanatophenyl α-D-mannopyranoside was coupled to the poly-L-lysine backbone through an isothiourea linkage. The structure of the 6-mer **503** is illustrated in Fig. 55. The affinity of these mannoclusters toward the mannose receptor increased steadily from 18-23 mM (dimer) to 0.5-2.6 nM (6-mer). As a consequence of its high affinity, **503** is a promising targeting device for cell-specific genes and delivery of drugs to liver endothelial cells or macrophages in bone marrow, lungs, spleen, and atherosclerotic plaques.

Another investigation targeting the mannose receptor expressed by the human dendritic cell (DC-SIGN) has been presented by Grandjean *et al.* but adding carbohydrate mimicry to the multivalent concept.[344,368] As with the strategy just described, quinic and shikimic acid derivatives used as mannose bioisosteres have been linked to dendritic L-lysine scaffolds to afford novel hyperbranched glycomimetics (Fig. 55). Fluorescein-labeled pseudoglycodendrimers with valencies of two to eight were tested by competitive-inhibition assays with mannan, which was evaluated by confocal microscopy using mannose receptors expressed in transfected COS-1-cells. Cells expressing mannose receptor-mediated uptake were assayed on monocyte-derived human dendritic cells by cytofluorometric analysis. The synthetic clusters were shown to be effective ligands against the dendritic cells, with an optimum affinity

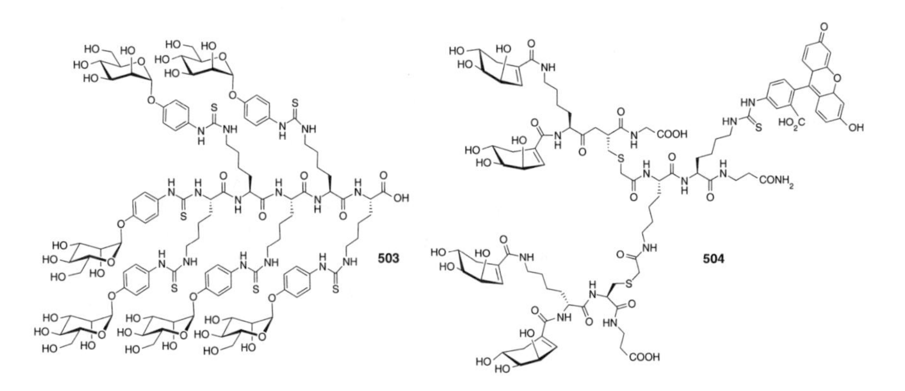

FIG. 55. Linear oligo-L-lysine bearing aryl mannosides (**503**) and the quinic acid bioisostere of mannose (**504**).[367,368]

toward clusters having a valency of four (**504**). However, the glycomimetics as well as the natural mannosides did not perform , although the results indicated that the mannose receptor could accommodate structures that diverged significantly from previously identified natural ligands and which could be further optimized using QSAR.[369,370] Additionally, monodisperse lysine dendrimers capped with 2–64 mono-, di-, and tri-α-D-mannopyranosyl residues did not induce dendritic cell maturation.[348]

As stated, DC-SIGN is a key mannoside receptor for exogenous pathogens that is used by viruses for entry into the lymph nodes. The inhibition of this process has thus been sought as an interesting strategy for blocking viral adhesion. Consequently, a series of mannosylated Boltorn® dendrimers was prepared (see later section) that were rather efficient in this respect.

An interesting extension of this strategy was proposed using various fluorescent probes: fluorescein, rhodamine, pyrene, and dansyl groups (**505**, Fig. 56). They were incorporated onto mannosylated dendrons prepared entirely by SPPS (Fmoc-chemistry) and used for the imaging studies of mannose receptor-mediated entry into dendritic cells by confocal fluorescence microscopy.[371] After pathogen capture, internalization, and digestion, peptide fragments were expressed on MHC molecules.

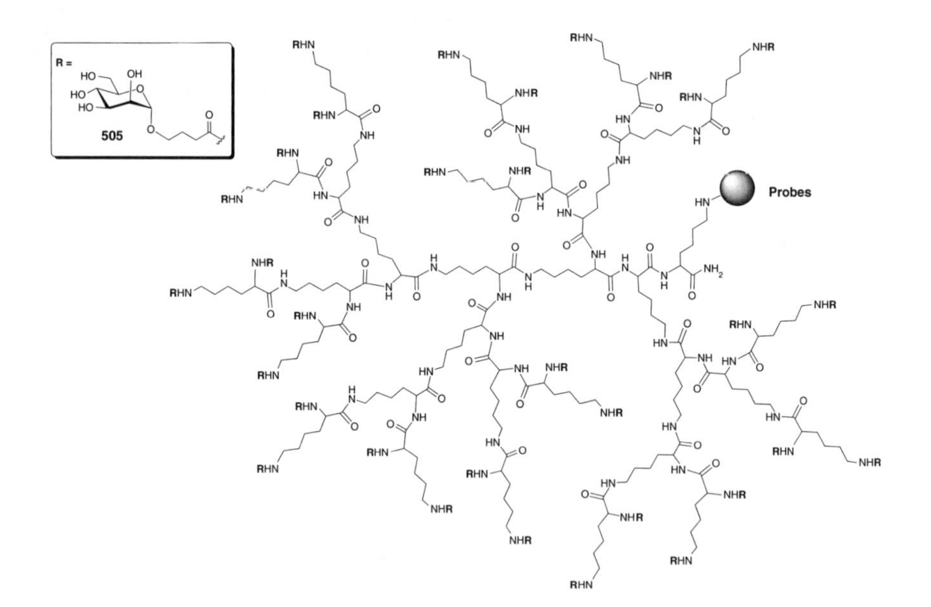

FIG. 56. Fluorescently labeled mannodendrimers for MHC capture and imaging.[371]

A T cell-specific immune response was thereby initiated. This sequence of biological events was exploited for vaccine design (see next). The heterobifunctional, high-affinity multivalent ligand was assembled on nontoxic, nonimmunogenic poly-L-lysine dendrimers to which the required number of mannosyl residues can be attached simultaneously via a 4-hydroxybutanoic acid linker, a substance occurring naturally in mammals. The synthesis of the model poly-L-lysine dendrons was initially performed on a commercial Tentagel resin preloaded with Fmoc-Sieber amide linker. The low loading level of the commercial resin (0.16 mmol/g) ensured that sufficient space was allowed for the large fourth and fifth-generation dendrons to be assembled without problems of steric hindrance. At the same time, the mild cleavage conditions required for this type of resin did not cause significant acid-promoted glycoside hydrolysis. The 4-(mannopyranosyloxy)butanoic acid was then introduced by classical peptide-coupling reagents.

The established protocol was next applied toward the fluorescently labeled analogues by introducing at the focal point a lysine residue possessing the acid-labile 4-methyltrityl (Mtt) protecting-group. The side chain of the focal ε-amine, protected by the Mtt group, was used for labeling, in conjunction with the considerably less acid-labile commercial Rink amide linker for maximum versatility. Thus, the Rink amide Tentagel resin was loaded with Fmoc-Lys(Mtt)-OH and the *N*-terminal Fmoc group was removed by classical treatment with 20% piperidine in DMF. After the reiterative insertion of bis-Fmoc lysine was terminated, the Mtt group was removed at low TFA concentrations with minimal losses of the dendron, and the newly freed amine was directly labeled onto the resin. Accordingly, treatment of the polymer-supported G(4) dendron with 3% TFA in CH_2Cl_2 led to complete removal of the Mtt group within a short time. Lower concentrations of TFA were inefficient. The free amine group of the glycodendrons was then treated with the appropriate fluorescent probes to provide desired mannosylated glycodendrons such as **505** (Fig. 56).

Spherical and hemispherical glycodendrimers containing a polylysine scaffold have been elaborated (Fig. 57).[372,373] Third-generation polylysine dendrimer **506** was prepared as a TFA salt from tris(2-ethylamino)amine (**8**) as a trivalent core and the stepwise condensation of diBoc-lysine, according to a published strategy. The *N*-Boc-protecting group was removed using standard conditions (2 M TFA) to provide the free polyamine. The polylysine dendrimer **506**, having 24 terminal amino groups on the surface, was then treated with the peracetylated cellobioside **507,** using BOP-mediated peptide coupling. Deprotection of the crude peracetylated glycodendrimer was achieved (Zemplén conditions: NaOMe, MeOH, 3 h), and the free dendritic glycopeptide **508** was obtained in 58% yield and fully characterized by ^1H NMR and MALDI-TOF MS. It was anticipated that, after random sulfation, the synthetic

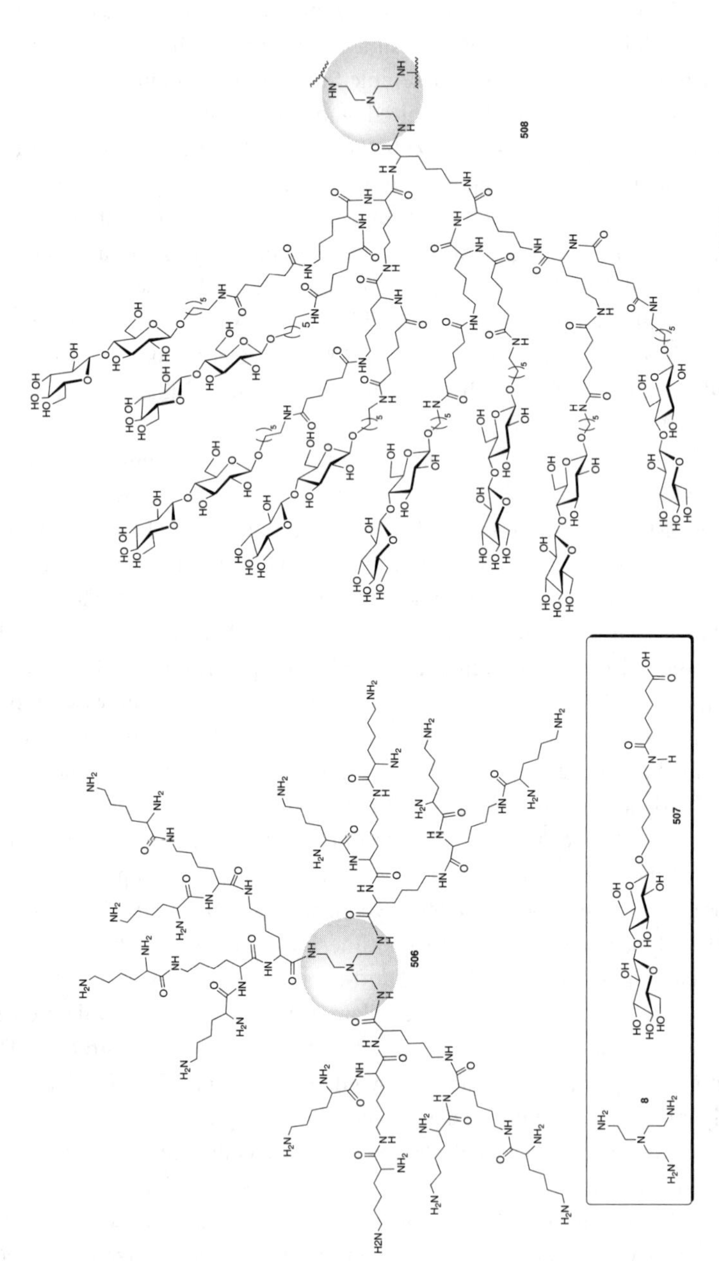

FIG. 57. Tris(2-ethylamino)amine (**8**) used as central core for the build up of poly-L-lysine cellobioside dendrimers.[372,373] (See Color Plate 15.)

cellobiosyl glycopeptide dendrimer would show key biological activities such as anti-HIV and blood anticoagulant activities.

Solid-phase combinatorial synthesis furnished a large (390,625-members) neoglycopeptide dendrimer library ending with *C*-fucoside derivatives.[374] A tetravalent dendrimer **509** (Fig. 58) showed the strongest binding affinity (IC_{50} of 0.6 μM) against the *P. aeruginosa* PA-IIL lectin, a virulence factor in cystic fibrosis (CF) patients. The optimized ligand combined multivalency with the presence of positive guanidine charges in proximity to the carbohydrate residues, and which happen to be better in comparison to a divalent analogue lacking the N-terminal lysine residues (IC_{50} of 5.0 μM). An improved 15,625-membered peptide dendrimer library was prepared analogously.[375] Dendrimer **510** (α-Fuc-CH_2CO-Lys-Pro-Leu)$_4$(Lys-Phe-Lys-Ile)$_2$Lys-His-Ile-NH_2 was the most potent ligand against the model plant lectin *Ulex europaeus* (UEA-I), with an IC_{50} of 11 μM and the bacterial lectin PA-IIL from *P. aeruginosa* (IC_{50} of 0.14 μM). Glycopeptide libraries have also been synthesized with divalent carbohydrate structures to optimize multivalent carbohydrate-binding protein interactions at subsite at the vicinity of the carbohydrate-recognition domain (CRD).[81]

As stated earlier, dendritic cells and macrophages are valuable antigen-presenting cells (APCs). These cells express MBPs and, as such, they constitute important entry mechanisms for vaccine targeting. Thus, instead of linking short peptide antigens to

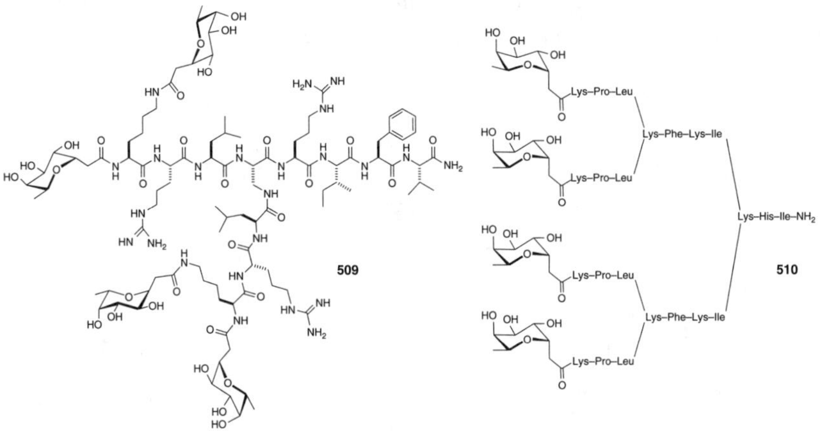

FIG. 58. Optimized *C*-fucosylated glycopeptides identified in a chemical library of 390,625 members produced by SPPS.[374,375]

classical immunogenic protein carriers, such as keyhole limpet hemocyanin (KLH) or tetanus toxoid and the like, a study was evaluated toward the possibility of using mannoside-capped polylysine glycodendrimers (**511**) constructed at the N-terminal of several immunogenic peptides (Fig. 59).[376] Peptide sequences from HIV-1 gp41 protein (541–555 bearing the LLSGIV motif capable of inhibiting viral fusion, 553-567),[377] SARS-CoV S2 (1081–1105, 1144–1187), and influenza hemagglutinin HA2 (1–25) were built on a Rink amide resin. The lysine moieties were then introduced as a G(3) lysine dendron followed by mannosylation, using 4-(mannopyranosyloxy)butanoic acid (**512**) at the terminal (ϵ) amino groups of the lysyl-peptide dendrimer (**513**). Preliminary data from the resulting vaccine candidate **511**, containing the LLSGIV motif, demonstrated that it could elicit a polyclonal antibody response in rabbit much stronger than the KLH constructs. It was concluded that the mannosylated dendron was stabilizing the peptide from proteolysis. N-Terminally mannosylated peptides carrying one to six mannose residues were also shown by Koning and coworkers to elicit an immune response, with efficiency up to 104-fold greater than peptide antigens alone.[378]

The capture of antigenic determinants by APCs that will ultimately appear in MHCs for T helper cell stimulation depends on several factors. If APCs are represented by B cells, multivalency becomes an issue, because both the recognition and the binding events are triggered by cell-surface immunoglobulins. A classical example of successful applications in this regard is presented by the high immunogenicity of bacterial capsular polysaccharides (CPS) which, on their own, can elicit protective antibodies.[379,380] However, the resulting antibodies are usually limited to low-affinity IgMs, and the immune responses lack memory effects because of the absence of T

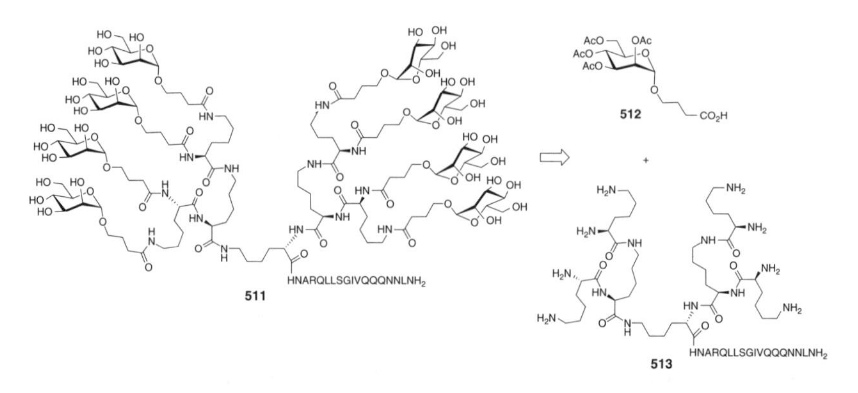

FIG. 59. Entirely synthetic anti-HIV vaccine candidate capable of APCs uptake.[376,377]

cell-dependent antigens normally associated to peptides. CPS are thus said to be T independent antigens. Alternatively, if the B cell antigens are not multivalent, such as those found in small peptides or oligosaccharides, they will greatly benefit from being mounted on dendritic scaffolds. Moreover, the key T cell epitopes, needed in the context of MHC complexes, can be typically short 15-mer peptides and yet, be too small for efficient uptake by APCs. Consequently, the design of dendritic glycoconjugates bearing multiples copies of both oligosaccharides (B cell epitopes) and short peptides (T cell epitopes) has been proposed as entirely synthetic vaccine candidates (**514**) (Fig. 60).

A few elegant applications of this principle have been recently proposed for triggering neutralizing antibodies against the V-3 loop glycoprotein of gp120 from HIV-1 isolates that recognize and bind to the high mannose oligosaccharide epitope $Man_9GlcNAc_2$ **183** (Fig. 24) that is present in numerous copies on gp120. The proposal was based on the observation that a swapped human antibody, named 2G12, was identified in a patient that successfully mounted a protective anti-HIV immune response (Fig. 61).[381]

Besides glycoclusters based on a central peptide, earlier presented in Section II.4, the group of Kunz was the first to fully demonstrate the feasibility and efficacy of eliciting complete immune responses with memory effects, using carbohydrate cancer antigens built on multiple antigen glycopeptide scaffolds.[382,383] Thus, the immunogenicity of synthetic multiple antigenic glycopeptides (MAGs) displaying four clustered T_N-epitopes anchored to an oligomeric branched lysine core was examined (Fig. 62). Conventional SPS of the peptidic structure was performed by Fmoc-methodology on Wang resin. The dendritic immunogen **516**, containing clusters of three consecutive T_N-epitopes linked to the poliovirus T cell epitope KLFAVWKITYKDT, was tested in mice as both a prophylactic and as a therapeutic cancer vaccine. As the analogous MAG **515**, harboring only four monomeric T_N-antigens that have been shown to increase the survival of tumor-bearing mice, added results indicated that **516** showed high immunogenicity and good protection against the development of T_N-expressing tumor cells.[382]

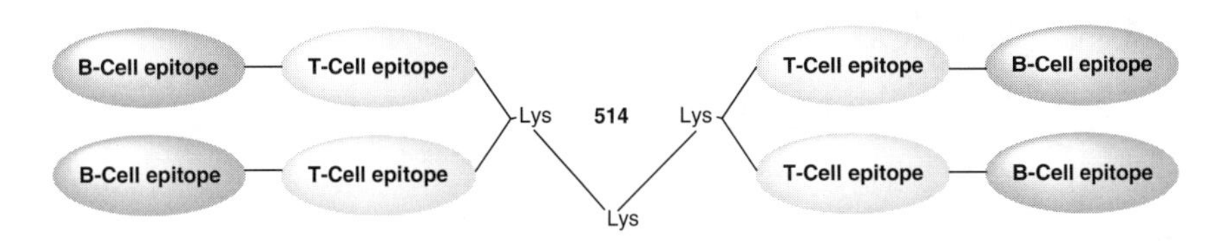

FIG. 60. Schematic representation of a fully synthetic carbohydrate-based vaccine.[379,380] (See Color Plate 16.)

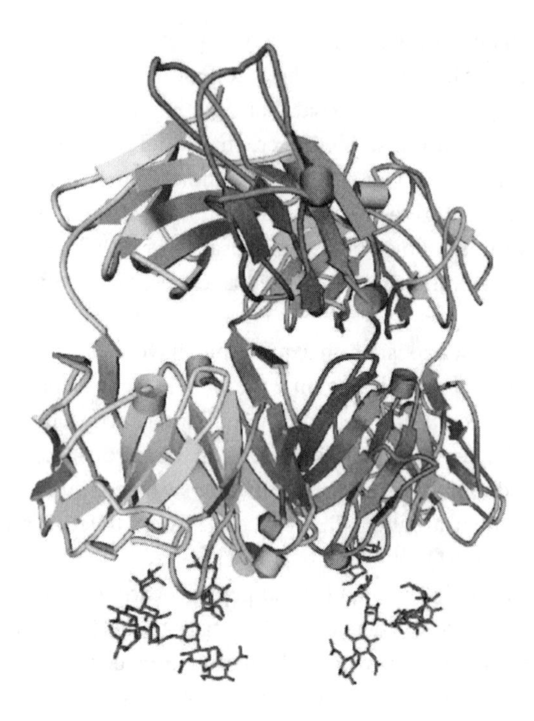

FIG. 61. Crystal structure of the HIV-1 neutralizing human antibody 2G12 bound to the oligomannoside $Man_9GlcNAc_2$ present on the "silent" face of the gp120 envelope glycoprotein (PDB 1OP5). (See Color Plate 17.)

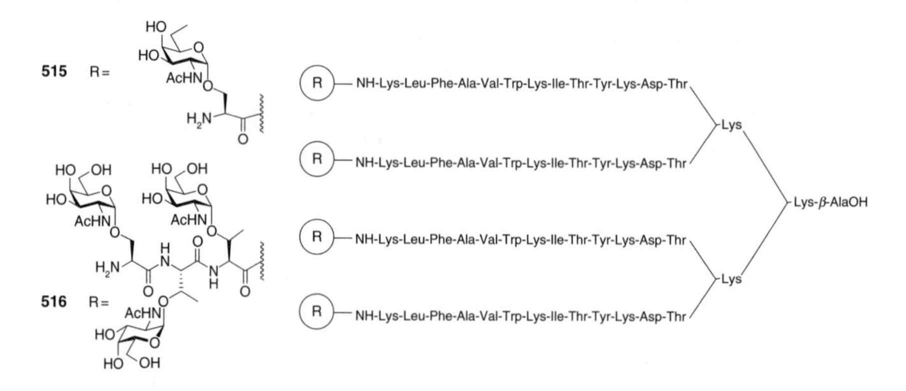

FIG. 62. Poliovirus T cell epitopes capped with monomeric (**515**) and trimeric T_N-antigens (**516**) that confered high murine immunogenicity.[382,383]

The N-linked pentasaccharide core $Man_3(GlcNAc)_2$ glycopeptide bearing the extracellular MMP inducer sequence **517** (emmprin 34–58) has been successfully linked to G(1) PAMAM aminodendrimer by thioester activation (Fig. 63). The resulting octameric 30 kDa construct was obtained in low yield, but was purified by preparative electrophoresis and fully characterized by MALDI-TOF mass spectrometry. The multivalent architecture was built to evaluate the requirement of emmprin multimerization for inducing MMP expression.[384]

The numerous successes achieved in the solid-phase syntheses (SPS) of glycopeptides dendrimers, coupled to their ease of preparation and purification, has also triggered chemists to apply the SPS nucleotide chemistry toward the synthesis of phosphodiester-linked glycodendrimers. As mentioned earlier in the glycoclusters section, the chemistry and geometry of phosphodiesters and triesters are obviously appealing factors for their use as glycodendrimer scaffolds. The following strategy was based on the synthesis of key phosphoramidite building blocks bearing either *N*-chloroacetyl or alcohol end groups for further branching.[385] DEG spacers were initially transformed into the mono *tert*-butyldiphenylsilyl derivative **518** which upon treatment with *N*,*N*-diisopropylphosphoramidous dichloride (Cl_2P-N-$(iPr)_2$, DIPEA, CH_2Cl_2), provided the essential building block *N*,*N*-diisopropyldiphosphoramidite (**519**) in 84% yield (Fig. 64). Alternatively, the mono *n*-chloroacetylated DEG derivative **520**, prepared from commercially available 2-(2-aminoethoxy)ethanol ($ClCH_2CO_2H$, EEDQ, 45 °C, 4 h, 82%) was similarly treated to provide phosphoramidite **521** (68%). Further couplings of **519** or **521** with alcohols **518** or **520** in the presence of 1*H*-tetrazole gave the corresponding phosphotriesters **522** or **523** after oxidation with *t*-butyl peroxide (70–71%). Coupling tris(*N*-chloroacetyl) derivative **524** (MeOH, Et_3N) with the thiolated *N*-acetylgalactosaminide **525** afforded a first-generation phosphoglycodendrimer (63%) having three α-D-GalNAc moieties (not

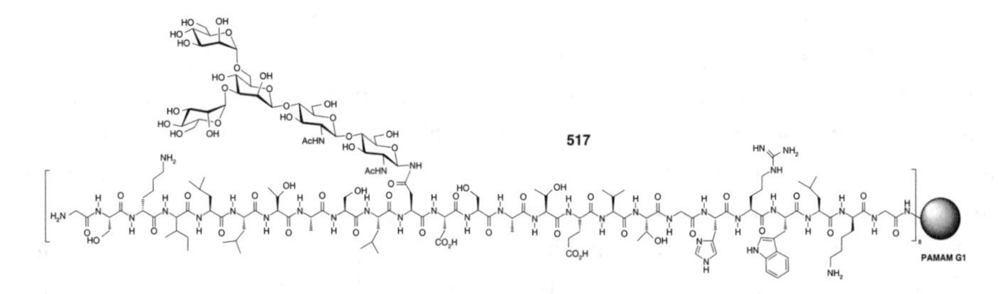

FIG. 63. $Man_3GlcNAc_2$ anchored to PAMAM-based matrix metalloproteinase (MMP) inducer sequence (emmprin 34-58).[384]

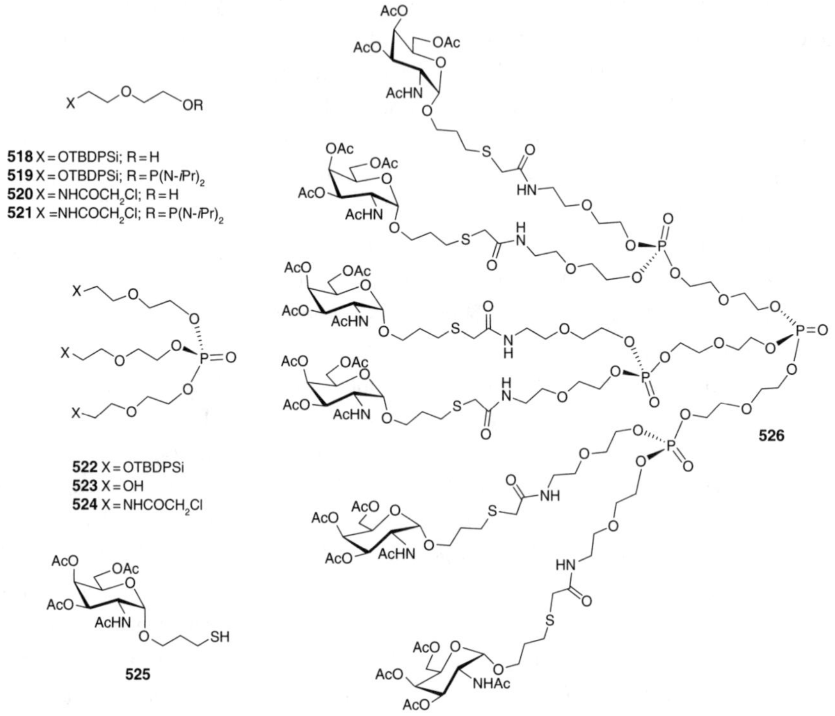

FIG. 64. An α-D-GalNAc phosphotriester dendron built using phosphoramidite chemistry (Kratzer and Roy, unpublished data).[385]

shown). Thiol **525** was prepared from allyl *N*-acetyl-α-D-galactosaminide by reaction with thioacetic acid (HSAc, MeOH, AIBN, reflux, 10 h, 77%) followed by Zemplén S-deacetylation.

tert-Butyldiphenylsilyl-protected phosphotriester **522** was deprotected by fluoride anion (Bu_4NF, THF, 78%) to give a triol **523**, which was further branched with building block **521** as before to provide a second-generation phosphotriester (57%) having six N-chloroacetyl residues after oxidation of the phosphite. Nucleophilic displacement of the hexavalent N-chloroacetylated precursor by **525** as before, provided hexavalent dendrimer **526** (52%). Further processing of the synthetic sequence, coupled with tethering strategies with various spacers, gave access to a family of dendrimers having many different valencies. All of the GalNAc-bearing phosphodendrimers thus produced were tested as inhibitors of the plant lectin *V. villosa* binding to asialoglycophorin, a natural ligand for hepatocytes receptors. The results indicated a 3–10-fold (hexamer) enhanced affinity over the monovalent ligand, thus supporting once again the glycoside cluster effect. Interestingly, sugar **525** is part of a key determinant known as the TF-antigen overlay expressed on breast cancer tissues and other melanoma cancers. Its structure and higher oligosaccharide homologues have formed the basis for the dendritic glycopeptides vaccines already described.

Additional examples of dendritic phosphodiesters have been highlighted by Dubber and Fréchet with mannoside and galactoside oligonucleotide conjugates, using a DNA synthesizer (Fig. 65).[386] The SPS of multivalent glycoconjugates enables the custom tailoring within hours of their valency and structural requirements toward the biological targets, as opposed to more cumbersome traditional approaches. A tetrameric mannoside (**530**) was thus prepared from simple precursors **527–529**, using phosphoramidite chemistry. A glycodendron was built with a fluorescein probe and another one with a thiolated oligonucleotide coded for the antisense inhibition of inducible nitric oxide synthase. Once again, the constructs, prepared on a 1 μmol scale, were satisfactorily characterized by standard ^{1}H NMR and MALDI-TOF MS.

b. Commercial Dendritic Scaffolds.— *(i) PAMAM Dendrimers.* Several dendrimer scaffolds having various surface functionalities and valencies, constructed from diverse building blocks and according to both general synthetic strategies (Fig. 66, see also Section V.1.b) are commercially available. Some of the most commonly used

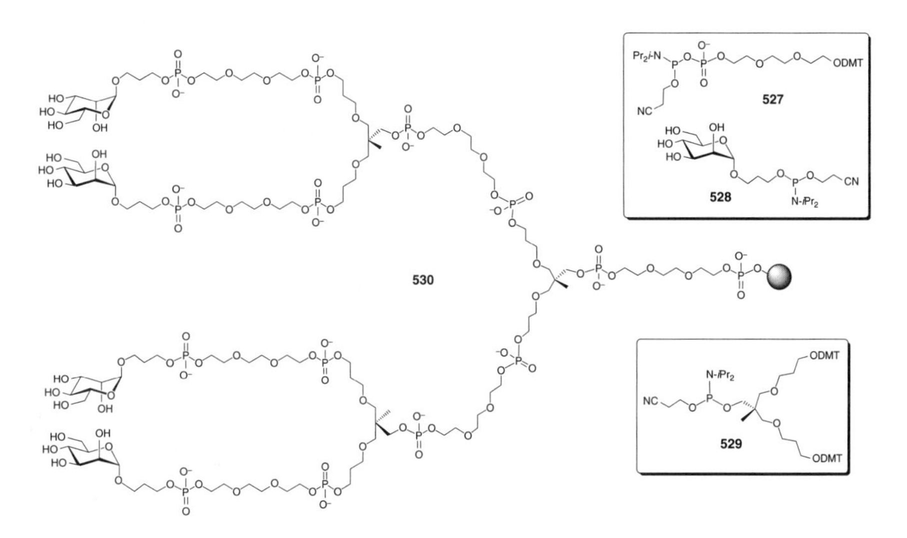

FIG. 65. Solid-phase synthesis of mannosylated dendrons for antisense gene delivery.[386]

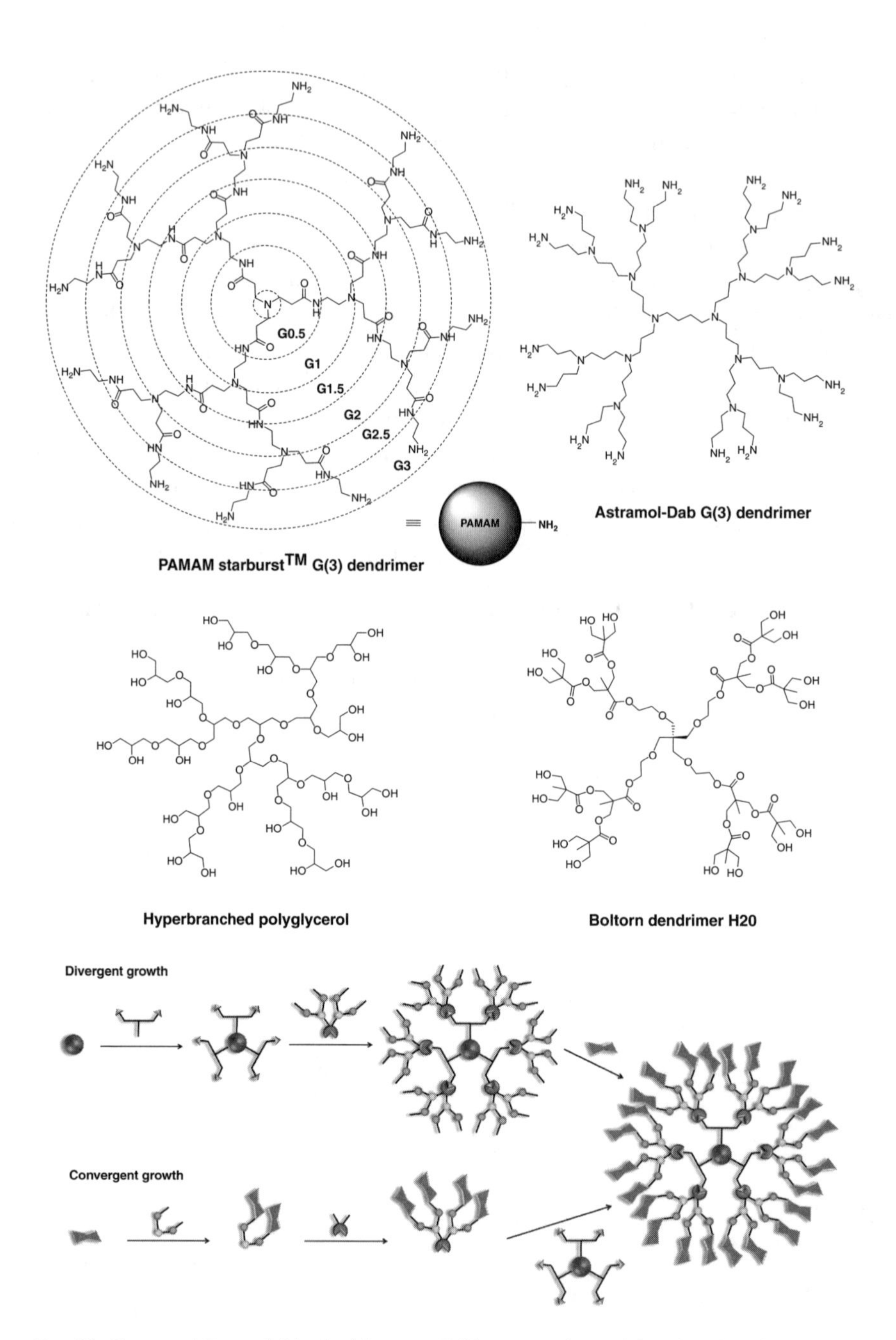

FIG. 66. Commercially available dendrimer scaffolds commonly used for glycodendrimer syntheses, and general synthetic strategies for synthesis of glycodendrimers. (See Color Plate 18.)

ones are illustrated in Fig. 66. Among these readily available and widely used dendritic scaffolds, poly(amidoamine) dendrimers (PAMAM, Dendritic Nanotechnologies) constitute candidates of choice that have been extensively exploited by many groups of glycochemists, notably for the rapid synthesis of highly branched glycodendrimers. Since the pioneering synthesis of poly(amidoamine) dendrimers proposed by Tomalia *et al.* using a divergent growth procedure, these attractive molecules have constituted an exciting new class of macromolecular and highly branched architectures with well-defined size, shape, and geometry. They have drawn much interest in several research areas.[284,288]

Accordingly, PAMAM-based dendrimers, having built-in surface amine functionalities, have been the first and most frequently used scaffolds for attachment of sugars. The very first example of saccharide-substituted PAMAM dendrimers was proposed by Okada and coworkers, who described the synthesis of ''sugar balls'' **532** via amide bond formation, starting from sugar lactones (**531**).[387] Although this process is a straightforward manipulation, it suffered from the disadvantage of sacrificing the reducing sugars, which alternatively served as extended linkers (Scheme 50).

Several other strategies have been used to adequately functionalize PAMAM dendrimers with carbohydrates, involving: (a) introduction of thiourea linkages formed by treating amino dendrimers with isothiocyanated saccharide derivatives, (b) direct amide linkages with sugar-bearing carboxylated or activated ester derivatives, (c) reductive amination, or (d) incorporation of chloro- or bromo-acetamido groups onto PAMAM dendrimers or saccaharides, to afford highly electrophilic species that can for instance react with thio or amino derivatives. All of these synthetic approaches are illustrated in the following section.

Thiourea Linkages. Attachment of saccharide units to the surface of PAMAM through thiourea linkages offers one of the most efficient ways to develop multivalent ligands quickly and efficiently for the study of protein–carbohydrate interactions.

SCHEME 50. ''Sugar-balls'' obtained by direct ring-opening of sugar lactones by PAMAM-ending polyamines.[387]

Historically, Lindhorst and Kieburg first published the simple and efficient coupling of different low-generation polyamine scaffolds, including PAMAM, to β-D-gluco-, α- and β-D-manno-, β-D-galacto-, β-cellobio-, and β-lacto-configured glycosyl isothiocyanates.[388] An improved version, using a hydrophobic aryl aglycone, later permitted the syntheses of the first four generations of monodispersed neoglycoconjugates having up to 32 mannoside units (**534**). *p*-Isothiocyanatophenyl α-D-mannopyranoside **533**[389] was used as key precursor (Scheme 51). The resulting glycodendrimers were evaluated as ligands for the phytohemagglutinins from Con A and *Pisum sativum* (pea lectin), using ELLA and turbidimetric analyses. The relative binding data indicated that incorporation of terminal α-substituted mannoside residues furnished glycodendrimers showing an up to 400-fold increase in binding capacities. Moreover, their ability to bind and form insoluble carbohydrate–lectin complexes was also demonstrated by radial double immunodiffusion and turbidimetric assays, and their capacity to precipitate selectively their homologous protein receptors from crude lectin mixtures made them convenient tools for the rapid and simple isolation of proteins.[390] The same procedure was successfully adapted by the same group a few years later, allowing the introduction of *p*-isothiocyanatophenyl β-D-lactoside onto PAMAM scaffolds to study their relative binding behavior toward the family of galectins, the influence of the generation, and the binding-site orientation of receptors.[391]

In 1997, a similar but simplified procedure, involving aqueous solutions that and avoiding protecting groups, for the synthesis, from methylthiourylene α-D-mannopyranoside or the aromatic derivative, of a G(2) PAMAM dendrimer containing up to six peripheral mannosyl residues was proposed.[392,393] Parallel investigations by Thompson and Schengrund involved the synthesis of glycodendrimers from poly(propylene)-imine and Starburst® (PAMAM) dendritic scaffolds, and provided potent inhibitors of CT.[394] They contained four to eight peripheral oligo-GM1 [βGal-(1→3)-βGalNAc-(1→4)-αNeu5Ac-(2→3)-βGal-(1→4)-βGlc-(1→1)Cer] group with IC_{50} of 14 nM against CT, that were covalently attached to the central core via thiourea linkages,

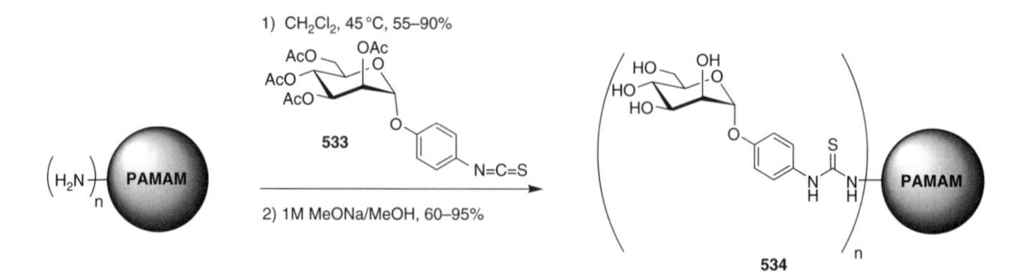

SCHEME 51. PAMAM-based mannosylated dendrimers incorporating the key aryl aglycone.[389]

using the PITC derivative of oligo-GM1. The IC_{50}s values were determined for the oligo-GM_1-PITC dendrimers, native GM_1, and the oligosaccharide moiety of GM_1 (oligo-GM_1), and studies revealed that the glycodendrimers inhibited binding of the CT to GM_1-coated wells at molar concentrations 5- to 15-fold lower than native GM_1 and more than 1000-fold lower than that of the free oligosaccharide.

G(2) and G(3) sialic acid-containing PAMAM glycoconjugates with *p*-isothiocyanatophenyl α-sialoside (**487**) (Fig. 50) as precursor have also been described by Zanini and Roy, and their lectin-binding properties evaluated.[395] Preliminary turbidimetric studies with G(2) and G(3) α-thiosialodendrimers, presenting 16 and 32 peripheral sugar units, respectively, demonstrated their ability to bind the slug lectin from LFA, showing well-organized precipitation patterns. Furthermore, their relative efficiency to inhibit the binding of HRPO-labeled LFA to human AGP (orosomucoid) was determined by a competitive ELLA. Interestingly, an increase in multivalency resulted in a steady increase of inhibitory potential, with IC_{50} values in the nanomolar range (2.89 and 1.13 for G(2) and G(3), respectively) corresponding to an ~210-fold jump in inhibitory potential over the monomeric analogue. Subsequently, Baker and coworkers evaluated similar sialic acid-conjugated G(4) PAMAM structures as a mean of preventing adhesion of three influenza A virus subtypes, and furnished one of the first documentations of the function of dendrimer conjugates as antiinfective agents *in vivo*.[396] HAI *in vitro* showed the glycodendrimers to inhibit some specific influenza subtype strains at concentrations 32–170 times lower than those of sialic acid monomers. The *in vivo* studies also demonstrated the ability of the glycodendrimers to protect against experimental infection by influenza A X-31 H3N2 virus in mice.

In 2001, Woller and Cloninger described the largest glycodendrimer then built, prepared on generation G(6)-PAMAM dendrimer and containing 256 mannoside residues.[397] When compared to methyl α-D-mannoside taken as the monomeric control, dendrimers G(1) and G(2) did not show any increase in activity toward the phytohemagglutinin Con A, and the G(3) dendrimer bound roughly one order of magnitude better than G(1), G(2), or methyl mannoside. This was suggestive of a glycoside cluster-effect (enhanced local concentration). As with G(1) and G(2) glycodendrimers, G(3) was too small for multivalent binding to occur (chelate effect). Dendrimers G(4) to G(6) showed a two orders of magnitude increase in activity against the tetrameric Con A, indicating that multivalent binding was occurring. It is also possible that the change in shape from circular [G(1)- to G(3)-PAMAM] to spherical [G(4) to G(6)] caused the observed binding enhancement.[398] Further studies led the same group to the efficient incorporation of more complex saccharide units such as αMan-(1→2)Man disaccharide **535**, manifesting the fact that dimannose functionalized G(3) and G(4)-PAMAM dendrimers **536**, having an average of 26 and 48 peripheral groups were very competent for the recruitment of cyanovirin-N (CV-N), an HIV-inactivating protein that blocks virus-to-cell fusion through high mannose-mediated interactions (Scheme 52).[399]

Finally, a subsequent investigation provided the first example of the use of glycodendrimers as model systems for studying carbohydrate–carbohydrate interactions (CCI).[400] More particularly, lactosylated PAMAM dendrimers with various peripheral carbohydrate numbers, depending on the generation, were synthesized to examine the CCI of lactose with mixed Langmuir monolayers containing GM_3 and dipalmitoyl phosphatidylcholine (DPPC). Thiourea ligation between glycosyl isothiocyanate derivatives and the aminated dendritic core, following standard protocols afforded glycodendrimers containing appropriate amounts of lactose, which were further capped with poly(ethylene)glycol chains or other carbohydrates. The results corroborated the concept that lactosyl dendrimers were engaged in a CCI with GM_3 in a Langmuir monolayer. This CCI was dependent on both the carbohydrate density and the density of glycolipid within the monolayer. Moreover, a specific CCI was observed only in the presence of calcium ions and when at least one-fifth of the monolayer was composed of GM_3.

Amide linkages. Strategies based on the direct formation of amide bonds via carboxylic acid or activated ester-containing saccharides and PAMAM dendrimers have afforded sophisticated dendritic architectures having promising biological properties. For instance, using a similar approach to the one just described, Aoi *et al.* described the synthesis of dendrimer-based star polymers (''oligoglycopeptide sugar balls'') by the original macromolecular design of living polymerization.[401] The actual dendritic oligoglycopeptides were thus synthesized by living oligomerization of glyco-*N*-carboxy anhydrides (glyco-NCAs) with poly(amidoamine) (PAMAM) dendrimer as a multifunctional macro initiator. This polymerization system was termed as ''radial-growth polymerization (RGP),'' since dendrimer-based living polymerizations

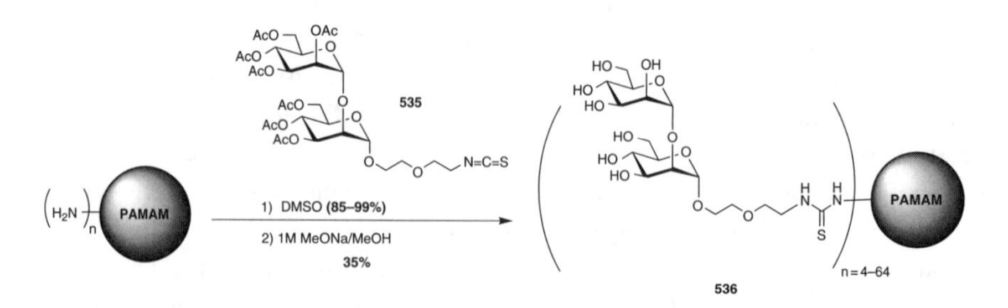

SCHEME 52. Manα1-2Man disaccharide attached to PAMAM via thiourea linkages.[399]

offered highly ordered star-shaped macromolecules with a number of arms, which should be different from conventional star polymers. The resulting sugar balls in this study are dendritic nanocapsules surrounded radially by oligoglycopeptide chains.

Roy *et al.* described the systematic preparation and biological evaluation of glycoPAMAMs containing up to 32 TF-antigen units [βGal-(1→3)-αGalNAc], known to be a cancer-related epitope and as an important antigen for the detection and immunotherapy of carcinomas, particularly relevant in breast cancer patients.[351,402,403] The synthetic sequence was based on the efficient acid functionalization of the allyled TF-antigen derivative **537**, accomplished using 3-mercaptopropanoic acid, to provide acid **538** in 83% yield. GlycoPAMAMs **539**, presenting up to 32 peripheral TF-antigens, were prepared by direct amide bond formation between unprotected **538** and PAMAM, using the efficient TBTU coupling reagent (2-(1*H*-benzotriazole-1-yl)-1,1,3,3-tetramethyluronium tetrafluoroborate) and DIPEA as a base (Scheme 53).

The protein-binding properties of these glycodendrimers were evaluated using peanut lectin from *Arachis hypogaea* and a mouse monoclonal IgG antibody. Based on bulk conjugates, the glycoPAMAMs with the highest carbohydrate density exhibited the strongest inhibitions, clearly indicating a cluster effect. All of these conjugates were antigenetically active and their IC_{50} values of 5.0, 2.4, 1.4, and 0.6 nmol, respectively, for G(0) to G(3) correspond to inhibitor abilities 460, 960, 1700, and 3800 times higher than that of the monomer **537** toward antibody-coating antigen interactions.

Based on an identical approach, mannosylated PAMAM dendrimers were constructed in order to evaluate their relative inhibitory properties against the type-1 fimbriated uropathogenic *E. coli*.[404] As an alternative to the synthetic strategy just described, the authors chose 3-aminopropyl α-D-mannopyranoside (**540**) pretreated with diglycolic anhydride. The resulting extended acid derivative **541** was then coupled to peripheral PAMAM amino groups with typical peptide coupling reagents such as BOP (benzotriazole-1-yloxy-tris(dimethylamino)phosphonium hexafluorophosphate) or TBTU (Scheme 54).

The deprotected glycodendrimers **542** were tested using a newly developed ELISA-based inhibition assay for their ability to inhibit the binding of recombinant type-1 fimbriated *E. coli* (FimH) to a monolayer of T24 cell lines derived from human

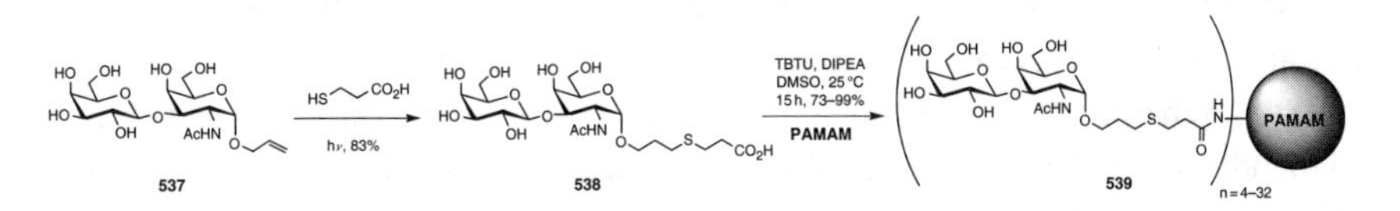

SCHEME 53. Antigenic PAMAM-based TF-antigens linked by amide coupling.[402]

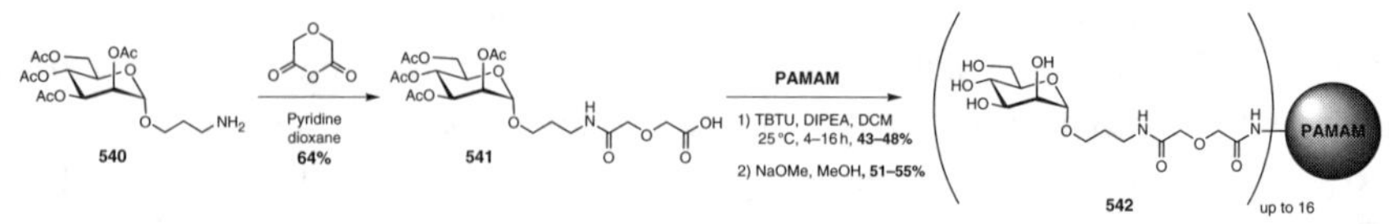

SCHEME 54. Amide-linked mannosylated PAMAM dendrimers.[404]

urinary bladder epithelium. The PAMAM mannodendrimers displayed potent affinity toward the target FimH, with IC_{50} of 37 and 19 μM for G(1) and G(2), respectively, although their relative potency per mannose was rather low.

The synthesis of mannose 6-phosphate-functionalized PAMAM dendrimers was also investigated and studied for their ability to bind goat-liver mannose-6-phopshate receptor (MPR 300).[405] The preparation of these glycodendrimers was achieved by the initial preparation of the amine-tethered Man-6-P derivative **544**. Since the amino function was present on the saccharide residue, the authors used preformed half-generation G(0.5) to G(3.5) PAMAM dendrimers **543** exhibiting 4, 8, 16, and 32 carboxylic acid groups, respectively. The resulting dendritic conjugates **545** were obtained by direct amide bond formation, using DIC–HOBt (1-hydroxybenzotriazole) or EDC (1-ethyl-3-(3-dimethylaminopropyl)carbodiimide)–NHS (*N*-hydroxysuccinimide) as coupling reagents (Scheme 55).

Preliminary biological studies suggested that the newly synthesized Man-6-phosphate-containing dendrimers could act as adsorbents in related affinity chromatography, with an interesting potential to bind the purified goat-liver mannose-6-phopshate receptor (MPR 300) protein. Additional conjugation of various nonprotected glycosides has been similarly described using *p*-nitrophenyl activated saccharide esters to conjugate disaccharides or sialic acid components onto PAMAM scaffolds.[406]

Reductive Amination. Sashiwa *et al.* have used reductive amination to provide dendronized chitosan–sialic acid hybrids containing different spacers (Scheme 56).[407–409] The sialic acid residues were successfully attached to each PAMAM dendrimer by reductive N-alkylation with *p*-formylphenyl α-sialoside **546**, using $NaBH_3CN$ in MeOH, to furnish hybrid chitosans **547**. However, since an excess of aldehyde must be used for high incorporation of sialic acid, the procedure also gave rise to double N-alkylation. The reaction thus gave complex sialic acid constituents, affording undesired mixed structures. To circumvent these drawbacks, N-methylamino derivatives (see Scheme 46) have been subsequently used for the reductive amination, and sialodendrimers have been prepared to target the immunoglobulin exposed Siglecs.[410]

Nucleophilic Substitution via Incorporation of Haloacetamido Groups. In 1996, Zanini and Roy described the design and synthesis of a novel family of symmetrical

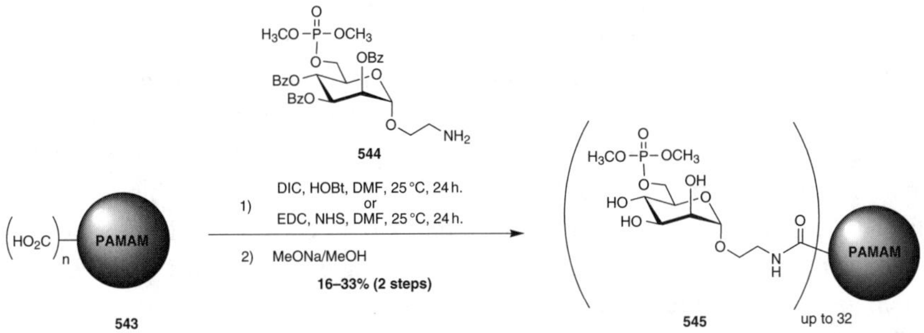

SCHEME 55. Synthesis of mannose-6-phopshate PAMAM conjugates.[405]

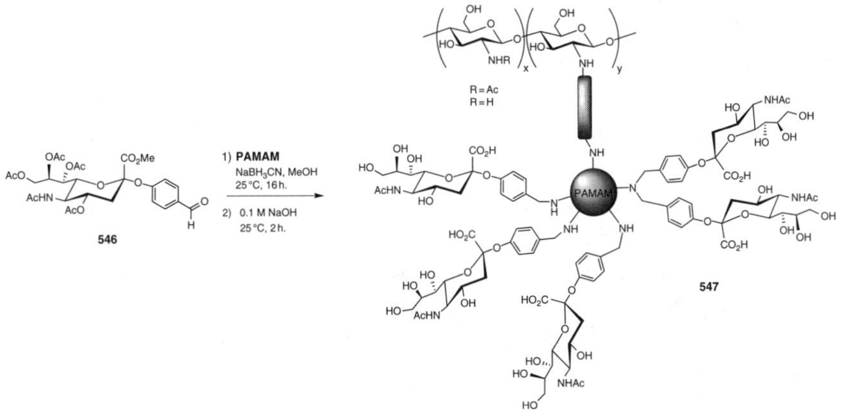

SCHEME 56. Synthesis of dendronized chitosan–sialic acid hybrids.[407–409]

dendrimers presenting even valencies between 2 and 16 residues, based on a 3,3′-iminobis(propylamine).[411] The synthetic approach was based on convergent assembly of suitable multibranched dendrimers containing N-chloroacetylated end groups to provide electrophilic species that could readily react with thiolated carbohydrate derivatives. Many synthetic benefits emanated from this strategy. First, the method was general, high-yielding, and readily amenable to such existing commercially available amine-terminated dendrimers as PAMAM that could be similarly treated with chloroacetic anhydride. Second, the synthesis of this dendritic family presented a viable alternative to PAMAM dendrimers, avoiding the need for large excesses of the reagents commonly used to ensure complete conversion. More importantly, these structures are not susceptible to base-catalyzed retro-Michael degradations, in contrast to classical PAMAMs, since the substituents are not β-carbonyl positioned. The authors thus accomplished the efficient preparation of dendritic α-thiosialosides **551–553**, starting from the AB_2-type core 3,3′-iminobis(propylamine) (**548**), which was subsequently transformed to afford various symmetrical N-chloroacetylated dendrons, such as tetramer **549** (Scheme 57). A slight excess of thiosialoside **550** allowed the synthesis of glycodendrimers under mild conditions. They were deprotected to afford the dendritic architectures containing between four (**551**, Scheme 56) and up to 16 sialoside residues (**552**, Fig. 67), respectively.

The same group conducted a systematic study of similar but tethered sialodendrimers with even valencies between 4 and 12 (**553**), based on the same AB_2 core **548**, and according to the same synthetic strategy.[412] The potential of these sialodendrimers to cross-link and precipitate LFA was confirmed by preliminary turbidimetric analysis. When tested in enzyme-linked lectin inhibition assays using human AGP (orosomucoid) as coating antigen, and HRPO-labeled LFA for detection, tetravalent dendron **551** showed IC_{50} values of 11.8 nM, while **552** presented a higher value of 425 nM. On the other hand, these assays indicated that for the corresponding tethered dendrimers, the inhibitory potency increased with a corresponding increase in valency. The highest IC_{50} value was obtained for **553** (8.22 nM) and constituted a 182-fold increases in inhibitory potential over the monovalent 5-acetamido-5-deoxy-D-*glycero*-α-D-*galacto*-2-nonulopyranosyl azide used as standard (IC_{50} of 1500 nM). The same group then extended this approach using solid-phase chemistry.[413]

Despite the demonstration that PAMAM dendrimers can exhibit up to 256 peripheral saccharide units, a general consensus highlights the fact that the most highly

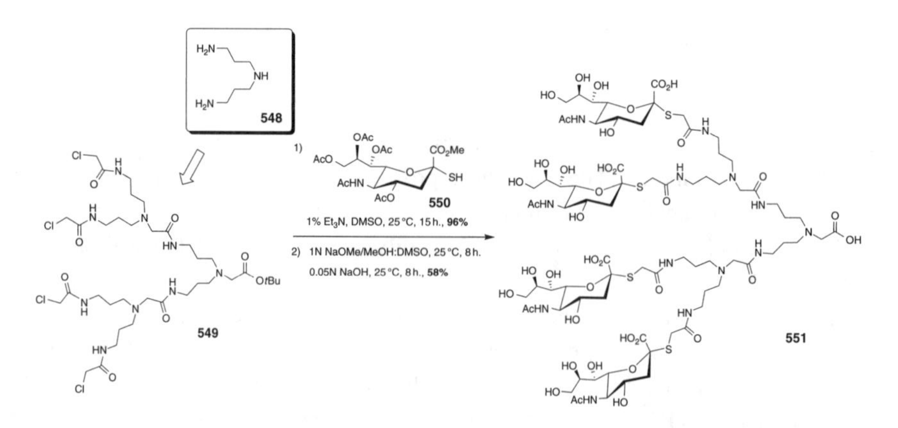

SCHEME 57. α-Thiosialodendrons built on base stable AB_2 amine core.[411]

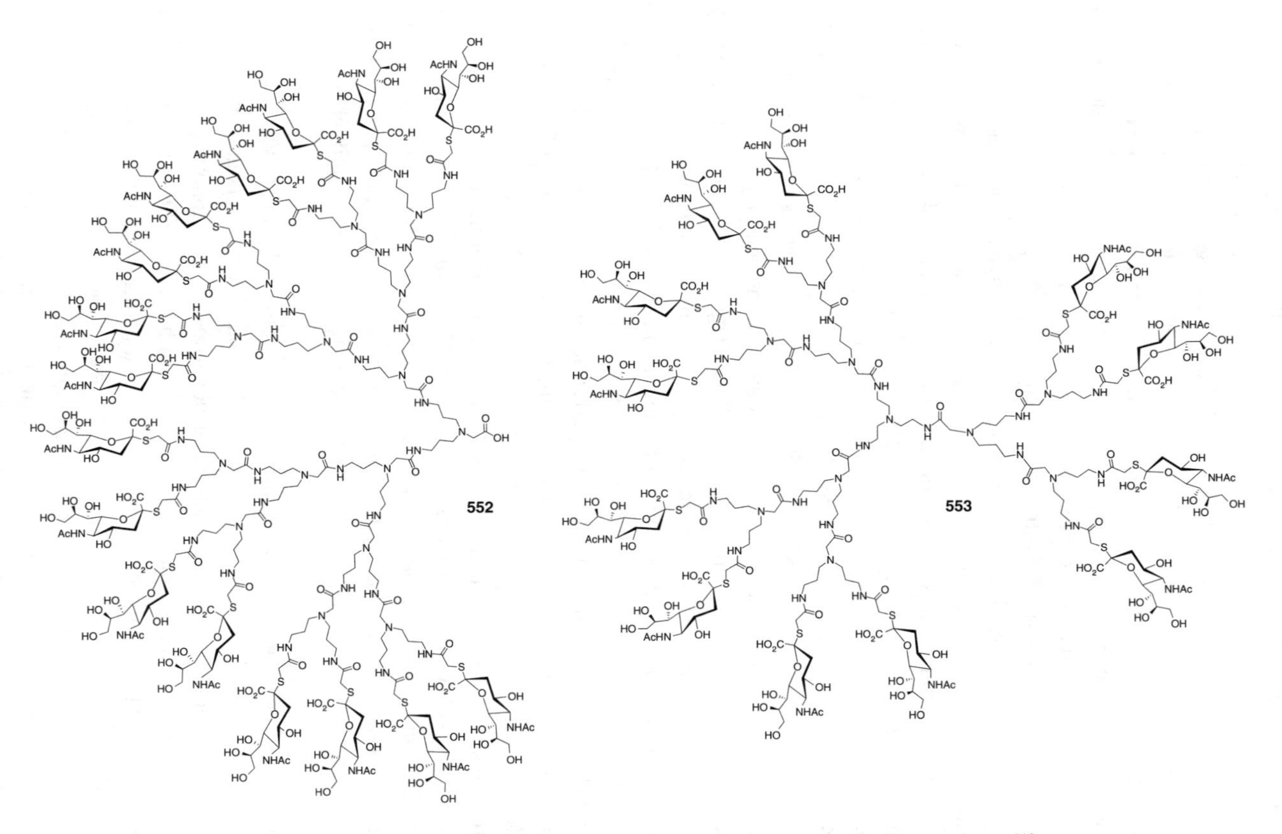

FIG. 67. α-Thiosialodendrons (**552**) and dendrimers **553** built on AB_2 amine scaffolds.[412]

functionalized scaffolds are not always the ones with the highest protein-binding activity. In fact, less-functionalized ligands may often present optimal potencies. A plausible explanation for this phenomenon has been suggested and highlights the fact that at high loading, the peripheral saccharide portions might become less accessible to proteins because of an increase in steric hindrance. Another hypothesis is based on the drawbacks of using large dendrimers bearing aromatic aglycones, which can intramolecularly undergo π-stacking, thus further tightening the saccharide units at the dendrimer surface. Furthermore, solubility problems are often encountered with PAMAM dendrimers bearing more than 64 arylated mannoside moieties.

Straightforward solutions to this problem have been addressed in the following publications describing the synthesis of sialoside dendrimers **554**, where the glycans are interspaced by TRIS residues used as ''dummy'' functionalities (Fig. 68).[335]

Aiming to validate these concepts, Wolfenden and Cloninger conducted the controlled preparation of a series of heterobifunctionalized mannoside and hydroxyl-G(3) to G(6)-PAMAM dendrimers **555** (Fig. 69).[414,415] A systematic study was addressed of the effect of carbohydrate loading on the activity of the dendrimer for a lectin and the influence of the conjugates' size. The degree of functionalization of mannose–hydroxyl groups via thiourea linkages was controlled by the stoichiometric amount of isothiocyanate derivatives of comparable reactivity. MS data allowed determination of the average number of mannose surface residues for each dendrimer generation. Hemagglutination assays were performed by adding rabbit erythrocytes to preincubated solutions of Con A and varying the concentrations of deprotected dendrimers. Comparison of the fourth through sixth PAMAM generations with different loading of surface mannose residues suggested that the binding efficiency was the highest for all generations at 30% to 50% loading. The most interesting observation was that the highest activity did not correlate with the maximum sugar loading, but rather occurred at slightly closer packing of the sugars as the generation increased, suggesting that unfavorable steric interactions precluded optimal binding at high carbohydrate density.

A follow-up investigation described the simultaneous anchoring of various sugars, such as mannose, galactose, and glucose residues via thiourea linkages, in order to quantify the effect that functionalization of dendrimers with monomers of varying affinities would have on its multivalent activity with lectins (**556**).As mentioned in earlier work, MALDI-TOF MS was used to determine the exact number of carbohydrate residues of each type on the resulting glycodendrimers. Both the change in MW after each sequential addition and the change in molecular weight after deacetylation were used for accurate measurement of the extent of functionalization. The association of the corresponding deprotected dendrimers with Con A was studied by

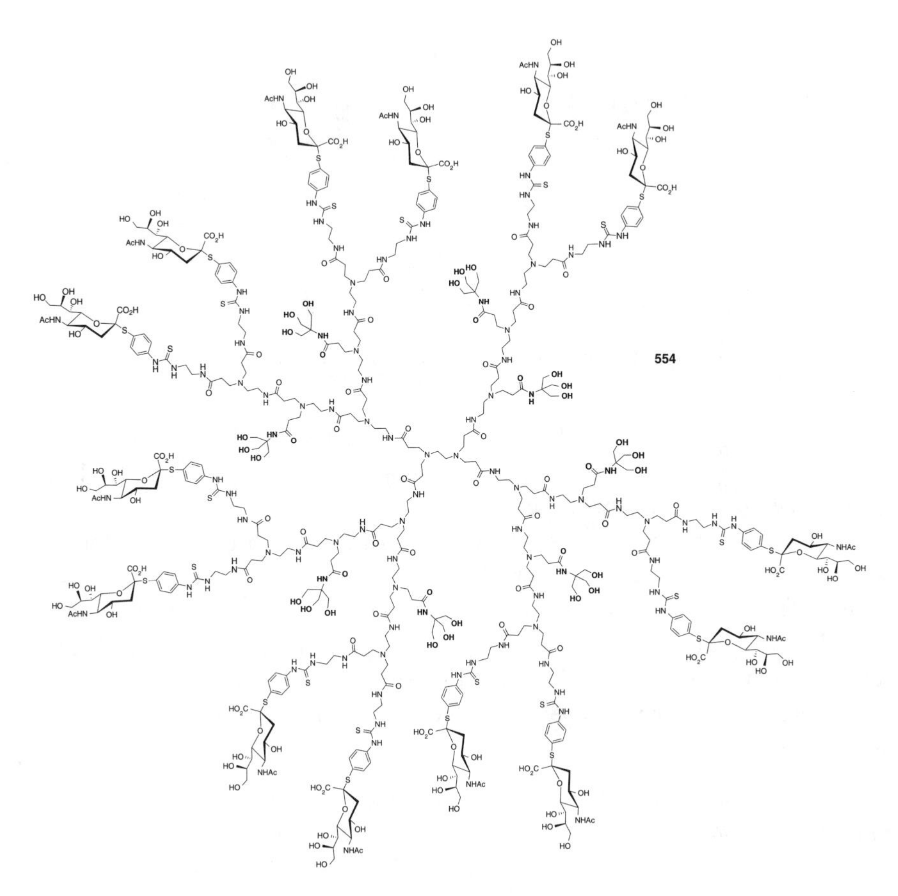

FIG. 68. Modified PAMAM scaffold bearing inter-spacing TRIS residues to promote better accessibility to surface saccharides.[335]

precipitation and hemagglutination assays. As expected, increasing the number of mannose residues, while decreasing the number of glucose residues, caused an increase in the relative affinity toward Con A. As with the dendrimers bearing 50% mannose: glucose loading, a linear relationship between Man:Glc loading and assay activity was observed for compounds of generations 4–6. However, the relative binding differences between mannosylated dendrimers and glucosylated dendrimers varied from one generation to another.

Alternatively, the G(3)-mannosylated PAMAM had a fivefold higher relative affinity toward Con A compared to the glucose-functionalized analogues. In summary, the reported results suggested that multivalency can be influenced in

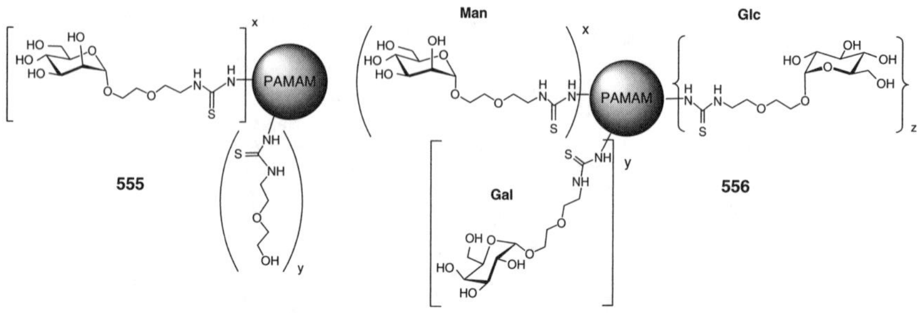

FIG. 69. Heterofunctional GlycoPAMAM dendrimers.[414,415]

predictable and therefore tunable ways. Monovalent differences are amplified by multivalent associations, and mixtures of low- and high-affinity ligands can be used to attenuate multivalent binding activities.

Given their commercial accessibility, PAMAMs have been the scaffold most widely used in modern investigations. They have been modified with a large variety of sugar derivatives and with varied sugar densities (Fig. 70). They have been used for several important binding phenomena and represent the most deeply studied scaffold for toxicity evaluations. In addition, scaffolded glycodendrimers were shown to be nonimmunogenic, a key property if they are to be used as bacterial or viral antiadhesins.

b. PPI Dendrimers.—A convenient reaction sequence for large-scale synthesis of PPI dendrimers [poly(propyleneimine) or DAB-*dendr*-$(NH_2)_x$] (Fig. 66) has been described by Meijer's group, who nicely adapted the original strategy proposed earlier by Vögtle's group (which suffered from cumbersome purifications).[286,416] By a repetition of double Michael addition of acrylonitrile into primary amines, followed by metal-catalyzed hydrogenation, the preparation of well-defined DAB-*dendr*-$(NH_2)_x$ containing up to 64 primary amine surface groups was efficiently achieved. As with PAMAM dendrimers, this family of commercial dendritic scaffolds has been widely functionalized. Obviously, saccharide-based PPI dendrimers have also been synthesized.

The first example of glycodendrimers using PPI scaffolds was proposed by Ashton *et al.* who described their use for the rapid and facile construction of high molecular weight G(1) to G(5) carbohydrate-coated dendrimers. A divergent approach was followed and glycodendrimers containing, respectively, 4, 8, 16, 32, and 64 D-galactose and lactose peripheral groups, were efficiently synthesized.[417] The saccharide residues were attached to the aminated dendritic scaffold was by amide

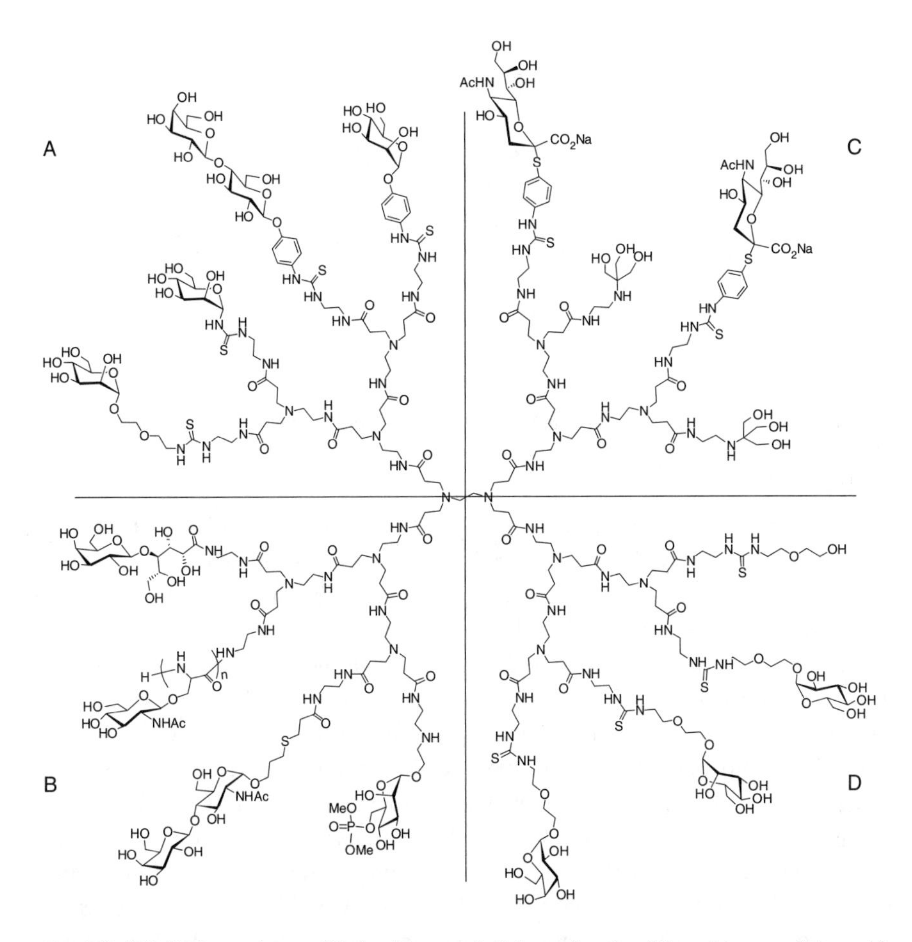

FIG. 70. PAMAM can be modified with varied linkage functionalities: thioureas (A); amides (B); heterobifunctional interspacing groups (C); and mixed sugars (D).

bonding through glycosides ending with the *N*-succinimidyl esters **557** and **558**. Highly globular glycodendrimers were obtained essentially quantitatively and were deacetylated under Zemplén conditions [to give G(4)-PPI glycodendrimers **559** and **560**] (Scheme 58).

In continuing efforts, the same authors constructed tris(galactoside)-modified DAB dendrimers by the accelerated convergent strategy. In this particular approach, the saccharides were connected first to a small TRIS branching derivative to form a cluster which served as a building block, once suitably functionalized.[417] N-Succinimidyl-activated ester **561** was then coupled onto G(1) and G(2)-PPI dendrimers with

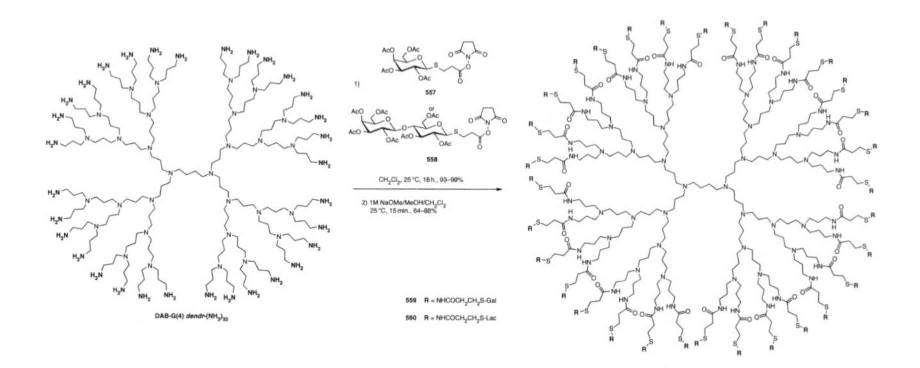

SCHEME 58. First poly(propyleneimine) glycodendrimers (PPI) coated with thioglycosides through amide bond formation.[417]

4 and 8 primary end groups, respectively, to afford glycoconjugates **562** and **563** containing respectively, 12 and 24 peripheral galactosides (Scheme 59).

The same group modified the linker by using different numbers of carbon atoms (1, 5, 10) to afford variations of the local saccharide concentrations at the dendrimer surface.[418] This study was aimed to determine the influence of this linker parameter on the glycodendrimer–protein interactions, the relationship between structure and water solubility, and to investigate amphiphilic properties.

In 2004, additional poly(sulfogalactosylated) dendrimers based on DAB scaffolds were investigated to study their antagonist properties against HIV-1 infection.[419] First, to assess the interaction of HIV-1 gp120 with its reported alternative glycolipid receptors, a series of glycodendrimers built from mimics of galactosyl ceramide (GalCer **564**) and its sulfated derivatives **565** were synthesized, analyzed by SPR as ligands for rgp120 IIIB, and evaluated for their ability to inhibit HIV-1 infection on CXCR4- and CCR5-expressing indicator cells. The synthesis was based on direct amine bond formation with carboxylated galactoside residues **567** or **568** and G(1) to G(5) DAB-dendrimer (Scheme 60), and afforded compounds **569** and **571**. Polysulfated dendrimers **570** were obtained by random sulfation of **568**, and the average number of sulfates was determined by MALDI-TOF MS analysis. The *in vitro* studies of their effectiveness at inhibiting infection of U373-MAGI-CCR5 cells by HIV-1 Ba-L indicated that the sulfated glycodendrimers were better inhibitors than those of the nonsulfated compounds, but not as effective as the sulfated dextran control **566**.

Further investigations led the same group to increase the degree of sulfation on galactosylated dendrimers, based on the hypothesis that glycosphingolipid SGalCer

SCHEME 59. Convergent synthesis of PPI glycodendrimers.[417]

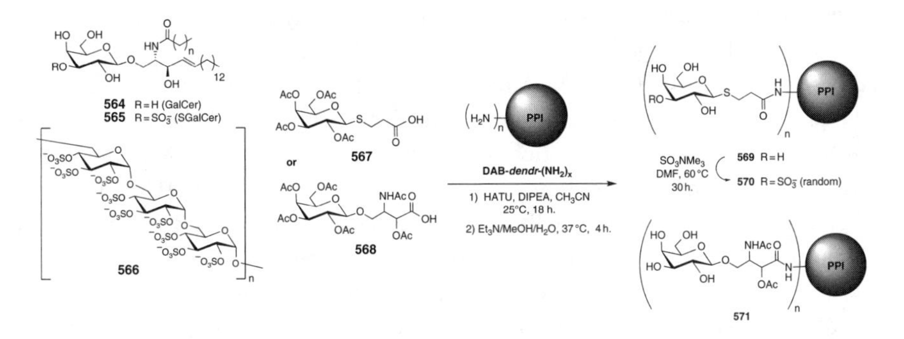

SCHEME 60. Galactosylated PPI dendrimers used as inhibitors of cell infectivity by HIV-1.[419]

565 was the best ligand for both recombinant and virus-associated gp120.[420] Therefore, the synthesis of G(5) DAB-dendrimer **570**, containing up to 64 peripheral galactoside groups and with an average of two sulfate groups per galactose residues (versus about 0.5 for the first example) was performed. The ability of **570** to inhibit infection of cultured indicator cells by HIV-1 was compared to that of dextran **566**. The results showed that **570** inhibited HIV-1 IIIB as well as dextran sulfate (a known potent inhibitor of HIV-1 infectivity) with EC_{50} values in the nanomolar range, and both were comparable in their ability to inhibit HIV infection of both X4 and R5 indicator cells. Furthermore, cytotoxicity studies revealed that neither the glycodendrimer nor **566** were toxic to the cells at the highest concentration (3 mg/mL) tested. Critical parameters such as the number and the position of sulfate groups (especially at C-3), which may constitute a key structural component of an efficient inhibitor, were clarified by this study.

An interesting application against HIV was developed, using G(5)-mannosylated PPI dendrimers (MPPI) as carriers for the controlled and targeted delivery of antiretroviral nonnucleoside reverse-transcriptase inhibitors, such as lamivudine (3TC) and efavirenz (EFV).[421,422] Numerous benefits over the standard treatment based on the free drug or its PPI-encapsulated version, were demonstrated *in vitro*. These nanocontainers ensured efficient entrapment of the bioactive drugs (~45%), which allowed their prolonged release profile for up to 144 h (vs. 24 h with PPI). The toxicity of these systems was generally found negligble. In cellular-uptake experiments, MPPI interacted with the lectin receptors present on the surface of MT2 cells or monocytes/macrophages, leading to cellular uptake of drugs 12 times higher than that of the free drugs. Furthermore, this phenomenon may be responsible for the significant, but preliminary anti-HIV activity displayed by MPPI. Despite these promising results, further studies are needed to consider those encapsulated systems as potent carriers for efficient control and targeting the delivery of anti-HIV drugs.

c. Hyperbranched Boltorn® Dendrimers.—Hyperbranched dendritic Boltorn® polymers (Perstorp Speciality Chemicals) are also useful scaffolds, which have been used for the multivalent presentation of various functional groups. Their selection is based on the fact that second, third, and fourth generation polymers (H20, H30, and H40, respectively) are commercially available at a low cost since they are prepared in bulk quantities by polymerization. They are, however, polydisperse.[423,424] Nevertheless, they can also be synthesized by a simple and iterative approach, using pentaerythritol (**572**) as the central core and 2,2-bis(hydroxymethyl)propanoic anhydride (bis-MPA anhydride, **573**) (Scheme 61). Deprotection in acidic media furnished G (1)-Boltorn® polymer (**574**) containing 8 peripheral hydroxyl functions, and the sequence can be repeated for subsequent generations, leading, for instance to G(3) BH30 (**575**).

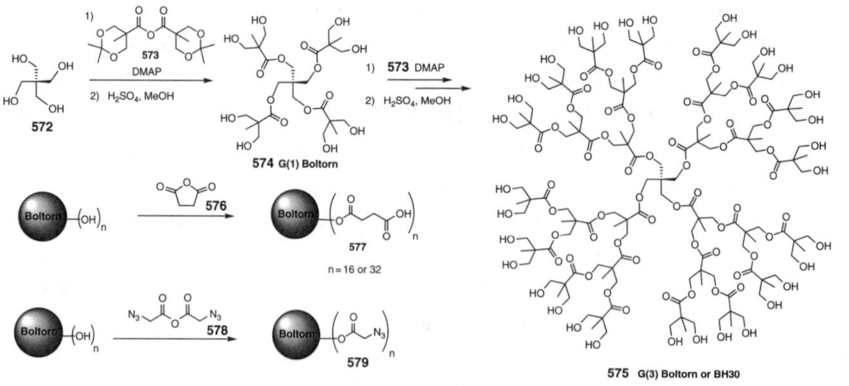

SCHEME 61. Idealized Boltorn® hyperbranched polyols.

To mimic the natural organization of high-mannose structures notably present in pathogens, mannosylated glycodendrimers obtained from these multivalent scaffolds have been synthesized. The commercial hyperbranched polymers of the second, third, and fourth generations were first functionalized with succinic anhydride (**576**) to furnish poly(acid) **577**, which was then capped with 2-aminoethyl α-D-mannopyranoside to give mannosylated dendrimers of various generations [**580** for G(3) (Fig. 71)]. Preliminary studies of these glycodendrimers, perfectly soluble under physiological conditions and nontoxic against several cell lines, evaluated their interactions with *Lens culinaris* lectin (LCA).[425] Data obtained by STD NMR

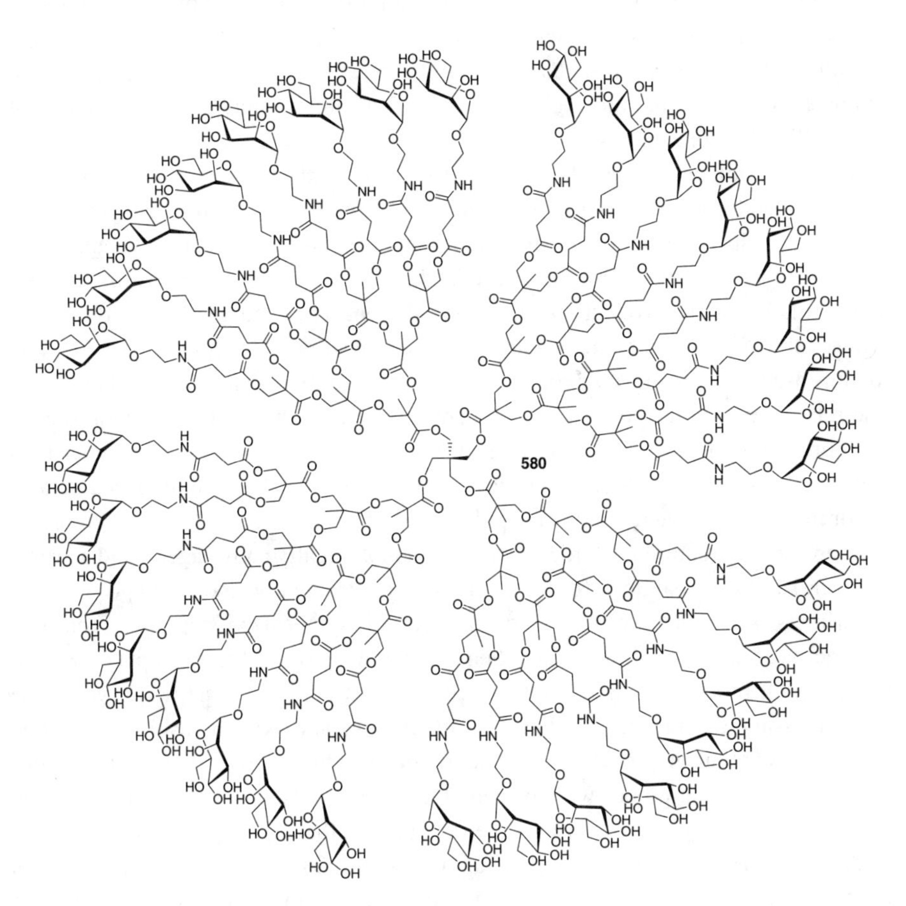

FIG. 71. Idealized mannosylated hyperbranched Boltorn® dendrimer.[425]

experiments and quantitative-precipitation assays demonstrated that glycodendrimers on the Boltorn® platform effectively interact with such biological receptors as plant and mammalian lectins.

The same group demonstrated that these mannodendrimers block the interaction between DC-SIGN and Ebola virus envelope glycoprotein.[426] Interestingly, dendrimer **580**, containing an average of 32 peripheral mannosides, was able to inhibit selectively the DC-SIGN-mediated Ebola infection in an efficient dose-dependent manner (IC_{50} of 337 nM), and showed no inhibitory effect in infection experiments using DC-SIGN-negative cell lines. These results showed that **580** was a potent inhibitor of Ebola infection mediated by DC-SIGN both in *cis* and *trans* (intra- and inter-cellular) presumably via the same inhibition mechanisms involving the interaction between the lectin and the viral envelope. In addition, a carbohydrate-dependent inhibitory effect was confirmed and a multivalent effect of two orders of magnitude demonstrated, since the monovalent manoside was able to inhibit this interaction at only millimolar concentrations.

A related application used ''click chemistry'' for synthesis of azido-functionalized Boltorn® dendrimer **579** (BH20) (Touaibia and Roy, unpublished data). Dendrimer **579** was obtained by treatment of hexadecahydroxylated BH20 (on average) with azidoacetic anhydride (**578**). IR analysis demonstrated complete hydroxyl-group transformation and introduction of azide function. The ''clicked'' hyperbranched dendrimer **582** was obtained under typical reaction conditions ($CuSO_4$ and sodium ascorbate) using nonprotected propargyl α-D-mannoside **581** (Scheme 62). The relative inhibitory potency of the mannosylated dendrimer for the inhibition of agglutination of *E. coli* by yeast mannan was approximately 400 times higher than that of the respective methyl mannoside (Benhamioud and Roy, unpublished data).

For most practical and biological applications of glycodendrimers, three functional units are usually required: a targeting moiety, a biologically active agent, and a probe. A general and facile strategy for functional-group introduction at defined positions on dendrimers is best achieved when dendrimers are synthesized stepwise. From this perspective, Sharpless, Hawker, and their group have provided an example of sophisticated, multifunctional materials that can be constructed in a stepwise, yet facile manner, using the efficient ''click'' methodology, and with a fashion-controlled strategy toward unsymmetrical glycodendrimers in which two distinct moieties (targeting and detection probe) were placed at the chain ends.[254] Dendritic block copolymers up to the fourth generation were coupled via the key step of Cu(I)-catalyzed azide–alkyne cycloaddition to prepare rapidly these dual-purpose and multifunctional materials without use of protecting groups. The orthogonal synthetic approach was based on 2,2-bis(hydroxymethyl)propanoic acid (bis-MPA), which constitute

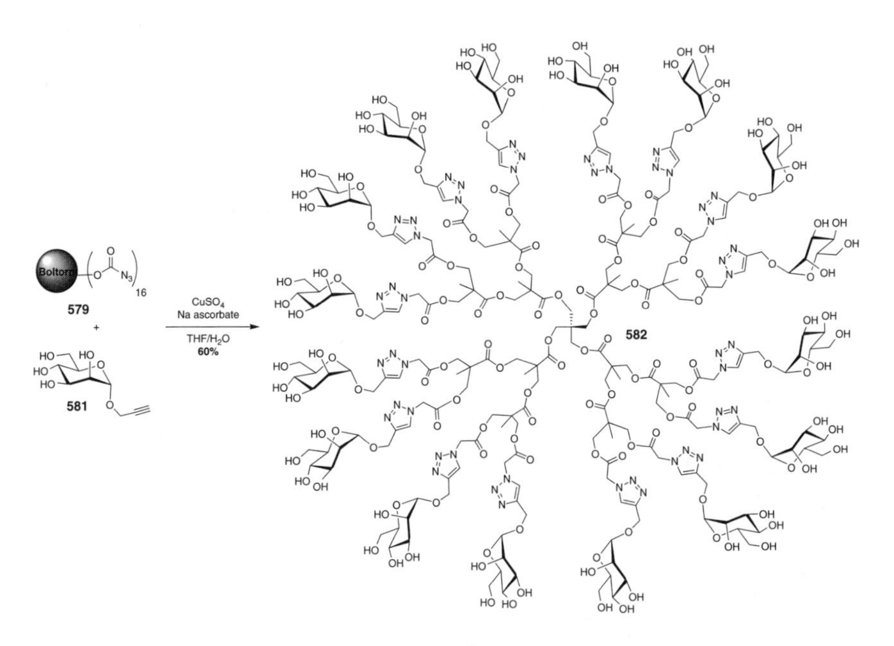

SCHEME 62. ''Clicked'' mannosylated Boltorn® H20 (Touaibia and Roy, unpublished data).

the main building block in synthesis of the commercial Boltorn® dendrimers. The corresponding bis-MPA anhydride (**573**) provided access to alkyne ester **583** and azide ester **584** by condensation with the appropriate alcohol (Scheme 63). Removal of the protecting groups and subsequent condensation by ''click chemistry'' allowed the generation growth of the core and afforded the asymmetrical structure **585** having a diol as the focal point. Two molecules of the functionalized coumarin-type fluorescent dye **586** were then introduced by esterification of the two free hydroxyl groups of **585** with pent-4-ynoic anhydride **283**, followed by the [3+2] cycloaddition. Acetal hydrolysis and subsequent introduction of the 16 alkynes via esterification, followed by [3+2] cycloaddition with an unprotected 2-azidoethyl α-D-mannopyranoside (**285a**) in THF–H_2O furnished the asymmetrical heterobifunctional dendrimer **587**.

The performance of the resulting mannosylated dendrimer **587** was investigated in a standard hemagglutination assay using the MBP Con A and rabbit red-blood cells. When compared to the activity of a single mannose, **587** exhibited 240-fold greater potency, corresponding to a relative activity of 15 per sugar moiety, thus demonstrating the synergistic benefit provided by the multivalent dendritic array of sugar groups.

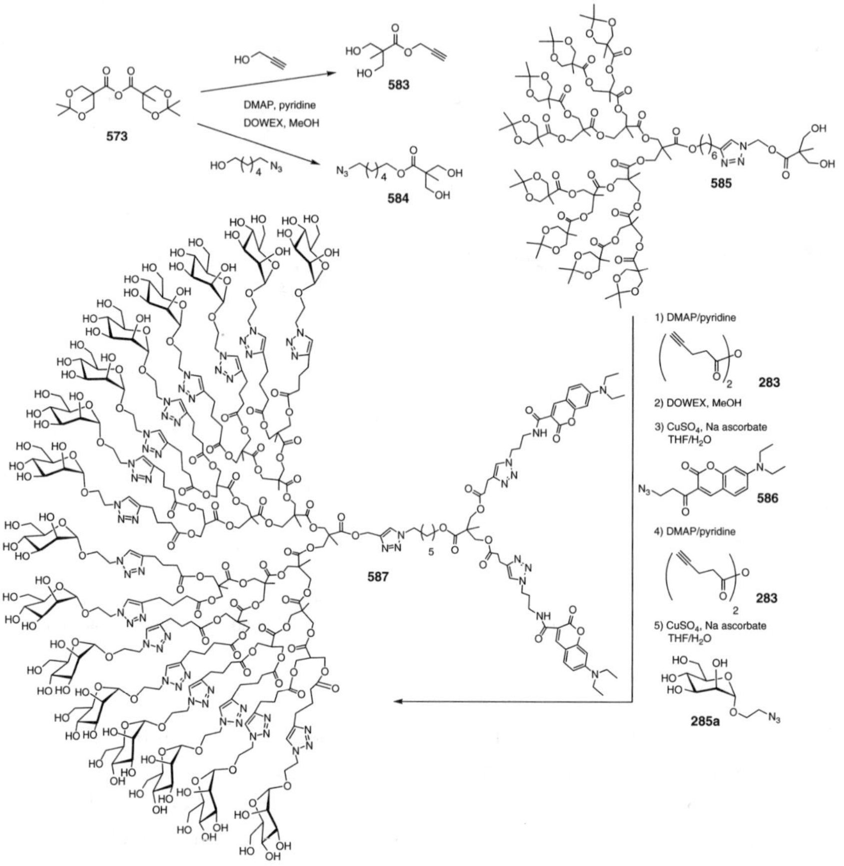

SCHEME 63. Synthesis of bifunctional glycodendrons with two distinct moieties containing targeting and detection probes.[254]

d. AB_2 and AB_3 Subunit-containing Glycodendrimers.— *(i) Aromatic AB_2 Systems.* As mentioned earlier, aromatic AB_2 building blocks constitute privileged components for rapid design of highly branched glycodendrimers possessing biological potential. As well as examples presented earlier, the approach has involved the design and synthesis of G(1) and G(2) glycodendrimers containing building blocks of 3,5-dihydroxybenzoic acid.[427] Homo- and heterobifunctional glycodendrimers ending with up to 16 fucoside and/or galactoside residues, installed via the ''click'' 1,3-dipolar cycloaddition, were synthesized via an ''outside in strategy,'' from a diazide derivative **588** of methyl 3,5-dihydroxybenzoate (Fig. 72) and alkyne-functionalized carbohydrates. The saccharides were selected based on the fact that the bacterium *P. aeruginosa* (responsible for chronic lung colonization and the major cause of morbidity and mortality in CF patients), expresses two lectins, PA-IL (LecA) and PA-IIL (LecB) specific for D-galactose and L-fucose residues, respectively. Polyazido dendritic scaffolds **589–593**, containing up to 16 peripheral azido functions were synthesized by successive formation of amide linkages between dendritic subunits. The polyazides were then totally functionalized with L-fucoside units to give **594–597** (Fig. 73) while the heterobifunctionalized dendrimer **598** bearing both D-galactoside and L-fucoside residues was also prepared, providing binding access to either PA-IL and PA-IIL, simultaneously. The sugar heterogeneity was incorporated in order to increase the biological activities of the dendritic architecture and afford potent new antiadhesin agents against *P. aeruginosa*.

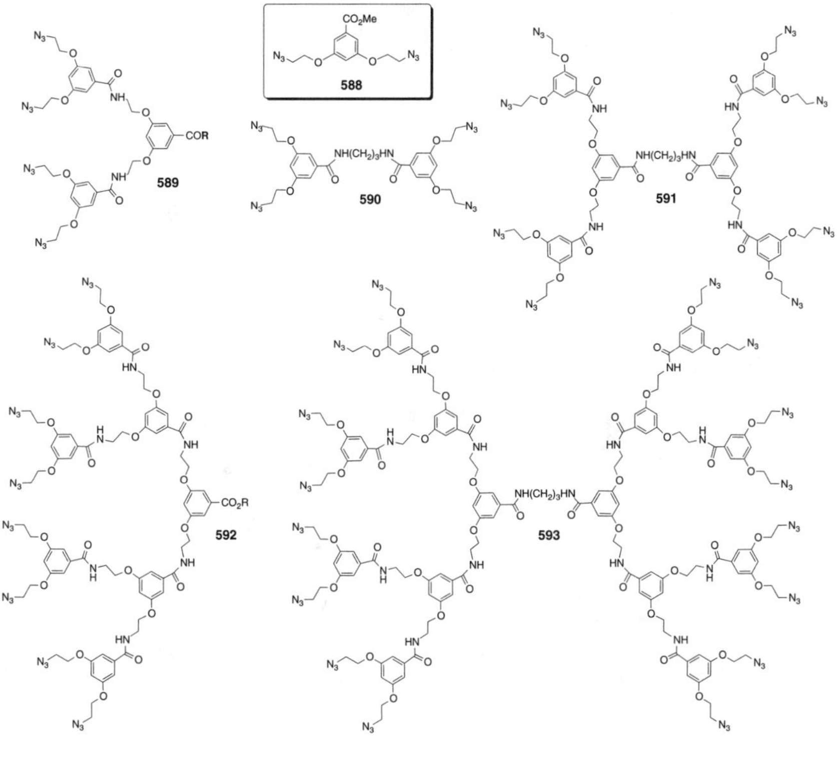

FIG. 72. AB_2-based polyazides used in ''click chemistry'' with propargylated β-D-galactosides and α-L-fucosides.[427]

FIG. 73. Galactosylated and/or fucosylated *P. aeruginosa* lectin ligands built on an AB_2 scaffold.[427] (See Color Plate 19.)

Turbidimetric assays indicated that glycodendrimers possessing a minimum of four fucoside residues on the same side showed rapid cross-linking abilities with tetrameric *P. aeruginosa* PA-IIL lectin, by forming insoluble complexes. As expected, heterodendrimer **598** containing 4 fucosides and 4 galactosides had the ability to recognize both binding-site domains of PA-IL and PA-IIL.

(ii) AB_3 Systems. One of the first examples of glycodendrimers based on AB_3 building blocks was proposed by Stoddart *et al.* who designed dense multivalent neoglycoconjugates in which the carbohydrates were located at the periphery of short peptidic chains emanating from an aromatic central core and TRIS-containing branching system.[229] The convergent approach was adopted, with preliminary construction of dendrons based on initial glycosylation of the three hydroxylmethyl groups of TRIS (**1**) with glucose (Scheme 64). The availability of the single free amino group in **1** following deprotection enabled further elaboration through formation of amide bond. This could be accomplished directly with triacid **599** or, preferably, to avoid steric problems through the intermediacy of the 3,3′-iminodipropanoic acid-derived dendron **240**. The glycine-derived glucosylated dendron **240** was thus connected with trifunctional scaffold **599** to afford glycodendrimer **600** decorated with up to 18 peripheral glucosides in a single step.

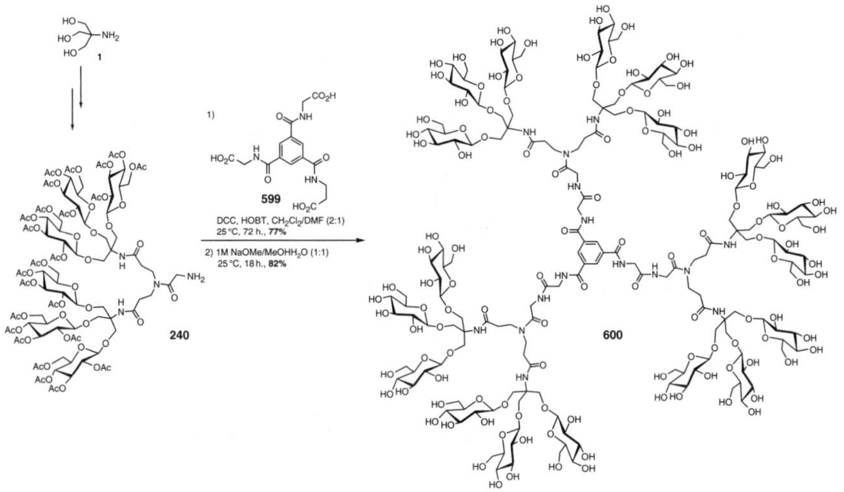

1) DCC, HOBT, CH_2Cl_2/DMF (2:1) 25 °C, 72 h., **77%**

2) 1M NaOMe/MeOHH$_2$O (1:1) 25 °C, 18 h., **82%**

SCHEME 64. Convergent strategy for the construction of glucosylated dendrimers using TRIS as the branching component.[229]

The authors then prepared even larger dendritic derivatives employing two different, yet closely related, convergent accelerated strategies.[304] The resulting D-glycopyranoside-containing glycodendrimers (**602**), bearing up to 36 saccharide residues, were obtained in good yields by a so-called 12 × 3 reaction sequence, involving three equivalents of dendritic wedge **601** (12-mer) and one equivalent of the trifunctional core **599** (Scheme 65). This advanced strategy circumvented the problem of incomplete dendrimer formation arising from limited reactivity of the core and accessibility impeded by the decreased interstitial space around it. High-generation glycodendrimers were thus obtained with very precise molecular sizes and shapes, and were completly monodisperse, despite the densely packed surface groups.

Furthermore, to avoid the anticipated limitation upon the growth of this type of dendrimer, common when saccharides contain protecting groups, Jayaraman and Stoddart focused on synthesis of glycodendrimers wherein the saccharides were totally unprotected during construction.[428]

e. Carbosilane Glycodendrimers.—In 1999, Matsuoka *et al.* synthesized the trisaccharide moieties present in globotriaosylceramide by using carbosilane-based

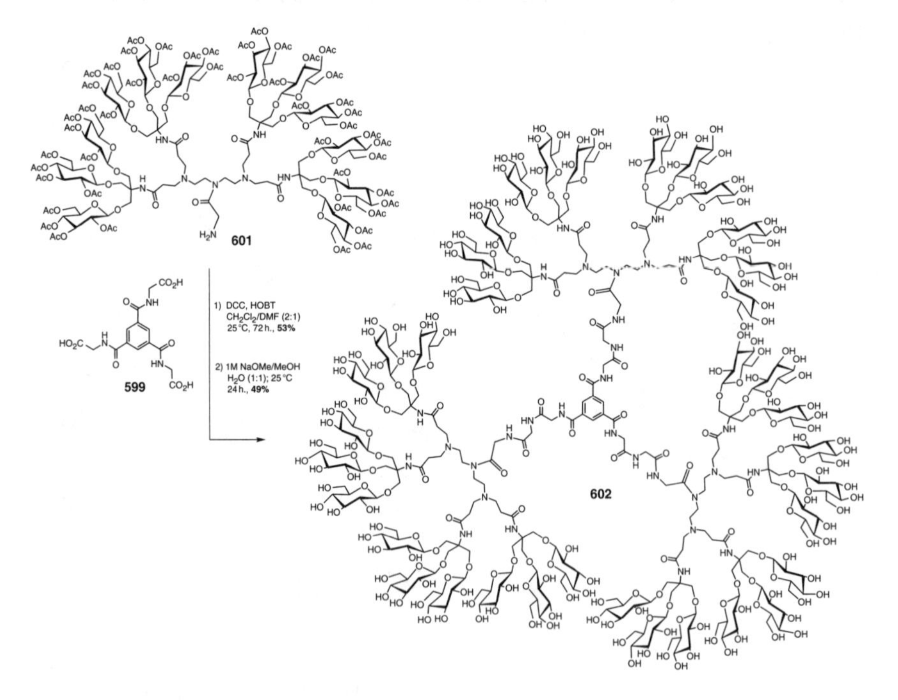

SCHEME 65. Advanced convergent strategy used by Ashton *et al.*[304]

dendrimers.[429] As well as an efficient methodology for constructing multivalent carbosilane scaffolds, the authors simultaneously developed a general synthetic strategy for the attachment of the carbohydrate moieties, employing an efficient SN2 substitution of an alkyl halide by the highly reactive thiolate anion. In addition to typical advantages encountered for dendrimers, multivalent carbosilane architectures offered (*i*) synthetic simplicity to extend the generation at will and thus provide access to derivatives of definite molecular weight and number of terminal functions, (*ii*) the neutral nature of dendritic scaffolds having chemical and biochemical stabilty, and (*iii*) biological inertness. These dendrimers (**603–606**), displaying unique shapes (fan-shape, ball-shape, and dumbbell-shape, respectively) were called ''SUPER TWIGs'' and are presented in Fig. 74.

The same group studied these ''SUPER TWIGs'' as therapeutic agents for treatment of infections by Stx-producing *E. coli* O157:H7, and demonstrated their inhibition potential.[430] SPR studies showed that dendrimers **604** and **606** presented very low K_D values of 4.2 and 1.4 μM, respectively, for STL-I, while the value for **603** was 30 times higher. This indicated that ''SUPER TWIGs'' **604** and **606** bound directly with high affinity to the STL-I B-subunit. In addition, six trisaccharides situated at an appropriate distance (span of ~30 Å from the central silicon core) in one molecule allowed full embrace of the predicted Gb_3-binding sites (CBPs 1 and 2 notably) and were sufficient for high-affinity binding. Furthermore, *in vivo* inhibitory effects of the ''SUPER TWIGs'' on the lethality of intravenously administred SLT-II in mice, indicated that **604** was a better candidate than **606** for this specific application. SLT-II was used in this study because it is more toxic than SLT-I both *in vitro* and *in vivo*, and clinically more significant. The assays indicated that **606** completely supressed the lethal effect of STL-II when administrated along with the toxin: **604**-treated mice survived more than 2 months without any pathological symptoms,

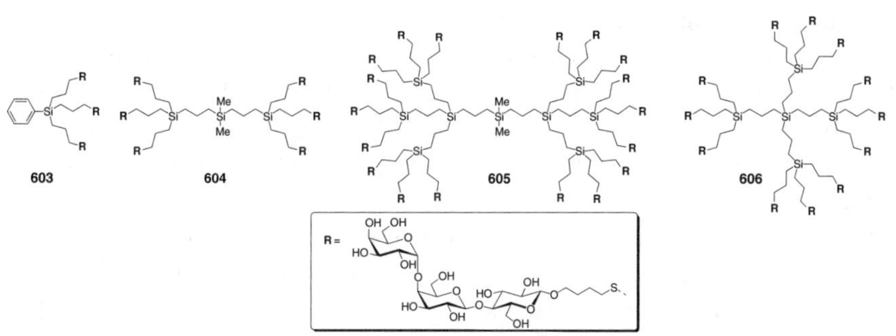

FIG. 74. Carbosilane glycodendrimers prepared by Matsuoka *et al.*[429]

whereas 100% of nontreated mice died within 5 days. The biological results suggested that **604** suppressed the lethality of SLT-II by diminishing the deposition of SLT-II in the brain and consequent fatal damage. Also, **604** protected mice from a challenge with a fatal dose of *E. coli* O157:H7, even when administrated after establishment of the infection. In the light of these results, a unique dual mechanism of action was proposed to explain this spectacular efficiency: (i) **604** bound to SLTs with high affinity and inhibited its Gb_3-dependent incorporation into target cells; (ii) it induced active uptake and subsequent degradation of SLTs by macrophages present in the reticuloendothelium.

A few years later, the same group identified the optimal dendritic structure, namely optimal valency and shape, required for SLTs neutralization.[431] They identified the 18-mer **605** as another potent SLTs neutralizer, functioning *in vivo*. Examination of additional structural features allowed the identification of crucial structural parameters. A dumbbell-shaped structure in which two clusters of trisaccharides, symmetrically located through their hydrophobic core and having an optimal length of at least 11 Å was first required. This indicated that grouping of the trisaccharides was more important than the number of trisaccharides. In the dumbbell shape, at least six trisaccharides needed to be involved for *in vivo* activity, and terminal trisaccharides with spacers had to be branched from the same terminal silicon atom to be clustered in high density. These structural requirements were essential for the appropriate induction of macrophage-dependent incorporation and degradation of SLT-II, further supporting the pivotal role for this mechanism in the *in vivo* SLTs-neutralizing activity of "SUPER TWIGs" having optimal structure. Interestingly, the data demonstrated the crucial implication of the CBP 3 present on the SLT-II B-subunit, and showed that it was an essential and sufficient site for high-affinity binding of **605**. In addition, the core length was a major structural parameter for perfectly embracing the three sites in a multiple way and, consequently, to provide an adequate hydrophobic volume for recognition by macrophages.

Considering the fact that galabiose is also known to bind SLTs with high affinity, Matsuoka *et al.* considered that enhancement of the binding ability by the clustering of galabiose would be advantageous as compared to clustering of globotriaose, both in terms of synthetic accessibility, and from the viewpoint of practical use for SLTs neutralization.

According to the synthetic strategy just described, a series of water-soluble carbosilane dendrimers were made, bearing peripheral galabiose trimers interspaced by 29 Å, adjusted to the binding-site distances (Fig. 75).[432] The binding affinity to SLT-I and SLT-II B subunits were determined by Scatchard plot analysis. Hexavalent compound **610** showed K_D values of 1.3 and 1.6 μM for SLT-I and SLT-II,

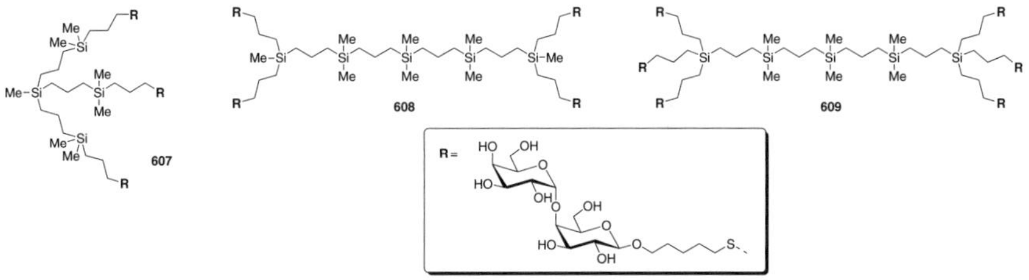

FIG. 75. Carbosilane glycodendrimers with pendant galabiose disaccharides.[432]

respectively, constituting one-tenth of the potency of the homologous **604**. Tri- and tetravalent analogues **607** and **608**, respectively, also presented weaker affinities (~60 and 10 μM, respectively, for both SLTs). On the other hand, further evaluations with ^{125}I-labeled SLTs binding and cytotoxic assays showed that multivalent **607–609** inhibited the binding of ^{125}I-labeled SLTs to vero cells, and presented very weak inhibitory effects in the cytotoxic assay as compared to the best dumbbell **604.**

In summary, results of *in vivo* experiments showed the effectiveness of carbosilane dendrimers having clustered **P^{k}** carbohydrate moieties, and the complete neutralization potency against STL-II was discovered when dumbbell-shaped dendrimers were identified as potent candidate inibitors. Although the precise mechanism of action remains to be elucidated, this type of inhibitor provided a new strategy for the detoxification of SLTs present in circulation.

The same research group further reported synthesis of αsialyl- (2→3)-lactosyl moieties [αNeu5Ac-(2→3)-βGal-(1→4)-βGlc-(1→] using a series of carbosilane dendrimer scaffolds and the products had interesting biological activities against various such human influenza virus strains as A/PR/8/34 (H1N1) and A/Aich/2/68 (H3N2). These dendrimers were thus uniformly functionalized with αsialyl-(2→3)-lactose residues, with different degrees of freedom of the sugar moities depending on the spacer length and core shape, and thus being responsible for a variety of 3D structures (Fig. 76). The inhibitory activity against the hemagglutination of influenza viruses to erythrocytes suggested that dumbbell amide **610**, having the longest spacer-arms and most carbohydrate epitopes, possessed the highest activity of the glycolibrary (IC_{50}s < 10 μM for both H1N1 and H3N2 viruses).[433]

Following this initiative, other complex glycosylylated architectures have been designed, finishing a series of large multigeneration glycodendrimers containing 27 [G(1), **615**], 81 [G(2), **616**], and 243 [G(3), **617**] terminal-modified D-xylose branches obtained by ''click chemistry'' from the corresponding polyalkynylated dendritic

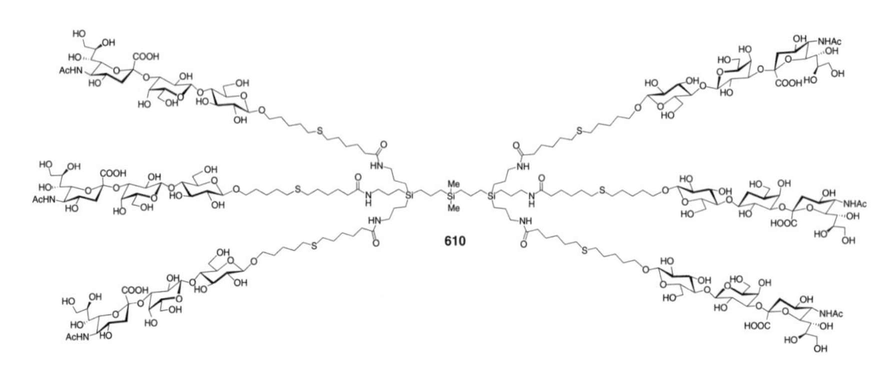

FIG. 76. Sialyloligosaccharide-capped carbosilane dendrimers successfully used as ligands against various strains of influenza virus hemagglutinins.[433]

cores (**612–614**) and 2-azidoethyl 2,3,4-tri-*O*-acetyl-β-D-xylopyranoside (**611**) (Fig. 77).[434] Characterization by ^{1}H- and ^{13}C NMR, elemental analysis, and IR spectroscopy confirmed their low polydispersity (1.04–1.05). The growth in size of the dendrimer from G(1)-27 to G(3)-243 and also upon functionalization within each generation, was shown by diffusion light scattering (DSL), DOSY NMR, and size exclusion chromatography (SEC), all of which satisfactorilly agreed with the expected size of the nanostructures.

f. Other Glycodendrimers.—Several other types of glycodendrimers have been synthesized from a wide variety of multivalent scaffolds.[435] Three of the most relevant exampes as described in the following sections.

Another interesting example illustrates the use of *scyllo*-inositol, the all-equatorial stereoisomer of *myo*-inositol. It has also been added to the growing selection of suitable scaffolds for synthesis of multivalent neoglycoconjugates.[436] The *N*-Boc-protected amino *scyllo*-inositol scaffold **618** constituted the key intermediate for subsequent glycosylation, as well as for chain elongation and multiplication directed toward the next dendrimeric generation (Scheme 66). Hence, conjugate 1,4-addition between triamine **618** and excess methyl acrylate, followed by amidation of the resulting esters with ethylenediamine afforded an amidoamine dendrimer under conditions previously seen for the synthesis of PAMAM dendrimers. Reiteration of this two-step sequence successfully doubled peripheral amino functionalities toward a G(2)-dendrimer **620**. Glycoconjugation of the dendrimers was then achieved by the thiourea linkages discussed previously based on mannosyl isothiocyanate **619**.

Another, architecture as yet exotic, has been proposed by Sakamoto and Müllen, who designed original glycodendrimers based on shape-persistent polyphenylenes building

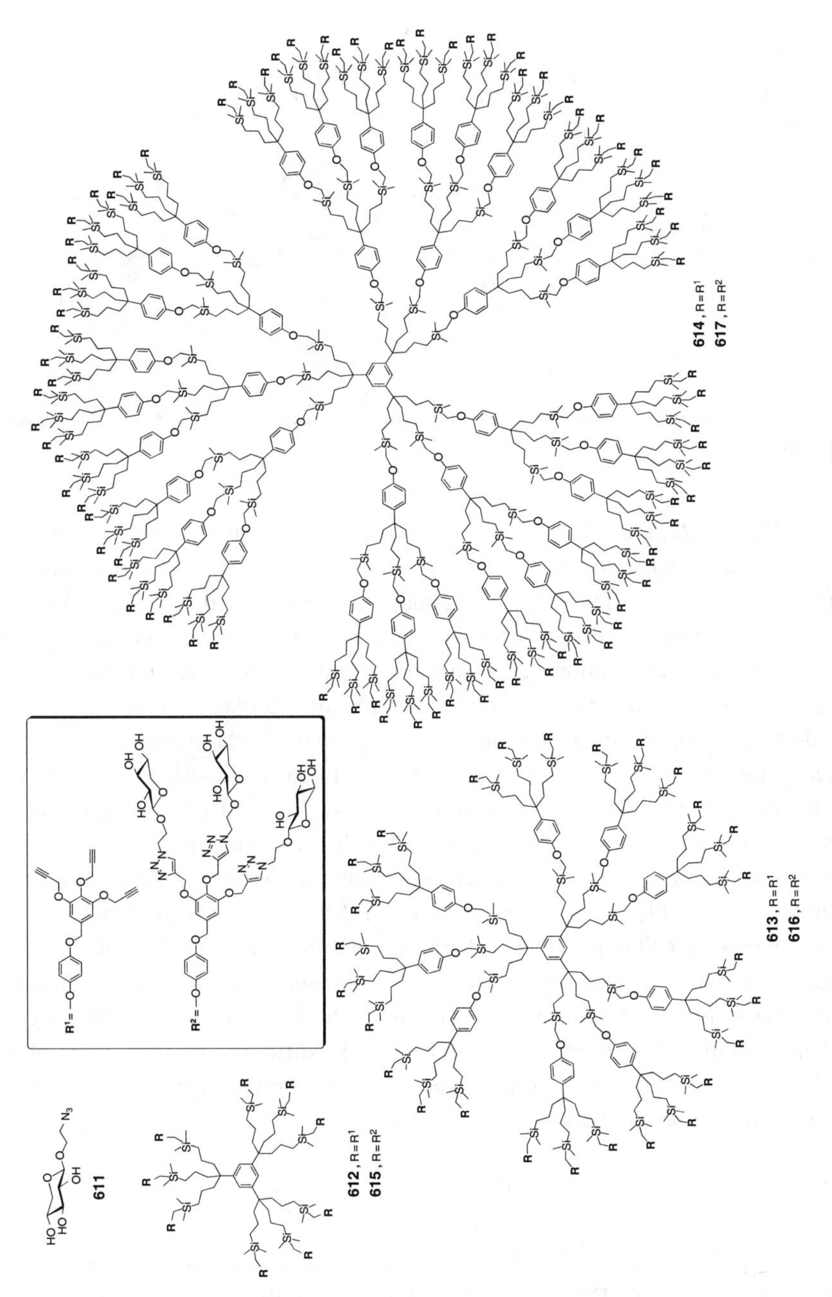

FIG. 77. D-Xylose-carbosilane dendrimers.[434]

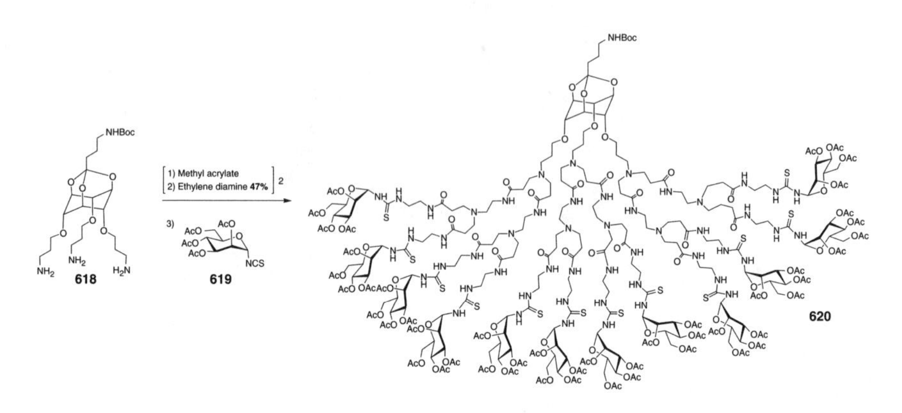

SCHEME 66. Synthesis of *scyllo*-inositol repeating unit toward the construction of novel mannosylated dendrimers.[436]

blocks (**621** and **622**, Fig. 78).[437] Both convergent and divergent routes were used to allow for sugar installation on not only the dendrimer surface but also within the hydrophobic internal scaffold, employing successive Schmidt glycosylation and the Diels–Alder reaction. All of the described glycodendrimers exhibited good solubility in weakly acidic aqueous solutions, and those possessing interior sugar moieties (reminescent of an active center situated inside a hydrophobic pocket of natural enzymes) were identified as interesting candidates for host–guest molecular recognition.

Finally, the group of Majoral prepared additional phosphorus-based glycodendrimers involving hydrazone groups, derived from D-xylose and containing up to 48 peripheral epitopes, in excellent yields and through very simple precipitation-based purification processes (Fig. 79).[438] Such multivalent derivatives were obtained via substitution of the chloride atoms of terminal –$P(S)Cl_2$ groups with a triacetylated phenolic derivative **623** of D-xylose under basic conditions [Cs_2CO_3 (4 eq./Cl atom)]. Consequently, G(1) to G(3) glycodendrimers, containing 12, 24, and 48 xyloside residues, respectively, were readily purified through filtration and extraction steps, and their uniformity was confirmed, notably by ^{31}P NMR spectroscopy. Controlled Zemplén O-deacetylation allowed full deprotection to furnish such valuable amphiphilic glycodendrimers as **624** without decomposition.

VI. CONCLUSION

This chapter has highlighted different strategies leading to a vast range of neoglycoconjugate architectures.[439,440] Since their discovery by Roy *et al.* in 1993, the field

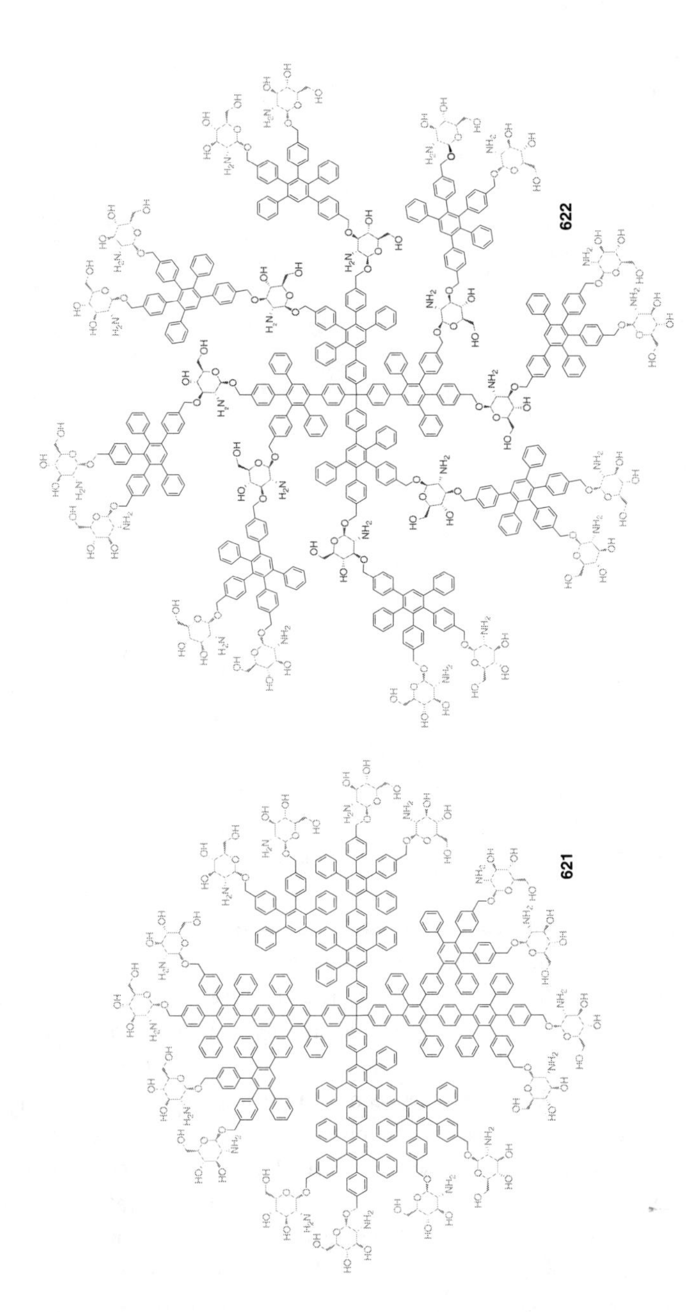

FIG. 78. Polyphenylenes building blocks used by Sakamoto and Müllen for glucosamine dendrimers.[437] (See Color Plate 20.)

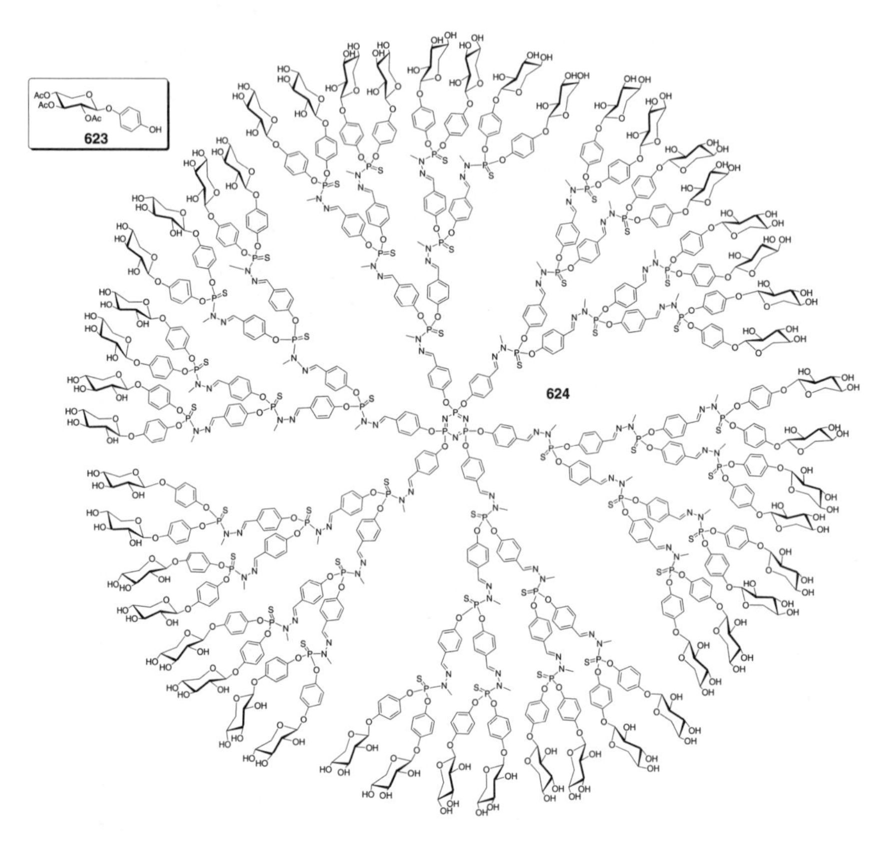

FIG. 79. ''Majoral-type'' glycodendrimers bearing peripheral D-xylose residues.[438]

of glycodendrimers has matured and expanded to unpredicted levels. The synthetic strategies leading to these nanostructures have also witnessed considerable progress, that including organometallic chemistry, chemoenzymatic build-up, silicon chemistry, and self-assembly. Most of the initial challenges in complete substitution by the carbohydrate, and structure determination, have been resolved. Undoubtedly, a fundamental contribution toward this goal has been the utilization of classical [1,3]-dipolar cycloaddition, using the soft and efficient copper-catalyzed, coupling reaction of azide to alkyne. This method, extensively used in this field, has proven extremely versatile. An additional enhancement has been ''accelerated convergent strategies,'' offering clear synthetic advantages and ease of purification. An important addition to the arsenal of sophisticated glycodendrimer methodologies is the recognition that most complex carbohydrate epitopes can be advantageously replaced by simpler saccharides, and most importantly by glycomimetics. In this way, by combining the multivalent scaffolding of carbohydrate ligands, together with their replacement by simpler but higher affinity mimetics, the research community has delivered very potent antiadhesin candidates against viral and bacterial infections, including HIV and Ebola viruses. Key to these discoveries and applications, glycodendrimers have permitted a better understanding and appreciation of the very complex nature of multivalent carbohydrate–protein interactions. The upcoming challenges must address the issues of fine-tuned specificity faced by numerous families of carbohydrate-binding proteins having common and conserved CRDs, such as those identified for instance in galectins and MBPs. Another challenge will be to determine the toxicities of glycodendrimers to allow development of safe therapeutic agents.

References

1. M. Mammen, S.-K. Choi, and G. M. Whitesides, Polyvalent interactions in biological systems: Implications for design and use of multivalent ligands and inhibitors, *Angew. Chem. Int. Ed. Engl.*, 37 (1998) 2754–2794.
2. D. Thompson, *On growth and form,* Cambridge University Press, London, 1987.
3. W. F. Ganong, *Review of medical physiology,* 15th edn. Prentice-Hall, New York, 1991.
4. K. Autumn, Y. A. Liang, S. T. Hsieh, W. Zesch, W. P. Chan, T. W. Kenny, R. Fearing, and R. J. Full, Adhesive force of a single gecko foot-hair, *Nature*, 405 (2000) 681–685.
5. S. Sirois, M. Touaibia, K.-C. Chou, and R. Roy, Glycosylation of HIV-1 gp120 V3 loop: Towards the rational design of a synthetic carbohydrate vaccine, *Curr. Med. Chem.*, 30 (2007) 3232–3242.
6. R. A. Dwek, Glycobiology: Toward understanding the function of sugars, *Chem. Rev.*, 96 (1996) 683–720.
7. C. R. Bertozzi and L. L. Kiessling, Chemical bioglycology, *Science*, 291 (2001) 2357–2364.
8. A. Varki, R. Cummings, J. Esko, H. Freeze, G. Hart, and J. Marth, *Essentials of glycobiology,* Cold Spring Harbor Laboratory Press, New York, 1999.
9. A. Varki, Biological roles of oligosaccharides: All of the theories are correct, *Glycobiology*, 3 (1997) 97–130.
10. R. A. Dwek, Biological importance of glycosylation, *Dev. Biol. Stand.*, 96 (1998) 43–47.

11. D. A. Mann, M. Kanai, D. J. Maly, and L. L. Kiessling, Probing low affinity and multivalent interactions with surface plasmon resonance: Ligands for Concanavalin A, *J. Am. Chem. Soc.*, 120 (1998) 10575–10582.
12. T. K. Dam, R. Roy, D. Pagé, and C. F. Brewer, Thermodynamic binding parameters of individual epitopes of multivalent carbohydrates to concanavalin A as determined by ''reverse'' isothermal titration microcalorimetry, *Biochemistry*, 41 (2002) 1359–1363.
13. J. L. Lundquist and E. J. Toone, The cluster glycoside effect, *Chem. Rev.*, 102 (2002) 555–578.
14. Y. C. Lee and R. T. Lee, Carbohydrate–protein interactions: Basis of glycobiology, *Acc. Chem. Res.*, 28 (1995) 321–327.
15. S. K. Choi, *Synthetic multivalent molecules: concepts and biomedical applications,* Wiley, New York, 2004.
16. H. J. Gabius, H. C. Siebert, S. André, J. Jiménez-Barbero, and H. Rüdiger, Chemical biology of the sugar code, *ChemBioChem*, 5 (2004) 740–764.
17. A. Imberty, Y. M. Chabre, and R. Roy, Glycomimetics and glycodendrimers as high affinity microbial anti-adhesins, *Chem. Eur. J.*, 14 (2008) 7490–7499.
18. T. Feizi, F. Fazio, W. Chai, and C.-H. Wong, Carbohydrate microarrays—a new set of technologies at the frontiers of glycomics, *Curr. Opin. Struct. Biol.*, 13 (2003) 637–645.
19. I. Shin, S. Park, and M.-R. Lee, Carbohydrate microarrays: An advanced technology for functional studies of glycans, *Chem. Eur. J.*, 11 (2005) 2894–2901.
20. J. L. De Paz and P. H. Seeberger, Recent advances in carbohydrate microarrays, *QSAR Comb. Sci.*, 25 (2006) 1027–1032.
21. D. M. Ratner and P. H. Seeberger, Carbohydrate microarrays as tools in HIV glycobiology, *Curr. Pharm. Des.*, 13 (2007) 173–183.
22. M. D. Disney and P. H. Seeberger, The use of carbohydrate microarrays to study carbohydrate–cell interactions and to detect pathogens, *Chem. Biol.*, 11 (2004) 1701–1707.
23. P. I. Kitov, J. M. Sadowska, G. Mulvey, G. D. Armstrong, H. Ling, N. S. Pannu, R. J. Read, and D. R. Bundle, Shiga-like toxins are neutralized by tailored multivalent carbohydrate ligands, *Nature*, 403 (2000) 669–672.
24. M. Mammen, G. Dahmann, and G. M. Whitesides, Effective inhibitors of hemagglutination by Influenza Virus synthesized from polymers having active ester groups. Insight into mechanisms of inhibition, *J. Med. Chem.*, 38 (1995) 4179–4190.

25. S. Howorka, J. Nam, H. Bayley, and D. Kahne, Stochastic detection of monovalent and bivalent protein–ligand interactions, *Angew. Chem. Int. Ed.*, 43 (2004) 842–846.
26. R. J. Pieters, Maximising multivalency effects in protein–carbohydrate interactions, *Org. Biomol. Chem.*, 7 (2009) 2013–2025.
27. K. J. Doores, D. P. Gamblin, and B. G. Davis, Exploring and exploiting the therapeutic potential of glycoconjugates, *Chem. Eur. J.*, 12 (2006) 656–665.
28. B. G. Davis, Synthesis of glycoproteins, *Chem. Rev.*, 102 (2002) 579–601.
29. T. Murata and T. Usui, Enzymatic synthesis of oligosaccharides and neoglycoconjugates, *Biosci. Biotechnol. Biochem.*, 70 (2006) 1049–1059.
30. H. Herzner, T. Reipen, M. Schultz, and H. Kunz, Synthesis of glycopeptides containing carbohydrate and peptide recognition motifs, *Chem. Rev.*, 100 (2000) 4495–4537.
31. D. Specker and V. Wittmann, Synthesis and application of glycopeptide and glycoprotein mimetics, *Top. Curr. Chem.*, 267 (2007) 65–107.
32. V. Verez-Bencomo, V. Fernández-Santana, E. Hardy, M. E. Toledo, M. C. Rodrıguez, L. Heynngnezz, A. Rodriguez, A. Baly, L. Herrera, M. Izquierdo, A. Villar, Y. Valdés, *et al.*, A synthetic conjugate polysaccharide vaccine against *Haemophilus influenzae* type b, *Science*, 305 (2004) 522–525.
33. M. Corti, L. Cantù, P. Brocca, and E. Del Favero, Self-assembly in glycolipids, *Curr. Opin. Colloid Interface Sci.*, 12 (2007) 148–154.
34. V. Sihorkar and S. P. Vias, Potential of polysaccharide anchored liposomes in drug delivery, targeting and immunization, *J. Pharm. Pharm. Sci.*, 4 (2001) 138–158.
35. R. Roy, Design and synthesis of glycoconjugates, in S. H. Khan and R. A. O'Neil, (Eds.) *Modern Methods in Carbohydrate Synthesis,* Harwood Academic, Amsterdam, 1996, pp. 378–402.
36. T. Ouchi and Y. Ohya, Drug delivery systems using carbohydrate recognition, in Y. C. Lee and R. T. Lee, (Eds.) *Neoglycoconjugates: Preparation and Applications,* Academic Press, San Diego, 1994, pp. 465–498.
37. N. V. Bovin and H.-J. Gabius, Polymer immobilized carbohydrate ligands versatile chemical tools for biochemistry and medical science, *Chem. Soc. Rev.*, 24 (1995) 413–421.
38. S. G. Spain, M. I. Gibson, and N. R. Cameron, Recent advances in the synthesis of well-defined glycopolymers, *J. Polym. Sci. Part A: Polym. Chem.*, 45 (2007) 2059–2072.
39. J. M. de la Fuente, A. G. Barrientos, T. C. Rojas, J. Rojo, J. Cañada, A. Fernández, and S. Penadés, Gold glyconanoparticles as water-soluble

polyvalent models to study carbohydrate interactions, *Angew. Chem. Int. Ed.*, 41 (2001) 1554–1557.

40. M. J. Hernáiz, J. M. de la Fuente, A. G. Barrientos, and S. Penadés, A model system mimicking glycosphingolipid clusters to quantify carbohydrate self-interactions by surface plasmon resonance, *Angew. Chem. Int. Ed.*, 40 (2001) 2258–2261.
41. A. G. Barrienyos, J. M. de la Fuente, T. C. Rojas, A. Fernández, and S. Penadés, Gold glyconanoparticles: Synthetic polyvalent ligands mimicking glycocalyx-like surfaces as tools for glycobiological studies, *Chem. Eur. J.*, 9 (2003) 1909–1921.
42. H. Otsuka, Y. Akiyama, Y. Nagasaki, and K. Kataoka, Quantitative and reversible lectin-induced association of gold nanoparticles modified with α-lactosyl-ω-mercapto-poly(ethyleneglycol), *J. Am. Chem. Soc.*, 123 (2001) 8226–8230.
43. C.-C. Lin, Y.-C. Yeh, C.-Y. Yang, C.-L. Chen, G.-F. Chen, C.-C. Chen, and Y.-C. Wu, Selective binding of mannose-encapsulated gold nanoparticles to type 1 pili in *Escherichia coli*, *J. Am. Chem. Soc.*, 124 (2002) 3508–3509.
44. B. Nolting, J.-J. Yu, G.-y. Liu, S.-J. Cho, S. Kauzlarich, and J. Gervay-Hague, Synthesis of gold glyconanoparticles and biological evaluation of recombinant gp120 interactions, *Langmuir*, 19 (2003) 6465–6473.
45. A. Carvalho de Souza, K. M. Halkes, J. D. Meedijk, A. J. Verkleij, J. F. G. Vliegenthart, and J. P. Kamerling, Gold glyconanoparticles as probes to explore the carbohydrate-mediated self-recognition of marine sponge cells, *ChemBioChem*, 6 (2005) 828–831.
46. S. G. Spain, L. Albertin, and N. R. Cameron, Facile *in situ* preparation of biologically active multivalent glyconanoparticles, *Chem. Commun.*, 40 (2006) 4198–4200.
47. J. M. de la Fuente and S. Penadés, Glyconanoparticles: types, synthesis and applications in glycoscience, biomedicine and material science, *Biochim. Biophys. Acta*, 1760 (2006) 636–651.
48. R. Kikkeri, B. Lepenies, A. Adibekian, P. Laurino, and P. H. Seeberger, *In vitro* and *in vivo* liver targeting with carbohydrate capped quantum dots, *J. Am. Chem. Soc.*, 131 (2009) 2110–2112.
49. B. Mukhopadhyay, M. B. Martins, R. Karamanska, D. A. Russell, and R. A. Field, Bacterial detection using carbohydrate-functionalised CdS quantum dots: A model study exploiting *E. coli* recognition of mannosides, *Tetrahedron Lett.*, 50 (2009) 886–889.
50. M. Meldal and C. W. Tomøe, Cu-catalyzed azide-alkyne cycloaddition, *Chem. Rev.*, 108 (2008) 2952–3015.

51. F. Santoyo-González and F. Hernández-Mateo, Azide-alkyne 1, 3-dipolar cycloadditions: A valuable tool in carbohydrate chemistry, *Top. Heterocycl. Chem.*, 7 (2007) 133–177.
52. S. Dedola, S. A. Nepogodiev, and R. A. Field, Recent applications of the CuI-catalysed Huisgen azide-alkyne 1,3-dipolar cycloaddition reaction in carbohydrate chemistry, *Org. Biomol. Chem.*, 5 (2007) 1006–1017.
53. R. J. Pieters, D. T. S. Rijkers, and R. M. J. Liskamp, Application of the 1,3-dipolar cycloaddition reaction in chemical biology: Approaches toward multivalent carbohydrates and peptides and peptide-based polymers, *QSAR Comb. Sci.*, 26 (2007) 1181–1190.
54. R. Roy, A decade of glycodendrimer chemistry, *Trends Glycosci. Glycotechnol.*, 15 (2003) 291–310.
55. M. Touaibia and R. Roy, Application of multivalent mannosylated dendrimers in glycobiology, in J. P. Kamerling, (Ed.), *Comprehensive Glycoscience, Vol. 3,* Elsevier, Amsterdam, 2007, pp. 821–870.
56. Y. M. Chabre and R. Roy, Recent trends in glycodendrimer syntheses and applications, *Curr. Top. Med. Chem.*, 8 (2008) 1237–1285.
57. F. Perez-Balderas, J. Morales-Sanfrutos, F. Hernández-Mateo, J. Isac-García, and F. Santoyo-González, Click multivalent homogeneous neoglycoconjugates—Synthesis and evaluation of their binding affinities, *Eur. J. Org. Chem.*, 2009 (2009) 2441–2453.
58. M. Ortega-Muñoz, F. Perez-Balderas, J. Morales-Sanfrutos, F. Hernández-Mateo, J. Isac-García, and F. Santoyo-González, Click multivalent heterogeneous neoglycoconjugates—Modular synthesis and evaluation of their binding affinities, *Eur. J. Org. Chem.*, 2009 (2009) 2454–2473.
59. Y. C. Lee, Synthesis of some cluster glycosides suitable for attachment to proteins or solid matrices, *Carbohydr. Res.*, 67 (1978) 509–514.
60. T. K. Lindhorst, S. Kötter, U. Krallmann-Wenzel, and S. Ehlers, Trivalent α-D-mannoside clusters as inhibitors of type-1 fimbriae-mediated adhesion of *Escherichia coli*: Structural variation and biotinylation, *J. Chem. Soc. Perkin Trans.*, 1 (2001) 823–831.
61. S. Kötter, U. Krallman-Wenzel, S. Ehlers, and T. B. Lindhorst, Multivalent ligands for the mannose-specific lectin on type 1 fimbriae of *Escherichia coli*: Syntheses and testing of trivalent α-D-mannoside clusters, *J. Chem. Soc. Perkin Trans.*, 1 (1998) 2193–2200.
62. A. Patel and T. B. Lindhorst, Multivalent glycomimetics: Synthesis of nonavalent mannoside clusters with variation of spacer properties, *Carbohydr. Res.*, 341 (2006) 1657–1668.

63. S. P. Gaucher, S. F. Pedersen, and J. A. Leary, Stereospecific synthesis and characterization of aminoglycoside ligands from diethylenetriamine, *J. Org. Chem.*, 64 (1999) 4012–4015.
64. W. Hayes, H. M. I. Osborn, S. D. Osborne, R. A. Rastall, and B. Romagnoli, One-pot synthesis of multivalent arrays of mannose mono- and disaccharides, *Tetrahedron*, 59 (2003) 7983–7996.
65. M. Dubber and T. K. Lindhorst, Exploration of reductive amination for the synthesis of cluster glycosides, *Synthesis*, 2 (2001) 327–330.
66. Y. Li, X. Zhang, S. Chu, K. Yu, and H. Guan, Synthesis of cluster mannosides via a Ugi four-component reaction and their inhibition against the binding of yeast mannan to concanavalin A, *Carbohydr. Res.*, 339 (2004) 873–879.
67. B. Liu and R. Roy, Facile synthesis of glycotope bioisosteres bearing β-D-galactoside moieties, *Tetrahedron*, 57 (2001) 6909–6913.
68. B. Liu and R. Roy, Synthesis of clustered xenotransplantation antagonists using palladium-catalyzed cross-coupling of prop-2-ynyl α-D-galactopyranoside, *J. Chem. Soc. Perkin Trans.*, 1 (2001) 773–779.
69. B. Liu and R. Roy, Olefin self-metathesis as a new entry into xenotransplantation antagonists bearing the Galili antigen, *Chem. Commun.* (2002) 594–595.
70. S. André, B. Liu, H.-J. Gabius, and R. Roy, First demonstration of differential inhibition of lectin binding by synthetic tri- and tetravalent glycoclusters from cross-coupling of rigidified 2-propynyl lactoside, *Org. Biomol. Chem.*, 1 (2001) 3909–3916.
71. S. Hanessian, D. Qiu, H. Prabhanjan, G. V. Reddy, and B. Lou, Synthesis of clustered D-GalNAc (Tn) and D-Galβ(1→3)GalNAc (T) antigenic motifs using a pentaerythritol scaffold, *Can. J. Chem.*, 74 (1996) 1738–1747.
72. H. C. Hansen, S. Haataja, J. Finne, and G. Magnusson, Di-, tri-, and tetravalent dendritic galabiosides that inhibit hemagglutination by *Streptococcus suis* at nanomolar concentration, *J. Am. Chem. Soc.*, 119 (1997) 6974–6979.
73. P. Langer, S. J. Ince, and S. V. Ley, Assembly of dendritic glycoclusters from monomeric mannose building blocks, *J. Chem. Soc. Perkin Trans.*, 1 (1998) 3913–3915.
74. H. Al-Mughaid and T. Bruce Grindley, Synthesis of a nonavalent mannoside glycodendrimer based on pentaerythritol, *J. Org. Chem.*, 71 (2006) 1390–1398.
75. T. K. Lindhorst, M. Dubber, U. Krallmann-Wenzel, and S. Ehlers, Cluster mannosides as inhibitors of type 1 fimbriae-mediated adhesion of *Escherichia coli*: Pentaerythritol derivatives as scaffolds, *Eur. J. Org. Chem.* (2000) 2027–2034.

76. M. Touaibia, T. C. Shiao, A. Papadopoulos, J. Vaucher, Q. Wang, K. Benhamioud, and R. Roy, Tri- and hexavalent mannoside clusters as potential inhibitors of type 1 fimbriated bacteria using pentaerythritol and triazole linkages, *Chem. Commun.* (2007) 380–382.
77. F. Himo, T. Lovell, R. Hilgraf, V. V. Rostovtsev, L. Noodleman, K. B. Sharpless, and V. V. Fokin, Copper(I)-catalyzed synthesis of azoles. DFT study predicts unprecedented reactivity and intermediates, *J. Am. Chem. Soc.*, 127 (2005) 210–216.
78. V. V. Rostovtsev, L. G. Green, V. V. Fokin, and K. B. Sharpless, A stepwise Huisgen cycloaddition process: Copper(I)-catalyzed regioselective ''ligation'' of azides and terminal alkynes, *Angew. Chem. Int. Ed.*, 41 (2002) 2596–2599.
79. P. Wu, A. K. Feldman, A. K. Nugent, C. J. Hawker, A. Scheel, B. Viot, J. Pyun, J. M. J. Fréchet, K. B. Sharpless, and V. V. Fokin, Efficiency and fidelity in a click-chemistry route to triazole dendrimers by the copper(I)-catalyzed ligation of azides and alkynes, *Angew. Chem. Int. Ed.*, 43 (2004) 3928–3932.
80. N. Sharon, Bacterial lectins, cell–cell recognition and infectious disease, *FEBS Lett.*, 217 (1987) 145–157.
81. P. Arya, K. M. K. Kutterer, H. Qin, J. Roby, M. L. Barnes, J. M. Kim, and R. Roy, Diversity of C-linked neoglycopeptides for the exploration of subsite-assisted carbohydrate binding interactions, *Bioorg. Med. Chem. Lett.*, 8 (1998) 1127–1132.
82. M. Touaibia, A. Wellens, T. C. Shiao, Q. Wang, S. Sirois, J. Bouckaert, and R. Roy, Mannosylated G(0) dendrimers with nanomolar affinities to *Escherichia coli* FimH, *ChemMedChem*, 2 (2007) 1190–1201.
83. S. Fortier, M. Touaibia, S. Lord-Dufour, J. Galipeau, R. Roy, and B. Annabi, Tetra- and hexavalent mannosides inhibit the pro-apoptotic, antiproliferative and cell surface clustering effects of concanavalin A: Impact on MT1-MMP functions in marrow-derived mesenchymal stromal cells, *Glycobiology*, 18 (2008) 195–204.
84. S. G. Gouin, A. Wellens, J. Bouckaert, and J. Kovensky, Synthetic multimeric heptyl mannosides as potent antiadhesives of uropathogenic *Escherichia coli*, *ChemMedChem*, 4 (2009) 749–755.
85. J. Bouckaert, J. Berglund, M. Schembri, E. De Genst, L. Cools, M. Wuhrer, C.-S. Hung, J. Pinkner, R. Slättegård, A. Zavialov, D. Choudhury, S. Langermann, *et al.*, Receptor binding studies disclose a novel class of high-affinity inhibitors of the *Escherichia coli* FimH adhesin, *Mol. Microbiol.*, 55 (2005) 441–455.

86. A. Wellens, C. Garofalo, H. Nguyen, N. Van Gerven, R. Slättegård, J.-P. Hernalsteens, L. Wyns, S. Oscarson, H. De Greve, S. Hultgren, and J. Bouckaert, Intervening with urinary tract infections using anti-adhesives based on the crystal structure of the FimH-oligomannose-3 complex, *PLoS ONE*, 3 (2008) e2040.
87. M. Weïwer, C.-C. Chen, M. M. Kemp, and R. J. Linhardt, Synthesis and biological evaluation of non-hydrolyzable 1,2,3-triazole-linked sialic acid derivatives as neuramidase inhibitors, *Eur. J. Org. Chem.* (2009) 2611–2620.
88. P. Albersheim and B. S. Valent, Host–pathogen interactions in plants—Plants, when exposed to oligosaccharides of fungal origin, defend themselves by accumulating antibiotics, *J. Cell. Biol.*, 78 (1978) 627–643.
89. J. Yariv, M. M. Rapport, and L. Graf, The interaction of glycosides and saccharides with antibody to the corresponding phenylazo glycosides, *Biochem. J.*, 85 (1962) 383–388.
90. R. Kaufman and R. S. Sidhu, Synthesis of aryl cluster glycosides by cyclotrimerization of 2-propynyl carbohydrate derivatives, *J. Org. Chem.*, 47 (1982) 4941–4947.
91. R. Roy, S. K. Das, R. Dominique, M. Corazon Trono, F. Hernández-Mateo, and F. Santoyo-González, Transition metal catalyzed neoglycoconjugate syntheses, *Pure Appl. Chem.*, 71 (1999) 565–571.
92. R. Roy, M. Corazon Trono, and D. Giguère, Effects of linker rigidity and orientation of mannoside clusters for multivalent interactions with proteins, in R. Roy, (Ed), *Glycomimetics: Modern Synthetic Methodologies, ACS Symp. Ser.*, 896 (2005) 137–150.
93. R. Dominique, B. Liu, S. K. Das, and R. Roy, Synthesis of ''molecular asterisks'' via sequential cross-metathesis Sonogashira and cyclotrimerization reactions, *Synthesis*, 6 (2000) 862–868.
94. A. Giannis and K. Sandhoff, Stereoselective synthesis of α-*C*-allyl-glycopyranosides, *Tetrahedron Lett.*, 26 (1985) 1479–1482.
95. S. K. Das and R. Roy, Mild Ruthenium-catalyzed intermolecular alkyne cyclotrimerization, *Tetrahedron Lett.*, 40 (1999) 4015–4018.
96. D. Pagé and R. Roy, Synthesis of divalent α-D-mannopyranosylated clusters having enhanced binding affinities towards concanavalin A and pea lectins, *Bioorg. Med. Chem. Lett.*, 6 (1996) 1765–1770.
97. J. Lehmann and U. P. Weitzel, Synthesis and application of α-D-mannosyl clusters as photoaffinity ligands for mannose-binding proteins: Concanavalin A as a model receptor, *Carbohydr. Res.*, 294 (1996) 65–94.

98. R. Roy, S. K. Das, F. Santoyo-González, F. Hernández-Mateo, T. K. Dam, and C. F. Brewer, Synthesis of ''sugar-rods'' with phytohemagglutinin cross-linking properties by using the palladium-catalyzed Sonogashira reaction, *Chem. Eur. J.*, 6 (2000) 1757–1762.
99. Z. Gan and R. Roy, Transition metal-catalyzed syntheses of ''rod-like'' thioglycoside dimers, *Tetrahedron Lett.*, 41 (2000) 1155–1158.
100. S. Sengupta and S. K. Sadhukhan, Synthetic studies on dendritic glycoclusters: A convergent palladium-catalyzed strategy, *Carbohydr. Res.*, 332 (2001) 215–219.
101. A. Dondoni, A. Marra, and M. G. Zampolli, Synthesis of all carbon linked glycoside clusters round benzene scaffold via Sonogashira–Heck–Cassar cross coupling of iodobenzenes with ethynyl *C*-glycosides, *Synlett*, 11 (2002) 1850–1854.
102. S. André, B. Liu, H.-J. Gabius, and R. Roy, First demonstration of differential inhibition of lectin binding by synthetic tri- and tetravalent glycoclusters from cross-coupling of rigidified 2-propynyl lactoside, *Org. Biomol. Chem.*, 1 (2003) 3909–3916.
103. Y. M. Chabre, C. Contino-Pépin, V. Placide, T. C. Shiao, and R. Roy, Expeditive synthesis of glycodendrimer scaffolds based on versatile TRIS and mannoside derivatives, *J. Org. Chem.*, 73 (2008) 5602–5605.
104. M. Sleiman, A. Varrot, J.-M. Raimundo, M. Gingras, and P. G. Goekjian, Glycosylated asterisks are among the most potent low valency inducers of Concanavalin A aggregation, *Chem. Commun.* (2008) 6507–6509.
105. E. Da Silva, A. N. Lazar, and A. W. Coleman, Biopharmaceutical application of calixarenes, *J. Drug Deliv. Sci.*, 14 (2004) 3–20.
106. A. Marra, M.-C. Schermann, A. Dondoni, A. Casnati, P. Minari, and R. Ungaro, Sugar calixarenes: Preparation of calix[4]arenes substituted at the lower and upper rims with *O*-glycosyl groups, *Angew. Chem. Int. Ed.*, 33 (1994) 2479–2481.
107. D. A. Fulton and J. F. Stoddart, Neoglycoconjugates based on cyclodextrins and calixarenes, *Bioconjug. Chem.*, 12 (2001) 655–672.
108. A. Casnati, F. Sansone, and R. Ungaro, Peptido- and glycocalixarenes: Playing with hydrogen bonds around hydrophobic cavities, *Acc. Chem. Res.*, 36 (2003) 246–254.
109. S. J. Meunier and R. Roy, Polysialosides scaffolded on *p-tert*-butylcalix[4] arene, *Tetrahedron Lett.*, 37 (1996) 5469–5472.
110. R. Roy and J. M. Kim, Amphiphilic *p-tert*-butylcalix[4]arene scaffolds containing exposed carbohydrate dendrons, *Angew. Chem. Int. Ed.*, 38 (1999) 369–372.

111. A. Dondoni, M. Kleban, and A. Marra, The assembly of Carbon-linked calix-arene-carbohydrate structures (*C*-calixsugars) by multiple Wittig olefination, *Tetrahedron Lett.*, 38 (1997) 7801–7804.
112. A. Dondoni, M. Kleban, X. Hu, A. Marra, and H. D. Banks, Glycoside-clustering round calixarenes toward the development of multivalent carbohydrate ligands. Synthesis and conformational analysis of calix[4]arene *O*- and *C*-glycoconjugates, *J. Org. Chem.*, 67 (2002) 4722–4733.
113. F. Pérez-Balderas and F. Santoyo-González, Synthesis of deeper calix-sugar-based on the Sonogashira reaction, *Synlett*, 11 (2001) 1699–1702.
114. A. Dondoni and A. Marra, C-glycoside clustering on calix[4]arene, adamantane, and benzene scaffolds through 1, 2, 3-triazole linkers, *J. Org. Chem.*, 71 (2006) 7546–7557.
115. A. Vecchi, B. Melai, A. Marra, C. Chiappe, and A. Dondoni, Microwave-enhanced ionothermal CuAAC for the synthesis of glycoclusters on a calix[4] arene platform, *J. Org. Chem.*, 73 (2008) 6437–6440.
116. A. Dondoni and A. Marra, Addressing the scope of the azide-nitrile cycloaddition in glycoconjugate chemistry. The assembly of *C*-glycoclusters on a calix[4] arene scaffold through tetrazole spacers, *Tetrahedron*, 63 (2007) 6339–6345.
117. U. Schädel, F. Sansone, A. Casnati, and R. Ungaro, Synthesis of upper rim calix [4]arene divalent glycoclusters via amide bond conjugation, *Tetrahedron*, 61 (2005) 1149–1154.
118. C. Saitz-Barria, A. Torres-Pinedo, and F. Santoyo-González, Synthesis of bridged thiourea calix-sugar, *Synlett*, 12 (1999) 1891–1894.
119. G. M. L. Consoli, F. Cunsolo, C. Geraci, T. Mecca, and P. Neri, Calix[8]arene-based glycoconjugates as multivalent carbohydrate-presenting systems, *Tetrahedron Lett.*, 44 (2003) 7467–7470.
120. Y. Ge, Y. Cai, and C. Yan, Synthesis of thiourea-bridged cluster glycoside calixarenes, *Synth. Commun.*, 35 (2005) 2355–2361.
121. G. M. L. Consoli, F. Cunsolo, C. Geraci, and V. Sgarlata, Synthesis and lectin binding ability of glycosamino acid-calixarenes exposing GlcNAc clusters, *Org. Lett.*, 6 (2004) 4163–4166.
122. A. Dondoni, A. Marra, M.-C. Scherrmann, A. Casnati, F. Sansone, and R. Ungaro, Synthesis and properties of *O*-glycosyl calix[4]arenes (Calixsugars), *Chem. Eur. J.*, 3 (1997) 1774–1782.
123. F. Sansone, E. Chierici, A. Casnati, and R. Ungaro, Thiourea-linked upper rim calix[4]arene neoglycoconjugates: Synthesis, conformations and binding properties, *Org. Biomol. Chem.*, 1 (2003) 1802–1809.

124. F. Sansone, L. Baldini, A. Casnati, and R. Ungaro, Conformationally mobile glucosylthioureidocalix[6]- and calix[8]arenes: Synthesis, aggregation and lectin binding, *Supramol. Chem.*, 20 (2008) 161–168.
125. D. Arosio, M. Fontanella, L. Baldini, L. Mauri, A. Bernardi, A. Casnati, F. Sansone, and R. Ungaro, A synthetic divalent Cholera Toxin glycocalix[4] arene ligand having higher affinity than natural GM1 oligosaccharide, *J. Am. Chem. Soc.*, 127 (2005) 3660–3661.
126. D. Vanden Broeck, C. Horvath, and M. J. S. De Wolf, *Vibrio cholerae*: Cholera toxin, *Int. J. Biochem. Cell Biol.*, 39 (2007) 1771–1775.
127. I. A. Velter, M. Politi, C. Podlipnik, and F. Nicotra, Natural and synthetic cholera toxin antagonists, *Mini-Rev. Med. Chem.*, 7 (2007) 159–170.
128. A. Bernardi, L. Carrettoni, A. Grosso Ciponte, D. Montib, and S. Sonnino, Second generation mimics of ganglioside GM1 as artificial receptors for Cholera Toxin: Replacement of the sialic acid moiety, *Bioorg. Med. Chem. Lett.*, 10 (2000) 2197–2200.
129. K. Křenek, M. Kuldová, K. Hulíková, I. Stibor, P. Lhoták, M. Dudič, J. Budka, H. Pelantová, K. Bezouška, A. Fišerová, and V. Křen, *N*-Acetyl-D-glucosamine substituted calix[4]arenes as stimulators of NK cell-mediated antitumor immune response, *Carbohydr. Res.*, 342 (2007) 1781–1792.
130. C. Geraci, G. M. L. Consoli, E. Galante, E. Bousquet, M. Pappalardo, and A. Spadaro, Calix[4]arene decorated with four Tn antigen glycomimetic units and P_3CS immunoadjuvant: Synthesis, characterization, and anticancer immunological evaluation, *Bioconjug. Chem.*, 19 (2008) 751–758.
131. S. André, F. Sansone, H. Kaltner, A. Casnati, J. Kopitz, H.-J. Gabius, and R. Ungaro, Calix[*n*]arene-based glycoclusters: Bioactivity of thiourea-linked galactose/lactose moieties as inhibitors of binding of medically relevant lectins to a glycoprotein and cell-surface glycoconjugates and selectivity among human adhesion/growth-regulatory galectins, *ChemBioChem*, 9 (2008) 1649–1661.
132. P. Maillard, J.-L. Guerquin-Kern, M. Momenteau, and S. Gaspard, Glycoconjugated tetrapyrrolic macrocycles, *J. Am. Chem. Soc.*, 111 (1989) 9125–9127.
133. I. Laville, S. Pigaglio, J.-C. Blais, F. Doz, B. Loock, P. Maillard, D. S. Grierson, and J. Blais, Photodynamic efficiency of diethylene glycol-linked glycoconjugated porphyrins in human retinoblastoma cells, *J. Med. Chem.*, 49 (2006) 2558–2567.
134. V. Sol, P. Branland, R. Granet, C. Kaldapa, B. Verneuil, and P. Krausz, Nitroglycosylated *meso*-arylporphyrins as photoinhibitors of gram positive bacteria, *Bioorg. Med. Chem. Lett.*, 8 (1998) 3007–3010.

135. J. P. C. Tomé, M. G. P. M. S. Neves, A. C. Tomé, J. A. S. Cavaleiro, A. F. Mendonça, I. N. Pegado, R. Duarte, and M. L. Valdeira, Synthesis of glycoporphyrin derivatives and their antiviral activity against herpes simplex virus types 1 and 2, *Bioorg. Med. Chem.*, 13 (2005) 3878–3888.
136. F.-C. Gong, X.-B. Zhang, C.-C. Guo, G.-L. Shen, and R.-Q. Yu, Amperometric metronidazole sensor based on the supramolecular recognition by metalloporphyrin incorporated in carbon paste electrode, *Sensors*, 3 (2004) 91–100.
137. T. G. Minehan and Y. Kishi, Total synthesis of the proposed structure of (+)-Tolyporphyn A *O*,*O*-diacetate, *Angew. Chem. Int. Ed.*, 38 (1999) 923–925.
138. T. G. Minehan, L. Cook-Blumberg, Y. Kishi, M. R. Prinsep, and R. E. Moore, Revised structure of Tolyporphyn A, *Angew. Chem. Int. Ed.*, 38 (1999) 926–928.
139. J. A. S. Cavaleiro, J. P. C. Tomé, and M. A. F. Faustino, Synthesis of glycoporphyrins, *Top. Heterocycl. Chem.*, 7 (2007) 179–248.
140. D. Oulmi, P. Maillard, J.-L. Guerquin-Kern, C. Huel, and M. Momenteau, Glycoconjugated porphyrins. 3. Synthesis of flat amphiphilic mixed *meso*-(glycosylated aryl)arylporphyrins and mixed *meso*-(glycosylated aryl)alkylporphyrins bearing some mono- and disaccharide groups, *J. Org. Chem.*, 60 (1995) 1554–1564.
141. Y. Mikata, Y. Onchi, K. Tabata, S.-i. Ogura, I. Okura, H. Ono, and S. Yano, Sugar-dependent photocytotoxic property of tetra- and octa-glycoconjugated tetraphenylporphyrins, *Tetrahedron Lett.*, 39 (1998) 4505–4508.
142. P. Pasetto, X. Chen, C. M. Drain, and R. W. Franck, Synthesis of hydrolytically stable porphyrin *C*- and *S*-glycoconjugates in high yields, *Chem. Commun.* (2001) 81–82.
143. A. A. Aksenova, Y. L. Sebyakin, and A. F. Mironov, Conjugates of porphyrins with carbohydrates, *Russ. J. Bioorg. Chem.*, 29 (2003) 201–219.
144. V. Sol, V. Chaleix, Y. Champavier, R. Granet, Y.-M. Huang, and P. Krausz, Glycosyl bis-porphyrin conjugates: Synthesis and potential application in PDT, *Bioorg. Med. Chem.*, 14 (2006) 7745–7760.
145. G. Fülling, D. Schröder, and B. Franck, Water-soluble porphyrin diglycosides with photosensitizing properties, *Angew. Chem. Int. Ed. Engl.*, 28 (1989) 1519–1521.
146. S.-i. Kawano, S.-i. Tamaru, N. Fujita, and S. Shinkai, Sol-gel polycondensation of tetraethyl orthosilicate (TEOS) in sugar-based porphyrin organogels: Inorganic conversion of a sugar-directed porphyrinic fiber library through sol-gel transcription processes, *Chem. Eur. J.*, 10 (2004) 343–351.

147. X. Chen, L. Hui, D. A. Foster, and C. M. Drain, Efficient synthesis and photodynamic activity of porphyrin–saccharide conjugates: Targeting and incapacitating cancer cells, *Biochemistry*, 43 (2004) 10918–10929.
148. T. Matsuo, H. Nagai, Y. Hisaeda, and T. Hayashi, Construction of glycosylated myoglobin by reconstitutional method, *Chem. Commun.* (2006) 3131–3133.
149. R. Ballardini, B. Colonna, M. T. Gandolfi, S. A. Kalovidouris, L. Orzel, F. M. Raymo, and J. F. Stoddart, Porphyrin-containing glycodendrimers, *Eur. J. Org. Chem.* (2003) 288–294.
150. R. Bonnett, *Chemical aspects of photodynamic therapy,* Gordon and Breach Science, Amsterdam, 2000.
151. S. Griegel, M. F. Rajewsky, T. Cisciolka, and H.-J. Gabius, Endogenous sugar receptor (lectin) profiles of human retinoblastoma and retinoblast cell lines analyzed by cytological markers, affinity chromatography and neoglycoprotein-targeted photolysis, *Anticancer Res.*, 9 (1989) 723–730.
152. P. Maillard, B. Loock, D. S. Grierson, I. Laville, J. Blais, F. Doz, L. Desjardins, D. Carrez, J.-L. Guerquin-Kern, and A. Croisy, *In vitro* phototoxicity of glycoconjugated porphyrins and chlorins in colorectal adenocarcinoma (HT29) and retinoblastoma (Y79) cell lines, *Photodiagnosis Photodyn. Ther.*, 4 (2004) 261–268.
153. R. G. Little, J. A. Anton, P. A. Loach, and J. A. Ibers, The synthesis of some substituted tetraarylporphyrins, *J. Heterocycl. Chem.*, 12 (1975) 343–349.
154. S. Ballut, A. Makky, B. Loock, J.-P. Michel, P. Maillard, and V. Rosilio, New strategy for targeting of photosensitizers. Synthesis of glycodendrimeric phenylporphyrins, incorporation into a liposome membrane and interaction with a specific lectin, *Chem. Commun.* (2009) 224–226.
155. R. J. Patch, H. Chen, and C. R. Pandit, Multivalent templated saccharides: Convenient syntheses of spacer-linked 1, $1'$-bis –and 1, $1'$, $1''$-tris-β-glycosides by the glycal epoxide glycosidation method, *J. Org. Chem.*, 62 (1997) 1543–1546.
156. D. Giguère, R. Patman, M.-A. Bellefleur, C. St-Pierre, S. Sato, and R. Roy, Carbohydrate triazoles and isoxazoles as inhibitors of galectins-1 and -3, *Chem. Commun.* (2006) 2379–2381.
157. K. Marotte, C. Préville, C. Sabin, M. Moumé-Pymbock, A. Imberty, and R. Roy, Synthesis and binding properties of divalent and trivalent clusters of the Lewis a disaccharide moiety to *Pseudomonas aeruginosa* lectin PA-IIL, *Org. Biomol. Chem.*, 5 (2007) 2953–2961.

158. A. R. M. Soares, J. P. C. Tomé, M. G. P. M. S. Neves, A. C. Tomé, J. A. S. Cavaleiro, and T. Torres, Synthesis of water-soluble phthalocyanines bearing four or eight D-galactose units, *Carbohydr. Res.*, 344 (2009) 507–510.
159. Z. Iqbal, M. Hanack, and T. Ziegler, Synthesis of an octasubstituted galactose zinc(II) phthalocyanine, *Tetrahedron Lett.*, 50 (2009) 873–875.
160. M. Köhn, J. M. Benito, C. O. Mellet, T. K. Lindhorst, and J. M. García Fernández, Functional evaluation of carbohydrate-centred glycoclusters by enzyme-linked lectin assay: Ligands for Concanavalin A, *ChemBioChem*, 5 (2004) 771–777.
161. M. Dubber and T. K. Lindhorst, Synthesis of carbohydrate-centered oligosaccharide mimetics equipped with a functionalized tether, *J. Org. Chem.*, 65 (2000) 5275–5281.
162. M. Dubber and T. K. Lindhorst, Trehalose-based octopus glycosides for the synthesis of carbohydrate-centered PAMAM dendrimers and thiourea-bridged glycoclusters, *Org. Lett.*, 3 (2001) 4019–4022.
163. M. Ortega-Muñoz, J. Morales-Sanfrutos, F. Perez-Balderas, F. Hernandez-Mateo, M. D. Giron-Gonzalez, N. Sevillano-Tripero, R. Salto-Gonzalez, and F. Santoyo-Gonzalez, Click multivalent neoglycoconjugates as synthetic activators in cell adhesion and stimulation of monocyte/macrophage cell lines, *Org. Biomol. Chem.*, 5 (2007) 2291–2301.
164. S. G. Gouin, E. Vanquelef, J. M. Garcia Fernández, C. Ortiz Mellet, F.-Y. Dupradeau, and J. Kovensky, Multi-mannosides based on a carbohydrate scaffold: Synthesis, force field development, molecular dynamics studies, and binding affinities for lectin Con A, *J. Org. Chem.*, 72 (2007) 9032–9045.
165. E. A. Merritt and W. G. J. Hol, AB5 toxins, *Curr. Opin. Struct. Biol.*, 5 (1995) 165–171.
166. G. L. Mulvey, P. Marcato, P. I. Kitov, J. Sadowska, D. R. Bundle, and G. D. Armstrong, Assessment in mice of the therapeutic potential of tailored, multivalent Shiga Toxin carbohydrate ligands, *J. Infect. Dis.*, 187 (2003) 640–649.
167. M. Singh, R. Sharma, and U. C. Banerjee, Biotechnological applications of cyclodextrins, *Biotechnol. Adv.*, 20 (2002) 341–359.
168. V. Lainé, A. Coste-Sarguet, A. Gadelle, J. Defaye, B. Perly, and F. Djedaïni-Pilard, Inclusion and solubilisation properties of 6-*S*-glycosyl-6-thio-derivatives of β-cyclodextrin, *J. Chem. Soc. Perkin Trans.*, 2 (1995) 1479–1487.
169. D. A. Fulton and J. F. Stoddart, An efficient synthesis of cyclodextrin-based carbohydrate cluster compounds, *Org. Lett.*, 2 (2000) 1113–1116.

170. T. Furiuke, R. Sadamoto, K. Niikura, K. Monde, N. Sakairi, and S.-I. Nishimura, Chemical and enzymatic synthesis of glycocluster having seven sialyl lewis X arrays using β-cyclodextrin as a key scaffold material, *Tetrahedron*, 61 (2005) 1737–1742.
171. J. J. García-López, F. Santoyo-González, A. Vargas-Berenguel, and J. J. Giménez-Martínez, Synthesis of cluster *N*-glycosides based on a β-cyclodextrin core, *Chem. Eur. J.*, 5 (1999) 1775–1784.
172. C. Ortiz-Mellet, J. M. Benito, J. M. García Fernández, H. Law, K. Chmurski, J. Defaye, M. L. O'Sullivan, and H. N. Caro, Cyclodextrin-scaffolded glycoclusters, *Chem. Eur. J.*, 4 (1998) 2523–2531.
173. R. Roy, F. Hernández-Mateo, and F. Santoyo-González, Synthesis of persialylated β-cyclodextrins, *J. Org. Chem.*, 65 (2000) 8743–8746.
174. J. J. García-López, F. Hernández-Mateo, J. Isac-García, J. M. Kim, R. Roy, F. Santoyo-González, and A. Vargas-Berenguel, Synthesis of per-glycosylated β-cyclodextrins having enhanced lectin binding affinity, *J. Org. Chem.*, 64 (1999) 522–531.
175. T. Furuike, S. Aiba, and S.-I. Nishimura, A highly practical synthesis of cyclodextrin-based glycoclusters having enhanced affinity with lectins, *Tetrahedron*, 56 (2000) 9909–9915.
176. J. M. Benito, M. Gómez-Garcia, C. Ortiz Mellet, I. Baussanne, J. Defaye, and J. M. García Fernández, Optimizing saccharide-directed molecular delivery to biological receptors: Design, synthesis, and biological evaluation of glycodendrimer-cyclodextrin conjugates, *J. Am. Chem. Soc.*, 126 (2004) 10355–10363.
177. C. Ortiz Mellet, J. Defaye, and J. M. García Fernández, Multivalent cyclooligosaccharides: Versatile carbohydrate clusters with dual role as molecular receptors and lectin ligands, *Chem. Eur. J.*, 8 (2002) 1983–1990.
178. I. Baussanne, J. M. Benito, C. Ortiz Mellet, J. M. Garcıa Fernandez, and J. Defaye, Synthesis and comparative lectin-binding affinity of mannosyl-coated β-cyclodextrin-dendrimer constructs, *Chem. Commun.* (2000) 1489–1490.
179. E. Benoist, A. Loussouarn, P. Remaud, J. C. Chatal, and J. F. Gestin, Convenient and simplified approaches to *N*-monoprotected triaminopropane derivatives. Key intermediates for bifunctional chelating agent synthesis, *Synthesis* (1998) 1113–1118.
180. P. Potier, Rhône-Poulenc Lecture. Search and discovery of new antitumour compounds, *Chem. Soc. Rev.*, 21 (1992) 113–119.
181. B. Westermann and S. Dörner, Synthesis of multivalent aminoglycoside mimics via the Ugi multicomponent reaction, *Chem. Commun.* (2005) 2116–2118.

182. K. Sato, N. Hada, and T. Takeda, Synthesis of new peptidic glycoclusters derived from β-alanine: Di- and trimerized glycoclusters and glycocluster-clusters, *Carbohydr. Res.*, 341 (2006) 836–845.
183. H. Kamitakahara, T. Suzuki, N. Nishigori, Y. Suzuki, O. Kanie, and C.-H. Wong, A lysoganglioside/poly-L-glutamic acid conjugate as a picomolar inhibitor of Influenza Hemagglutinin, *Angew. Chem. Int. Ed.*, 37 (1998) 1524–1528.
184. H. A. Shaikh, F. D. Sönnichsen, and T. K. Lindhorst, Synthesis of glycocluster peptides, *Carbohydr. Res.*, 343 (2008) 1665–1674.
185. V. Wittmann and S. Seeberger, Combinatorial solid-phase synthesis of multivalent cyclic neoglycopeptides, *Angew. Chem. Int. Ed.*, 39 (2000) 4348–4352.
186. O. Renaudet and P. Dumy, A fully solid-phase synthesis of biotinylated glycoclusters, *Open Glycosci.*, 1 (2008) 1–7.
187. Z. Zhang, J. Liu, C. L. M. J. Verlinde, W. G. J. Hol, and E. Fan, Large cyclic peptides as cores of multivalent ligands: Application to inhibitors of receptor binding by cholera toxin, *J. Org. Chem.*, 69 (2004) 7737–7740.
188. I. J. Krauss, J. G. Joyce, A. C. Finnefrock, H. C. Song, V. Y. Dudkin, X. Geng, J. D. Warren, M. Chastain, J. W. Shiver, and S. J. Danishefsky, Fully synthetic carbohydrate HIV antigens designed on the logic of the 2G12 Antibody, *J. Am. Chem. Soc.*, 129 (2007) 11042–11044.
189. L. M. Likhosherstov, O. S. Novikova, V. A. Derevitskaja, and N. K. Kochetkov, A new simple synthesis of amino sugar β -D-glycosylamines, *Carbohydr. Res.*, 146 (1986) C1–C5.
190. S. T. Cohen-Anisfeld and P. T. Lansbury, A practical, convergent method for glycopeptide synthesis, *J. Am. Chem. Soc.*, 115 (1993) 10531–10537.
191. J. Wang, H. Li, G. Zou, and L.-X. Wang, Novel template-assembled oligosaccharide clusters as epitope mimics for HIV-neutralizing antibody 2G12. Design, synthesis, and antibody binding study, *Org. Biomol. Chem.*, 5 (2007) 1529–1540.
192. S. D. Burke, Q. Zhao, M. C. Schuster, and L. L. Kiessling, Synergistic formation of soluble lectin clusters by a templated multivalent saccharide ligand, *J. Am. Chem. Soc.*, 122 (2000) 4518–4519.
193. S. D. Burke and Q. Zhao, Synthesis and study of C_3-symmetric hydropyran cyclooligolides with oriented aryl and alcohol appendages at 10 Å spacing, *J. Org. Chem.*, 65 (2000) 1489–1500.
194. J. Kim, Y. Ahn, K. M. Park, Y. Kim, Y. H. Ko, D. H. Oh, and K. Kim, Carbohydrate wheels: Cucurbituril-based carbohydrate clusters, *Angew. Chem. Int. Ed.*, 46 (2007) 7393–7395.

195. M. Touaibia and R. Roy, First synthesis of ''Majoral-type'' glycodendrimers bearing covalently bound α-D-mannopyranoside residues onto a hexachlorocyclotriphosphazene core, *J. Org. Chem.*, 73 (2008) 9292–9302.
196. B. König, T. Fricke, A. Waβmann, U. Krallmann-Wenzel, and T. K. Lindhorst, α-Mannosyl clusters scaffolded on azamacrocycles: Synthesis and inhibitory properties in the adhesion of type 1 fimbriated *Escherichia coli* to guinea pig erythrocytes, *Tetrahedron Lett.*, 39 (1998) 2307–2310.
197. H. Stephan, A. Röhrich, S. Noll, J. Steinbach, R. Kirchner, and J. Seidel, Carbohydratation of 1,4,8,11-tetraazacyclotetradecane (cyclam): Synthesis and binding properties toward concanavalin A, *Tetrahedron Lett.*, 48 (2007) 8834–8838.
198. E. Fan, Z. Zhang, W. E. Minke, Z. Hou, C. L. M. J. Verlinde, and W. G. J. Hol, High-affinity pentavalent ligands of *Escherichia coli* heat-labile enterotoxin by modular structure-based design, *J. Am. Chem. Soc.*, 122 (2000) 2663–2664.
199. E. A. Meritt, Z. Zhang, J. C. Pickens, M. Ahn, W. G. J. Hol, and E. Fan, Characterization and crystal structure of a high-affinity pentavalent receptor-binding inhibitor for cholera toxin and *E. coli* heat-labile enterotoxin, *J. Am. Chem. Soc.*, 124 (2002) 8818–8824.
200. Z. Zhang, J. C. Pickens, W. G. J. Hol, and E. Fan, Solution- and solid-phase syntheses of guanidine-bridged, water-soluble linkers for multivalent ligand design, *Org. Lett.*, 6 (2004) 1377–1380.
201. Z. Zhang, E. A. Merritt, M. Ahn, C. Roach, Z. Hou, C. L. M. J. Verlinde, W. G. J. Hol, and E. Fan, Solution and crystallographic studies of branched multivalent ligands that inhibit the receptor-binding of cholera toxin, *J. Am. Chem. Soc.*, 124 (2002) 12991–12998.
202. G. Pourceau, A. Meyer, J.-J. Vasseur, and F. Morvan, Combinatorial and automated synthesis of phosphodiester galactosyl cluster on solid support by click chemistry assisted by microwaves, *J. Org. Chem.*, 73 (2008) 6014–6017.
203. F. Morvan, A. Meyer, A. Jochum, C. Sabin, Y. Chevolot, A. Imberty, J.-P. Praly, J.-J. Vasseur, E. Souteyrand, and S. Vidal, Fucosylated pentaerythrityl phosphodiester oligomers (PePOs): Automated synthesis of DNA-based glycoclusters and binding to *Pseudomonas aeruginosa* lectin (PA-IIL), *Bioconjug. Chem.*, 18 (2007) 1637–1643.
204. F. J. Feher, K. D. Wyndham, and D. J. Knauer, Synthesis, characterization and lectin binding study of carbohydrate functionalized silsesquioxanes, *Chem. Commun.* (1998) 2393–2394.
205. Y. Gao, A. Eguchi, K. Kakehi, and Y. C. Lee, Efficient preparation of glycoclusters from silsesquioxanes, *Org. Lett.*, 6 (2004) 3457–3460.

206. H. W. Kroto, J. R. Heath, S. C. O'Brien, R. F. Curl, and R. E. Smalley, C60: Buckminsterfullerene, *Nature*, 318 (1985) 162–163.
207. F. Giacalone and N. Martin, Fullerene polymers: Synthesis and properties, *Chem. Rev.*, 106 (2006) 5136–5190.
208. A. W. Jensen, S. R. Wilson, and D. I. Schuster, Biological applications of fullerenes, *Bioorg. Med. Chem.*, 4 (1996) 767–779.
209. T. Da Ros and M. Prato, Medicinal chemistry with fullerenes and fullerene derivatives, *Chem. Commun.* (1999) 663–669.
210. S. Bosi, T. Da Ros, G. Spalluto, and M. Prato, Fullerene derivatives: An attractive tool for biological applications, *Eur. J. Med. Chem.*, 38 (2003) 913–923.
211. S. Marchesan, T. Da Ros, G. Spalluto, J. Balzarini, and M. Prato, Anti-HIV properties of cationic fullerene derivatives, *Bioorg. Med. Chem. Lett.*, 15 (2005) 3615–3618.
212. A. Vasella, P. Uhlmann, C. A. A. Waldraff, F. Diederich, and C. Thielgen, Fullerene sugars: Preparation of enantiomerically pure, spiro linked C-glycosides of C_{60}, *Angew. Chem. Int. Ed. Engl.*, 31 (1992) 1388–1390.
213. A. Yashiro, Y. Nishida, M. Ohno, S. Eguchi, and K. Kobayashi, Fullerene glycoconjugates: A general synthetic approach via cycloaddition of per-*O*-acetyl glycosyl azides to [60]fullerene, *Tetrahedron Lett.*, 39 (1998) 9031–9034.
214. A. Dondoni and A. Marra, Synthesis of [60]fulleropyrrolidine glycoconjugates using 1, 3-dipolar cycloaddition with C-glycosyl azomethine ylides, *Tetrahedron Lett.*, 43 (2002) 1649–1652.
215. C. I. C. Jordao, A. S. F. Farinha, R. F. Enes, A. C. Tomé, A. M. S. Silva, J. A. S. Cavaleiro, C. I. V. Ramos, M. G. Santana-Marques, F. A. Almeida Paz, J. M. de la Torre Ramirez, M. D. L. de la Torre, and M. Nogueras, Synthesis of [60]fullerene-glycopyranosylaminopyridin-4-one conjugates, *Tetrahedron*, 64 (2008) 4427–4437.
216. M. R. Banks, J. I. G. Cadogan, I. Gosney, P. K. G. Hodgson, P. R. R. Langridge-Smith, J. R. A. Millar, and A. T. Taylor, Aziridino[2′,3′:1,2][60]fullerene, *J. Chem. Soc. Chem. Commun.* (1995) 885–886.
217. H. Ito, T. Tada, M. Sudo, Y. Ishida, T. Hino, and K. Saigo, [60]Fullerenoacetyl chloride as a versatile precursor for fullerene derivatives: Efficient ester formation with various alcohols, *Org. Lett.*, 5 (2003) 2643–2645.
218. Y. Wang, J. Cao, D. I. Schuster, and S. R. Wilson, A superior synthesis of [6, 6]-methanofullerenes: The reaction of sulfonium ylides with C60, *Tetrahedron Lett.*, 36 (1995) 6843–6846.

219. S. Tanimoto, S. Sakai, S. Matsumura, D. Takahashi, and K. Toshima, Target-selective photo-degradation of HIV-1 protease by a fullerene-sugar hybrid, *Chem. Commun.* (2008) 5767–5769.
220. Y.-Z. An, C.-H. B. Chen, J. L. Anderson, D. S. Sigman, C. S. Foote, and Y. Rubin, Sequence-specific modification of Guanosine in DNA by a C_{60}-linked deoxyoligonucleotide: Evidence for a non-singlet oxygen mechanism, *Tetrahedron*, 52 (1996) 5179–5189.
221. H. Kato, A. Yashiro, A. Mizuno, Y. Nishida, K. Kobayashi, and H. Shinohara, Syntheses and biological evaluations of α-D-mannosyl [60]fullerenols, *Bioorg. Med. Chem. Lett.*, 11 (2001) 2935–2939.
222. Y. Nishida, A. Mizuno, H. Kato, A. Yashiro, T. Ohtake, and K. Kobayashi, Stereo- and biochemical profiles of the 5-6- and 6-6-junction isomers of α-D-mannosyl [60]fullerenes, *Chem. Biodivers.*, 1 (2004) 1452–1464.
223. Y. Mikata, S. Takagi, M. Tanahashi, S. Ishii, M. Obata, Y. Miyamoto, K. Wakita, T. Nishisaka, T. Hirano, T. Ito, M. Hoshino, C. Ohtsuki, *et al.*, Detection of 1270 nm emission from singlet oxygen and photocytotoxic property of sugar-pendant [60]fullerene, *Bioorg. Med. Chem. Lett.*, 13 (2003) 3289–3292.
224. H. Kato, N. Kaneta, S. Nii, K. Kobayashi, N. Fukui, H. Shinohara, and Y. Nishida, Preparation and supramolecular properties of unadulterated glycosyl liposomes from a bis(α-D-mannopyranosyl)-[60]fullerene conjugate, *Chem. Biodivers.*, 2 (2005) 1232–1241.
225. C. Bingel, Cyclopropanierung von fullerenen, *Chem. Ber.*, 126 (1993) 1957–1959.
226. X. Camps and A. Hirsch, Efficient cyclopropanation of C_{60} starting from malonates, *J. Chem. Soc. Perkin Trans.*, 1 (1997) 1595–1596.
227. F. Cardullo, F. Diederich, L. Echegoyen, T. Habicher, N. Jayaraman, R. M. Leblanc, J. F. Stoddart, and S. Wang, Stable Langmuir and Langmuir–Blodgett films of fullerene–glycodendron conjugates, *Langmuir*, 14 (1998) 1955–1959.
228. I. Lamparth and A. Hirsch, Water-soluble malonic acid derivatives of C60 with a defined three-dimensional structure, *J. Chem. Soc. Chem. Commun.* (1994) 1727–1728.
229. P. R. Ashton, S. E. Boyd, C. L. Brown, N. Jayaraman, S. A. Nepogodiev, and J. F. Stoddart, A convergent synthesis of carbohydrate-containing dendrimers, *Chem. Eur. J.*, 2 (1996) 1115–1128.
230. J.-F. Nierengarten, T. Habicher, R. Kessinger, F. Cardullo, F. Diederich, V. Gramlich, J.-P. Gisselbrecht, C. Boudon, and M. Gross, Macrocyclization on the fulleren core: Direct regio- and diasterodelective multi-functionalization

of [60]fullerene, and synthesis of fullerene-dendrimer derivatives, *Helv. Chim. Acta*, 80 (1997) 2238–2276.
231. H. Kato, C. Böttcher, and A. Hirsch, Sugar balls: synthesis and supramolecular assembly of [60]fullerene glycoconjugates, *Eur. J. Org. Chem.* (2007) 2659–2666.
232. G. R. Newkome, R. K. Behera, C. N. Moorefield, and G. R. Baker, Cascade polymers: Syntheses and characterization of one-directional arborols based on adamantane, *J. Org. Chem.*, 56 (1991) 7162–7167.
233. M. Sawamura, H. Likura, and E. Nakamura, The first pentahaptofullerene metal complexes, *J. Am. Chem. Soc.*, 118 (1996) 12850–12851.
234. H. Isobe, H. Mashima, H. Yorimitsu, and E. Nakamura, Synthesis of fullerene glycoconjugates through sulfide connection in aqueous media, *Org. Lett.*, 5 (2003) 4461–4463.
235. H. Isobe, K. Cho, N. Solin, D. B. Werz, P. H. Seeberger, and E. Nakamura, Synthesis of fullerene glycoconjugates via a copper-catalyzed Huisgen cycloaddition reaction, *Org. Lett.*, 9 (2007) 4611–4614.
236. D. Tasis, N. Tagmatarchis, A. Bianco, and M. Prato, Chemistry of carbon nanotubes, *Chem. Rev.*, 106 (2006) 1105–1136.
237. A. Bianco, K. Kostarelos, and M. Prato, Applications of carbon nanotubes in drug delivery, *Curr. Opin. Chem. Biol.*, 9 (2005) 674–679.
238. C. R. Martin and P. Kohli, The emerging field of nanotube biotechnology, *Nat. Rev. Drug Discov.*, 2 (2003) 29–37.
239. M. Prato, K. Kostarelos, and A. Bianco, Functionalized carbon nanotubes in drug design and discovery, *Acc. Chem. Res.*, 41 (2008) 60–68.
240. G. Pastorin, K. Kostarelos, M. Prato, and A. Bianco, Functionalized carbon nanotubes: Towards the delivery of therapeutic molecules, *J. Biomed. Nanotechnol.*, 1 (2005) 133–142.
241. D. Pantarotto, R. Singh, D. McCarthy, M. Erhardt, J.-P. Briand, M. Prato, K. Kostarelos, and A. Bianco, Functionalized carbon nanotubes for plasmid DNA gene delivery, *Angew. Chem. Int. Ed.*, 43 (2004) 5242–5246.
242. N. W. Shi Kam, T. C. Jessop, P. A. Wender, and H. Dai, Nanotube molecular transporters: Internalization of carbon nanotube–protein conjugates into mammalian cells, *J. Am. Chem. Soc.*, 126 (2004) 6850–6851.
243. A. Bianco, K. Kostarelos, C. D. Partidos, and M. Prato, Biomedical applications of functionalised carbon nanotubes, *Chem. Commun.* (2005) 571–577.
244. L. Lacerda, A. Bianco, M. Prato, and K. Kostarelos, Carbon nanotubes as nanomedicines: From toxicology to pharmacology, *Adv. Drug Deliv. Rev.*, 58 (2006) 1460–1470.

245. Y. Lin, S. Taylor, H. Li, K. A. Shiral Fernando, L. Qu, W. Wang, L. Gu, B. Zhou, and Y.-P. Sun, Advances toward bioapplications of carbon nanotubes, *J. Mater. Chem.*, 14 (2004) 527–541.
246. A. Star, D. W. Steuerman, J. R. Heath, and J. F. Stoddart, Starched carbon nanotubes, *Angew. Chem. Int. Ed.*, 41 (2002) 2508–2512.
247. O.-K. Kim, J. Je, J. W. Baldwin, S. Kooi, P. E. Pehrsson, and L. J. Buckley, Solubilization of single-wall carbon nanotubes by supramolecular encapsulation of helical amylose, *J. Am. Chem. Soc.*, 125 (2003) 4426–4427.
248. H. Dodziuk, A. Ejchart, W. Anczewski, H. Ueda, E. Krinichnaya, G. Dolgonos, and W. Kutner, Water solubilization, determination of the number of different types of single-wall carbon nanotubes and their partial separation with respect to diameters by complexation with η-cyclodextrin, *Chem. Commun.* (2003) 986–987.
249. X. Chen, G. S. Lee, A. Zettl, and C. R. Bertozzi, Biomimetic engineering of carbon nanotubes by using cell surface mucin mimics, *Angew. Chem. Int. Ed.*, 43 (2004) 6112–6116.
250. X. Chen, U. C. Tam, J. L. Czlapinsli, G. S. Lee, D. Rabuka, A. Zettl, and C. R. Bertozzi, Interfacing carbon nanotubes with living cells, *J. Am. Chem. Soc.*, 128 (2006) 6292–6293.
251. T. Hasegawa, T. Fujisawa, M. Numata, M. Umeda, T. Matsumodo, T. Kimura, S. Okumura, K. Sakurai, and S. Shinkai, Single-walled carbon nanotubes acquire a specific lectin-affinity through supramolecular wrapping with lactose-appended schizophyllan, *Chem. Commun.* (2004) 2150–2151.
252. H. Dohi, S. Kikuchi, S. Kuwahara, T. Sugai, and H. Shinohara, Synthesis and spectroscopic characterization of single-wall carbon nanotubes wrapped by glycoconjugate polymer with bioactive sugars, *Chem. Phys. Lett.*, 428 (2006) 98–101.
253. P. Wu, X. Chen, N. Hu, U. C. Tam, O. Blixt, A. Zettl, and C. R. Bertozzi, Biocompatible carbon nanotubes generated by functionalization with glycodendrimers, *Angew. Chem. Int. Ed.*, 47 (2008) 5022–5025.
254. P. Wu, M. Malkoch, J. N. Hunt, R. Vestberg, E. Kaltgrad, M. G. Finn, V. V. Folkin, K. B. Sharpless, and C. J. Hawker, Multivalent, bifunctional dendrimers prepared by click chemistry, *Chem. Commun.* (2005) 5775–5777.
255. S. Niyogi, M. A. Hamon, H. Bu, B. Zhao, P. Bhowmik, R. Sen, M. E. Itkis, and R. C. Haddon, Chemistry of single-walled carbon nanotubes, *Acc. Chem. Res.*, 35 (2002) 1105–1113.
256. J. Chen, A. M. Rao, S. Lyuksyutov, M. E. Itkis, M. A. Hamon, H. Hu, R. W. Cohn, P. C. Eklund, D. T. Colbert, R. E. Smalley, and R. C. Haddon,

Dissolution of full-length single-walled carbon nanotubes, *J. Phys. Chem. B*, 105 (2001) 2525–2528.
257. J. Liu, A. G. Rinzler, H. Dai, H. Hafner, R. K. Bradley, P. J. Boul, A. Lu, T. Iverson, K. Shelimov, C. B. Huffman, F. Rodriguez-Macias, Y.-S. Shon, *et al.*, Fullerene pipes, *Science*, 280 (1998) 1253–1256.
258. F. Pompeo and D. E. Resasco, Water solubilization of single-walled carbon nanotubes by functionalization with glucosamine, *Nano Lett.*, 2 (2002) 369–373.
259. K. Matsuura, K. Hayashi, and N. Kimizuka, Lectin-mediated supramolecular junctions of galactose-derivatized single-walled carbon nanotubes, *Chem. Lett.*, 32 (2003) 212–213.
260. L. Gu, T. Elkin, X. Jiang, H. Li, Y. Lin, L. Qu, T.-R. J. Tzeng, R. Joseph, and Y.-P. Sun, Single-walled carbon nanotubes displaying multivalent ligands for capturing pathogens, *Chem. Commun.* (2005) 874–876.
261. H. Wang, L. Gu, Y. Lin, F. Lu, M. J. Meziani, P. G. Luo, W. Wang, L. Cao, and Y.-P. Sun, Unique aggregation of anthrax (*Bacillus anthracis*) spores by sugar-coated single-walled carbon nanotubes, *J. Am. Chem. Soc.*, 128 (2006) 13364–13365.
262. L. Gu, P. G. Luo, H. Wang, M. J. Meziani, Y. Lin, L. M. Veca, L. Cao, F. Lu, X. Wang, R. A. Quinn, W. Wang, P. Zhang, *et al.*, Single-walled carbon nanotubes as a unique scaffold for the multivalent display of sugars, *Biomacromolecules*, 9 (2008) 2408–2418.
263. C. Gao, S. Muthukrishnan, W. Li, J. Yuan, Y. Xu, and A. H. E. Müller, Linear and hyperbranched glycopolymer-functionalized carbon nanotubes: Synthesis, kinetics and characterization, *Macromolecules*, 40 (2007) 1803–1815.
264. J.-M. Lehn, Toward self-organization and complex matter, *Science*, 295 (2002) 2400–2403.
265. C. H. M. Amijs, G. P. M. van Klink, and G. van Koten, Metallasupramolecular architectures, an overview of functional properties and applications, *Dalton Trans.*, 327 (2006) 308–327.
266. N. C. Gianneschi, M. S. Masar, and C. A. Mirkin, Development of a coordination chemistry-based approach for functional supramolecular structures, *Acc. Chem. Res.*, 38 (2005) 825–837.
267. S. Sakai and T. Sasaki, Multivalent carbohydrate ligands assembled on a metal template, *J. Am. Chem. Soc.*, 116 (1994) 1587–1588.
268. S. Sakai, Y. Shigemasa, and T. Sasaki, A self-adjusting carbohydrate ligand for GalNac specific lectins, *Tetrahedron Lett.*, 38 (1997) 8145–8148.

269. S. Sakai, Y. Shigemasa, and T. Sasaki, Iron(II)-assisted assembly of trivalent GalNAc clusters and their interactions with GalNAc-specific lectins, *Bull. Chem. Soc. Jpn.*, 72 (1999) 1313–1319.
270. T. Hasegawa, T. Yonemura, K. Matsuura, and K. Kobayashi, Tris-bipyridine ruthenium complex-based glyco-clusters: Amplified luminescence and enhanced lectin affinities, *Tetrahedron Lett.*, 42 (2001) 3989–3992.
271. T. Hasegawa, T. Yonemura, K. Matsuura, and K. Kobayashi, Artificial metalloglycoclusters: Compact saccharide shell to induce high lectin affinity as well as strong luminescence, *Bioconjug. Chem.*, 14 (2003) 728–737.
272. S. Kojima, T. Hasegawa, T. Yonemura, K. Sasaki, K. Yamamoto, Y. Makimura, T. Takahashi, T. Suzuki, Y. Susuki, and K. Kobayashi, Ruthenium complexes carrying a disialo complex-type oligosaccharide: Enzymatic synthesis and its application to a luminescent probe to detect influenza viruses, *Chem. Commun.* (2003) 1250–1251.
273. E. C. Constable and S. Mundwiler, Metal-ion control of molecular recognition-sugar-functionalised 2,2′:6′, 2″-terpyridines, *Polyhedron*, 18 (1999) 2433–2444.
274. E. C. Constable, B. Kariuki, and A. Mahmood, New approaches to sugar-functionalised 2,2′:6′,2″-terpyridines based upon tetrafluorophenoxy spacers; crystal and molecular structures of 4′-(tetrafluoro-4-hydroxyphenyl)-2,2′:6′, 2″-terpyridine and 4′-(4-methoxytetrafluorophenyl)-2, 2′:6′,2″’-terpyridine, *Polyhedron*, 22 (2003) 687–698.
275. S. Orlandi, R. Annunziata, M. Benaglia, F. Cozzi, and L. Manzoni, Synthesis of some oligopyridine–galactose conjugates and their metal complexes: A simple entry to multivalent sugar ligands, *Tetrahedron*, 61 (2005) 10048–10060.
276. M. Gottschaldt, D. Koth, D. Müller, I. Klette, S. Rau, H. Görls, B. Schäfer, R. P. Baum, and S. Yano, Synthesis and structure of novel sugar-substituted bipyridine complexes of rhenium and 99m-technetium, *Chem. Eur. J.*, 13 (2007) 10273–10280.
277. R. Roy and J. M. Kim, Cu(II)-Self-assembling bipyridyl-glycoclusters and dendrimers bearing the Tn-antigen cancer marker: Syntheses and lectin binding properties, *Tetrahedron*, 59 (2003) 3881–3893.
278. R. Kikkeri, L. H. Hossain, and P. H. Seeberger, Supramolecular one-pot approach to fluorescent glycodendrimers, *Chem. Commun.* (2008) 2127–2129.
279. R. Kikkeri, I. Garcia-Rubio, and P. H. Seeberger, Ru(II)-carbohydrate dendrimers as photoinduced electron transfer lectin biosensors, *Chem. Commun.* (2009) 235–237.

280. G. Thoma, A. G. Katopodis, N. Voelcker, R. O. Duthaler, and M. B. Streiff, Novel glycodendrimers self-assemble to nanoparticles which function as polyvalent ligands *in vitro* and *in vivo*, *Angew. Chem. Int. Ed.*, 41 (2002) 3195–3198.
281. G. Thoma, M. B. Streiff, A. G. Katopodis, R. O. Duthaler, N. H. Voelcker, C. Ehrhardt, and C. Masson, Non-covalent polyvalent ligands by self-assembly of small glycodendrimers: A novel concept for the inhibition of polyvalent carbohydrate-protein interactions *in vitro* and *in vivo*, *Chem. Eur. J.*, 12 (2006) 99–117.
282. G. R. Newkome, C. N. Moorefield, and F. Vögtle, *Dendrimers and dendrons: concepts, synthesis, applications,* Wiley-VCH, New York, 2001.
283. J. M. J. Fréchet and D. Tomalia, *Dendrimers and other dendritic polymers,* Wiley, New York, 2001.
284. R. Esfand and D. A. Tomalia, Poly(amidoamine) (PAMAM) dendrimers: From biomimicry to drug delivery and biomedical applications, *Drug Discov. Today*, 6 (2001) 427–436.
285. D. A. Tomalia, A. M. Naylor, and W. A. Goddard, Starburst dendrimers-molecular level control of size, shape, surface chemistry, topology, and flexibility from atoms to macroscopic matter, *Angew. Chem. Int. Ed. Engl.*, 29 (1990) 138–175.
286. E. Buhleier, W. Wehner, and F. Vögtle, ''Cascade''- and ''nonskid-chain-like'' syntheses of molecular cavity topologies, *Synthesis* (1978) 155–158.
287. G. R. Newkome, Z.-q. Yao, G. R. Baker, and V. K. Gupta, Micelles. Part 1. Cascade molecules: A new approach to micelles. A [27]-arborol, *J. Org. Chem.*, 50 (1985) 2004–2006.
288. D. A. Tomalia, H. Baker, J. Dewald, M. Hall, G. Kallos, S. Martin, J. Roeck, J. Ryder, and P. Smith, A new class of polymers: Starburst-dendritic macromolecules, *Polym. J.*, 17 (1985) 117–132.
289. R. Haag and F. Kratz, Polymer therapeutics: Concepts and applications, *Angew. Chem. Int. Ed. Engl.*, 45 (2006) 1198–1215.
290. R. Duncan and L. Izzo, Dendrimer biocompatibility and toxicity, *Adv. Drug Deliv. Rev.*, 57 (2005) 2215–2237.
291. U. Boas, J. B. Christensen, and P. M. H. Heegaard, *Dendrimers in medicine and biotechnology: new molecular tools,* RSC Publishing, Cambridge, 2006.
292. R. Roy, D. Zanini, S. J. Meunier, and A. Romanowska, Solid phase synthesis of dendritic sialoside inhibitors of influenza A virus haemagglutinin, *J. Chem. Soc. Chem. Commun.* (1993) 1869–1872.
293. N. Röckendorf and T. K. Lindhorst, Glycodendrimers, *Top. Curr. Chem.*, 217 (2001) 201–238.

294. M. J. Cloninger, Biological applications of dendrimers, *Curr. Opin. Chem. Biol.*, 6 (2002) 742–748.
295. K. Bezouska, Design, functional evaluation and biomedical applications of carbohydrate dendrimers (glycodendrimers), *Rev. Mol. Biotechnol.*, 90 (2002) 269–290.
296. W. B. Turnbull and J. F. Stoddart, Design and synthesis of glycodendrimers, *Rev. Mol. Biotechnol.*, 90 (2002) 231–255.
297. S. A. Nepogodiev and J. F. Stoddart, Glycodendrimers: Chemical aspects, *Adv. Macromol. Carbohydr. Res.*, 2 (2003) 191–239.
298. Y. Li, Y. Cheng, and T. Xu, Design, synthesis and potent pharmaceutical applications of glycodendrimers: A mini review, *Curr. Drug Discov. Technol.*, 4 (2007) 246–254.
299. C. J. Hawker and J. M. J. Fréchet, Preparation of polymers with controlled molecular architecture. A new convergent approach to dendritic macromolecules, *J. Am. Chem. Soc.*, 112 (1990) 7638–7647.
300. C. J. Hawker, K. L. Wooley, and J. M. J. Fréchet, Unimolecular micelles and globular amphiphiles: Dendritic macromolecules as novel recyclable solubilization agents, *J. Chem. Soc. Perkin Trans. 1*, 12 (1993) 1287–1297.
301. K. L. Wooley, C. J. Hawker, and J. M. J. Fréchet, Hyperbranched macromolecules via a novel double-stage convergent growth approach, *J. Am. Chem. Soc.*, 113 (1991) 4252–4261.
302. K. L. Wooley, C. J. Hawker, and J. M. J. Fréchet, A ''branched-monomer approach'' for the rapid synthesis of dendrimers, *Angew. Chem. Int. Ed. Engl.*, 33 (1994) 82–85.
303. H. Ihre, A. Hult, J. M. J. Fréchet, and I. Gitsov, Double-stage convergent approach for the synthesis of functionalized dendritic aliphatic polyesters based on 2,2-bis(hydroxymethyl)propionic acid, *Macromolecules*, 31 (1998) 4061–4068.
304. P. R. Ashton, S. E. Boyd, C. L. Brown, N. Jayaraman, and J. F. Stoddart, A convergent synthesis of a carbohydrate-containing dendrimer, *Angew. Chem. Int. Ed. Engl.*, 36 (1997) 732–735.
305. F. Zeng and S. C. Zimmerman, Rapid synthesis of dendrimers by an orthogonal coupling strategy, *J. Am. Chem. Soc.*, 118 (1996) 5326–5327.
306. P. Antoni, D. Nyström, C. J. Hawker, A. Hult, and M. Malkoch, A chemoselective approach for the accelerated synthesis of well-defined dendritic architectures, *Chem. Commun.* (2007) 2249–2251.
307. W. E. Minde, C. Roach, W. G. J. Hol, and C. L. Verlinde, Structure-based exploration of the ganglioside GM1 binding sites of *Escherichia coli* heat-labile

enterotoxin and Cholera Toxin for the discovery of receptor anatgonists, *Biochemistry*, 38 (1999) 5684–5692.

308. J. C. Pickens, E. A. Merritt, M. Ahn, C. L. M. J. Verlinde, W. G. J. Hol, and E. Fan, Anchor-based design of improved cholera toxin and *E. coli* heat-labile enterotoxin receptor binding antagonists that display multiple binding modes, *Chem. Biol.*, 9 (2002) 215–224.
309. I. Vrasidas, N. J. de Mol, R. M. J. Liskamp, and R. J. Pieters, Synthesis of lactose dendrimers and multivalency effects in binding to the cholera toxin B subunit, *Eur. J. Org. Chem.* (2001) 4685–4692.
310. I. Vrasidas, J. Kemmink, R. M. J. Liskamp, and R. J. Pieters, Synthesis and Cholera Toxin binding properties of a lactose-2-aminothiazoline conjugate, *Org. Lett.*, 4 (2002) 1807–1808.
311. D. Arosio, I. Vrasidas, P. Valentini, R. M. J. Liskamp, R. J. Pieters, and A. Bernardi, Synthesis and Cholera Toxin binding properties of multivalent GM1 mimics, *Org. Biomol. Chem.*, 2 (2004) 2113–2124.
312. A. V. Pukin, H. M. Branderhorst, C. Sisu, C. A. G. M. Weijers, M. Gilbert, R. M. J. Liskamp, G. M. Visser, H. Zuilhof, and R. J. Pieters, Strong inhibition of Cholera Toxin by multivalent GM1 derivatives, *ChemBioChem*, 8 (2007) 1500–1503.
313. C. Sisu, A. J. Baron, H. M. Branderhorst, S. D. Connell, C. A. G. M. Weijers, R. de Vries, E. D. Hayes, A. V. Pukin, M. Gilbert, R. J. Pieters, H. Zuilhof, G. M. Visser, *et al.*, The influence of ligand valency on aggregation mechanisms for inhibiting bacterial toxins, *ChemBioChem*, 10 (2009) 329–337.
314. H. M. Branderhorst, R. M. J. Liskamp, G. M. Visser, and R. J. Pieters, Strong inhibition of cholera toxin binding by galactose dendrimers, *Chem. Commun.*, (2007) 5043–5047.
315. I. Vrasidas, S. André, P. Valentini, C. Böck, M. Lensch, H. Kaltner, R. M. J. Liskamp, H.-J. Gabius, and R. J. Pieters, Rigidified multivalent lactose molecules and their interactions with mammalian galectins: A route to selective inhibitors, *Org. Biomol. Chem.*, 1 (2003) 803–810.
316. M.-G. Baek and R. Roy, Simultaneous binding of mouse monoclonal antibody and streptavidin to heterobifunctional dendritic L-lysine core bearing T-antigen tumor marker and biotin, *Bioorg. Med. Chem.*, 9 (2001) 3005–3011.
317. R. R. Kale, C. M. McGannon, C. Fuller-Schaefer, D. M. Hatch, M. J. Flager, S. D. Gamage, A. A. Weiss, and S. S. Iyer, Differentiation between structurally homologous Shiga 1 and Shiga 2 toxins by using synthetic glycoconjugates, *Angew. Chem. Int. Ed.*, 47 (2008) 1265–1268.

318. D. M. Hatch, A. A. Weiss, R. R. Kale, and S. S. Iyer, Biotinylated bi- and tetra-antennary glycoconjugates for *Escherichia coli* detection, *ChemBioChem*, 9 (2008) 2433–2442.
319. K. Sadalapure and T. K. Lindhorst, A general entry into glycopeptide ''dendrons''. *Angew. Chem. Int. Ed.*, 39 (2000) 2010–2013.
320. A. Nelson and J. F. Stoddart, Synthesis of lactoside glycodendrons using photoaddition and reductive amination methodologies, *Carbohydr. Res.*, 339 (2004) 2069–2075.
321. C. D. Heidecke and T. K. Lindhorst, Iterative synthesis of spacered glycodendrons as oligomannoside mimetics and evaluation of their antiadhesive properties, *Chem. Eur. J.*, 13 (2007) 9056–9067.
322. M. Jayaraman and J. M. J. Fréchet, A convergent route to novel aliphatic polyether dendrimers, *J. Am. Chem. Soc.*, 120 (1998) 12996–12997.
323. K. Elsner, M. M. K. Boysen, and T. K. Lindhorst, Synthesis of new polyether glycodendrons as oligosaccharide mimetics, *Carbohydr. Res.*, 342 (2007) 1715–1725.
324. G. R. Newkome, Z.-q. Yao, G. R. Baker, V. K. Gupta, P. S. Russo, and M. J. Saunders, Chemistry of micelles series. Part 2. Cascade molecules. Synthesis and characterization of a benzene [9]3-arborol, *J. Am. Chem. Soc.*, 108 (1986) 849–850.
325. S.-K. Wang, P. H. Liang, R. D. Astronomo, T.-L. Hsu, D. R. Burton, and C.-H. Wong, Targeting the carbohydrates on HIV-1: Interaction of oligomannose dendrons with human monoclonal antibody 2G12 and DC-SIGN, *Proc. Natl. Acad. Sci. (USA)*, 105 (2008) 3690–3695.
326. R. Roy, W. K. C. Park, Q. Wu, and S.-N. Wang, Synthesis of hyperbranched dendritic lactosides, *Tetrahedron Lett.*, 36 (1995) 4377–4380.
327. S. J. Meunier, Q. Wang, S.-N. Wang, and R. Roy, Synthesis of hyperbranched glycodendrimers incorporating α-thiosialosides based on a gallic acid core, *Can. J. Chem.*, 75 (1997) 1472–1482.
328. H. Sashiwa, Y. Shigemasa, and R. Roy, Chemical modification of chitosan. 10.1. Synthesis of dendronized chitosan-sialic acid acid hybrid using convergent grafting of preassembled dendrons built on gallic acid and tri(ethylene glycol) backbone, *Macromolecules*, 34 (2001) 3905–3909.
329. J. A. F. Joosten, N. T. H. Tholen, F. Ait El Maate, A. J. Brouwer, G. Wilma van Esse, D. T. S. Rijkers, R. M. J. Liskamp, and R. J. Pieters, High-yielding microwave-assisted synthesis of triazole-linked glycodendrimers by copper-catalyzed [3+2] cycloaddition, *Eur. J. Org. Chem.* (2005) 3182–3185.

330. E. Fernandez-Megia, J. Correa, I. Rodriguez-Meizoso, and R. Riguera, A click approach to unprotected glycodendrimers, *Macromolecules*, 39 (2006) 2113–2120.
331. E. Fernandez-Megia, J. Correa, and R. Riguera, ''Clickable'' PEG-dendritic block copolymers, *Biomacromolecules*, 7 (2006) 3104–3111.
332. R. Roy, D. Zanini, S. J. Meunier, and A. Romanowska, Syntheses and antigenic properties of sialic acid-based dendrimers, in P. Kovàc, (Ed.), *Synthetic Oligosaccharides: Indispensable Probes for the Life Sciences, ACS Symp. Ser.*, 560, 1994, pp. 104–119.
333. V. M. Krishnamurthy, V. Semetey, P. J. Bracher, N. Shen, and G. M. Whitesides, Dependence of effective molarity on linker length for an intramolecular protein–ligand system, *J. Am. Chem. Soc.*, 129 (2007) 1312–1320.
334. V. M. Krishnamurthy, L. A. Estroff, and G. M. Whitesides, Multivalency in ligand design, in W. Jahnke and D. A. Erlanson, (Eds.) *Fragment-Based Approaches in Drug Discovery, Vol. 34,* Wiley-VCH, Weinheim, 2006, pp. 11–53.
335. J. D. Reuter, A. Myc, M. M. Hayes, Z. Gan, R. Roy, D. Qin, R. Yin, L. T. Piehler, R. Esfand, D. A. Tomalia, and J. R. Baker, Jr., Inhibition of viral adhesion and infection by sialic-acid-conjugated dendritic polymers, *Bioconjug. Chem.*, 10 (1999) 271–278.
336. N. Nagahori, R. T. Lee, S.-I. Nishimura, D. Pagé, R. Roy, and Y. C. Lee, Inhibition of adhesion of type 1 Fimbriated *Escherichia coli* to highly mannosylated ligands, *ChemBioChem*, 3 (2002) 836–844.
337. R. G. Denkewalter, J. Kolc, and W. J. Lukasavage, Macromolecular highly branched homogeneous compound based on lysine units (1981) US Patent, 4289872.
338. L. Crespo, G. Sanclimens, M. Pons, E. Giralt, M. Royo, and F. Albericio, Peptide and amide bond-containing dendrimers, *Chem. Rev.*, 105 (2005) 1663–1681.
339. P. Niederhafner, J. Šebestík, and J. Ježek, Peptide dendrimers, *J. Pept. Sci.*, 11 (2005) 757–788.
340. P. Niederhafner, J. Šebestík, and J. Ježek, Glycopeptide dendrimers. Part I, *J. Pept. Sci.*, 14 (2008) 2–43.
341. P. Niederhafner, J. Šebestík, and J. Ježek, Glycopeptide dendrimers. Part II, *J. Pept. Sci.*, 14 (2008) 44–65.

342. P. Niederhafner, M. Reiniš, J. Šebestík, and J. Ježek, Glycopeptide dendrimers, Part III—A review: Use of glycopeptide dendrimers in immunotherapy and diagnosis of cancer and viral diseases, *J. Pept. Sci.*, 14 (2008) 556–587.
343. A. R. P. M. Valentjin, G. A. van der Marel, L. A. J. M. Sliedregt, T. J. C. van Berkel, E. A. L. Biessen, and J. H. van Boom, Solid-phase synthesis of lysine-based cluster galactosides with high affinity for the asialoglycoprotein receptor, *Tetrahedron*, 53 (1997) 759–770.
344. C. Grandjean, C. Rommens, H. Gras-Masse, and O. Melnyk, Convergent synthesis of fluorescein-labelled lysine-based cluster glycosides, *Tetrahedron Lett.*, 40 (1999) 7235–7238.
345. J. Ježek, J. Velek, P. Veprek, V. Velkova, T. Trnka, J. Pecka, M. Ledvina, J. Vondrasek, and M. Pisacka, Solid phase synthesis of glycopeptide dendrimers with Tn antigenic structure and their biological activities. Part I, *J. Pept. Sci.*, 5 (1999) 46–55.
346. N. Frison, M. E. Taylor, E. Soilleux, M.-T. Bousser, R. Mayer, M. Monsigny, K. Drickamer, and A.-C. Roche, Oligolysine-based oligosaccharide clusters: selective recognition and endocytosis by the mannose receptor and dendritic cell-specific intercellular adhesion molecule 3 (ICAM-3)-grabbing nonintegrin, *J. Biol. Chem.*, 278 (2003) 23922–23929.
347. H. Baigude, K. Katsuraya, K. Okuyama, S. Tokunaga, and T. Uryu, Synthesis of sphere-type monodispersed oligosaccharide-polypeptide dendrimers, *Macromolecules*, 36 (2003) 7100–7106.
348. B. W. Greatex, S. J. Brodie, R. H. Furneaux, S. M. Hook, W. T. McBurney, G. F. Painter, T. Rades, and P. M. Rendle, The synthesis and immune stimulating action of mannose-capped lysine-based dendrimers, *Tetrahedron*, 65 (2009) 2939–2950.
349. D. Zanini and R. Roy, Chemoenzymatic synthesis and lectin binding properties of dendritic *N*-acetyllactosamine, *Bioconjug. Chem.*, 8 (1997) 187–192.
350. R. Roy, M.-G. Baek, and K. Rittenhouse-Olson, Synthesis of *N,N′*-bis(acrylamido)acetic acid base-T antigen glycodendrimers and their mouse monoclonal IgG antibody binding properties, *J. Am. Chem. Soc.*, 123 (2001) 1809–1816.
351. M.-G. Baek and R. Roy, Glycodendrimers: Novel glycotope isosteres unmasking sugar coding. Case study with T-antigen markers from Breast Cancer MUC1 glycoprotein, *Rev. Mol. Biotechnol.*, 90 (2002) 291–309.
352. M. M. Palcic, H. Li, D. Zanini, R. S. Bhella, and R. Roy, Chemoenzymatic synthesis of dendritic sialyl Lewis^x, *Carbohydr. Res.*, 305 (1998) 433–442.

353. M. Otter, M. M. Barett-Bergshoeff, and D. C. J. Rijken, Binding of tissue-type plasminogen activator by the mannose receptor, *J. Biol. Chem.*, 266 (1991) 13931–13935.
354. I. P. Fraser, H. Koziel, and R. A. B. Ezekowitz, The serum mannose-binding protein and the macrophage mannose receptor are pattern recognition molecules that link innate and adaptive immunity, *Semin. Immunol.*, 10 (1998) 363–372.
355. M. K. Bijsterbosch, W. Donker, H. Van de Bilt, S. Van Weely, T. J. Van Berkel, and J. M. Aerts, Quantitative analysis of the targeting of mannose-terminal glucocerebrosidase. Predominant uptake by liver endothelial cells, *Eur. J. Biochem.*, 237 (1996) 344–349.
356. B. Friedman, K. Vaddi, C. Preston, E. Mahon, J. R. Cataldo, and J. M. A. McPherson, Comparison of the pharmacological properties of carbohydrate remodeled recombinant and placental-derived β-Glucocerebrosidase: Implications for clinical efficacy in treatment of Gaucher disease, *Blood*, 93 (1999) 2807–2816.
357. G. M. Baratt, D. Nolibe, A. Yapo, J. F. Petit, and J. P. Tenu, Use of mannosylated liposomes for *in vivo* targeting of a macrophage activator and control of artificial pulmonary metastases, *Ann. Inst. Pasteur/Immunol.*, 138 (1987) 437–450.
358. S. Gac, J. Coudane, M. Boustta, M. Domurado, and M. Vert, Synthesis, characterization and *in vivo* behavior of a norfloxacin-poly(L-lysine citramide imide) conjugate bearing mannosyl residues, *J. Drug Targeting*, 7 (2000) 393–406.
359. P. Chakraborty, A. N. Bhaduri, and P. K. Das, Neoglycoproteins as carriers for receptor-mediated drug targeting in the treatment of experimental visceral leishmaniasis, *J. Protozool.*, 37 (1990) 358–364.
360. W. W. Liang, X. Shi, D. Deshpande, C. J. Malanga, and Y. Rojanasakul, Oligonucleotide targeting to alveolar macrophages by mannose receptor-mediated endocytosis, *Biochim. Biophys. Acta*, 1279 (1996) 227–234.
361. T. Ferkol, F. Mularo, J. Hilliard, S. Lodish, J. C. Perales, A. Ziady, and M. Konstan, Transfer of the human α-antitrypsin gene into pulmonary macrophages *in vivo*, *Am. J. Respir. Cell. Mol. Biol.*, 18 (1998) 591–601.
362. S. S. Diebold, H. Lehrmann, M. Kursa, E. Wagner, M. Cotten, and M. Zenke, Efficient gene delivery into human dendritic cells by adenovirus polyethylenimine and mannose polyethylenimine transfection, *Hum. Gene Ther.*, 10 (1999) 775–786.
363. M. Nishikawa, S. Takamura, F. Yamashita, Y. Takahura, D. K. Meijer, M. Hashida, and J. P. Swart, Pharmacokinetics and *in vivo* gene transfer of plasmid DNA complexed with mannosylated poly(L-lysine) in mice, *J. Drug Target.*, 8 (2000) 29–38.

364. J. Van Bergen, F. Ossendorp, R. Jordens, A. M. Mommaas, J. W. Drijfhout, and F. Koning, Get into the groove! Targeting antigens to MHC class II, *Immunol. Rev.*, 172 (1999) 87–96.
365. M. Fukasawa, Y. Shimizu, K. Shikata, M. Nakata, R. Sakakibara, N. Yamamoto, M. Hatanaka, and T. Mizuochi, Liposome oligomannose-coated with neoglycolipid, a new candidate for a safe adjuvant for induction of CD8+ cytotoxic T lymphocytes, *FEBS Lett.*, 441 (1998) 353–356.
366. V. Apostolopoulos, N. Barnes, G. A. Pietersz, and I. F. McKenzie, *Ex vivo* targeting of the macrophage mannose receptor generates anti-tumor CTL responses, *Vaccine*, 18 (2000) 3174–3184.
367. E. A. L. Biessen, F. Noorman, M. E. van Teijlingen, J. Kuiper, M. Barrett-Bergshoeff, M. K. Bijsterbosch, D. C. Rijken, and T. J. C. Van Berkel, Lysine-based cluster mannosides that inhibit ligand binding to the human mannose receptor at nanomolar concentration, *J. Biol. Chem.*, 271 (1996) 28024–28030.
368. C. Grandjean, G. Angyalosi, E. Loing, E. Adriaenssens, O. Melnyk, V. Pancré, C. Auriault, and H. Gras-Masse, Novel hyperbranched glycomimetics recognized by the human mannose receptor: Quinic or shikimic acid derivatives as mannose bioisosteres, *ChemBioChem*, 2 (2001) 747–757.
369. M. C. Schuster, D. A. Mann, T. J. Buchholz, K. M. Johnson, W. D. Thomas, and L. L. Kiessling, Parallel synthesis of glycomimetic libraries: Targeting a C-type lectin, *Org. Lett.*, 5 (2003) 1407–1410.
370. M. J. Borrock and L. L. Kiessling, Non-carbohydrate inhibitors of the lectin DC-SIGN, *J. Am. Chem. Soc.*, 129 (2007) 12780–12785.
371. E. A. B. Kantchev, C.-C. Chang, S.-F. Cheng, A.-C. Roche, and D.-K. Chang, Direct solid-phase synthesis and fluorescence labeling of large, monodisperse mannosylated dendrons in a peptide synthesizer, *Org. Biomol. Chem.*, 6 (2008) 1377–1385.
372. S. Hana, H. Baigude, K. Hattori, T. Yoshida, and T. Uryu, Synthesis of new spherical and hemispherical oligosaccharides with polylysine core scaffold, *Carbohydr. Polym.*, 68 (2007) 26–34.
373. S. Hana, T. Yoshida, and T. Uryu, Synthesis of a new polylysine-dendritic oligosaccharide with alkyl spacer having peptide linkage, *Carbohydr. Polym.*, 69 (2007) 436–444.
374. E. M. V. Johansson, E. Kolomiets, F. Rosenau, K.-E. Jaeger, T. Darbre, and J.-L. Reymond, Combinatorial variation of branching length and multivalency in a large (390, 625 member) glycopeptide dendrimer library: Ligands for Fucose-specific lectins, *New J. Chem.*, 31 (2007) 1291–1299.

375. E. Kolomiets, E. M. V. Johansson, O. Renaudet, T. Darbre, and J. L. Reymond, Neoglycopeptide dendrimer libraries as a source of lectin binding ligands, *Org. Lett.*, 9 (2007) 1465–1468.
376. E. A. B. Kantchev, C.-C. Chang, and D.-K. Chang, Direct Fmoc/tert-Bu solid phase synthesis of octamannosyl polylysine dendrimer-peptide conjugates, *Biopolymers*, 84 (2006) 232–240.
377. C. T. Wild, D. C. Shugars, and T. K. Greenwell, Peptides corresponding to a predictive α-helical domain of human immunodeficiency virus type 1 gp41 are potent inhibitors of virus infection, *Proc. Natl. Acad. Sci. (USA)*, 91 (1994) 9770–9774.
378. M. C. A. A. Tan, A. M. Mommaas, R. Drijfhout, J. J. Jordens, M. Onderwater, D. Vervoerd, A. A. Mulder, A. N. van der Heiden, D. Scheidegger, L. C. J. M. Oomen, T. H. M. Ottenhoff, J. J. Neefjes, and F. Koning, Mannose receptor-mediated uptake of antigens strongly enhances HLA class II-restricted antigen presentation by cultured dendritic cells, *Eur. J. Immunol.*, 27 (1997) 2426–2435.
379. R. Roy, New trends in carbohydrate-based vaccines, *Drug Discov. Today*, 1 (2004) 327–336.
380. R. Roy, Carbohydrate-based vaccines, in R. Roy, (Ed.), *Vol. 989,* ACS Symposium Series, ACS, Washington, DC, 2008.
381. E. O. Saphire, P. W. Parren, R. Pantophlet, M. B. Zwick, G. M. Morris, P. M. Rudd, R. A. Dwek, R. L. Stanfield, D. R. Burton, and I. A. Wilson, Crystal structure of a neutralizing human IGG against HIV-1: A template for vaccine design, *Science*, 293 (2001) 1155–1159.
382. C. Brocke and H. Kunz, Synthesis of tumor-associated glycopeptides antigens, *Bioorg. Med. Chem.*, 10 (2002) 3085–3112.
383. S. Dziadek and H. Kunz, Synthesis of tumor-associated glycopeptides antigens for the development of tumor-selective vaccines, *Chem. Rec.*, 3 (2004) 308–321.
384. C. Ozawa, H. Hojo, Y. Nakahara, H. Katayama, K. Nabeshima, T. Akahane, and Y. Nakahara, Synthesis of glycopeptide dendrimer by a convergent method, *Tetrahedron*, 63 (2007) 9685–9690.
385. R. Roy, Recent development in the rational design of multivalent glycoconjugates, *Top. Curr. Chem.*, 187 (1997) 241–274.
386. M. Dubber and J. M. J. Fréchet, Solid-phase synthesis of multivalent glycoconjugates on a DNA synthesizer, *Bioconjug. Chem.*, 14 (2003) 239–246.
387. K. Aoi, K. Itoh, and M. Okada, Globular carbohydrate macromolecules ''sugar balls'' Synthesis of novel sugar-persubstituted poly(amido amine) dendrimers, *Macromolecules*, 28 (1995) 5391–5393.

388. T. K. Lindhorst and C. Kieburg, Glycocoating of oligovalent amines: Synthesis of thiourea-bridged cluster glycosides from glycosyl isothiocyanates, *Angew. Chem. Int. Ed. Engl.*, 35 (1996) 1953–1956.
389. D. Pagé and R. Roy, Synthesis and biological properties of mannosylated starburst Poly(amidoamine) dendrimers, *Bioconjug. Chem.*, 8 (1997) 714–723.
390. D. Pagé and R. Roy, Glycodendrimers as novel biochromatography adsorbents, *Int. J. Bio-Chromatogr.*, 3 (1997) 231–244.
391. S. André, P. J. C. Ortega, M. A. Perez, R. Roy, and H.-J. Gabius, Lactose-containing starburst dendrimers: Influence of dendrimer generation and binding-site orientation of receptors (plant/animal lectins and immunoglobulins) on binding properties, *Glycobiology*, 9 (1999) 1253–1261.
392. C. Kieburg and T. K. Lindhorst, Glycodendrimer synthesis without using protecting groups, *Tetrahedron Lett.*, 38 (1997) 3885–3888.
393. T. K. Lindhorst, C. Kieburg, and U. Krallmann-Wenzel, Inhibition of the type 1 fimbriae-mediated adhesion of *Escherichia coli* to erythrocytes by multiantennary α-mannosyl clusters: the effect of multivalency, *Glycoconjug. J.*, 15 (1998) 605–613.
394. J. P. Thompson and C.-L. Schengrund, Oligosaccharide-derivatized dendrimers: Defined multivalent inhibitors of the adherence of the cholera toxin B subunit and the heat labile enterotoxin of *E. coli* to GM1, *Glycoconjug. J.*, 14 (1997) 837–845.
395. D. Zanini and R. Roy, Practical synthesis of starburst PAMAM α-thiosialodendrimers for probing multivalent carbohydrate-lectin binding properties, *J. Org. Chem.*, 63 (1998) 3486–3491.
396. J. J. Landers, Z. Cao, I. Lee, L. T. Piehler, P. P. Myc, A. Myc, T. Hamouda, A. T. Galecki, and J. R. Baker, Jr., Prevention of influenza pneumonitis by sialic acid-conjugated dendritic polymers, *J. Infect. Dis.*, 186 (2002) 1222–1230.
397. E. K. Woller and M. J. Cloninger, Mannose functionalization of a sixth generation dendrimer, *Biomacromolecules*, 2 (2001) 1052–1054.
398. E. K. Woller and M. J. Cloninger, The lectin-binding properties of six generations of mannose-functionalized dendrimers, *Org. Lett.*, 4 (2002) 7–10.
399. S. L. Mangold, J. R. Morgan, G. C. Strohmeyer, A. M. Gronenborn, and M. J. Cloninger, Cyanovirin-N binding to Manα1-2Man functionalized dendrimers, *Org. Biomol. Chem.*, 3 (2005) 2354–2358.
400. N. Seah, P. V. Santacroce, and A. Basu, Probing the lactose·GM3 carbohydrate–carbohydrate interaction with glycodendrimers, *Org. Lett.*, 11 (2009) 559–562.

401. K. Aoi, K. Tsutsumiuchi, A. Yamamoto, and M. Okada, Globular carbohydrate macromolecule ''sugar balls'' 3. ''Radial-growth polymerization'' of sugar-substituted α-amino acid N-carboxyanhydrides (glyco-NCAs) with a dendritic initiator, *Tetrahedron*, 53 (1997) 15415–15427.
402. M.-G. Baek and R. Roy, Synthesis and protein binding properties of T-antigen containing GlycoPAMAM dendrimers, *Bioorg. Med. Chem.*, 10 (2002) 11–17.
403. M.-G. Baek, K. Rittenhouse-Olson, and R. Roy, Synthesis and antibody binding properties of glycodendrimers bearing the tumor related T-antigen, *Chem. Commun.* (2001) 257–258.
404. C. C. M. Appeldoorn, J. A. F. Joosten, F. A. el Maate, U. Dobrindt, J. Hacker, R. M. J. Liskamp, A. S. Khan, and R. J. Pieters, Novel multivalent mannose compounds and their inhibition of the adhesion of type 1 fimbriated uropathogenic *E. coli*, *Tetrahedron: Asymmetry*, 16 (2005) 361–372.
405. O. Srinivas, S. Radhika, N. M. Bandaru, S. K. Nadimpalli, and N. Jayaraman, Synthesis and biological evaluation of mannose-6-phosphate-coated multivalent dendritic cluster glycosides, *Org. Biomol. Chem.*, 3 (2005) 4252–4257.
406. D. E. Tsvetkov, P. E. Cheshev, A. B. Tuzikov, A. A. Chinarev, G. V. Pazynina, M. A. Sablina, A. S. Gambaryan, N. V. Bovin, R. Rieben, A. S. Shashkov, and N. E. Nifantiev, Neoglycoconjugates based on dendrimer Poly(aminoamides), *Russ. J. Bioorg. Chem.*, 28 (2002) 470–486.
407. H. Sashiwa, Y. Shigemasa, and R. Roy, Chemical modification of chitosan. 3. Hyperbranched chitosan–sialic acid dendrimer hybrid with tetraethylene glycol spacer, *Macromolecules*, 33 (2000) 6913–6915.
408. H. Sashiwa, Y. Shigemasa, and R. Roy, Highly convergent synthesis of dendrimerized chitosan-sialic acid hybrid, *Macromolecules*, 34 (2001) 3211–3214.
409. H. Sashiwa, H. Yajima, R. Roy, and S.-I. Aiba, Chitosan–dendrimer hybrid, *Adv. Chitin Sci.*, 7 (2003) 196–200.
410. S. A. Kalovidouris, O. Blixt, A. Nelson, S. Vidal, W. B. Turnbull, J. C. Paulson, and J. F. Stoddart, Chemically defined sialoside scaffolds for investigation of multivalent interactions with sialic acid binding proteins, *J. Org. Chem.*, 68 (2003) 8485–8493.
411. D. Zanini and R. Roy, Novel dendritic α-sialosides: Synthesis of glycodendrimers based on a 3,3'-iminobis(propylamine) core, *J. Org. Chem.*, 61 (1996) 7348–7354.
412. D. Zanini and R. Roy, Synthesis of new α-thiosialodendrimers and their binding properties to the sialic acid specific lectin from *Limax flavus*, *J. Am. Chem. Soc.*, 119 (1997) 2088–2095.

413. M. Llinares and R. Roy, Multivalent neoglycoconjugates: Solid-phase synthesis of N-linked α-sialodendrimers, *Chem. Commun.,* (1997) 2119–2120.
414. M. L. Wolfenden and M. J. Cloninger, Mannose/glucose-functionalized dendrimers to investigate the predictable tunability of multivalent interactions, *J. Am. Chem. Soc.*, 127 (2005) 12168–12169.
415. M. L. Wolfenden and M. J. Cloninger, Carbohydrate-functionalized dendrimers to investigate the predictable tunability of multivalent interactions, *Bioconjug. Chem.*, 17 (2006) 958–966.
416. E. M. M. de Brabander-van den Berg and E. W. Meijer, Poly(propylene imine) dendrimers: Large-scale synthesis by hetereogeneously catalyzed hydrogenations, *Angew. Chem. Int. Ed. Engl.*, 32 (1993) 1308–1311.
417. P. R. Ashton, S. E. Boyd, C. L. Brown, S. A. Nepogodiev, E. W. Meijer, H. W. I. Peerlings, and J. F. Stoddart, Synthesis of glycodendrimers by modification of poly(propyleneimine) dendrimers, *Chem. Eur. J.*, 3 (1997) 974–984.
418. H. W. I. Peerlings, S. A. Nepogodiev, J. F. Stoddart, and E. W. Meijer, Synthesis of spacer-armed glucodendrimers based on the modification of poly(propylene imine) dendrimers, *Eur. J. Org. Chem.* (1998) 1879–1886.
419. R. D. Kensinger, B. C. Yowler, A. J. Benesi, and C.-L. Schengrund, Synthesis of novel, multivalent glycodendrimers as ligands for HIV-1 gp120, *Bioconjug. Chem.*, 15 (2004) 349–358.
420. R. D. Kensinger, B. J. Catalone, F. C. Krebs, B. Wigdahl, and C.-L. Schengrund, Novel polysulfated galactose-derivatized dendrimers as binding antagonists of Human Immunodeficiency virus type 1 infection, *Antimicrob. Agents Chemother.*, 48 (2004) 1614–1623.
421. T. Dutta, H. B. Agashe, M. Garg, P. Balasubramanium, M. Kabra, and N. K. Jain, Poly(propyleneimine) dendrimer based nanocontainers for targeting of efavirenz to human monocytes/macrophages *in vitro*, *J. Drug. Target.*, 15 (2007) 89–98.
422. T. Dutta and N. K. Jain, Targeting potential and anti-HIV activity of lamivudine loaded mannosylated poly(propyleneimine) dendrimer, *Biochim. Biophys. Acta Gen. Subj.*, 1770 (2007) 681–686.
423. H. Magnusson, E. Malmström, and A. Hult, Structure builup in hyperbranched polymers from 2, 2-bis(hydroxymethyl)propionic acid, *Macromolecules*, 33 (2000) 3099–3104.
424. A. Hult, M. Johansson, and E. Malmström, Hyperbranched polymers, *Adv. Polym. Sci.*, 143 (1999) 1–34.

425. E. Arce, P. M. Nieto, V. Diaz, R. Garcia Castro, A. Bernad, and J. Rojo, Glycodendritic structures based on Boltorn hyperbranched polymers and their interactions with *Lens culinaris* lectin, *Bioconjug. Chem.*, 14 (2003) 817–823.
426. F. Lasala, E. Arce, J. R. Otero, J. Rojo, and R. Delgado, Mannosyl glycodendritic structure inhibits DC-SIGN-mediated Ebola virus infection in cis and in trans, *Antimicrob. Agents Chemother.*, 47 (2003) 3970–3972.
427. I. Deguise, D. Lagnoux, and R. Roy, Synthesis of glycodendrimers containing both fucoside and galactoside residues and their binding properties to PA-IL and PA-IIL lectins from *Pseudomonas aeruginosa*, *New J. Chem.*, 31 (2007) 1321–1331.
428. N. Jayaraman and J. F. Stoddart, Synthesis of carbohydrate-containing dendrimers. 5. Preparation of dendrimers using unprotected carbohydrates, *Tetrahedron Lett.*, 38 (1997) 6767–6770.
429. K. Matsuoka, M. Terabatake, Y. Esumi, D. Terunuma, and H. Kuzuhara, Synthetic assembly of trisaccharide moieties of globotriaosyl ceramide using carbosilane dendrimers as cores. A new type of functional glyco-material, *Tetrahedron Lett.*, 40 (1999) 7839–7842.
430. K. Nishikawa, K. Matsuoka, E. Kita, N. Okabe, M. Mizugushi, K. Hino, S. Miyazawa, C. Yamasaki, J. Aoki, S. Takashima, Y. Yamakawa, M. Nishijima, *et al.*, A therapeutic agent with oriented carbohydrates for treatment of infections by Shiga toxin-producing *Escherichia coli* O157:H7, *Proc. Natl. Acad. Sci. (USA)*, 99 (2002) 7669–7674.
431. K. Nishikawa, K. Matsuoka, M. Watanabe, K. Igai, K. Hino, K. Hatano, A. Yamada, N. Abe, D. Terunuma, H. Kuzuhara, and Y. Natori, Identification of the optimal structure required for a Shiga Toxin neutralizer with oriented carbohydrates to function in the circulation, *J. Infect. Dis.*, 191 (2005) 2097–2106.
432. A. Yamada, K. Hatano, K. Matsuoka, T. Koyama, Y. Esumi, H. Koshino, K. Hino, K. Nishikawa, Y. Natori, and D. Terunuma, Syntheses and Vero toxin-binding activities of carbosilane, *Tetrahedron*, 62 (2006) 5074–5083.
433. H. Oka, T. Onaga, T. Koyama, C.-T. Guo, Y. Suzuki, Y. Esumi, K. Hatano, D. Terunuma, and K. Matsuoka, Sialyl $\alpha(2\rightarrow3)$ lactose clusters using carbosilane dendrimer core scaffolds as influenza hemagglutinin blockers, *Bioorg. Med. Chem. Lett.*, 18 (2008) 4405–4408.
434. J. Camponovo, C. Hadad, J. Ruiz, E. Cloutet, S. Gatard, J. Muzart, S. Bouquillon, and D. Astruc, Click glycodendrimers containing 27, 81, and 243 modified xylopyranoside termini, *J. Org. Chem.*, 74 (2009) 5071–5074.
435. D. Zanini and R. Roy, Architectonic neoglycoconjugates: Effects of shapes and valencies in multiple carbohydrate-protein interactions, in Y. Chapleur, (Ed.),

Carbohydrate mimics, concepts and methods, Wiley-VCH, Weinheim, 1998, pp. 385–415.

436. N. Y. Lee, W. J. Jang, S. H. Yu, J. U. Im, and S. K. Chung, Syntheses of glycodendrimers having scyllo-inositol as the scaffold, *Tetrahedron Lett.*, 46 (2005) 6063–6066.
437. J. Sakamoto and K. Müllen, Sugars within a hydrophobic scaffold: Glycodendrimers from polyphenylenes, *Org. Lett.*, 6 (2004) 4277–4280.
438. C. Hadad, J.-P. Majoral, J. Muzart, A.-M. Caminade, and S. Bouquillon, First phosphorus D-xylose-derived glycodendrimers, *Tetrahedron Lett.*, 50 (2009) 1902–1905.
439. C. P. Stowell and Y. C. Lee, Neoglycoproteins. The preparation and application of synthetic glycoproteins, *Adv. Carbohydr. Chem. Biochem.*, 37 (1980) 225–281.
440. F. Nicotra, L. Cipolla, F. Peri, B. La Ferla, and C. Radaelli, Chemoselective neoglycosylation, *Adv. Carbohydr. Chem. Biochem.*, 61 (2007) 353–398.

AUTHOR INDEX

Page numbers in roman type indicate that the listed author is cited on that page; page numbers in italic denote the page where the literature citation is given.

D

E

H

L

M

S

W

X

Y

Z

SUBJECT INDEX

N

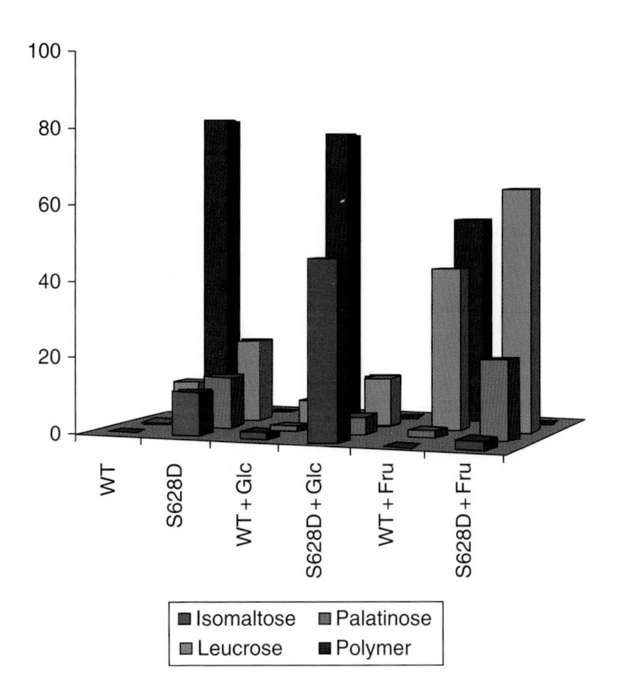

Plate 1 Product spectra of the wild-type (WT) GTFR and mutant S628D enzymes (200 U/L) incubated (7 days at 30 °C) with sucrose (146 mM) and different acceptor substrates (292 mM, Glc: glucose; Fru: fructose). Yields are given in percentage (mol/mol Glc). Yields of higher oligosaccharides (DP > 5) and hydrolysis products are not shown. (See Fig. 3 in the Seibel–Buchholz chapter, p. 110.)

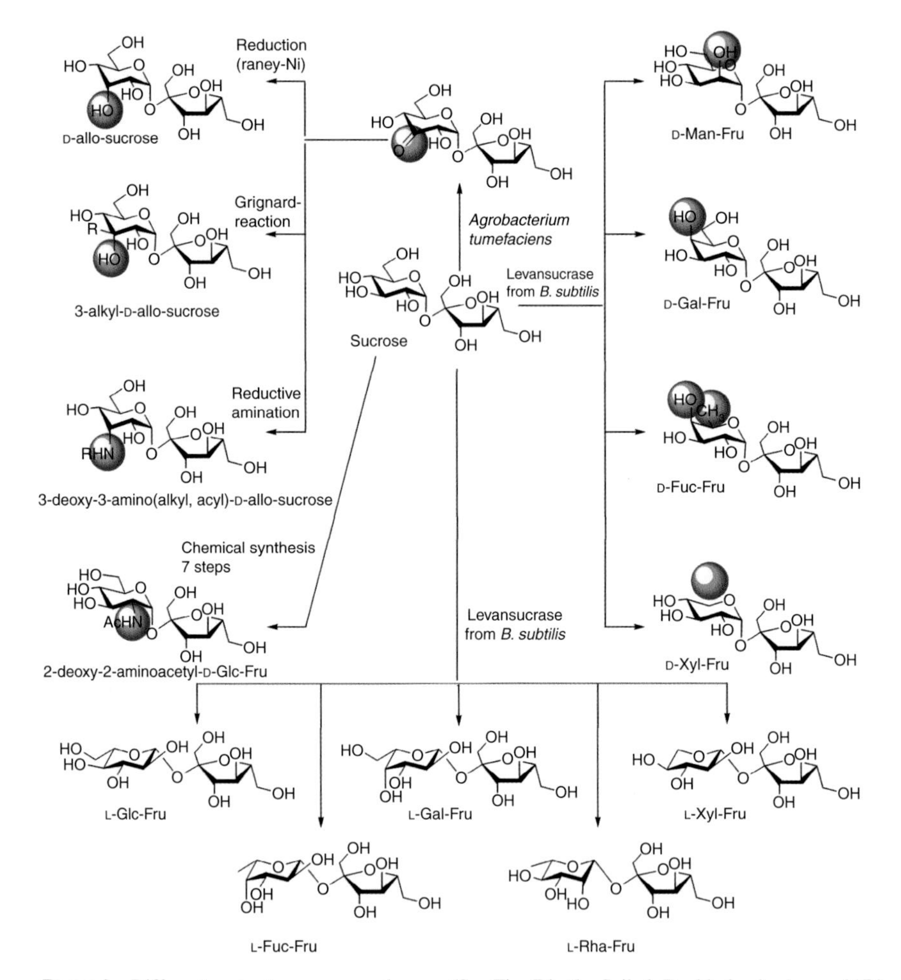

PLATE 2 Different routes to sucrose analogues. (See Fig. 7 in the Seibel–Buchholz chapter, p. 117.)

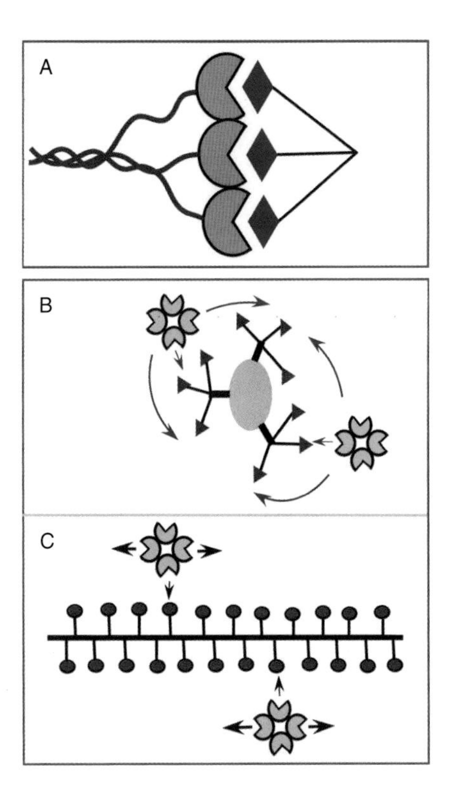

Plate 3 Schematic representations of (A) face-to-face binding of a lectin with three subsites (green) to a trivalent carbohydrate (blue); (B) binding of a nonavalent glycoprotein (orange/black/pink) to two lectin molecules (green); (C) binding of a linear glycoprotein (black/red) to two lectin molecules (green). (See Fig. 1 in the Dam–Brewer chapter, p. 141.)

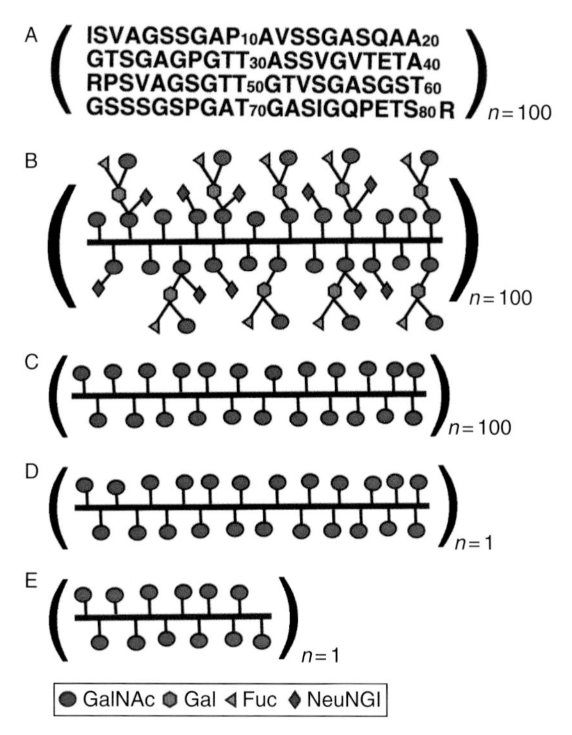

PLATE 4 Structural representations of (A) the amino acid sequence of the 100-repeat 81-residue polypeptide O-glycosylation domain of intact PSM; (B) the fully carbohydrate-decorated form (described in the text) of the 100-repeat 81-residue polypeptide O-glycosylation domain of PSM (Fd-PSM); (C) the 100-repeat 81-residue polypeptide O-glycosylation domain of PSM containing only peptide-linked αGalNAc residues (Tn-PSM); (D) the single 81-residue polypeptide O-glycosylation domain of PSM containing peptide-linked αGalNAc residues (81-mer Tn-PSM); (E) the 38/40-residue polypeptide(s) derived from the 81-residue polypeptide O-glycosylation domain of PSM containing peptide-linked αGalNAc residues (38/40-mer Tn-PSM). The number of glycan chains in Fd-PSM and Tn-PSM is ~2300. The number of αGalNAc residues in 81-mer Tn-PSM is ~23, while the number of αGalNAc residues in 38/40-mer Tn-PSM is ~11–12. (See Fig. 2 in the Dam–Brewer chapter, p. 144.)

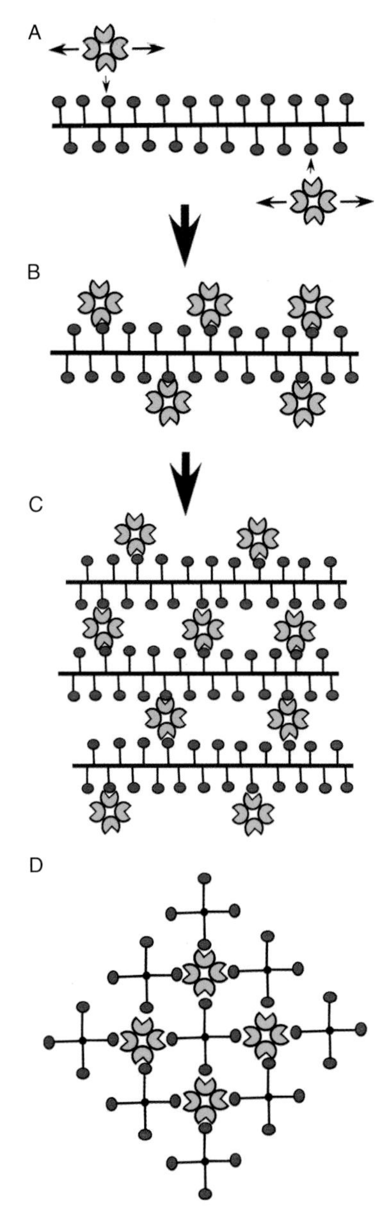

Plate 5 Schematic representations of (A) SBA or VML binding at low density to Tn-PSM; (B) SBA or VML binding at higher density to Tn-PSM; (C) SBA and VML binding at higher density to Tn-PSM and initiating crosslinking of the complexes; (D) SBA crosslinked complexes with Tn-PSM under saturation-binding conditions. The view is end on of the polypeptide chains of Tn-PSM in Fig. 3C. αGalNAc residues extend out from the polypeptide chains of Tn-PSM in three dimensions. Lectin tetramers are bound to four separate Tn-PSM chains, with staggered binding down the length of the mucin chains. (See Fig. 3 in the Dam–Brewer chapter, p. 151.)

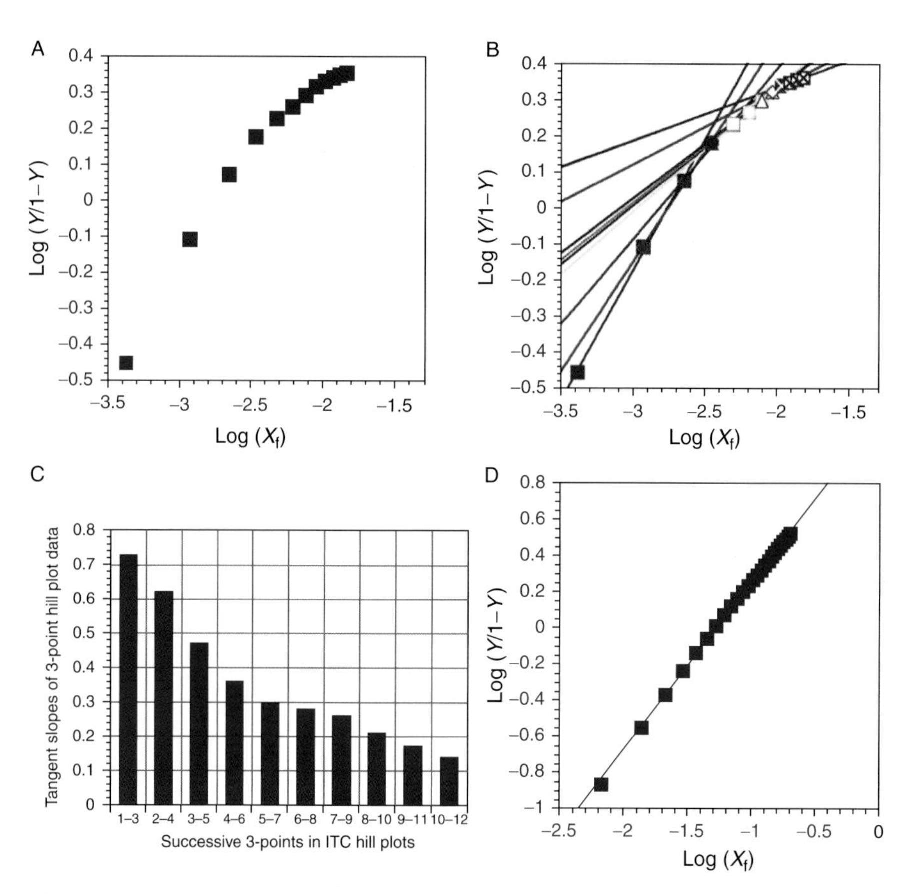

Plate 6 (A). Hill plot of ITC data for SBA binding to 38/40-mer Tn-PSM. (B). Tangent slopes of progressive three-point intervals of the Hill plot for SBA binding to 38/40-mer Tn-PSM in (A). (C). Bar graphs of the three-point tangent slopes of the ITC data Hill plots of SBA binding to 38/40-mer Tn-PSM in (B). (D). Hill plot of the ITC data of SBA binding to αGalNAc1-*O*-Ser (3.24 mM). The slope value is 0.93. (See Fig. 4 in the Dam–Brewer chapter, p. 154.)

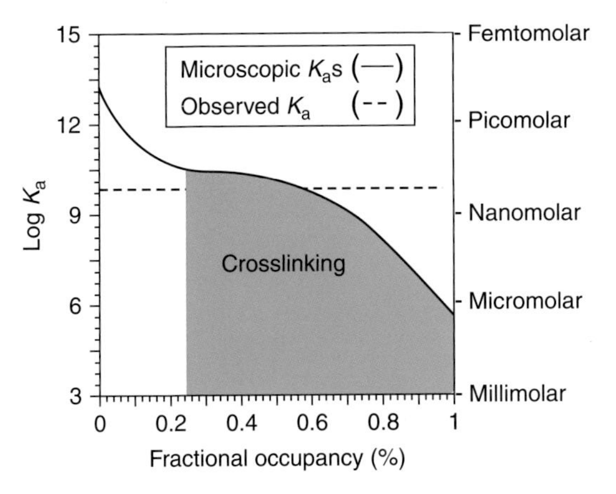

Plate 7 Hypothetical plot (solid line) of the microscopic K_d values of SBA binding to Tn-PSM from 0% to 100% occupancy (0–540 αGalNAc residues). The dashed line represents the observed K_d value of 0.2 nM for SBA binding to Tn-PSM derived from the ITC data in Table I. Crosslinking begins at about 25% occupancy of Tn-PSM by SBA. (See Fig. 5 in the Dam–Brewer chapter, p. 157.)

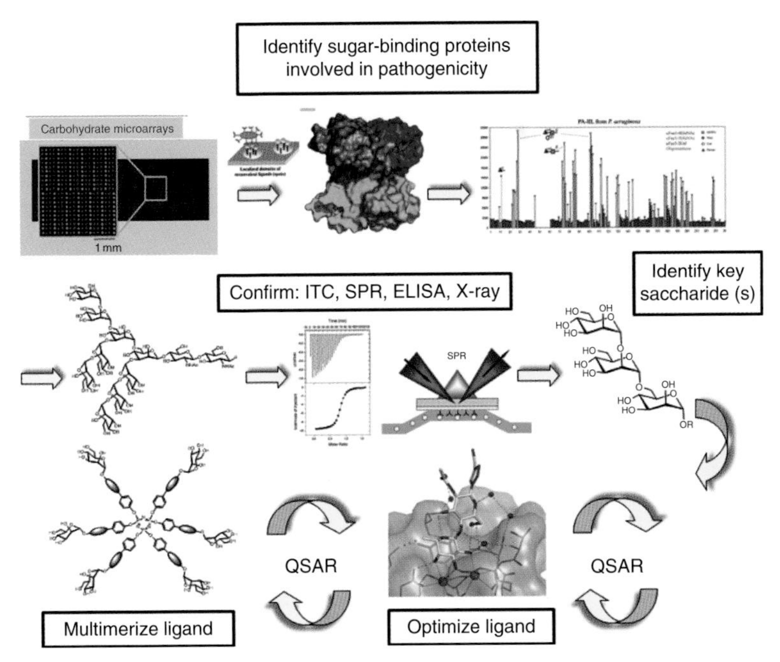

Plate 8 Steps involved in the discovery of optimized multivalent glycodendrimers. (See Fig. 1 in the Chabre–Roy chapter, p. 172.)

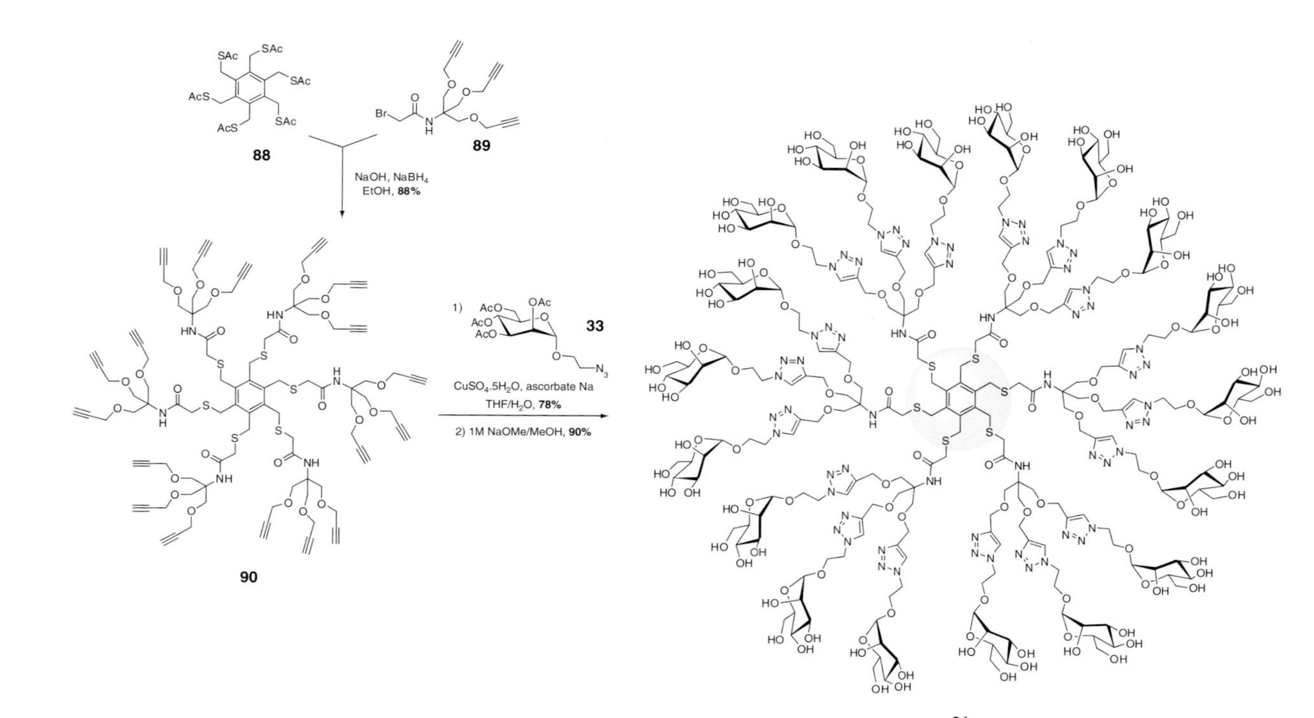

PLATE 9 Persulfurated glycoclusters bearing 18 α-D-mannopyranoside residues. (See Scheme 13 in Chabre–Roy chapter, p. 199.)

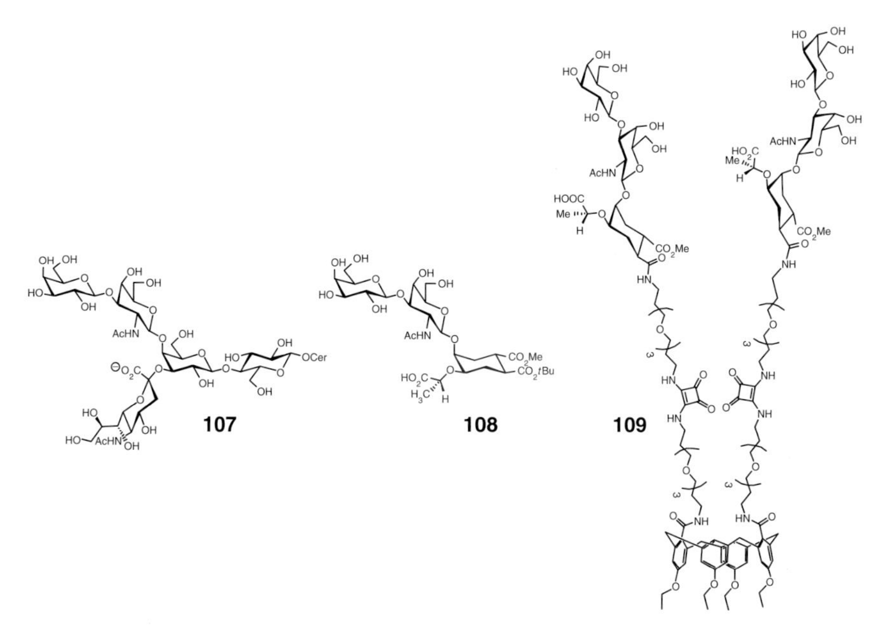

Plate 10 Divalent GM1 mimic having high affinity against cholera toxin. (See Fig. 9 in Chabre–Roy chapter, p. 204.)

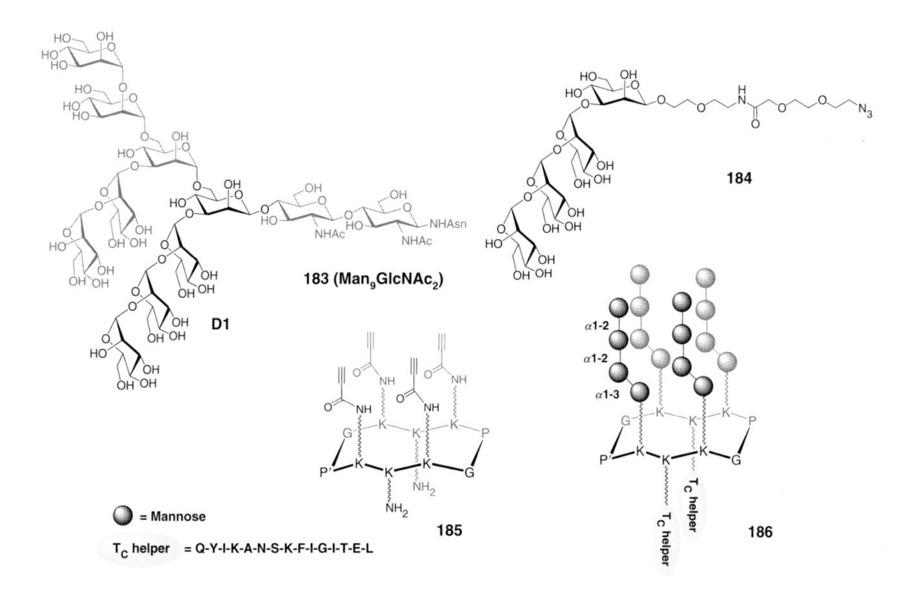

Plate 11 Cyclic peptide vaccine candidate bearing the minimally epitopic D1 branch of the Man_9 $GlcNAc_2$ antigen of HIV-1 gp120 recognized by the protective human antibody 2G12. (See Fig. 24 in Chabre–Roy chapter, p. 232.)

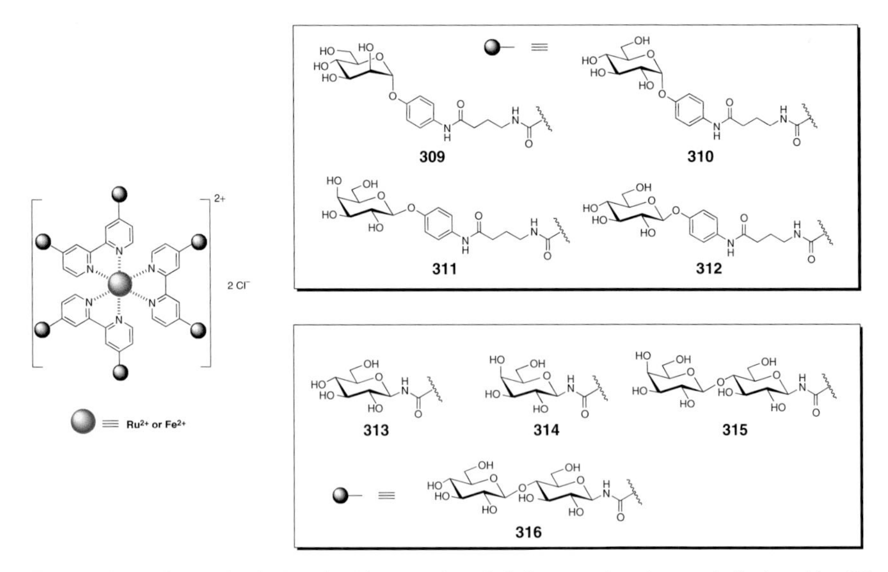

Plate 12 Self-assembled glycodendrimers using chelating metal cations and divalent bipyridine scaffolds. (See Fig. 36 in Chabre–Roy chapter, p. 267.)

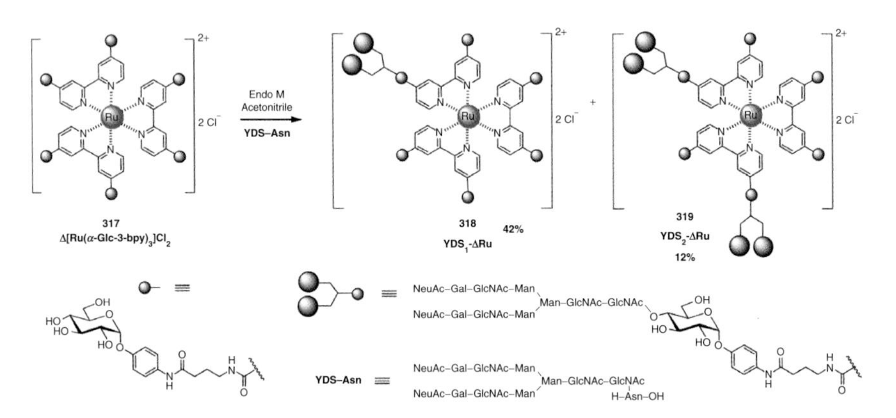

Plate 13 Enzymatic transglycosylation of a preformed bisglucobipyridine core self-assembled as a hexavalent cluster around a ruthenium cation. (See Scheme 32 in Chabre–Roy chapter, p. 269.)

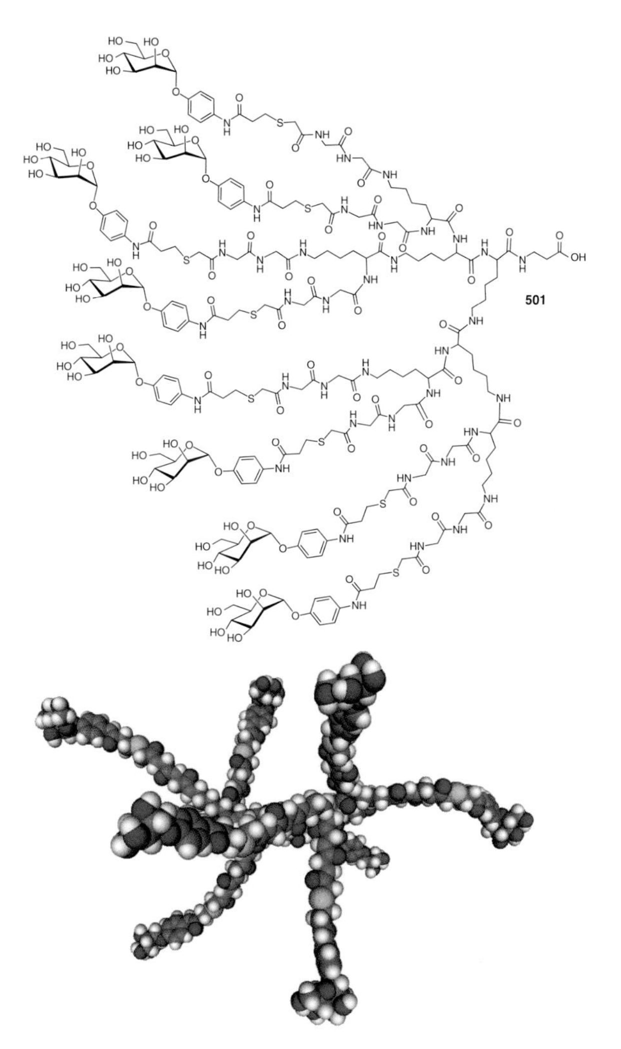

PLATE 14 Mannosylated dendron based on a poly-L-lysine scaffold. This construct leads to subnanomolar inhibitory potency against uropathogenic *E. coli*. (See Fig. 53 in Chabre–Roy chapter, p. 311.)

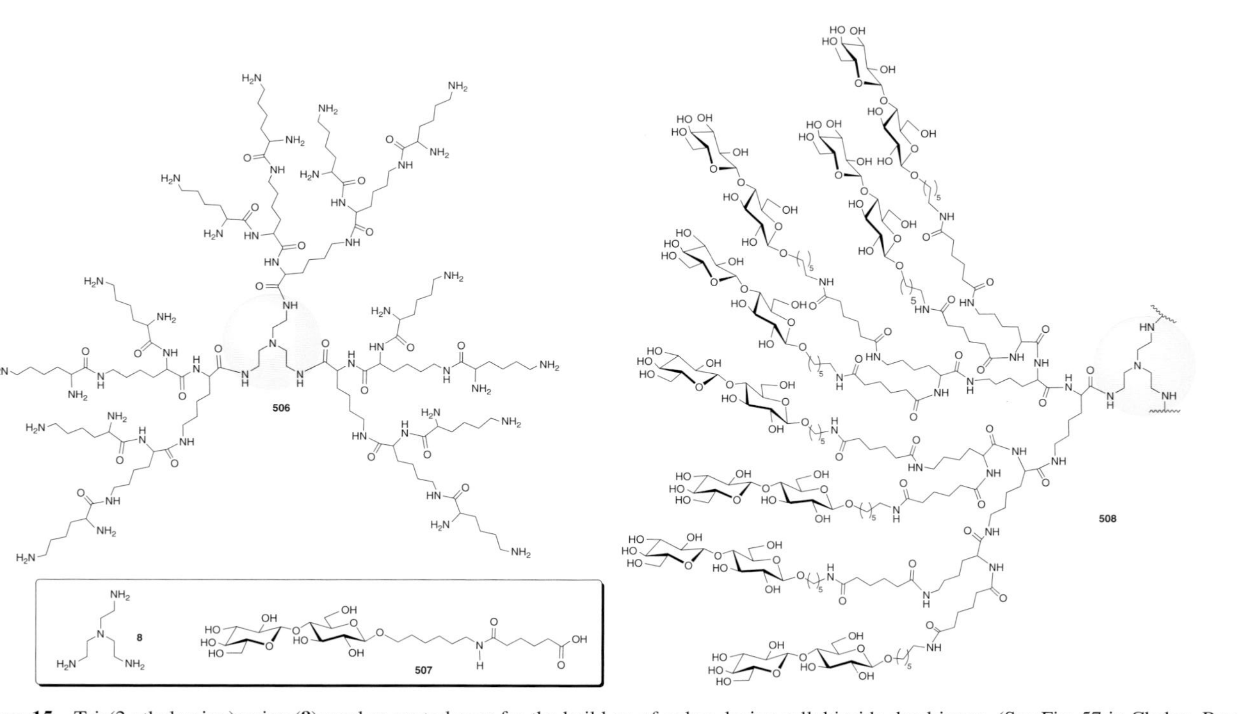

Plate 15 Tris(2-ethylamino)amine (**8**) used as central core for the build up of poly-L-lysine cellobioside dendrimers. (See Fig. 57 in Chabre–Roy chapter, p. 316.)

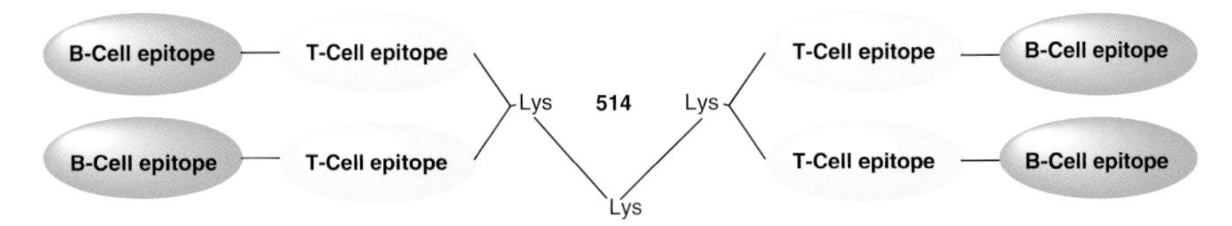

Plate 16 Schematic representation of a fully synthetic carbohydrate-based vaccine. (See Fig. 60 in Chabre–Roy chapter, p. 319.)

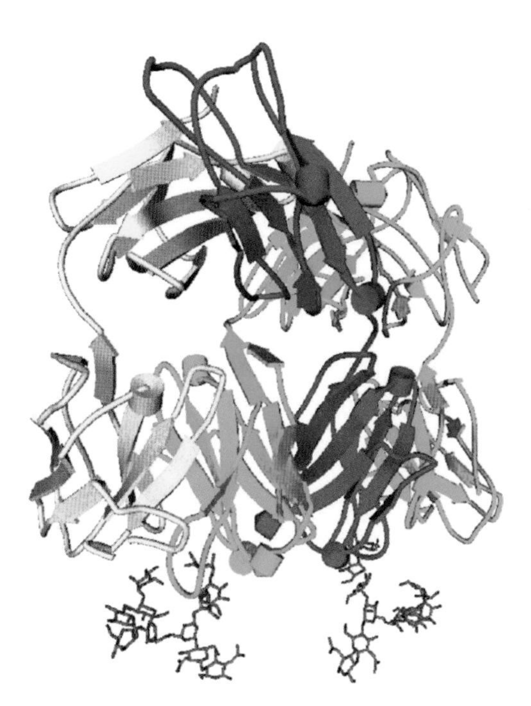

Plate 17 Crystal structure of the HIV-1 neutralizing human antibody 2G12 bound to the oligomannoside $Man_9GlcNAc_2$ present on the ''silent'' face of the gp120 envelope glycoprotein (PDB 1OP5). (See Fig. 61 in Chabre–Roy chapter, p. 320.)

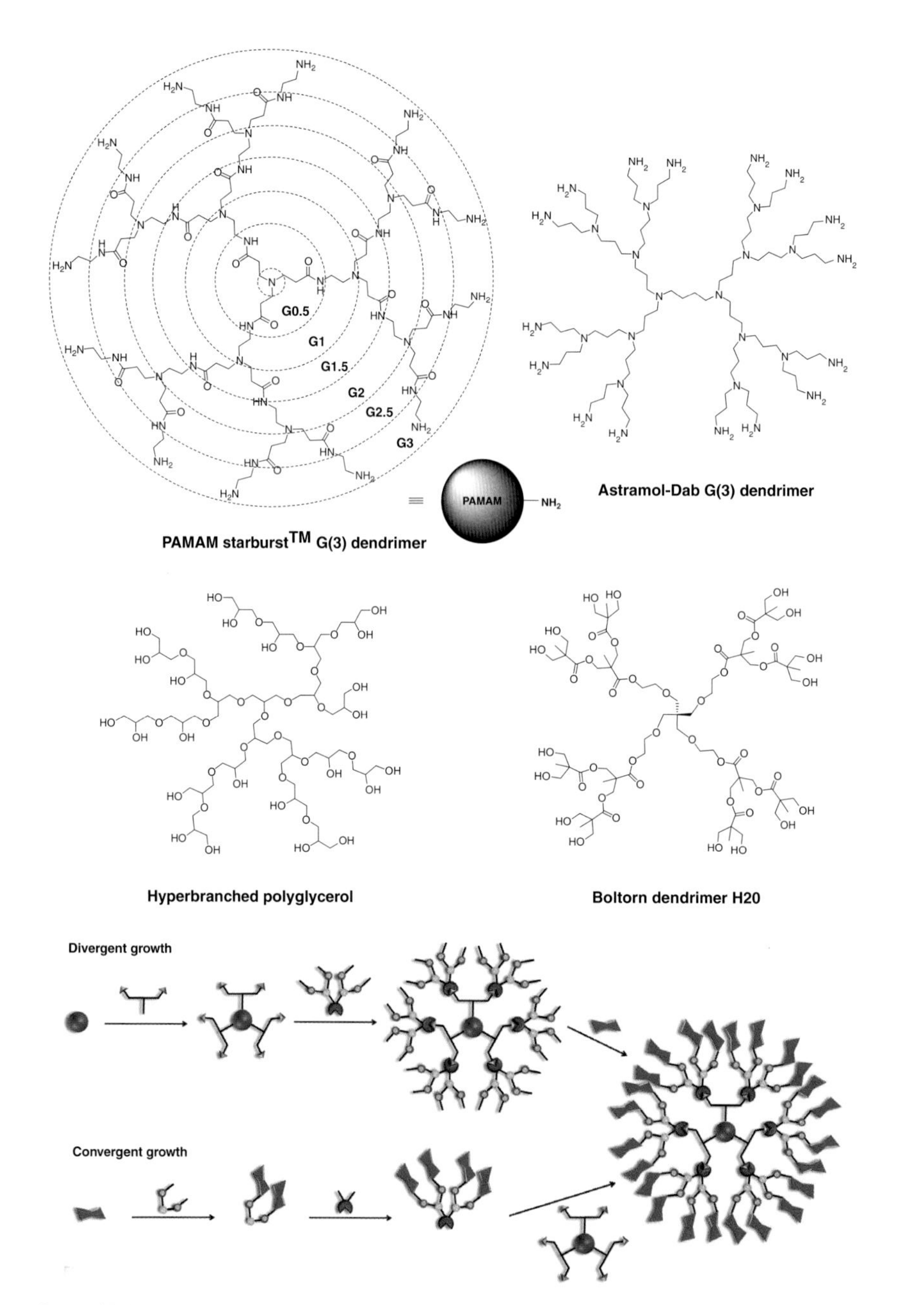

PLATE 18 Commercially available dendrimer scaffolds commonly used for glycodendrimer syntheses, and general synthetic strategies for synthesis of glycodendrimers. (See Fig. 66 in Chabre–Roy chapter, p. 324.)

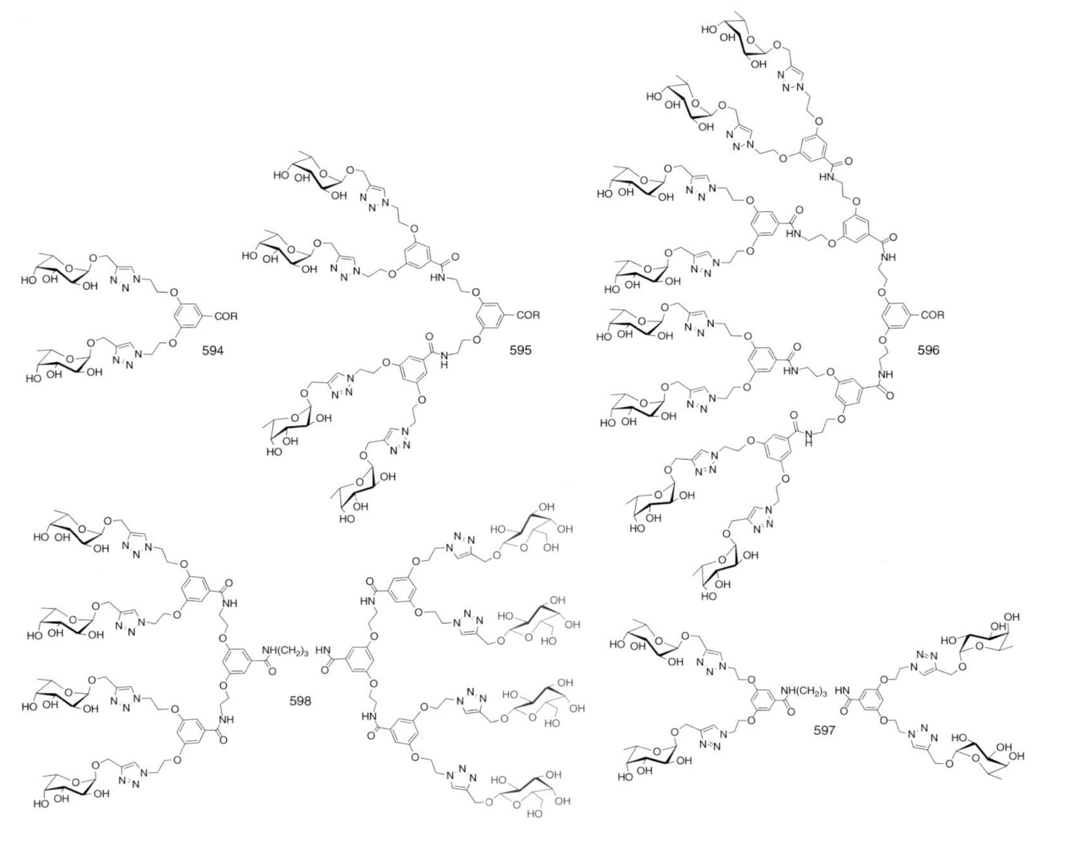

PLATE 19 Galactosylated and/or fucosylated *P. aeruginosa* lectin ligands built on an AB_2 scaffold. (See Fig. 73 in Chabre–Roy chapter, p. 346.)

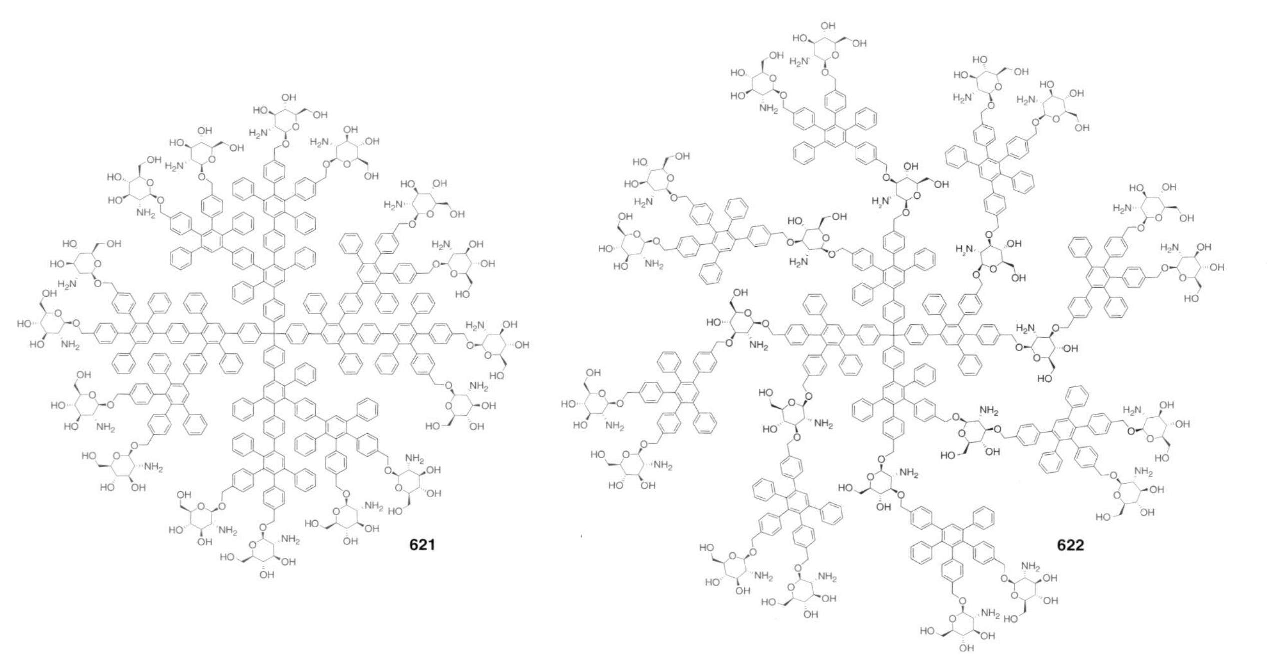

PLATE 20 Polyphenylenes building blocks used by Sakamoto and Müllen for glucosamine dendrimers. (See Fig. 78 in Chabre–Roy chapter, p. 355.)